1# Test covers: page 1-50, 70-75

FUNDAMENTALS OF COLLEGE GEOMETRY

Fundamentals of COLLEGE GEOMETRY

SECOND EDITION

Edwin M. Hemmerling

Department of Mathematics

Bakersfield College

JOHN WILEY & SONS, New York • Chichester • Brisbane • Toronto

10

Library of Congress Catalogue Card Number: 75-82969

SBN 471 37034 7

Printed in the United States of America

Preface

Before revising *Fundamentals of College Geometry,* extensive questionnaires were sent to users of the earlier edition. A conscious effort has been made in this edition to incorporate the many fine suggestions given the respondents to the questionnaire. At the same time, I have attempted to preserve the features that made the earlier edition so popular.

The postulational structure of the text has been strengthened. Some definitions have been improved, making possible greater rigor in the development of the theorems. Particular stress has been continued in observing the distinction between equality and congruence. Symbols used for segments, intervals, rays, and half-lines have been changed in order that the symbols for the more common segment and ray will be easier to write. However, a symbol for the interval and half-line is introduced, which will still logically show their relations to the segment and ray.

Fundamental space concepts are introduced throughout the text in order to preserve continuity. However, the postulates and theorems on space geometry are kept to a minimum until Chapter 14. In this chapter, particular attention is given to mensuration problems dealing with geometric solids.

Greater emphasis has been placed on utilizing the principles of deductive logic covered in Chapter 2 in deriving geometric truths in subsequent chapters. Venn diagrams and truth tables have been expanded at a number of points throughout the text.

There is a wide variance throughout the United States in the time spent in geometry classes. Approximately two fifths of the classes meet three days a week. Another two fifths meet five days each week. The student who studies the first nine chapters of this text will have completed a well-rounded minimum course, including all of the fundamental concepts of plane and space geometry.

Each subsequent chapter in the book is written as a complete package, none of which is essential to the study of any of the other last five chapters, yet each will broaden the total background of the student. This will permit the instructor considerable latitude in adjusting his course to the time available and to the needs of his students.

Each chapter contains several sets of summary tests. These vary in type to include true-false tests, completion tests, problems tests, and proofs tests. A key for these tests and the problem sets throughout the text is available.

January 1969 *Edwin M. Hemmerling*

Preface to First Edition

During the past decade the entire approach to the teaching of geometry has been undergoing serious study by various nationally recognized professional groups. This book reflects many of their recommendations.

The style and objectives of this book are the same as those of my *College Plane Geometry*, out of which it has grown. Because I have added a significant amount of new material, however, and have increased the rigor employed, it has seemed desirable to give the book a new title. In *Fundamentals of College Geometry*, the presentation of the subject has been strengthened by the early introduction and continued use of the language and symbolism of sets as a unifying concept.

This book is designed for a semester's work. The student is introduced to the basic structure of geometry and is prepared to relate it to everyday experience as well as to subsequent study of mathematics.

The value of the precise use of language in stating definitions and hypotheses and in developing proofs is demonstrated. The student is helped to acquire an understanding of deductive thinking and a skill in applying it to mathematical situations. He is also given experience in the use of induction, analogy, and indirect methods of reasoning.

Abstract materials of geometry are related to experiences of daily life of the student. He learns to search for undefined terms and axioms in such areas of thinking as politics, sociology, and advertising. Examples of circular reasoning are studied.

In addition to providing for the promotion of proper attitudes, understandings, and appreciations, the book aids the student in learning to be critical in his listening, reading, and thinking. He is taught not to accept statements blindly but to think clearly before forming conclusions.

The chapter on coordinate geometry relates geometry and algebra. Properties of geometric figures are then determined analytically with the aid of algebra and the concept of one-to-one correspondence. A short chapter on trigonometry is given to relate ratio, similar polygons, and coordinate geometry.

Illustrative examples which aid in solving subsequent exercises are used liberally throughout the book. The student is able to learn a great deal of the material without the assistance of an instructor. Throughout the book he is afforded frequent opportunities for original and creative thinking. Many of the generous supply of exercises include developments which prepare for theorems that appear later in the text. The student is led to discover for himself proofs that follow.

The summary tests placed at the end of the book include completion, true-false, multiple-choice items, and problems. They afford the student and the instructor a ready means of measuring progress in the course.

Bakersfield, California, 1964 *Edwin M. Hemmerling*

Contents

FUNDAMENTALS OF COLLEGE GEOMETRY

|1|

Basic Elements of Geometry

1.1. Historical background of geometry. Geometry is a study of the properties and measurements of figures composed of points and lines. It is a very old science and grew out of the needs of the people. The word geometry is derived from the Greek words *geo*, meaning "earth," and *metrein*, meaning "to measure." The early Egyptians and Babylonians (4000–3000 B.C.) were able to develop a collection of practical rules for measuring simple geometric figures and for determining their properties.

These rules were obtained inductively over a period of centuries of trial and error. They were not supported by any evidence of logical proof. Applications of these principles were found in the building of the Pyramids and the great Sphinx.

The irrigation systems devised by the early Egyptians indicate that they had an adequate knowledge of geometry as it may be applied in land surveying. The Babylonians were using geometric figures in tiles, walls, and decorations of their temples.

From Egypt and Babylonia the knowledge of geometry was taken to Greece. From the Greek people we have gained some of the greatest contributions to the advancement of mathematics. The Greek philosophers studied geometry not only for utilitarian benefits derived but for the esthetic and cultural advantages gained. The early Greeks thrived on a prosperous sea trade. This sea trade brought them not only wealth but also knowledge from other lands. These wealthy citizens of Greece had considerable time for fashionable debates and study on various topics of cultural interest because they had slaves to do most of their routine work. Usually theories and concepts brought back by returning seafarers from foreign lands made topics for lengthy and spirited debate by the Greeks.

Thus the Greeks became skilled in the art of logic and critical thinking. Among the more prominent Greeks contributing to this advancement were Thales of Miletus (640–546 B.C.), Pythagoras, a pupil of Thales (580?–500 B.C.), Plato (429–348 B.C.), Archimedes (287–212 B.C.), and Euclid (about 300 B.C.).

Euclid, who was a teacher of mathematics at the University of Alexandria, wrote the first comprehensive treatise on geometry. He entitled his text "Elements." Most of the principles now appearing in a modern text were present in Euclid's "Elements." His work has served as a model for most of the subsequent books written on geometry.

1.2. Why study geometry? The student beginning the study of this text may well ask, "What is geometry? What can I expect to gain from this study?"

Many leading institutions of higher learning have recognized that positive benefits can be gained by all who study this branch of mathematics. This is evident from the fact that they require study of geometry as a prerequisite to matriculation in those schools.

A study of geometry is an essential part of the training of the successful engineer, scientist, architect, and draftsman. The carpenter, machinist, tinsmith, stonecutter, artist, and designer all apply the facts of geometry in their trades. In this course the student will learn a great deal about geometric figures such as lines, angles, triangles, circles, and designs and patterns of many kinds.

One of the most important objectives derived from a study of geometry is making the student be more critical in his listening, reading, and thinking. In studying geometry he is led away from the practice of blind acceptance of statements and ideas and is taught to think clearly and critically before forming conclusions.

There are many other less direct benefits the student of geometry may gain. Among these one must include training in the exact use of the English language and in the ability to analyze a new situation or problem into its basic parts, and utilizing perseverence, originality, and logical reasoning in solving the problem. An appreciation for the orderliness and beauty of geometric forms that abound in man's works and of the creations of nature will be a by-product of the study of geometry. The student should also develop an awareness of the contributions of mathematics and mathematicians to our culture and civilization.

1.3. Sets and symbols. The idea of "set" is of great importance in mathematics. All of mathematics can be developed by starting with sets.

The word "set" is used to convey the idea of a collection of objects, usually with some common characteristic. These objects may be pieces of furniture

in a room, pupils enrolled in a geometry class, words in the English language, grains of sand on a beach, etc. These objects may also be distinguishable objects of our intuition or intellect, such as points, lines, numbers, and logical possibilities. The important feature of the set concept is that the collection of objects is to be regarded as a single entity. It is to be treated as a whole. Other words that convey the concept of set are "group," "bunch," "class," "aggregate," "covey," and "flock."

There are three ways of specifying a set. One is to give a rule by which it can be determined whether or not a given object is a *member* of the set; that is, the set is described. This method of specifying a set is called the *rule method.* The second method is to give a complete list of the members of the set. This is called the *roster method.* A third method frequently used for sets of real numbers is to graph the set on the number line. The members of a set are called its *elements.* Thus "members" and "elements" can be used interchangeably.

It is customary to use braces { } to surround the elements of a set. For example, {1, 3, 5, 7} means the set whose members are the odd numbers 1, 3, 5, and 7. {Tom, Dick, Harry, Bill} might represent the members of a vocal quartet. A capital letter is often used to name or refer to a set. Thus, we could write $A = \{1, 3, 5, 7\}$ and $B =$ {Tom, Dick, Harry, Bill}.

A set may contain a finite number of elements, or an infinite number of elements. A finite set which contains no members is the *empty* or *null* set. The symbol for a null set is $\emptyset$ or { }. Thus, {even numbers ending in 5} = $\emptyset$. A set with a definite number* of members is a *finite* set. Thus, {5} is a finite set of which 5 is the only element. When the set contains many elements, it is customary to place inside the braces a description of the members of the set, e.g. {citizens of the United States}. A set with an infinite number of elements is termed an *infinite set.* The natural numbers 1, 2, 3, ... form an infinite set. {0, 2, 4, 6, . . .} means the set of all nonnegative even numbers. It, too, is an infinite set.

In mathematics we use three dots (. . .) in two different ways in listing the elements of a set. For example

	Rule	**Roster**
1.	{integers greater than 10 and less than 100} Here the dots . . . mean "and so on up to and including."	{11, 12, 13, . . . , 99}
2.	{integers greater than 10} Here the dots . . . mean "and so on indefinitely."	{11, 12, 13, . . .}

*Zero is a definite number.

To symbolize the notion that 5 is an element of set A, we shall write $5 \in A$. If 6 is not a member of set A, we write $6 \notin A$, read "6 is not an element of set A."

Exercises

In exercises 1–12 it is given:

$A = \{1, 2, 3, 4, 5\}$. $B = \{6, 7, 8, 9, 10\}$.
$C = \{1, 2, 3, \ldots, 10\}$. $D = \{2, 4, 6, \ldots\}$.
$E = \emptyset$. $F = \{0\}$.
$G = \{5, 3, 2, 1, 4\}$. $H = \{1, 2, 3, \ldots\}$.

1. How many elements are in C? in E?
2. Give a rule describing H.
3. Do E and F contain the same elements?
4. Do A and G contain the same elements?
5. What elements are common to set A and set C?
6. What elements are common to set B and set D?
7. Which of the sets are finite?
8. Which of the sets are infinite?
9. What elements are common to A and B?
10. What elements are either in A or C or in both?
11. Insert in the following blank spaces the correct symbol $\in$ or $\notin$.
    (*a*) 3 ___ A (*b*) 3 ___ D (*c*) 0 ___ F
    (*d*) 0 ___ E (*e*) $\frac{3}{2}$ ___ H (*f*) 1002 ___ D
12. Give a rule describing F.

13–20. Use the roster method to describe each of the following sets.

Example. {whole numbers greater than 3 and less than 9}
Solution. {4, 5, 6, 7, 8}

13. {days of the week whose names begin with the letter T}
14. {even numbers between 29 and 39}
15. {whole numbers that are neither negative or positive}
16. {positive whole numbers}
17. {integers greater than 9}
18. {integers less than 1}
19. {months of the year beginning with the letter J}
20. {positive integers divisible by 3}

21–28. Use the rule method to describe each of the following sets.

Example. {California, Colorado, Connecticut}
Solution. {member states of the United States whose names begin with the letter C}

21. $\{a, e, i, o, u\}$
22. $\{a, b, c, \ldots, z\}$
23. {red, orange, yellow, green, blue, violet}
24. { }
25. $\{2, 4, 6, 8, 10\}$
26. $\{3, 4, 5, \ldots, 50\}$
27. $\{-2, -4, -6, \ldots\}$
28. $\{-6, -4, -2, 0, 2, 4, 6\}$

1.4. Relationships between sets. Two sets are *equal* if and only if they have the same elements. The equality between sets A and B is written $A = B$. The inequality of two sets is written $A \neq B$. For example, let set A be {whole numbers between $1\frac{1}{2}$ and $6\frac{3}{4}$} and let set B be {whole numbers between $1\frac{2}{3}$ and $6\frac{5}{6}$}. Then $A = B$ because the elements of both sets are the same: 2, 3, 4, 5, and 6. Here, then, is an example of two equal sets being described in two different ways. We could write {days of the week} or {Sunday, Monday, Tuesday, Wednesday, Thursday, Friday, Saturday} as two ways of describing equal sets.

Often several sets are parts of a larger set. The set from which all other sets are drawn in a given discussion is called the *universal set.* The universal set, which may change from discussion to discussion, is often denoted by the letter U. In talking about the set of girls in a given geometry class, the universal set U might be all the students in the class, or it could be all the members of the student body of the given school, or all students in all schools, and so on.

Schematic representations to help illustrate properties of and operations with sets can be formed by drawing *Venn diagrams* (see Figs. 1.1*a* and 1.1*b*).

Here, points within a rectangle represent the elements of the universal set. Sets within the universal set are represented by points inside circles enclosed by the rectangle.

We shall frequently be interested in relationships between two or more sets. Consider the sets A and B where

$$A = \{2, 4, 6\} \qquad \text{and} \qquad B = \{1, 2, 3, 4, 5, 6\}.$$

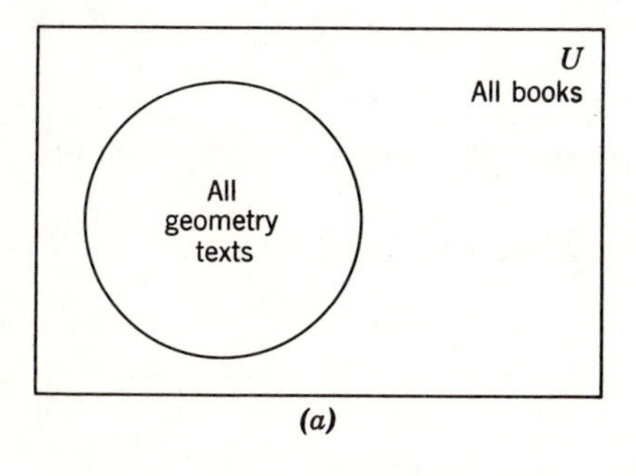

(a)

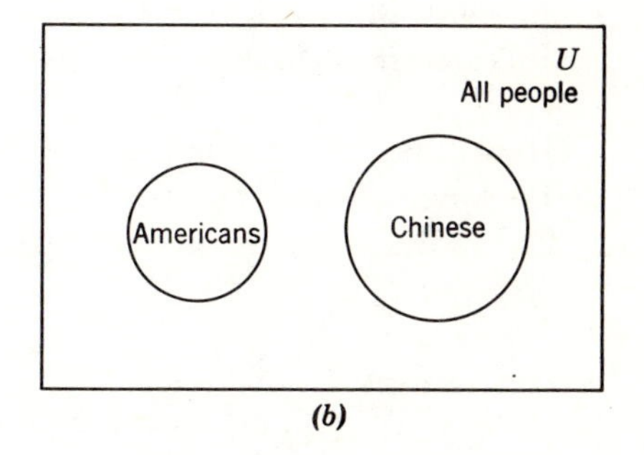

(b)

Fig. 1.1.

Definition: The set A is a *subset* of set B if, and only if, every element of set A is an element of set B. Thus, in the above illustration *A is a subset of B*. We write this relationship $A \subset B$ or $B \supset A$. In the illustration there are more elements in B than in A. This can be shown by the Venn diagram of Fig. 1.2. Notice, however, that our definition of subset does not stipulate it must contain fewer elements than does the given set. The subset can have exactly the same elements as the given set. In such a case, the two sets are equal and each is a subset of the other. Thus, any set is a subset of itself.

Illustrations

(*a*) Given $A = \{1, 2, 3\}$ and $B = \{1, 2\}$. Then $B \subset A$.

(*b*) Given $R = \{\text{integers}\}$ and $S = \{\text{odd integers}\}$. Then $S \subset R$.

(*c*) Given $C = \{\text{positive integers}\}$ and $D = \{1, 2, 3, 4, \ldots\}$. Then $C \subset D$ and $D \subset C$, and $C = D$.

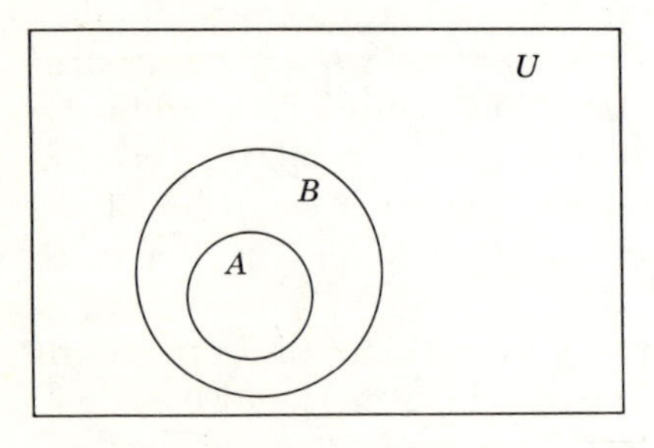

Fig. 1.2. $A \subset B$.

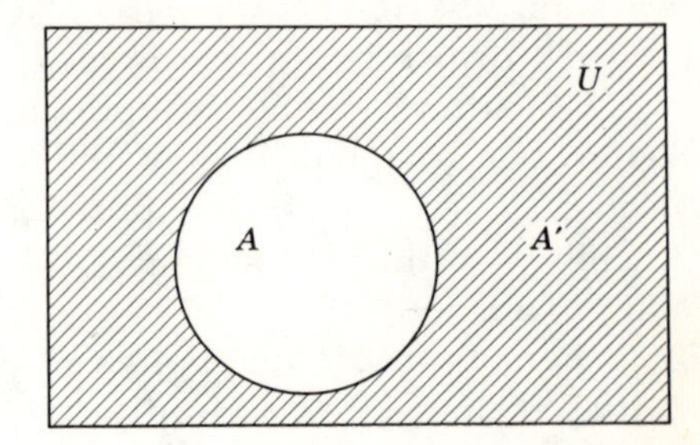

Fig. 1.3.

When A is a subset of a universal set U, it is natural to think of the set composed of all elements of U that are *not in A*. This set is called the *complement of A* and is denoted by A'. Thus, if U represents the set of integers and A the set of negative integers, then A' is the set of nonnegative integers, i.e., $A' = \{0, 1, 2, 3, \ldots\}$. The shaded area of Fig. 1.3 illustrates A'.

1.5. Operations on sets. We shall next discuss two methods for generating new sets from given sets.

Definition: The *intersection* of two sets P and Q is the set of all elements that belong to both P and Q.

The intersection of sets P and Q is symbolized by $P \cap Q$ and is read "P intersection Q" or "P cap Q."

Illustrations:

(*a*) If $A = \{1, 2, 3, 4, 5\}$ and $B = \{2, 4, 6, 8, 10\}$, then $A \cap B = \{2, 4\}$.

(*b*) If $D = \{1, 3, 5, \ldots\}$ and $E = \{2, 4, 6, \ldots\}$, then $D \cap E = \emptyset$.

When two sets have no elements in common they are said to be *disjoint sets* or *mutually exclusive sets.*

(*c*) If $F = \{0, 1, 2, 3, \ldots\}$ and $G = \{0, -2, -4, -6, \ldots\}$, then $F \cap G = \{0\}$.
(*d*) Given A is the set of all bachelors and B is the set of all males. Then $A \cap B = A$. Here A is a subset of B.

Care should be taken to distinguish between the set whose sole member is the number zero and the null set (see *b* and *c* above). They have quite distinct and different meanings. Thus $\{0\} \neq \emptyset$. The null set is empty of any elements. Zero is a number and can be a member of a set. The null set is a subset of all sets.

The intersection of two sets can be illustrated by a Venn diagram. The shaded area of Fig. 1.4 represents $A \cap B$.

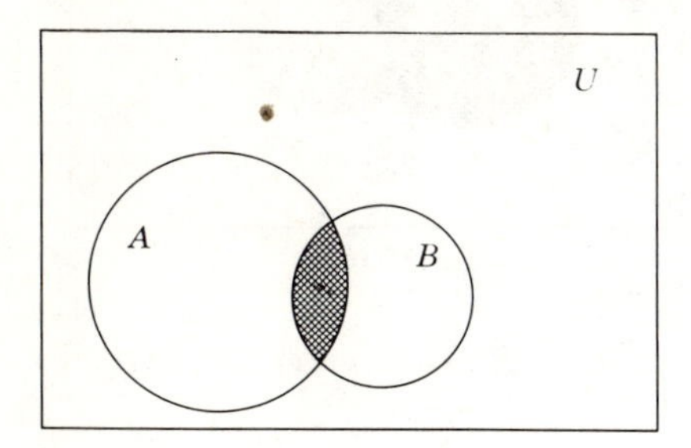

Fig. 1.4. $A \cap B$.

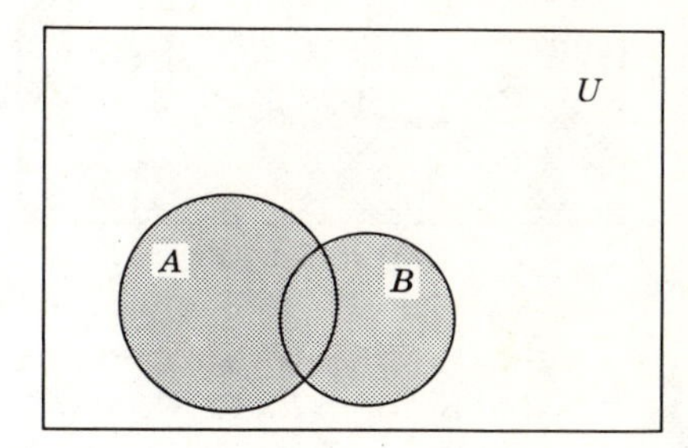

Fig. 1.5. $A \cup B$.

Definition: The *union* of two sets P and Q is the set of all elements that belong to either P or Q or that belong to both P and Q.

The union of sets P and Q is symbolized by $P \cup Q$ and is read "P union Q" or "P cup Q." The shaded area of Fig. 1.5 represents the Venn diagram of $A \cup B$.

Illustrations:

(*a*) If $A = \{1, 2, 3\}$ and $B = \{1, 3, 5, 7\}$, then $A \cup B = \{1, 2, 3, 5, 7\}$.
Note. Individual elements of the union are listed only once.
(*b*) If $A = \{\text{whole even numbers between } 2\frac{1}{2} \text{ and } 5\}$ and $B = \{\text{whole numbers between } 3\frac{1}{4} \text{ and } 6\frac{1}{2}\}$, then $A \cup B = \{4, 5, 6\}$ and $A \cap B = \{4\}$.
(*c*) If $P = \{\text{all bachelors}\}$ and $Q = \{\text{all men}\}$, then $P \cup Q = Q$.

Example. Draw a Venn diagram to illustrate $(R' \cap S')'$ in the figure.

Solution

(*a*) Shade R'.
(*b*) Add a shade for S'.
$R' \cap S'$ is represented by the region common to the area slashed up to the

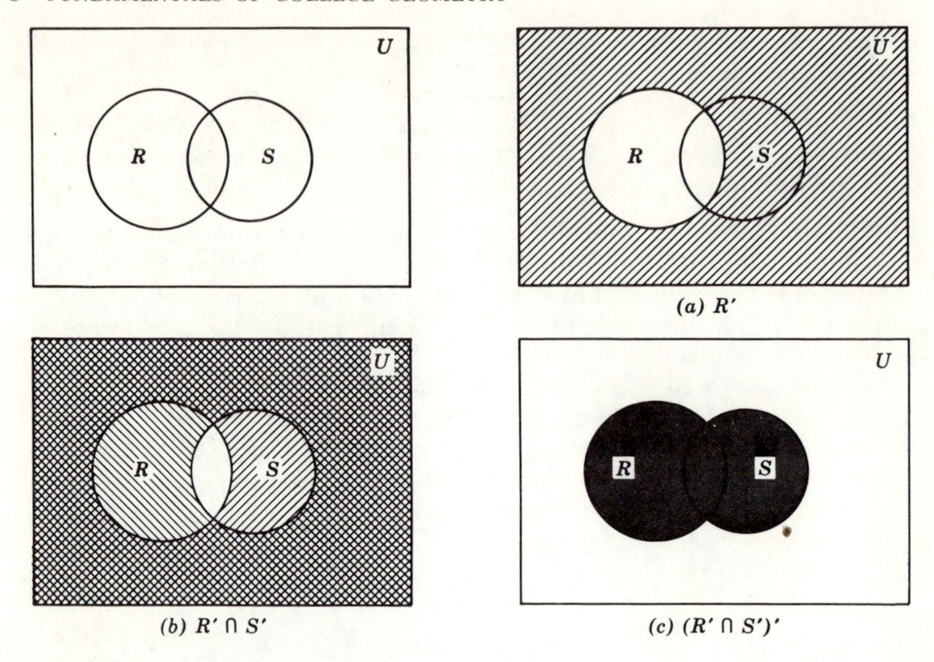

(a) R'

(b) $R' \cap S'$

(c) $(R' \cap S')'$

right and the area slashed down to the right. $(R' \cap S')'$ is all the area in U that is not in $R' \cap S'$.

(*c*) The solution is shaded in the last figure.
We note that $(R' \cap S')' = R \cup S$.

Exercises

1. Let $A = \{2, 3, 5, 6, 7, 9\}$ and $B = \{3, 4, 6, 8, 9, 10\}$.
   (*a*) What is $A \cap B$? (*b*) What is $A \cup B$?
2. Let $R = \{1, 3, 5, 7, \ldots\}$ and $S = \{0, 2, 4, 6, \ldots\}$.
   (*a*) What is $R \cap S$? (*b*) What is $R \cup S$?
3. Let $P = \{1, 2, 3, 4, \ldots\}$ and $Q = \{3, 6, 9, 12, \ldots\}$.
   (*a*) What is $P \cap Q$? (*b*) What is $P \cup Q$?
4. $(\{1, 3, 5, 7, 9\} \cap \{2, 3, 4, 5\}) \cup \{2, 4, 6, 8\} = ?$
5. Simplify: $\{4, 7, 8, 9\} \cup (\{1, 2, 3, \ldots\} \cap \{2, 4, 6, \ldots\})$.
6. Consider the following sets.
   $A = \{$students in your geometry class$\}$.
   $B = \{$male students in your geometry class$\}$.
   $C = \{$female students in your geometry class$\}$.
   $D = \{$members of student body of your school$\}$.
   What are (*a*) $A \cap B$; (*b*) $A \cup B$; (*c*) $B \cap C$; (*d*) $B \cup C$; (*e*) $A \cap D$; (*f*) $A \cup D$?

7. In the following statements P and Q represent sets. Indicate which of the following statements are true and which ones are false.
   (*a*) $P \cap Q$ is always contained in P.
   (*b*) $P \cup Q$ is always contained in Q.
   (*c*) P is always contained in $P \cup Q$.
   (*d*) Q is always contained in $P \cup Q$.
   (*e*) $P \cup Q$ is always contained in P.
   (*f*) $P \cap Q$ is always contained in Q.
   (*g*) P is always contained in $P \cap Q$.
   (*h*) Q is always contained in $P \cap Q$.
   (*i*) If $P \supset Q$, then $P \cap Q = P$.
   (*j*) If $P \supset Q$, then $P \cap Q = Q$.
   (*k*) If $P \subset Q$, then $P \cup Q = P$.
   (*l*) If $P \subset Q$, then $P \cup Q = Q$.
8. What is the solution set for the statement $a + 2 = 2$, i.e., the set of all solutions, of statement $a + 2 = 2$?
9. What is the solution set for the statement $a + 2 = a + 4$?
10. Let D be the set of ordered pairs (x, y) for which $x + y = 5$, and let E be the set of ordered pairs (x, y) for which $x - y = 1$. What is $D \cap E$?

11–30. Copy figures and use shading to illustrate the following sets.

11. $R \cup S$.
12. $R \cap S$.
13. $(R \cap S)'$.
14. $(R \cup S)'$.
15. R'.
16. S'.
17. $(R')'$.
18. $R' \cup S'$.
19. $R' \cap S'$.
20. $(R' \cap S')'$.
21. $R \cup S$.
22. $R \cap S$.
23. $R' \cap S'$.
24. $R' \cup S'$.
25. $R \cup S$.
26. $R \cap S$.
27. $R' \cap S'$.
28. $R' \cup S'$.
29. $R' \cup S$.
30. $R \cup S'$.

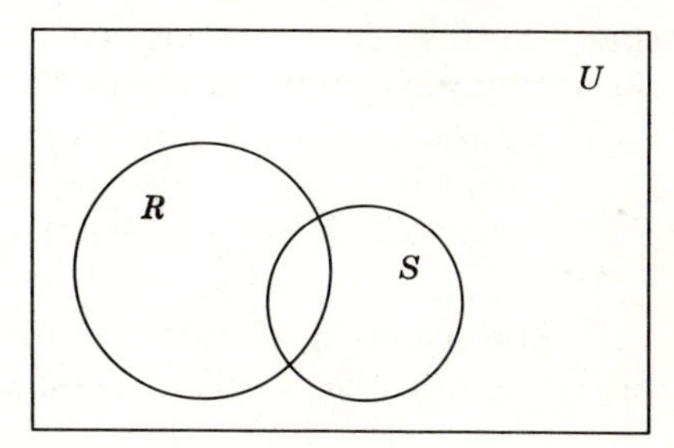

Exs. 11–20.

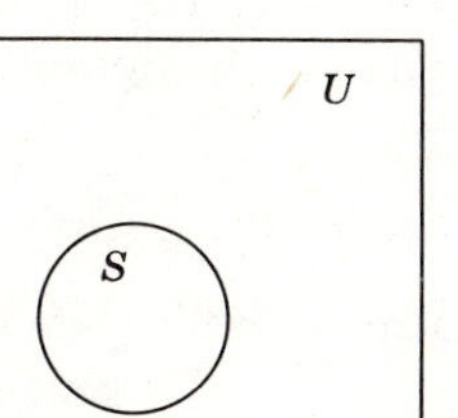

Exs. 21–24.

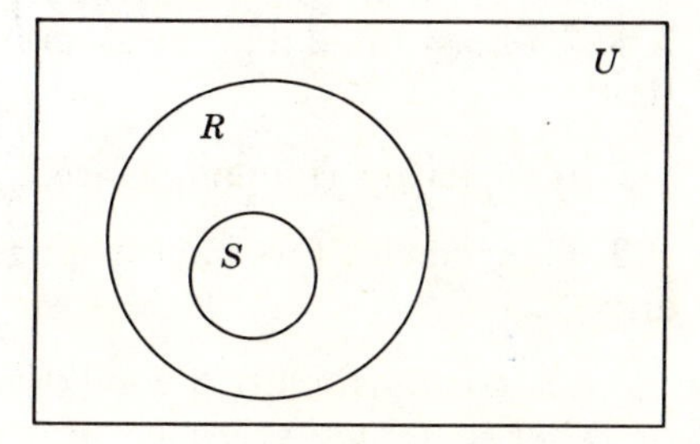

Exs. 25–30.

1.6. Need for definitions. In studying geometry we learn to prove statements by a process of deductive reasoning. We learn to analyze a problem in terms of what data are given, what laws and principles may be accepted as true and, by careful, logical, and accurate thinking, we learn to select a solution to the problem. But before a statement in geometry can be proved, we must agree on certain definitions and properties of geometric figures. It is necessary that the terms we use in geometric proofs have exactly the same meaning to each of us.

Most of us do not reflect on the meanings of words we hear or read during the course of a day. Yet, often, a more critical reflection might cause us to wonder what really we have heard or read.

A common cause for misunderstanding and argument, not only in geometry but in all walks of life, is the fact that the same word may have different meanings to different people.

What characteristics does a good definition have? When can we be certain the definition is a good one? No one person can establish that his definition for a given word is a correct one. What is important is that the people participating in a given discussion agree on the meanings of the word in question and, once they have reached an understanding, no one of the group may change the definition of the word without notifying the others.

This will especially be true in this course. Once we agree on a definition stated in this text, we cannot change it to suit ourselves. On the other hand, there is nothing sacred about the definitions that will follow. They might well be improved on, as long as everyone who uses them in this text agrees to it.

A good definition in geometry has two important properties:

1. The words in the definition must be simpler than the word being defined and must be clearly understood.

2. The definition must be a reversible statement.

Thus, for example, if "right angle" is defined as "an angle whose measure is 90," it is assumed that the meaning of each term in the definition is clear and that:

1. If we have a right angle, we have an angle whose measure is 90.

2. Conversely, if we have an angle whose measure is 90, then we have a right angle.

Thus, the converse of a good definition is always true, although the converse of other statements are *not necessarily* true. The above statement and its converse can be written, "An angle is a right angle if, and only if, its measure is 90. The expression "if and only if" will be used so frequently in this text that we will use the abbreviation "iff" to stand for the entire phrase.

1.7. Need for undefined terms. There are many words in use today that are difficult to define. They can only be defined in terms of other equally undefinable concepts. For example, a "straight line" is often defined as a line "no part of which is curved." This definition will become clear if we can define the word curved. However, if the word curved is then defined as a line "no part of which is straight," we have no true understanding of the definition of the word "straight." Such definitions are called "*circular definitions*." If we define a straight line as one extending without change in direction, the word "direction" must be understood. In defining mathematical terms, we start with undefined terms and employ as few as possible of those terms that are in daily use and have a common meaning to the reader.

In using an undefined term, it is assumed that the word is so elementary that its meaning is known to all. Since there are no easier words to define the term, no effort is made to define it. The dictionary must often resort to "defining" a word by either listing other words, called *synonyms*, which have the same (or almost the same) meaning as the word being defined or by describing the word.

We will use three undefined geometric terms in this book. They are: point, straight line, and plane. We will resort to synonyms and descriptions of these words in helping the student to understand them.

1.8. Points and lines. Before we can discuss the various geometric figures as sets of points, we will need to consider the nature of a point. What is a point? Everyone has some understanding of the term. Although we can represent a point by marking a small dot on a sheet of paper or on a blackboard, it certainly is not a point. If it were possible to subdivide the marker, then subdivide again the smaller dots, and so on indefinitely, we still would not have a point. We would, however, approach a condition which most of us assign to that of a point. Euclid attempted to do this by defining a point as that which has position but no dimension. However, the words "position" and "dimension" are also basic concepts and can only be described by using circular definitions.

We name a point by a capital letter printed beside it, as point "A" in Fig. 1.6. Other geometric figures can be defined in terms of sets of points which satisfy certain restricting conditions.

We are all familiar with lines, but no one has seen one. Just as we can represent a point by a marker or dot, we can represent a line by moving the tip of a sharpened pencil across a piece of paper. This will produce an approximation for the meaning given to the word "line." Euclid attempted to define a line as that which has only one dimension. Here, again, he used

Fig. 1.6. *Fig. 1.7.*

the undefined word "dimension" in his definition. Although we cannot define the word "line," we recognize it as a set of points.

On page 11, we discussed a "straight line" as one no part of which is "curved," or as one which extends without change in directions. The failures of these attempts should be evident. However, the word "straight" is an abstraction that is generally used and commonly understood as a result of many observations of physical objects. The line is named by labeling two points on it with capital letters or by one lower case letter near it. The straight line in Fig. 1.7 is read "line AB" or "line l." Line AB is often written "$\overleftrightarrow{AB}$." In this book, unless otherwise stated, when we use the term "line," we will have in mind the concept of a *straight* line.

If $B \in l$, $A \in l$, and $A \neq B$, we say that l is the line which *contains* A and B. Two points determine a line (see Fig. 1.7). Thus $\overleftrightarrow{AB} = \overleftrightarrow{BA}$.

Two straight lines can intersect in only one point. In Fig. 1.6, $\overleftrightarrow{AB} \cap \overleftrightarrow{AC} = \{A\}$. What is $\overleftrightarrow{AB} \cap \overleftrightarrow{BC}$?

If we mark three points R, S, and T (Fig. 1.8) all on the same line, we see that $\overleftrightarrow{RS} = \overleftrightarrow{ST}$. Three or more points are *collinear* iff they belong to the same line.

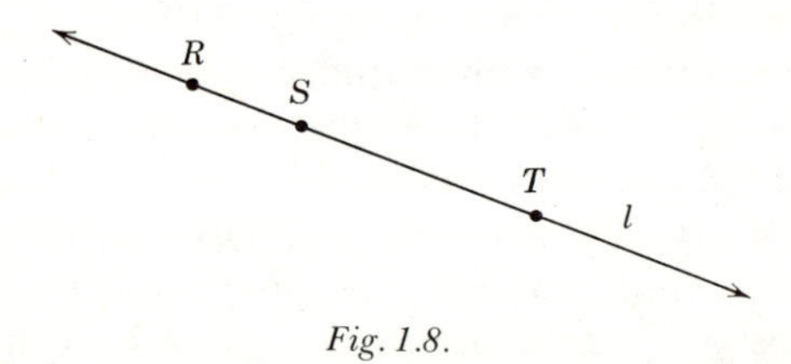

Fig. 1.8.

1.9. Solids and planes. Common examples of solids are shown in Fig. 1.9.

The geometric solid shown in Fig. 1.10 has six faces which are smooth and flat. These faces are subsets of *plane surfaces* or simply *planes*. The surface of a blackboard or of a table top is an example of a plane surface. A plane can be thought of as a set of points.

Definition. A set of points, all of which lie in the same plane, are said to be *coplanar*. Points D, C, and E of Fig. 1.10 are coplanar. A plane can be named by using two points or a single point in the plane. Thus, Fig. 1.11

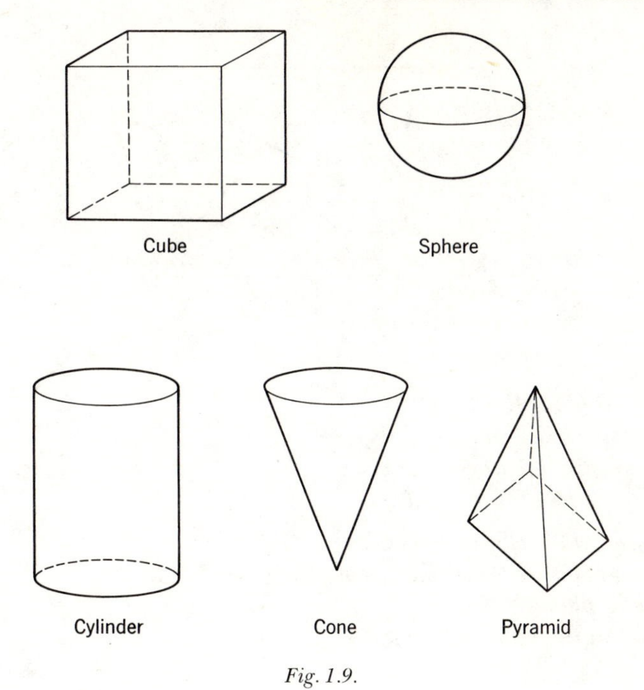

Fig. 1.9.

represents plane *MN* or plane *M*. We can think of the plane as being made up of an infinite number of points to form a surface possessing no thickness but having infinite length and width.

Two lines lying in the same plane whose intersection is the null set are said to be *parallel lines*. If line l is parallel to line m, then $l \cap m = \emptyset$. In Fig. 1.10, $\overleftrightarrow{AB}$ is parallel to $\overleftrightarrow{DC}$ and $\overleftrightarrow{AD}$ is parallel to $\overleftrightarrow{BC}$.

The drawings of Fig. 1.12 and Fig. 1.13 illustrate various combinations of points, lines, and planes.

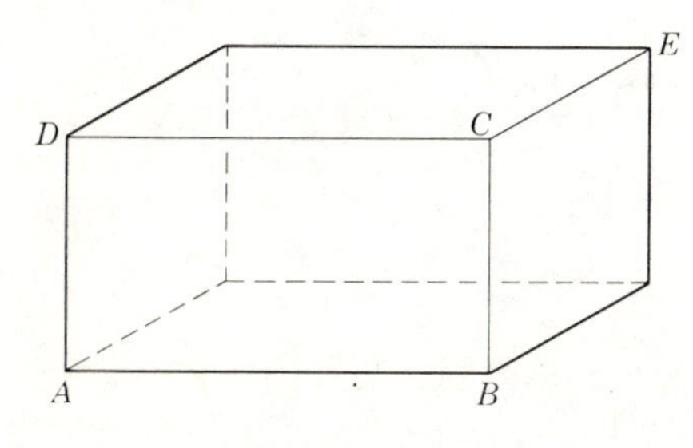

Fig. 1.10.

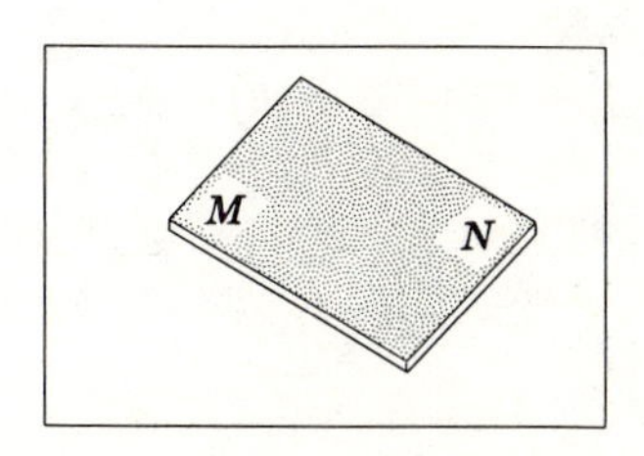

Fig. 1.11.

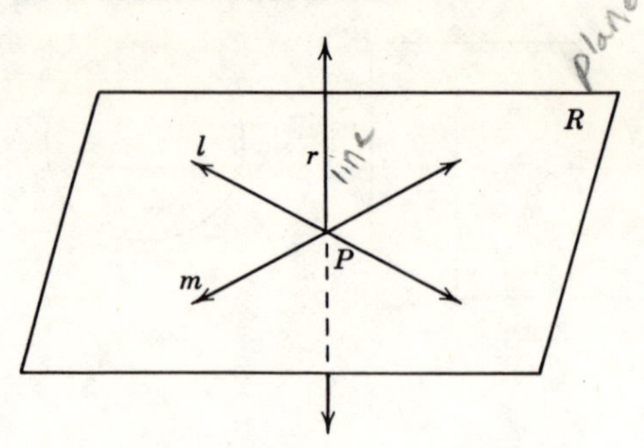

Fig. 1.12.

Line r intersects plane R.
Plane R contains line l and m.
Plane R passes through lines l and m.
Plane R does not pass through line r.

Plane MN and plane RS intersect in $\overleftrightarrow{AB}$.
Plane MN and plane RS both pass through $\overleftrightarrow{AB}$.
$\overleftrightarrow{AB}$ lies in both planes.
$\overleftrightarrow{AB}$ is contained in planes MN and RS.

Exercises

1. How many points does a line contain?
2. How many lines can pass through a given point?
3. How many lines can be passed through two distinct points?
4. How many planes can be passed through two distinct points?

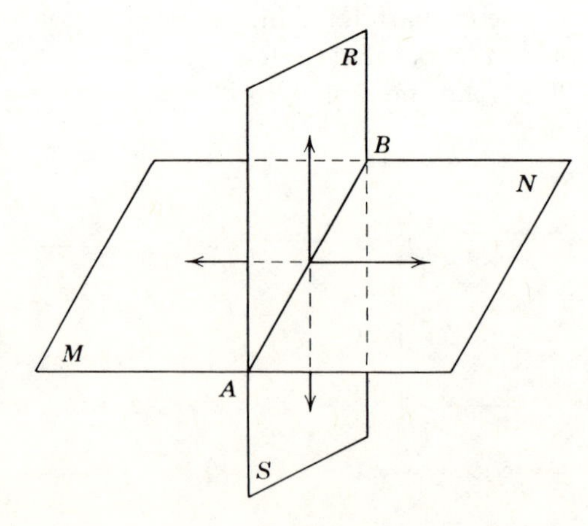

Fig. 1.13.

5. Can a line always be passed through any three distinct points?
6. Can a plane always be passed through any three distinct points?
7. Can two planes ever intersect in a single point?
8. Can three planes intersect in the same straight line?

9–17. Refer to the figure and indicate which of the following statements are true and which are false.

9. Plane AB intersects plane CD in line l.
10. Plane AB passes through line l.
11. Plane AB passes through $\overleftrightarrow{EF}$.
12. Plane CD passes through $\overleftrightarrow{EF}$.
13. $P \in$ plane CD.
14. (plane AB) $\cap$ (plane CD) $= \overleftrightarrow{EF}$.
15. $l \cap \overleftrightarrow{EF} = G$.
16. (plane CD) $\cap\ l = G$.
17. (plane AB) $\cap\ \overleftrightarrow{EF} = \overleftrightarrow{EF}$.

18–38. Draw pictures (if possible) that illustrate the situations described.

18. l and m are two lines and $l \cap m = \{P\}$.
19. l and m are two lines, $P \in l, R \in l, S \in m$ and $\overleftrightarrow{RS} \neq \overleftrightarrow{PR}$.
20. $C \notin \overleftrightarrow{AB}$, and $A \neq B$.
21. $R \in \overleftrightarrow{ST}$.

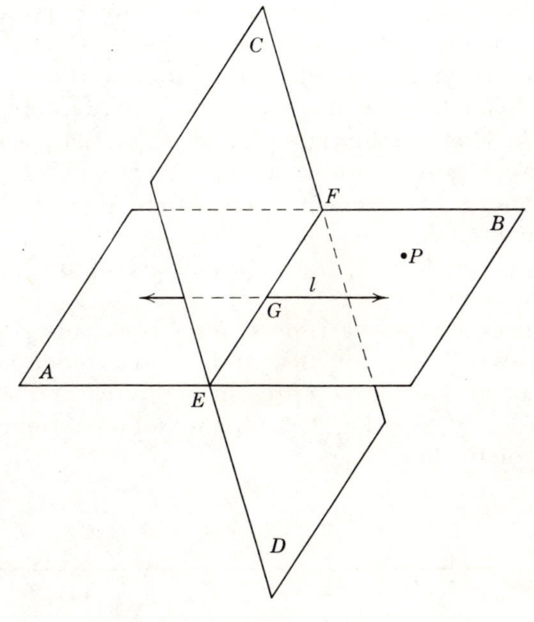

Exs. 9–17.

22. r and s are two lines, and $r \cap s = \emptyset$.
23. r and s are two lines, and $r \cap s \neq \emptyset$.
24. $P \notin \overleftrightarrow{KL}, P \in l$, and $l \cap \overleftrightarrow{KL} = \emptyset$.
25. R, S, and T are three points and $T \in (\overleftrightarrow{RT} \cap \overleftrightarrow{ST})$.
26. r and s are two lines, $A \neq B$, and $\{A, B\} \subset (r \cap s)$.
27. P, Q, R, and S are four points, $Q \in \overleftrightarrow{PR}$, and $R \in \overleftrightarrow{QS}$.
28. P, Q, R, and S are four noncollinear points, $Q \in \overleftrightarrow{PR}$, and $Q \in \overleftrightarrow{PS}$.
29. A, B, and C are three noncollinear points, A, B, and D are three collinear points, and A, C, and D are three collinear points.
30. l, m, and n are three lines, and $P \in (m \cap n) \cap l$.
31. l, m, and n are three lines, $A \neq B$, and $\{A, B\} \subset (l \cap m) \cap n$.
32. l, m, and n are three lines, $A \neq B$, and $\{A, B\} = (l \cap m) \cup (n \cap m)$.
33. A, B, and C are three collinear points, C, D, and E are three noncollinear points, and $E \in \overleftrightarrow{AB}$.
34. (plane RS) $\cap$ (plane MN) $= \overleftrightarrow{AB}$.
35. (plane AB) $\cap$ (plane CD) $= \emptyset$.
36. line $l \subset$ plane AB. line $m \subset$ plane CD. $l \cap m = \{P\}$.
37. (plane AB) $\cap$ (plane CD) $= l$. line $m \in$ plane CD. $l \cap m = \emptyset$.
38. (plane AB) $\cap$ (plane CD) $= l$. line $m \in$ plane CD. $l \cap m \neq \emptyset$.

1.10. Real numbers and the number line. The first numbers a child learns are the counting or natural numbers, e.g., $\{1, 2, 3, \ldots\}$. The natural numbers are infinite; that is, given any number, however large, there is always another number larger (add 1 to the given number). These numbers can be represented by points on a line. Place a point O on the line $X'X$ (Fig. 1.14). The point O will divide the line into two parts. Next, let A be a point on $X'X$ to the right of O. Then, to the right of A, mark off equally spaced points B, C, D, ... For every positive whole number there will be exactly one point to the right of point O. Conversely, each of these points will represent only one positive whole number.

In like manner, points R, S, T, ... can be marked off to the left of point O to represent negative whole numbers.

The distance between points representing consecutive integers can be divided into halves, thirds, fourths, and so on, indefinitely. Repeated division would make it possible to represent all positive and negative fractions with points on the line. Note Fig. 1.15 for a few of the numbers that might be assigned to points on the line.

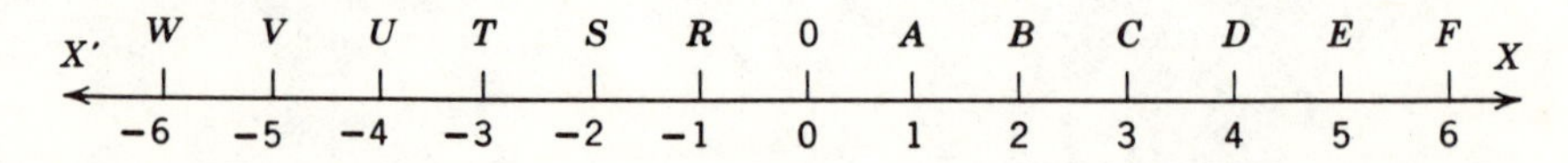

Fig. 1.14.

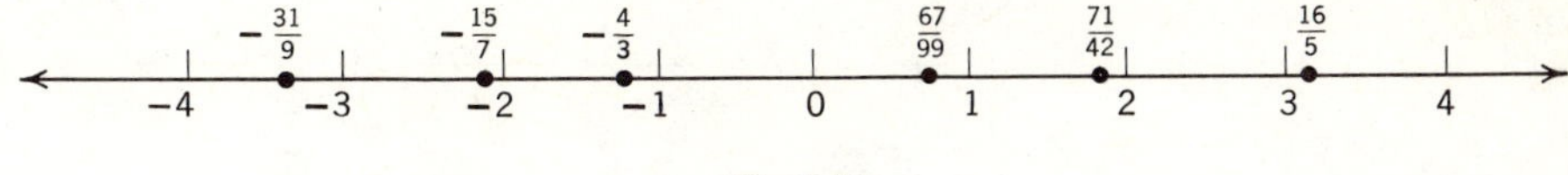

Fig. 1.15.

We have now expanded the points on the line to represent all real rational numbers.

Definition: A *rational number* is one that can be expressed as a quotient of integers.

It can be shown that every quotient of two integers can be expressed as a repeating decimal or decimal that terminates, and every such decimal can be written as an equivalent indicated quotient of two integers. For example, $13/27 = 0.481481\ldots$ and $1.571428571428\ldots = 11/7$ are rational numbers.

The rational numbers form a very large set, for between any two rational numbers there is a third one. Therefore, there are an infinite number of points representing rational numbers on any given scaled line. However, the rational numbers still do not completely fill the scaled line.

Definition: An *irrational number* is one that cannot be expressed as the quotient of two integers (or as a repeating or terminating decimal).

Examples of irrational numbers are $\sqrt{2}$, $-\sqrt{3}$, $\sqrt[3]{5}$, and π. Approximate locations of some rational and irrational numbers on a scaled line are shown in Fig. 1.16.

The union of the sets of rational and irrational numbers form the set of *real numbers.* The line that represents all the real numbers is called the *real number line.* The number that is paired with a point on the number line is called the *coordinate* of that point.

We summarize by stating that the real number line is made up of an infinite set of points that have the following characteristics.

1. *Every point on the line is paired with exactly one real number.*
2. *Every real number can be paired with exactly one point on the line.*

When, given two sets, it is possible to pair each element of each set with exactly one element of the other, the two sets are said to have a *one-to-one correspondence.* We have just shown that there is a one-to-one correspondence between the set of real numbers and the set of points on a line.

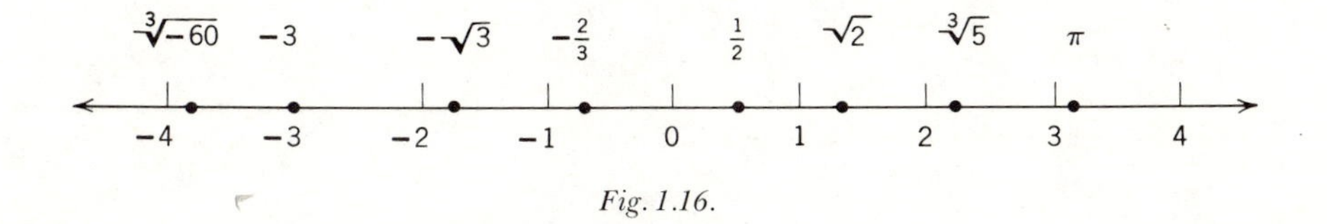

Fig. 1.16.

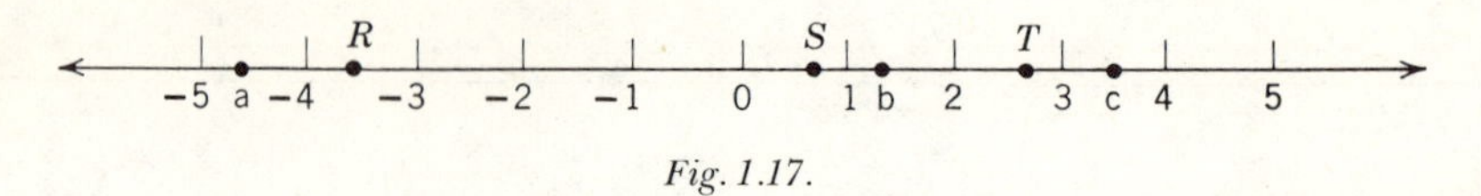

Fig. 1.17.

1.11. Order and the number line. All of us at one time or another engage in comparing sizes of real numbers. Symbols are often used to indicate the relative sizes of real numbers. Consider the following.

Symbol	*Meaning*
$a = b$	a equals b
$a \neq b$	a is not equal to b
$a > b$	a is greater than b
$a < b$	a is less than b
$a \geqslant b$	a is either greater than b or a is equal to b
$a \leqslant b$	a is either less than b or a is equal to b

It should be noted that $a > b$ and $b < a$ have exactly the same meaning; that is, if a is more than b, then b is less than a.

The number line is a convenient device for visualizing the *ordering* of real numbers. If $b > a$, the point representing the number b will be located to the right of the point on the number line representing the number a (see Fig. 1.17). Conversely, if point S is to the right of point R, then the number which is assigned to S must be larger than that assigned to R. In the figure, $b < c$ and $c > a$.

When we write or state $a = b$ we mean simply that a and b are different names for the same number. Thus, points which represent the same number on a number line must be identical.

1.12. Distance between points. Often in the study of geometry, we will be concerned with the "distance between two points." Consider the number line of Fig. 1.18 where points A, P, B, C, respectively represent the integers $-3, 0, 3, 6$. We note that A and B are the same distance from P, namely 3.

Next consider the distance between B and C. While the coordinates differ in these and the previous two cases, it is evident that the distance between the points is represented by the number 3.

How can we arrive at a rule for determining distance between two points? We could find the distance between two points on a scaled line by subtracting

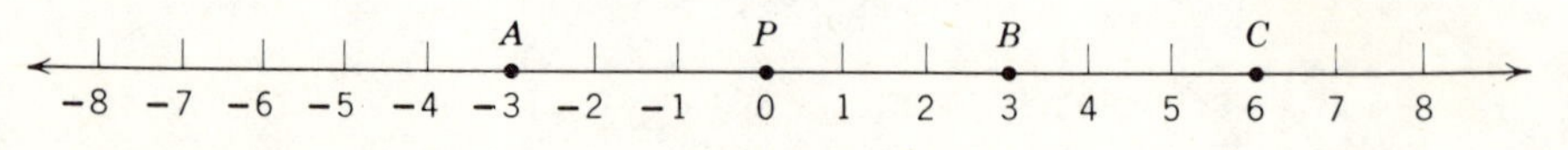

Fig. 1.18.

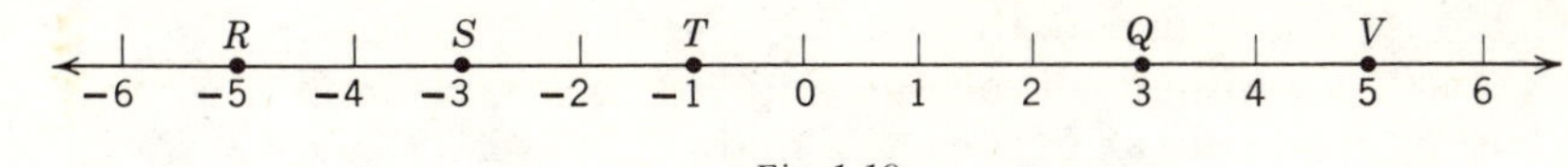

Fig. 1.19.

the smaller number represented by these two points from the larger. Thus, in Fig. 1.19:

The distance from T to V $= 5-(-1) = 6$.
The distance from S to T $= (-1)-(-3) = 2$.
The distance from Q to R $= 3-(-5) = 8$.

Another way we could state the above rule could be: "Subtract the coordinate of the left point from that of the point to the right." However, this rule would be difficult to apply if the coordinates were expressed by place holders a and b. We will need to find some way of always arriving at a number that is positive and is associated with the difference of the coordinates of the point. To do this we use the symbol $|\ |$. The symbol $|x|$ is called the *absolute value of x*. In the study of algebra the absolute value of any number x is defined as follows.

$$|x| = x \text{ if } x \geqslant 0$$
$$|x| = -x \text{ if } x < 0$$

Consider the following illustrations of the previous examples.

Column 1	*Column 2*
$\|3\| = 3$	$\|-3\| = 3$
$\|5-(-1)\| = \|6\| = 6$	$\|(-1)-(+5)\| = \|-6\| = 6$
$\|(-1)-(-3)\| = \|2\| = 2$	$\|(-3)-(-1)\| = \|-2\| = 2$
$\|3-(-5)\| = \|8\| = 8$	$\|(-5)-(+3)\| = \|-8\| = 8$

Thus, we note that to find the distance between two points we need only to subtract the coordinates in either order and then take the absolute value of the difference. *If a and b are the coordinates of two points, the distance between the points can be* expressed either by $|a-b|$ or $|b-a|$.

Exercises

1. What is the coordinate of B? of D?
2. What point lies halfway between B and D?
3. What is the coordinate of the point 7 units to the left of D?

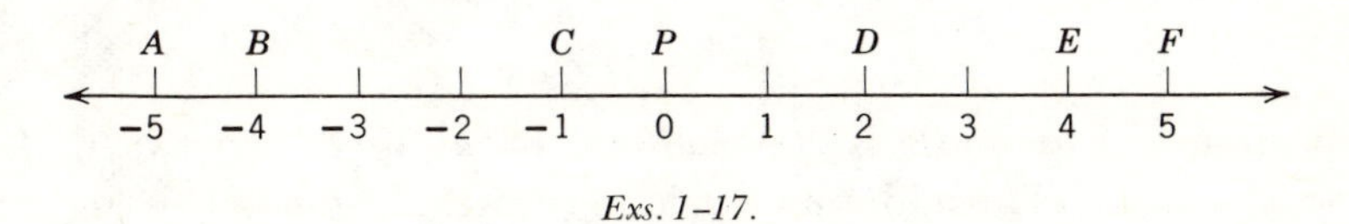

Exs. 1–17.

4. What is the coordinate of the point 3 units to the right of C?
5. What is the coordinate of the point midway between C and F?
6. What is the coordinate of the point midway between D and F?
7. What is the coordinate of the point midway between C and E?
8. What is the coordinate of the point midway between A and C?

9–16. Let a, b, c, d, e, f, p represent the coordinates of points A, B, C, D, E, F, P, respectively. Determine the values of the following.

9. $e-p$	10. $b-p$	11. $b-c$
12. $\lvert d-b \rvert$	13. $\lvert e-d \rvert$	14. $\lvert d-f \rvert$
15. $\lvert c-d \rvert$	16. $\lvert a-c \rvert$	17. $\lvert a-e \rvert$

18–26. Evaluate the following.

18. $\lvert -1 \rvert + \lvert 2 \rvert$	19. $\lvert -3 \rvert + \lvert -4 \rvert$	20. $\lvert -8 \rvert - \lvert -3 \rvert$
21. $\lvert -4 \rvert - \lvert -6 \rvert$	22. $\lvert -3 \rvert \times \lvert 3 \rvert$	23. $2\lvert -4 \rvert$
24. $\lvert -4 \rvert^2$	25. $\lvert 2 \rvert^2 + \lvert -2 \rvert^2$	26. $\lvert 2 \rvert^2 - \lvert -2 \rvert^2$

1.13. Segments. Half-lines. Rays. Let us next consider that part of the line between two points on a line.

Definitions: The part of line AB between A and B, together with points A and B, is called *segment AB* (Fig. 1.19*a*). Symbolically it is written $\overline{AB}$. The points A and B are called the *endpoints* of $\overline{AB}$. The number that tells how far it is from A to B is called the *measure* (or *length*) of $\overline{AB}$. In this text we will use the symbol $m\overline{AB}$ to mean the length of $\overline{AB}$.

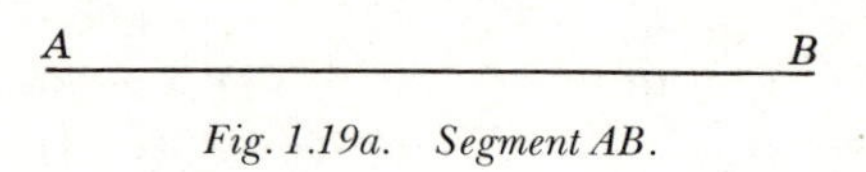

Fig. 1.19a. Segment AB.

The student should be careful to recognize the differences between the meanings of the symbols $\overline{AB}$ and $m\overline{AB}$. The first refers to a geometric figure; the second to a number.

Definition: B is *between* A and C (see Fig. 1.20) if, and only if, A, B, and C are distinct points on the same line and $m\overline{AB} + m\overline{BC} = m\overline{AC}$. Using the equal sign implies simply that the name used on the left ($m\overline{AB} + m\overline{BC}$) and the name used on the right of the equality sign ($m\overline{AC}$) are but two different names for the same number.

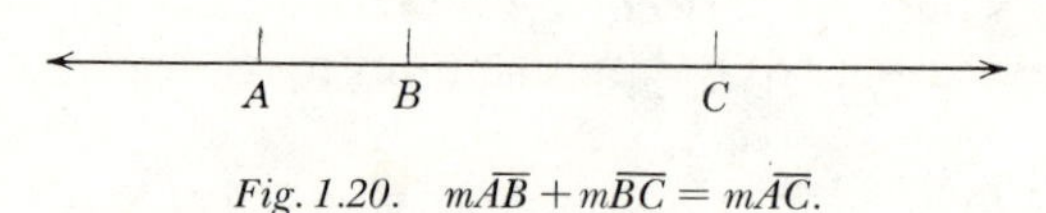

Fig. 1.20. $m\overline{AB} + m\overline{BC} = m\overline{AC}$.

Definition: A point B is the *midpoint* of $\overline{AC}$ iff B is between A and C and $m\overline{AB} = m\overline{BC}$. The midpoint is said to *bisect* the segment (see Fig. 1.21).

Fig. 1.21. $m\overline{AB} = m\overline{BC}$.

A line or a segment which passes through the midpoint of a second segment bisects the segment. If, in Fig. 1.22, M is the midpoint of $\overline{AB}$, then $\overleftrightarrow{CD}$ bisects $\overline{AB}$.

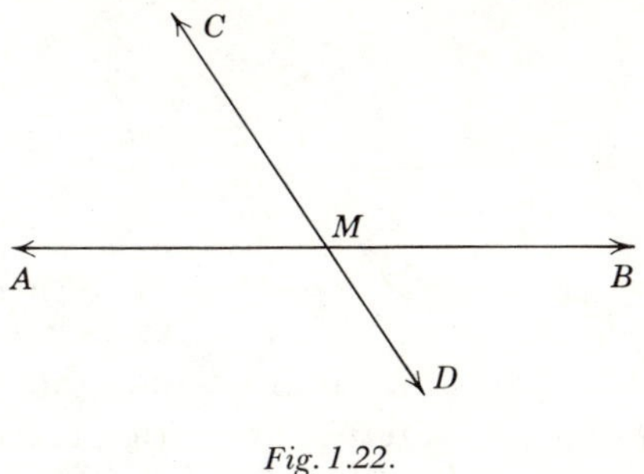

Fig. 1.22.

Definition: The set consisting of the points between A and B is called an *open segment* or the *interval* joining A and B. It is designated by the symbol $\overset{\circ\!-\!\circ}{AB}$.

Definition: For any two distinct points A and B, the figure $\{A\} \cup \{\overset{\circ\!-\!\circ}{AB}\}$ is called a *half-open segment*. It is designated by the symbol $\overset{-\!\circ}{AB}$. Open segments and half-open segments are illustrated in Fig. 1.23.

Every point on a line divides that line into two parts. Consider the line l through points A and B (Fig. 1.24*a*).

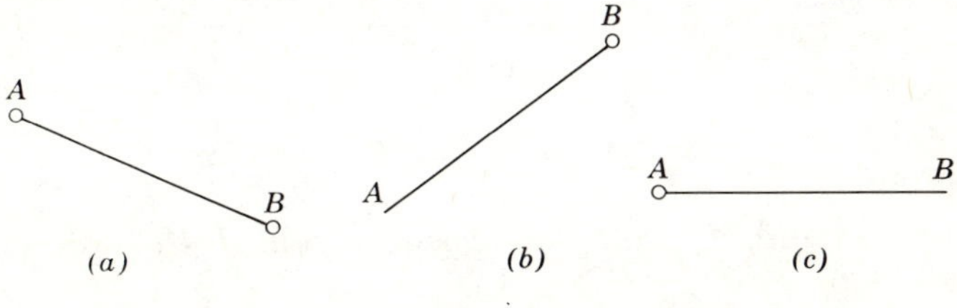

Fig. 1.23. (*a*) $\overset{\circ\!-\!\circ}{AB}$ (*b*) $\overset{-\!\circ}{AB}$ (*c*) $\overset{\circ\!-}{AB}$.

Definition: If A and B are points of line l, then the set of points of l which are on the same side of A as is B is the *half-line from A through B* (Fig. 1.24*b*).

The symbol for the half-line from A through B is $\overset{\circ\!\!\rightarrow}{AB}$ and is read "half-line AB." The arrowhead indicates that the half-line includes *all* points of the line on the same side of A as is B. The symbol for the half-line from B through A (Fig. 1.24*c*) is $\overset{\circ\!\!\rightarrow}{BA}$. Note that A is not an element of $\overset{\circ\!\!\rightarrow}{AB}$. Similarly, B does not belong to $\overset{\circ\!\!\rightarrow}{BA}$.

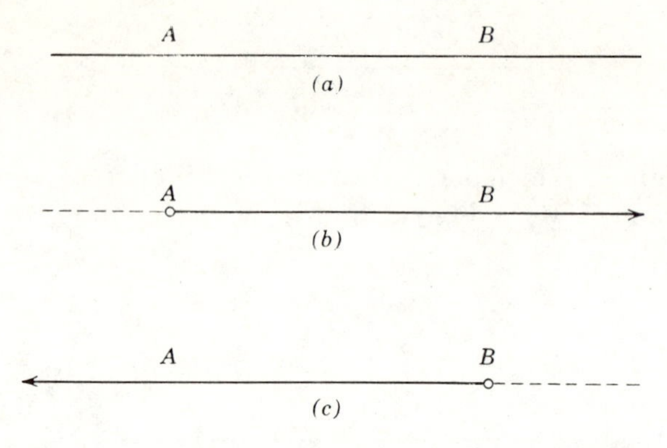

Fig. 1.24. (a) Line AB. (b) Half-line AB. Half-line BA.

Definition: If A and B are points of line l, then the set of points consisting of A and all the points which are on the same side of A as is B is the *ray from A through B*. The point A is called the *endpoint* of ray AB.

The symbol for the ray from A through B is $\overrightarrow{AB}$ (Fig. 1.25*a*) and is read "ray AB." The symbol for the ray from B through A (Fig. 1.25*b*) is $\overrightarrow{BA}$.

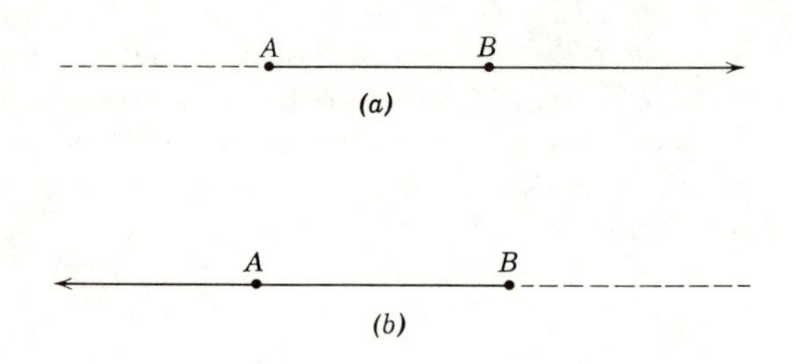

Fig. 1.25. (a) Ray AB. (b) Ray BA.

Definition: $\overrightarrow{BA}$ and $\overrightarrow{BC}$ are called *opposite rays* iff A, B, and C are collinear points and B is between A and C (Fig. 1.26).

It will be seen that points A and B of Fig. 1.26 determine nine geometric figures: $\overleftrightarrow{AB}$, $\overline{AB}$, $\overset{\circ\!-\!\circ}{AB}$, $\overset{\circ\!\rightarrow}{AB}$, $\overrightarrow{AB}$, $\overrightarrow{BA}$, $\overset{\circ\!\rightarrow}{BA}$, the ray opposite $\overrightarrow{AB}$, and the ray opposite $\overrightarrow{BA}$. The union of $\overrightarrow{BA}$ and $\overrightarrow{BC}$ is $\overleftrightarrow{BC}$ (or $\overleftrightarrow{AC}$). The intersection of $\overset{\circ\!\rightarrow}{BA}$ and $\overset{\circ\!\rightarrow}{AB}$ is $\overset{\circ\!-\!\circ}{AB}$.

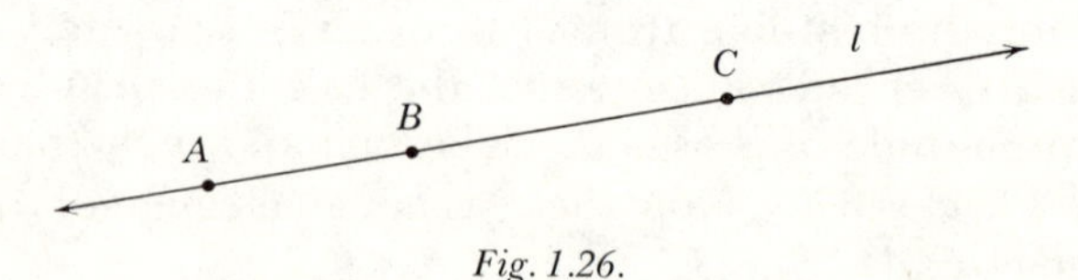

Fig. 1.26.

Exercises

1–12. Given: A, B, C, D are collinear and C is the midpoint of $\overline{AD}$.

1. Does C bisect $\overline{AD}$?
2. Are B, C, and D collinear?
3. Does $\overleftrightarrow{BC}$ pass through A?
4. Does $m\overline{AB} + m\overline{BC} = m\overline{AC}$?
5. Is C between A and B?

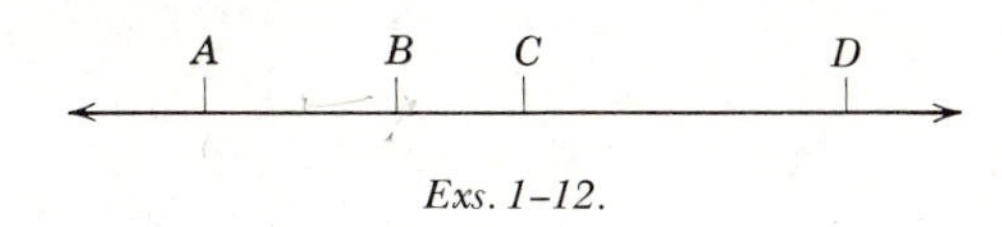

Exs. 1–12.

6. Are $\overrightarrow{CA}$ and $\overrightarrow{CD}$ opposite rays?
7. Is $C \in \overrightarrow{BD}$?
8. What is $\overrightarrow{CA} \cap \overrightarrow{BD}$?
9. What is $\overset{\circ\!\rightarrow}{BA} \cap \overset{\circ\!\rightarrow}{BD}$?
10. What is $\overline{AB} \cup \overset{\circ\!\rightarrow}{BC}$?
11. What is $\overline{AB} \cup \overrightarrow{BC}$?
12. What is $\overset{\circ\!\rightarrow}{CB} \cap \overset{\circ\!\rightarrow}{AD}$?

13–32. Draw pictures (if possible) that illustrate the situations described in the following exercises.

13. B is between A and C, and C is between A and D.
14. A, B, C, and D are four collinear points, A is between C and D, and D is between A and B.
15. $R \in \overleftrightarrow{ST}$ and $R \notin \overrightarrow{ST}$.
16. $\overset{\circ\!-\!\circ}{PQ} \subset \overset{\circ\!-\!\circ}{RS}$.
17. $\overset{\circ\!-\!\circ}{QP} \subset \overrightarrow{RG}$.
18. $B \in \overset{\circ\!-\!\circ}{AC}$ and C is between B and D.
19. $\overleftrightarrow{PQ} = \overset{\circ\!\rightarrow}{PR} \cup \overset{\circ\!\rightarrow}{PQ}$
20. $T \in \overset{\circ\!-\!\circ}{RS}$ and $S \in \overset{\circ\!-\!\circ}{RT}$.
21. $\overleftrightarrow{PQ} = \overrightarrow{PR} \cup \overrightarrow{PQ}$.
22. $\overset{\circ\!-\!\circ}{AB} \cap \overset{\circ\!-\!\circ}{CD} = \{E\}$.
23. $\overset{\circ\!\rightarrow}{PQ}$, $\overset{\circ\!\rightarrow}{PR}$, and $\overset{\circ\!\rightarrow}{PS}$ are three half-lines, and $\overset{\circ\!-\!\circ}{QR} \cap \overset{\circ\!\rightarrow}{PS} \neq \emptyset$.
24. $\overset{\circ\!\rightarrow}{PQ}$, $\overset{\circ\!\rightarrow}{PR}$, and $\overset{\circ\!\rightarrow}{PS}$ are three half-lines, and $\overset{\circ\!-\!\circ}{QR} \cap \overset{\circ\!\rightarrow}{PS} = \emptyset$.
25. $\overleftrightarrow{PQ} = \overset{\circ\!\rightarrow}{PR} \cup \overrightarrow{PQ}$.
26. $\overset{\circ\!\rightarrow}{PQ} = \overline{PQ} \cup \overset{\circ\!\rightarrow}{QR}$.
27. $\overset{\circ\!\rightarrow}{PQ} = \overset{\circ\!-\!\circ}{PQ} \cup \overrightarrow{QR}$.
28. P, Q, and R are three collinear points, $P \in \overset{\circ\!-\!\circ}{QR}$, and $R \notin \overset{\circ\!-\!\circ}{PQ}$.
29. l, m, and n are three distinct lines, $l \cap m = \emptyset$, $m \cap n = \emptyset$.
30. l, m, and n are three distinct lines, $l \cap m = \emptyset$, $m \cap n = \emptyset$, $l \cap n \neq \emptyset$.
31. $R \in \overset{\circ\!-\!\circ}{KL}$ and $L \in \overset{\circ\!-\!\circ}{RH}$.
32. $D \in \overset{\circ\!-\!\circ}{JK}$ and $F \in \overset{\circ\!-\!\circ}{DK}$.

1.14. Angles. The figure drawn in Fig. 1.27 is a representation of an angle.

Definitions: An *angle* is the union of two rays which have the same endpoint. The rays are called the *sides* of the angle, and their common endpoint is called the *vertex* of the angle.

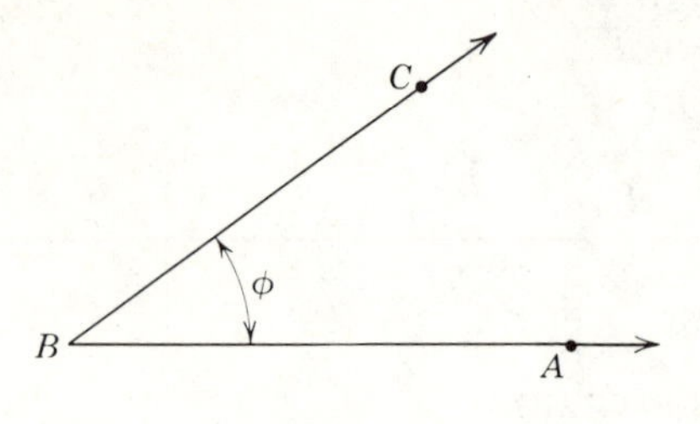

Fig. 1.27.

The symbol for angle is $\angle$; the plural, $\measuredangle$s. There are three common ways of naming an angle: (1) by three capital letters, the middle letter being the vertex and the other two being points on the sides of the angle, as $\angle ABC$; (2) by a single capital letter at the vertex if it is clear which angle is meant, as $\angle B$; and (3) by a small letter in the interior of the angle. In advanced work in mathematics, the small letter used to name an angle is usually a Greek letter, as $\angle\phi$. The student will find the letters of the Greek alphabet in the appendix of this book.

The student should note that the sides of an angle are infinitely long in two directions. This is because the sides of an angle are rays, not segments. In Fig. 1.28, $\angle AOD$, $\angle BOE$, and $\angle COF$ all refer to the same angle, $\angle O$.

1.15. Separation of a plane. A point separates a line into two half-lines. In a similar manner, we can think of a line separating a plane U into two

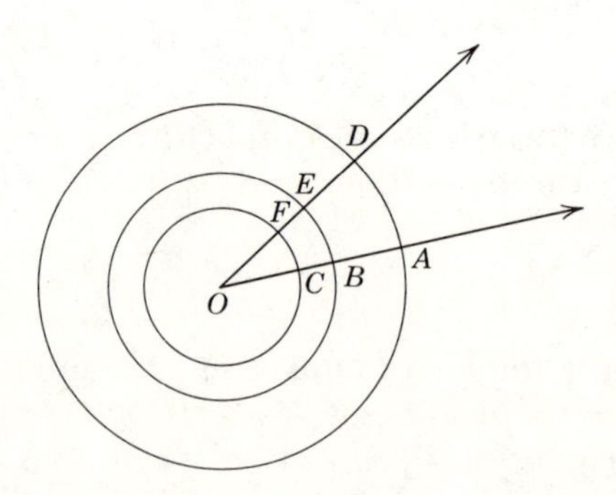

Fig. 1.28.

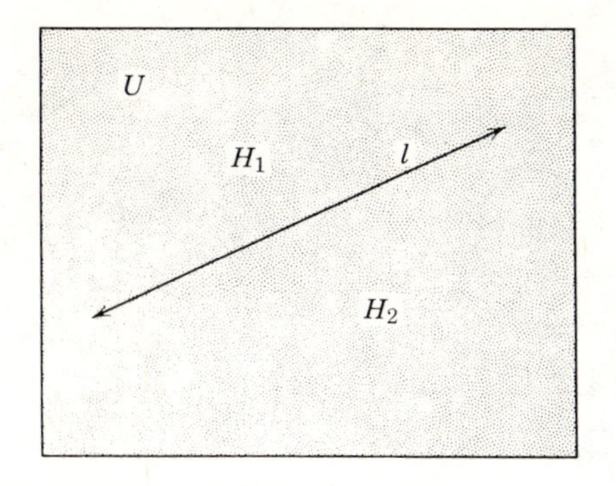

Fig. 1.29.

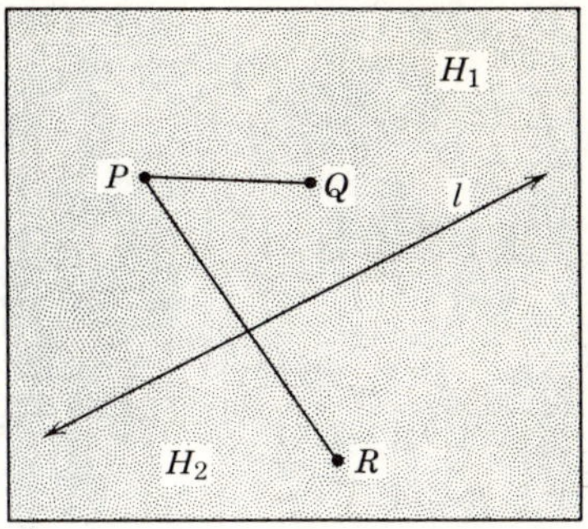

Fig. 1.30.

half-planes H_1 and H_2 (Fig. 1.29). The two sets of points H_1 and H_2 are called *sides* (or *half-planes*) of line l. The line l is called the *edge* of each half-plane. Notice that a half-plane does not contain points of its edge; that is, l does not lie in either of the two half-planes. We can write this fact as $H_1 \cap l = \emptyset$ and $H_2 \cap l = \emptyset$. A half-plane together with its edge is called a *closed half-plane*. The plane $U = H_1 \cup l \cup H_2$.

If two points P and Q of plane U lie in the same half-plane, they are said to lie on *the same side* of the line l which divides the plane into the half-planes. In this case $\overline{PQ} \cap l = \emptyset$. If P lies in one half-plane of U and R in the other (Fig. 1.30), they lie on *opposite sides* of l. Here $\overline{PR} \cap l \neq \emptyset$.

1.16. Interior and exterior of an angle. Consider $\angle ABC$ (Fig. 1.31) lying in plane U. Line AB separates the plane into two half-planes, one of which contains C. Line BC also separates the plane into two half-planes, one of which contains A. The intersection of these two half-planes is the interior of the $\angle ABC$.

Definitions: Consider an $\angle ABC$ lying in plane U. The *interior* of the angle is the set of all points of the plane on the same side of $\overleftrightarrow{AB}$ as C and on the same side of $\overleftrightarrow{BC}$ as A. The *exterior* of $\angle ABC$ is the set of all points of U that do not lie on the interior of the angle or on the angle itself.

A check of the definitions will show that in Fig. 1.31, point P is in the interior of $\angle ABC$; points Q, R, and S are in the exterior of the angle.

1.17. Measures of angles. We will now need to express the "size" of an angle in some way. Angles are usually measured in terms of the degree unit.

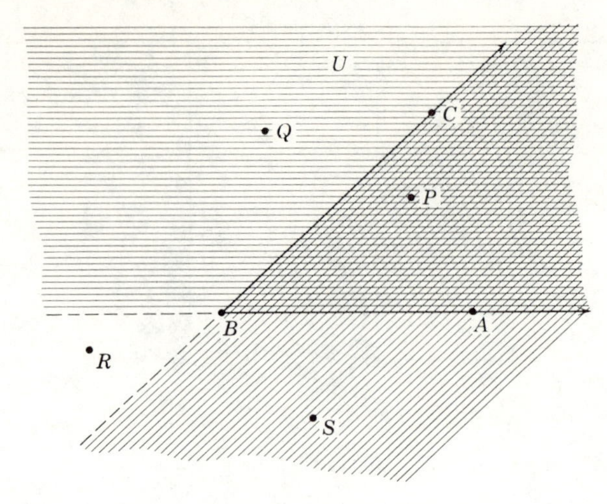

Fig. 1.31.

Definition: To each angle there corresponds exactly one real number r between 0 and 180. The number r is called the *measure* or *degree measure* of the angle.

While we will discuss circles, radii, and arcs at length in Chapter 7, it is assumed that the student has at least an intuitive understanding of the terms. Thus, to help the student better to comprehend the meaning of the term we will state that if a circle is divided into 360 equal arcs and radii are drawn to any two consecutive points of division, the angle formed at the center by these radii has a measure of one degree. It is a one-degree angle. The symbol for degree is °. The degree is quite small. We gain a rough idea of the "size" of a one-degree angle when we realize that, if in Fig. 1.32 (not drawn to scale), $\overline{BA}$ and $\overline{BC}$ are each 57 inches long and $\overline{AC}$ is one inch long, then $\angle ABC$ has a measure of approximately one.

We can describe the measure of angle ABC three ways:

The measure of $\angle ABC$ is 1.

$m\angle ABC = 1$.

$\angle ABC$ is a one-degree angle.

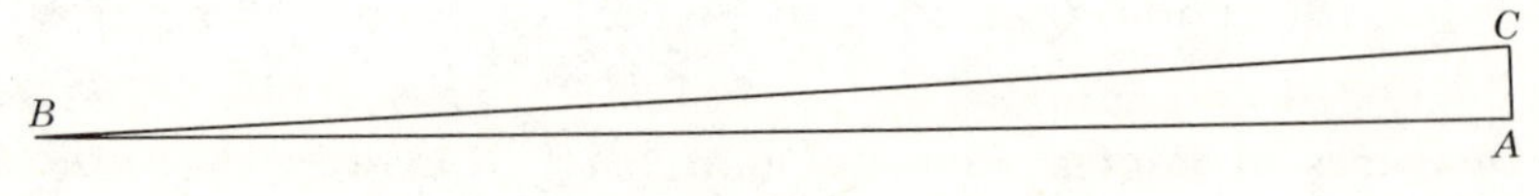

Fig. 1.32.

Just as a ruler is used to estimate the measures of segments, the measure of an angle can be found roughly with the aid of a protractor. (Fig. 1.33).

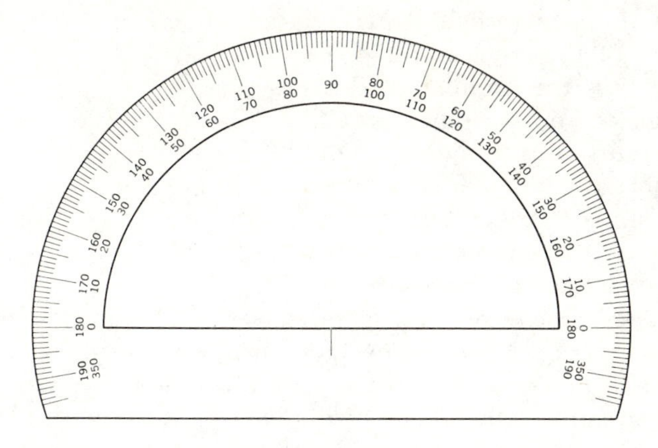

Fig. 1.33. A protractor.

Thus, in Fig. 1.34, we indicate the angle measures as:

$$m\angle AOB = 20 \qquad m\angle COD = |86-50| \text{ or } |50-86| = 36$$
$$m\angle AOD = 86 \qquad m\angle DOF = |150-86| \text{ or } |86-150| = 64$$
$$m\angle AOF = 150 \qquad m\angle BOE = |110-20| \text{ or } |20-110| = 90$$

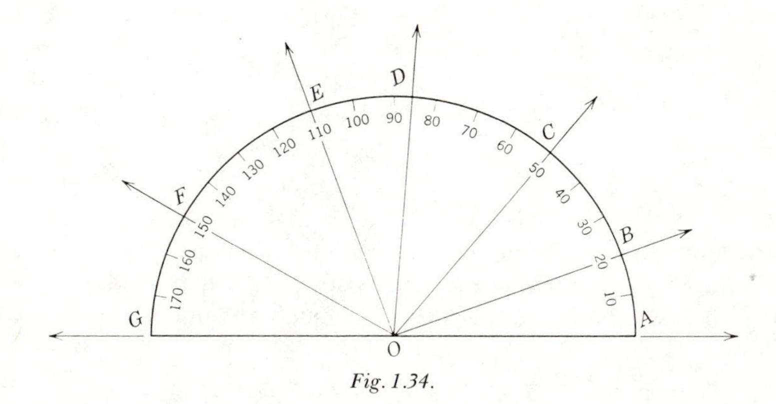

Fig. 1.34.

The reader should note that the measure of an angle is merely the absolute value of the difference between numbers corresponding to the sides of the angle. Hence, as such, it is merely a number and no more. We should *not express the measure of an angle as*, let us say, *30 degrees*. However, we will always indicate in a diagram the measure of an angle by inserting the number with a degree sign in the interior of the angle (see Fig. 1.35). The number 45 is the number of degrees in the angle. The number itself is called the measure of the angle. By defining the measure of the angle as a number, we make it unnecessary to use the word degree or to use the symbol for degree in expressing the measure of the angle.

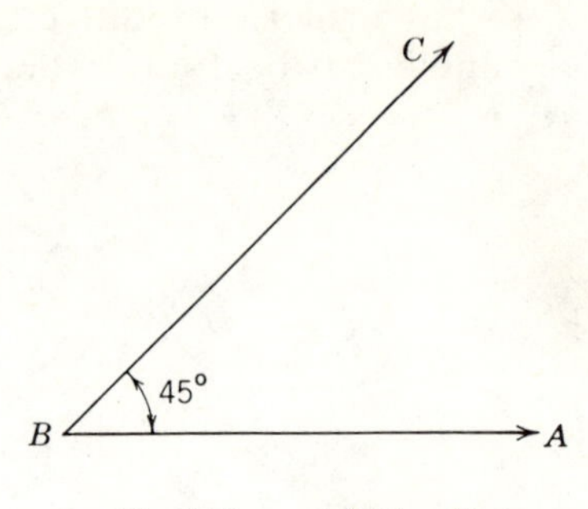

Fig. 1.35. $m\angle ABC = 45$.

In using the protractor, we restrict ourselves to angles whose measures are no greater than 180. This will exclude the measures of a figure such as $\angle ABC$ illustrated in Fig. 1.36. While we know that angles can occur whose measures are greater than 180, they will not arise in this text. Hence $\angle ABC$ in such a figure will refer to the angle with the smaller measure. The study of angles whose measures are greater than 180 will be left to the more advanced courses in mathematics.

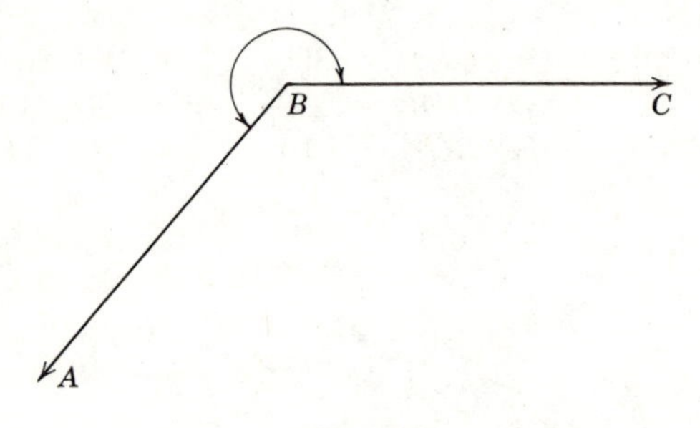

Fig. 1.36.

The student may wonder about the existence of an angle whose measure is 0. We will assume that such an angle exists when the two sides of the angle coincide. You will note that the interior of such an angle is the empty set, $\emptyset$.

Exercises (A)

1. Name the angle formed by $\overrightarrow{MD}$ and $\overrightarrow{MC}$ in three different ways.
2. Name $\angle\alpha$ in four additional ways.
3. Give three additional ways to name $\overleftrightarrow{DM}$.

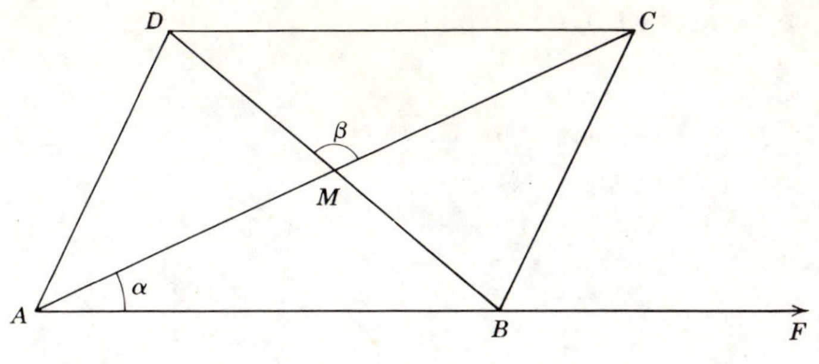

Exs. 1–10.

4. Name the two sides of $\angle FBC$.
5. What is $\angle ABD \cap \angle DBC$?
6. What is $\angle AMD \cap \angle BMC$?
7. Name three angles whose sides are pairs of opposite rays.
8. What is $\overleftrightarrow{AC} \cap \overleftrightarrow{BD}$?
9. What is $\overrightarrow{\mathrm{MA}} \cup \overrightarrow{MD}$?
10. What is $\overrightarrow{MA} \cup \overset{\circ\!\!\longrightarrow}{MD}$?

11–20. Draw (if possible) pictures that illustrate the situations described in each of the following.

11. l is a line. $\overline{PQ} \cap l = \emptyset$.
12. l is a line. $\overline{PQ} \cap l \neq \emptyset$.
13. l is a line. $\overline{PQ} \cap l \neq \emptyset$. $\overset{\circ\!-\!\circ}{PQ} \cap l = \emptyset$.
14. l is a line. $\overline{PQ} \cap l = \emptyset$. $\overline{PR} \cap l = \emptyset$.
15. l is a line. $\overline{PQ} \cap l = \emptyset$. $\overline{PR} \cap l \neq \emptyset$.
16. l is a line. $\overline{PQ} \cap l = \emptyset$. $QR \cap l = \emptyset$. $\overline{PR} \cap l \neq \emptyset$.
17. l is a line. $\overline{PQ} \cap l = \emptyset$. $\overline{QR} \cap l \neq \emptyset$. $\overset{\circ\!-\!\circ}{PR} \cap l = \emptyset$.
18. l is a line which separates plane U into half-planes H_1 and H_2. $\overleftrightarrow{PQ} \cap l = \emptyset$, $P \in H_1, Q \in H_2$.
19. l determines the two half-planes h_1 and h_2. $R \in l, S \notin l, \overset{\circ\!\!\longrightarrow}{RS} \subset h_1$.
20. l determines the two half-planes h_1 and h_2. $R \in l, S \notin l, \overrightarrow{RS} \subset h_1$.

Exercises (B)

21. Draw two angles whose interiors have no points in common.
22. Indicate the measure of the angle in three different ways.
23. By using a protractor, draw an angle whose measure is 55. Label the angle $\angle KTR$.

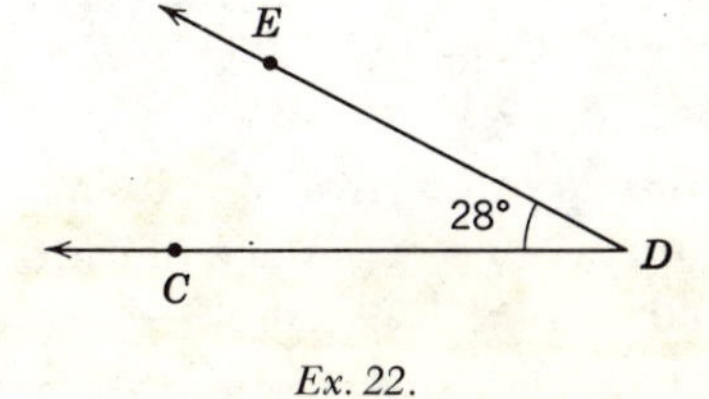

Ex. 22.

24. Find the value of each of the following:

(*a*) $m\angle AJC$. (*d*) $m\angle DJB$. (*g*) $m\angle HJC + m\angle FJE$.
(*b*) $m\angle CJE$. (*e*) $m\angle BJF$. (*h*) $m\angle HJB - m\angle FJD$.
(*c*) $m\angle HJC$. (*f*) $m\angle CJD + m\angle GJD$. (*i*) $m\angle DJG - m\angle BJC$.

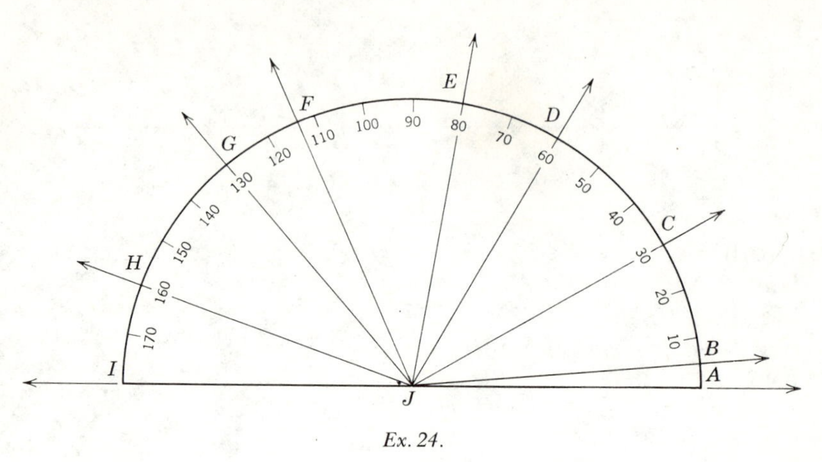

Ex. 24.

25. Draw $\overline{AB} \subset l$ such that $m(\overline{AB}) = 4$ inches. At A draw $\overrightarrow{AC}$ such that $m\angle BAC = 63$. At B draw $\overrightarrow{BD}$ such that $m\angle ABD = 48$. Label the point where the rays intersect as K. That is, $\overrightarrow{AC} \cap \overrightarrow{BD} = \{K\}$. With the aid of a protractor find $m\angle AKB$.

26. Complete:

(*a*) $m\angle KPL + m\angle LPM = m\angle$.
(*b*) $m\angle MPN + m\angle LPM = m\angle$.
(*c*) $m\angle KPM - m\angle LPM = m\angle$.
(*d*) $m\angle KPN - m\angle MPN = m\angle$.

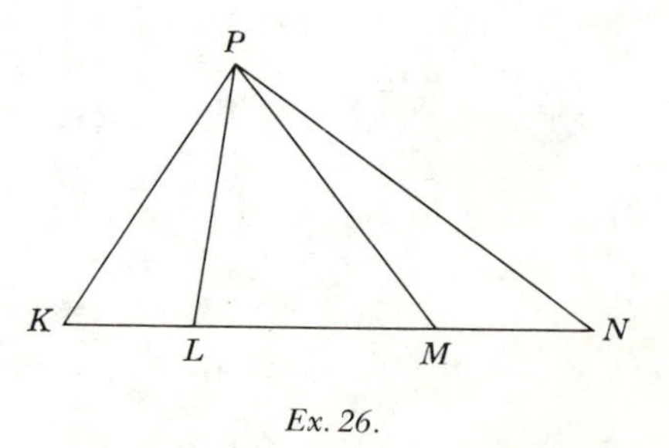

Ex. 26.

27. With the aid of a protractor draw an angle whose measure is 70. Call it $\angle RST$. Locate a point P in the interior of $\angle RST$ such that $m(\overrightarrow{SP} \cup \overrightarrow{ST}) = 25$. What is $m\angle PSR$?

28. With the aid of a protractor draw $\angle ABC$ such that $m\angle ABC = 120$. Locate a point P in the exterior of $\angle ABC$ such that $B \in \overleftrightarrow{PC}$. Find the value of $m(\overrightarrow{BP} \cup \overrightarrow{BA})$.

1.18. Kinds of angles. Two angles are said to be *adjacent angles* iff they have the same vertex, a common side, and the other two sides are contained in opposite closed half-planes determined by the line which contains the common side. The rays not common to both angles are called *exterior sides* of the two adjacent angles. In Fig. 1.37 $\angle AOB$ and $\angle BOC$ are adjacent angles. $\overrightarrow{OB}$ lies in the interior of $\angle AOC$.

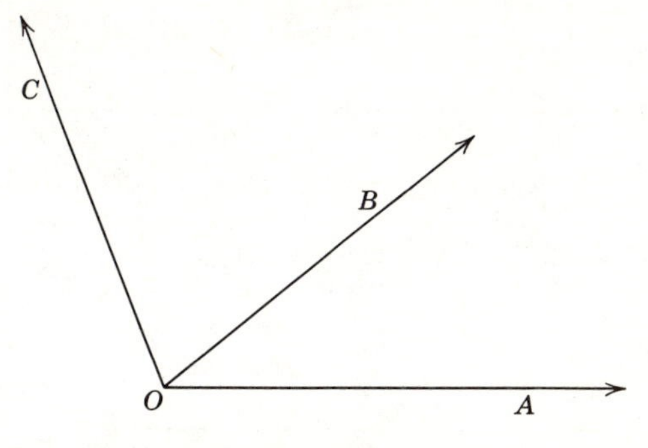

Fig. 1.37. Adjacent ∠S.

The pairs of nonadjacent angles formed when two lines intersect are termed *vertical angles.* In Fig. 1.38 $\angle\alpha$ and $\angle\alpha'$ are vertical angles and so are $\angle\beta$ and $\angle\beta'$.

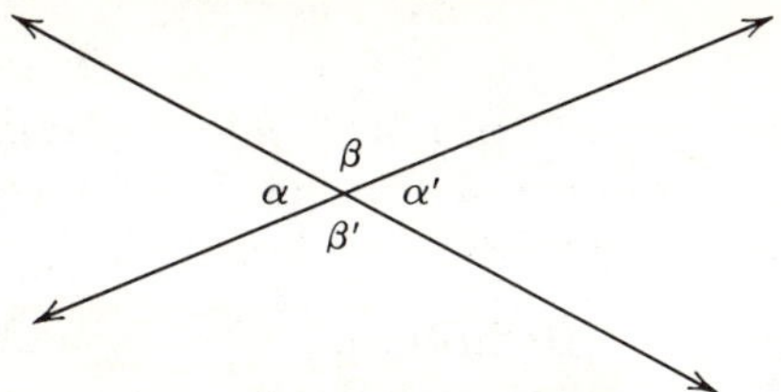

Fig. 1.38. $\angle\alpha$ and $\angle\alpha'$ are vertical ∠S.

As the measure of an angle increases from 0 to 180 the following kinds of angles are formed: acute angle, right angle, obtuse angle, and straight angle (see Fig. 1.39).

Definitions: An angle is an *acute angle* iff it has a measure less than 90. An angle is a *right angle* iff it has a measure of 90. An angle is an *obtuse angle* iff its measure is more than 90 and less than 180. An angle is a *straight angle* iff its measure is equal to 180.

Actually, our definition for the straight angle lacks rigor. Since we defined an angle as the "union of two rays which have a common endpoint," we know that the definition should be a reversible statement. Therefore, we would have to conclude that every union of two rays which have the same endpoint would produce an angle. Yet we know that $\overrightarrow{BC} \cup \overrightarrow{BA}$ is $\overleftrightarrow{AC}$. We are, in effect, then saying that a straight angle is a straight line. This we know is not true. An angle is not a line.

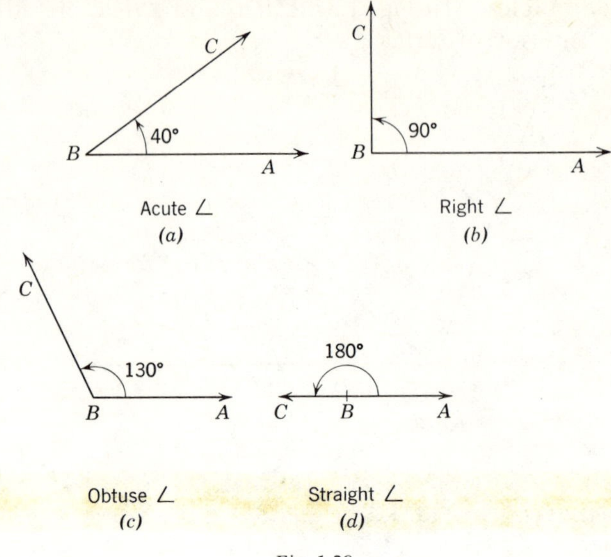

Fig. 1.39.

However, since the term "straight angle" is quite commonly used to represent such a figure as illustrated in Fig. 1.39*d*, we will follow that practice in this book. Some texts call the figure a linear pair.*

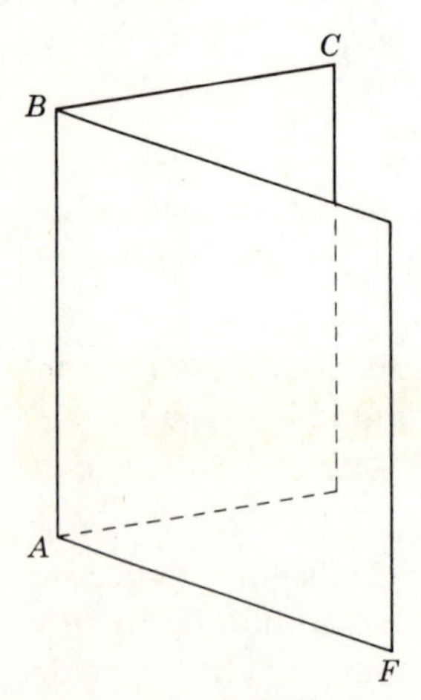

Fig. 1.40. Dihedral angle.

Definition: If A, B, and C are collinear and A and C are on opposite sides of B, then $\overrightarrow{BA} \cup \overrightarrow{BC}$ is called a *straight angle* with B its vertex and $\overrightarrow{BA}$ and $\overrightarrow{BC}$ the sides.

Definition: A *dihedral angle* is formed by the union of two half-planes with the same edge. Each half-plane is called a *face* of the angle (see Fig. 1.40). Dihedral angles will be studied in Chapter 14.

1.19. Congruent angles. Congruent segments. A common concept in daily life is that of size and comparative sizes. We frequently speak of two things having the same size. The word "congruent" is used in geometry to define what we intuitively speak of as "having the same size

*Many textbooks, also, will define an angle as a *reflex angle* iff its measure is more than 180 but less than 360. We will have no occasion to use such an angle in this text.

and the same shape." Congruent figures can be thought of as being duplicates of each other.

Definitions: Plane angles are *congruent* iff they have the same measure. Segments are *congruent* iff they have the same measure. Thus, if we know that $m\overline{AB} = m\overline{CD}$, we say that $\overline{AB}$ and $\overline{CD}$ are congruent, that $\overline{AB}$ is congruent to $\overline{CD}$ or that $\overline{CD}$ is congruent to $\overline{AB}$. Again, if we know that $m\angle ABC = m\angle RST$, we can say that $\angle ABC$ and $\angle RST$ are congruent angles, $\angle ABC$ is congruent to $\angle RST$, or that $\angle RST$ is congruent to $\angle ABC$.

The symbols we have used thus far in expressing the equality of measures between line segments or between angles is rather cumbersome. To overcome this, mathematicians have invented a new symbol for congruence. The symbol for "is congruent to" is $\cong$. Thus, the following are equivalent statements.*

$$m\overline{AB} = m\overline{CD} \qquad \overline{AB} \cong \overline{CD}$$
$$m\angle ABC = m\angle RST \qquad \angle ABC \cong \angle RST$$

Definition: The *bisector* of an angle is the ray whose endpoint is the vertex of the angle and which divides the angle into two congruent angles. The ray BD of Fig. 1.41 *bisects,* or is the *angle bisector* of, $\angle ABC$ iff D is in the interior of $\angle ABC$ and $\angle ABD \cong \angle DBC$.

1.20. Perpendicular lines and right angles. Consider the four figures shown in Fig. 1.42. They are examples of representations of right angles and perpendicular lines.

Definition: Two lines are *perpendicular* iff they intersect to form a right angle. Rays and segments are said to be perpendicular to each other iff the lines of which they are subsets are perpendicular to each other.

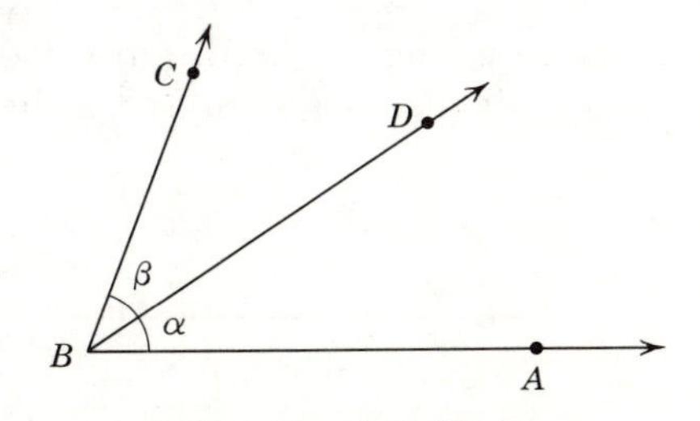

Fig. 1.41. Angle bisector.

---

*Many texts will also use the symbol $AB = CD$ to mean that the measures of the segments are equal. Your instructor may permit this symbolism. However, in this text, we will not use this symbolism for congruence of segments until Chapter 8. By that time, surely, the student will not confuse a geometric figure with that of its measure.

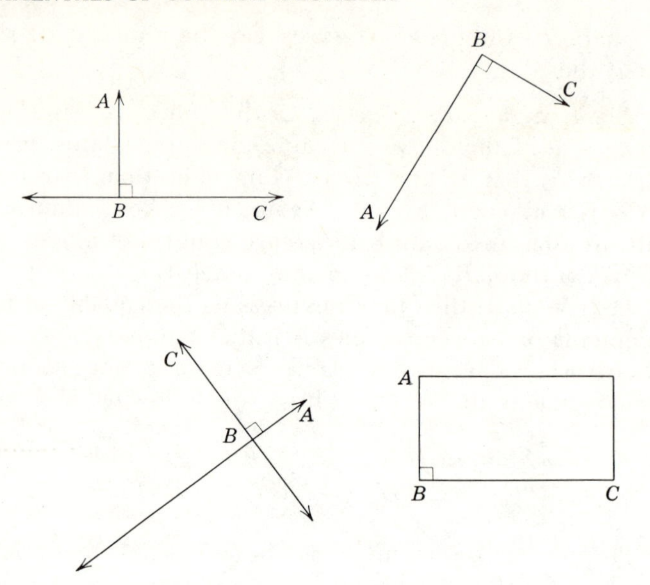

Fig. 1.42. Perpendicular lines.

The symbol for perpendicular is $\perp$. The symbol may also be read "perpendicular to." A right angle of a figure is usually designated by placing a square corner mark ⊾ where the two sides of the angle meet. The foot of the perpendicular to a line is the point where the perpendicular meets the line. Thus, B is the foot of the perpendiculars in Fig. 1.42.

A line, ray, or segment is perpendicular to a plane if it is perpendicular to every line in the plane that passes through its foot. In Fig. 1.43, $\overline{PQ} \perp \overleftrightarrow{AQ}$, $\overleftrightarrow{PQ} \perp \overleftrightarrow{AQ}$, $\overline{PQ} \perp \overrightarrow{QB}$.

1.21. Distance from a point to a line. The distance from a point to a line is the measure of the perpendicular segment from the point to the line.

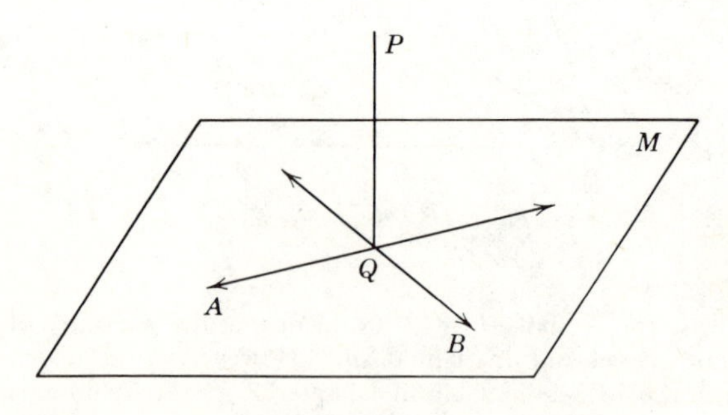

Fig. 1.43.

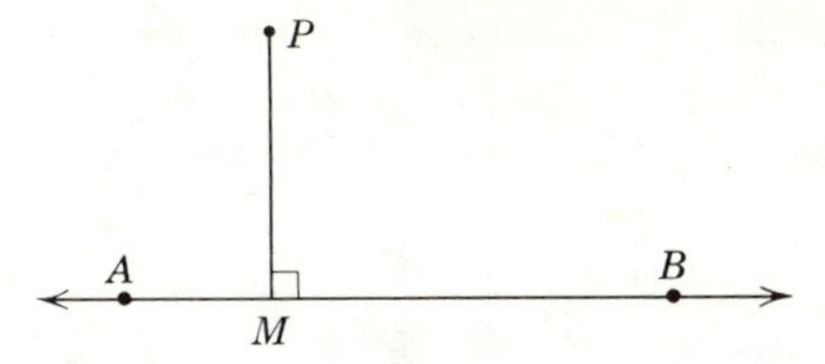

Fig. 1.44. Distance from point to line

Thus, in Fig. 1.44, the measure of $\overline{PM}$ is the distance from point P to $\overleftrightarrow{AB}$. In Chapter 9, we will prove that *the perpendicular distance is the shortest distance from a point to a line.*

1.22. Complementary and supplementary angles. Two angles are called *complementary angles* iff the sum of their measures is 90. Complementary angles could also be defined as two angles the sum of whose measures equals the measure of a right angle. In Fig. 1.45 $\angle\alpha$ and $\angle\beta$ are complementary angles. Each is the complement of the other. Angle α is the complement of $\angle\beta$; and $\angle\beta$ is the complement of $\angle\alpha$.

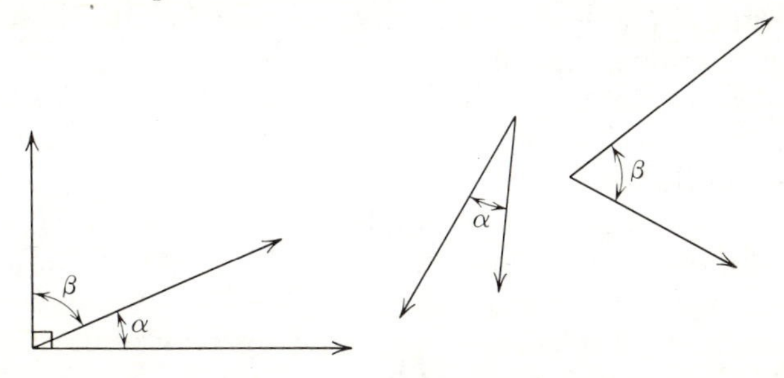

Fig. 1.45. Complementary ∠S.

Angles are *supplementary angles* iff the sum of their measures is 180. We could also say supplementary angles are two angles the sum of whose measures is equal to the measure of a straight angle. In Fig. 1.46 $\angle\alpha$ and $\angle\beta$ are supplementary angles. Angle α is the supplement of $\angle\beta$; and $\angle\beta$ is the supplement of $\angle\alpha$.

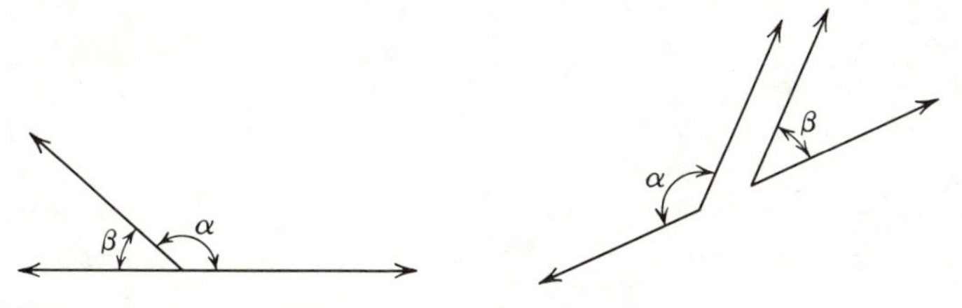

Fig. 1.46. Supplementary ∠S.

1.23. Trigangles. Kinds of triangles. The union of the three segments $\overline{AB}$, $\overline{BC}$, and $\overline{AC}$ is called a *triangle* iff A, B, and C are three noncollinear points. The symbol for triangle is $\triangle$ (plural $\triangle\!\!\!\triangle$). Thus, in Fig. 1.47, $\triangle ABC = \overline{AB} \cup \overline{BC} \cup \overline{AC}$.

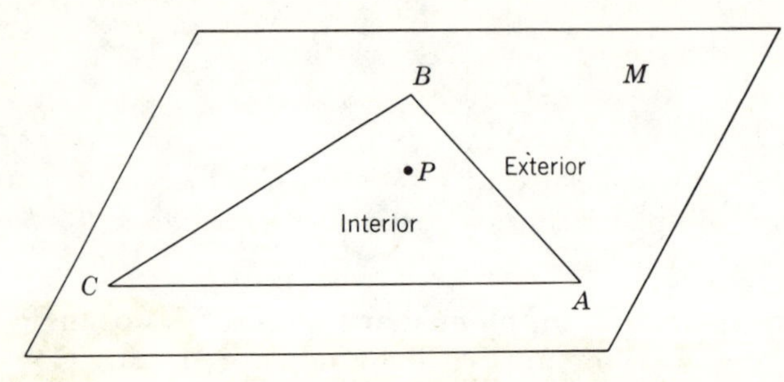

Fig. 1.47.

Each of the noncollinear points is called a *vertex* of the triangle, and each of the line segments is a *side* of the triangle. Angle ABC, $\angle ACB$, and $\angle CAB$ are called the interior angles or simply the *angles* of the triangle. In Fig. 1.47, A, B, and C are vertices of $\triangle ABC$; $\overline{AB}, \overline{BC}$, and $\overline{CA}$ are sides of $\triangle ABC$. Angle C is opposite side AB; $\overline{AB}$ is opposite $\angle C$. The sides AC and BC are said to include $\angle C$. Angle C and $\angle A$ include side CA.

A point P lies in the *interior* of a triangle iff it lies in the interior of each of the angles of the triangle. Every triangle separates the points of a plane into three subsets: the triangle itself, the interior of the triangle and the exterior of the triangle. The *exterior* of a triangle is the set of points of the plane of the triangle that are neither elements of the triangle nor of its interior. Thus, exterior of $\triangle ABC = [(\text{interior of } \triangle ABC) \cup \triangle ABC]'$.

The set of triangles may be classified into three subsets by comparing the sides of the $\triangle$ (Fig. 1.48). A triangle is *scalene* iff it has no two sides that are congruent. A triangle is *isosceles* iff it has two sides that are congruent.

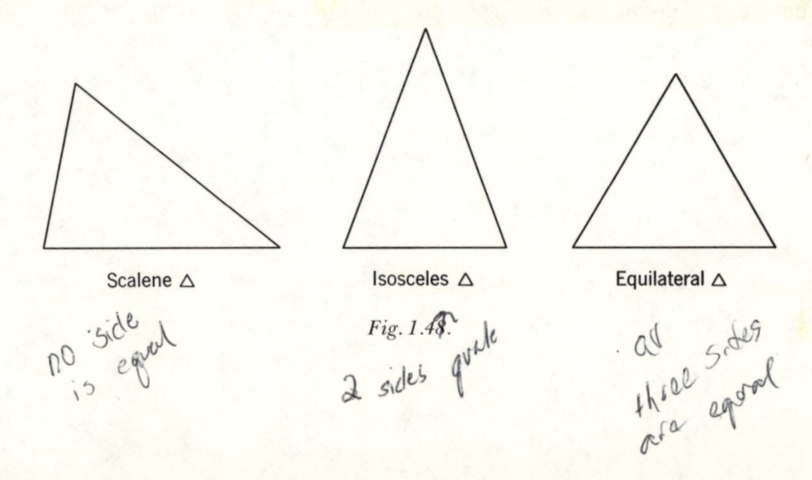

Fig. 1.48.

A triangle is *equilateral* iff it has three congruent sides. The parts of an isosceles triangle are labeled in Fig. 1.49. In the figure $\overline{AC} \cong \overline{BC}$. Sometimes, the congruent sides are called *legs* of the triangle. Angle A, opposite $\overline{BC}$, and angle B, opposite $\overline{AC}$, are called the *base angles* of the isosceles triangle. Side $\overline{AB}$ is the *base* of the triangle. Angle C, opposite the base, is the vertex angle.

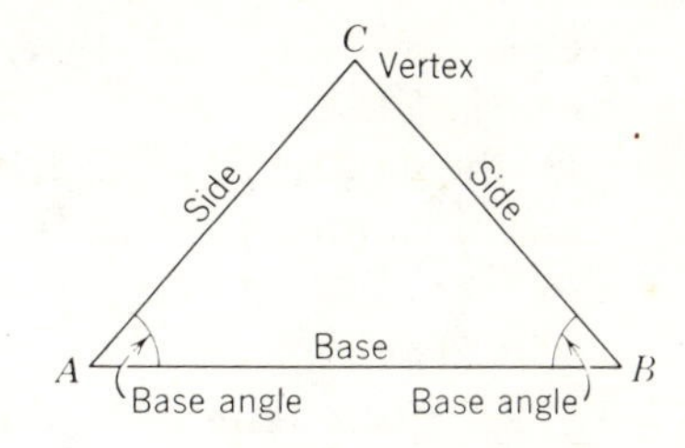

Fig. 1.49. Isosceles triangle.

The set of triangles may also be classified into four subsets, according to the kind of angles the ⟁s contain (Fig. 1.50). A triangle is an *acute* triangle iff it has three acute angles. A triangle is an *obtuse* triangle iff it has one obtuse angle. A triangle is a *right* triangle iff it has one right angle. The sides that form the right angle of the triangle are termed *legs* of the triangle; and the side opposite the right angle is called the *hypotenuse*. In Fig. 1.51, $\overline{AB}$ and $\overline{BC}$ are the legs and $\overline{AC}$ is the hypotenuse of the right triangle. A triangle is *equiangular* iff it has three congruent angles.

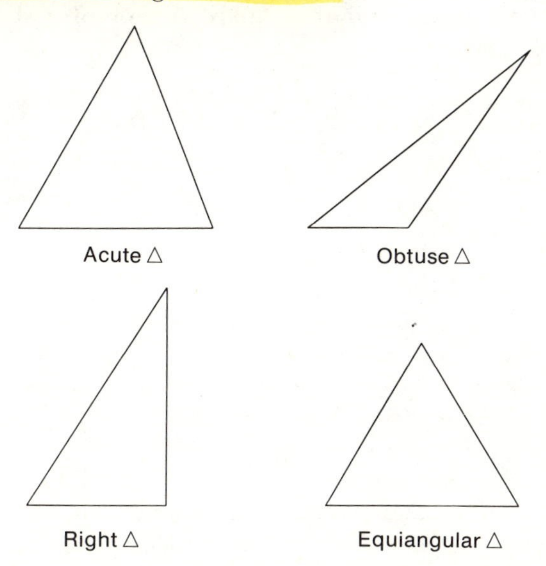

Fig. 1.50.

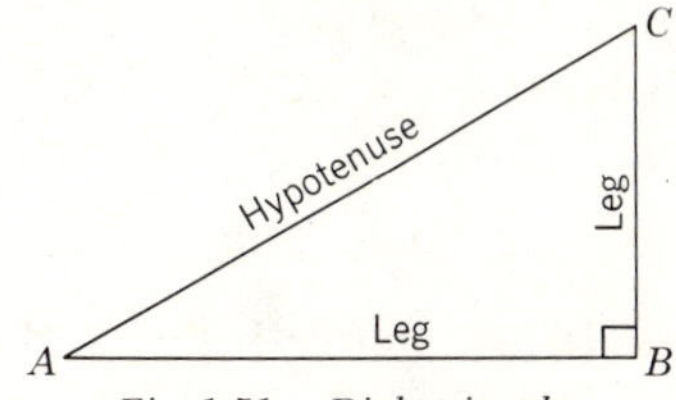

Fig. 1.51. Right triangle.

Exercises

1. Using a protractor and ruler, construct a triangle ABC with $m\overline{AB} = 4''$, $m\angle A = 110$, and $m\angle B = 25$. Give two names for this kind of triangle.
2. In the figure for Ex. 2, what side is common to ⚠ ADC and BDC? What vertices are common to the two ⚠?

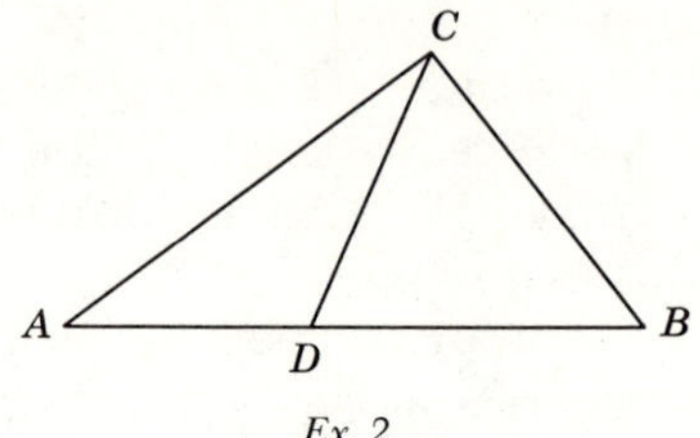

Ex. 2.

3–12. State the kind of triangle each of the following seems to be (*a*) according to the sides and (*b*) according to the angles of the triangles. (If necessary, use a ruler to compare the length of the sides and the square corner of a sheet of paper to compare the angles.

3. $\triangle RST$.
4. $\triangle MNT$.

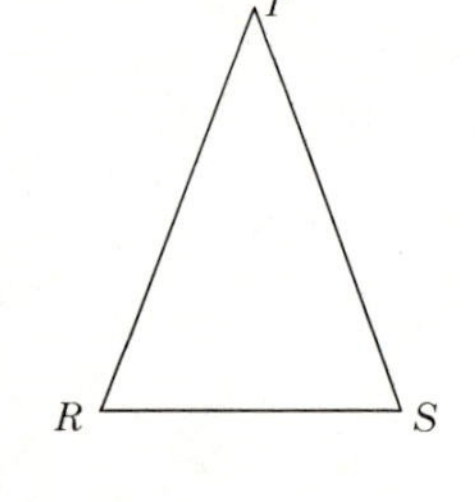

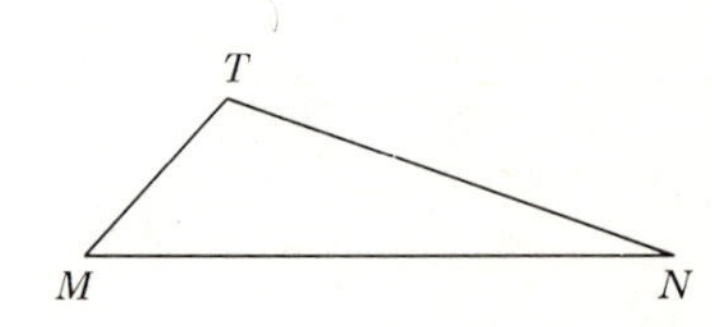

Exs. 3, 4.

5. $\triangle ABC$.
6. $\triangle DEF$.

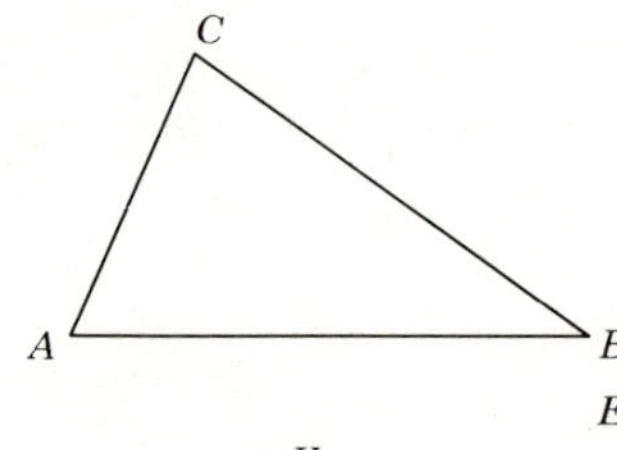

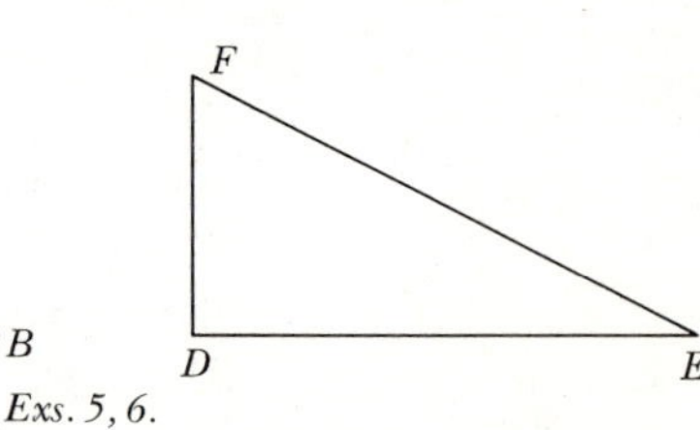

Exs. 5, 6.

7. $\triangle GHK$.
8. $\triangle ABC$.
9. $\triangle ADC$.
10. $\triangle BDC$.
11. $\triangle AEC$.
12. $\triangle ABE$.

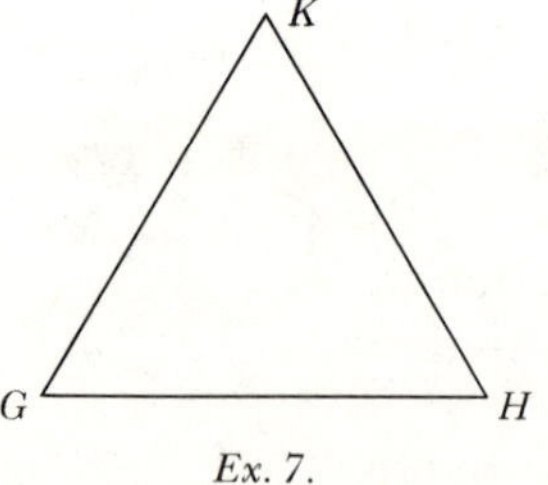

Ex. 7.

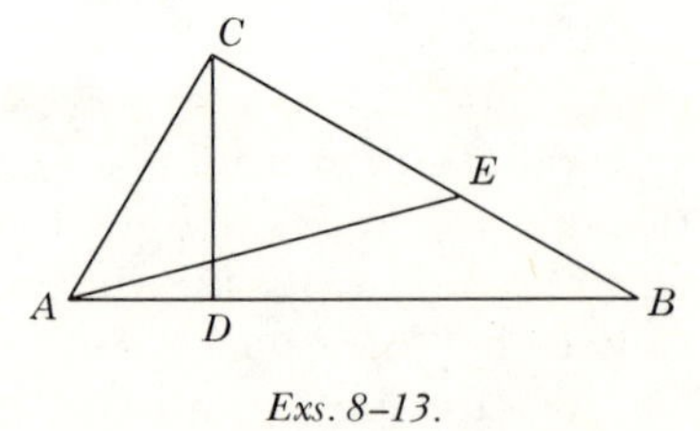

Exs. 8–13.

13. In the figure for Exs. 8 through 13, indicate two pairs of perpendicular lines.

14–16. Name a pair of complementary angles in each of the following diagrams.

17. Tell why $\angle\alpha$ and $\angle\beta$ are complementary angles.

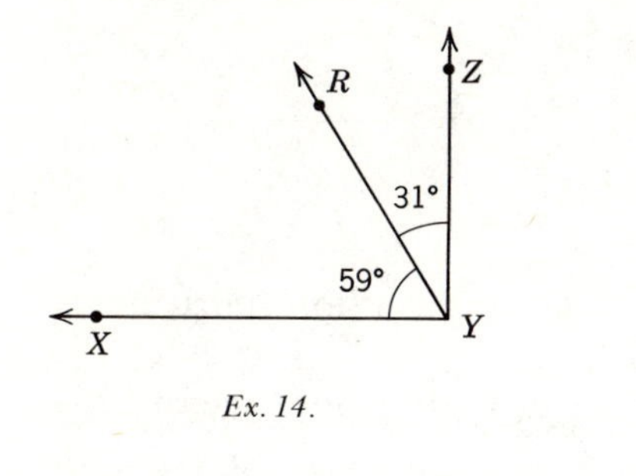

Ex. 14.

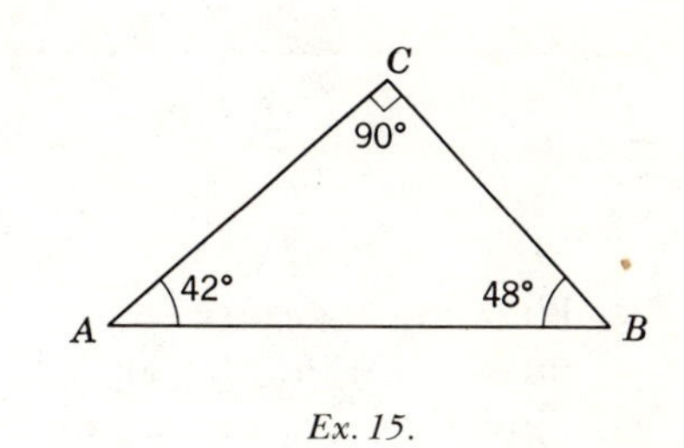

Ex. 15.

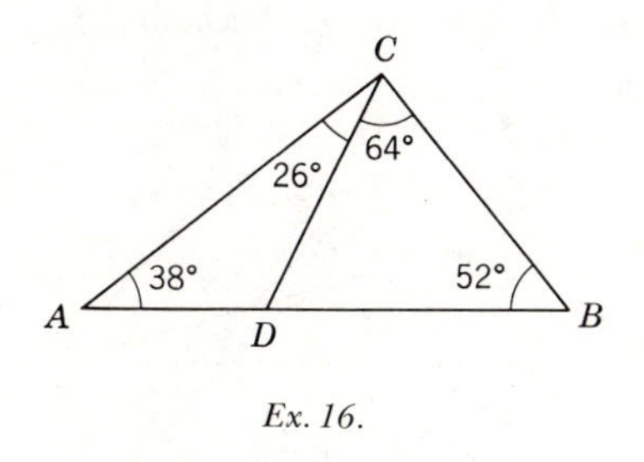

Ex. 16.

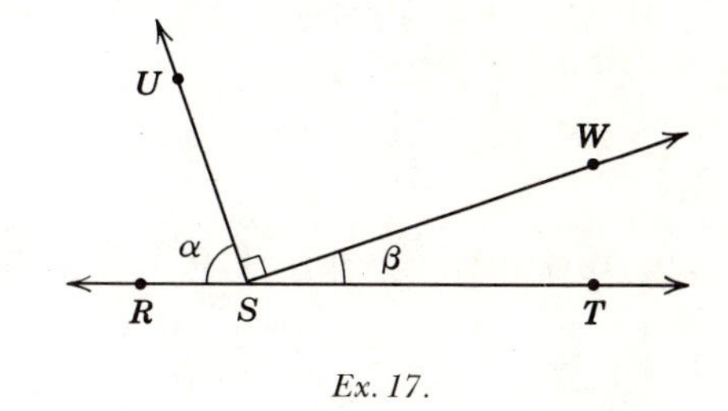

Ex. 17.

18–20. Name a pair of supplementary angles in each of the following diagrams.

21–23. Name two pairs of adjacent angles in each of the following figures.

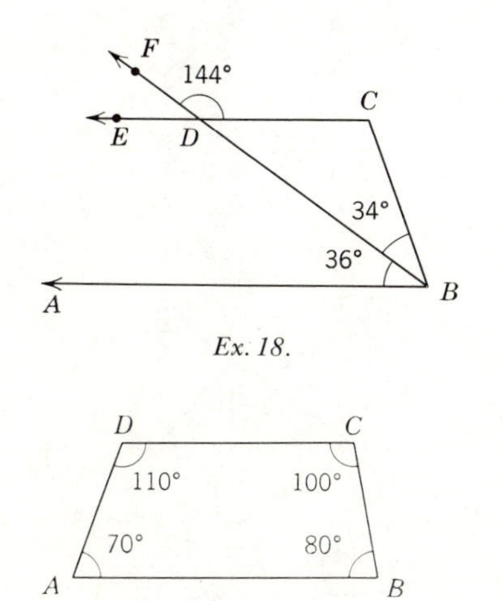

Ex. 18.

Ex. 20.

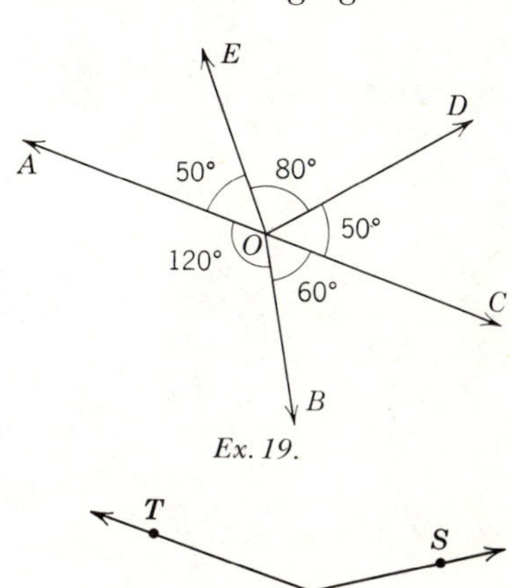

Ex. 19.

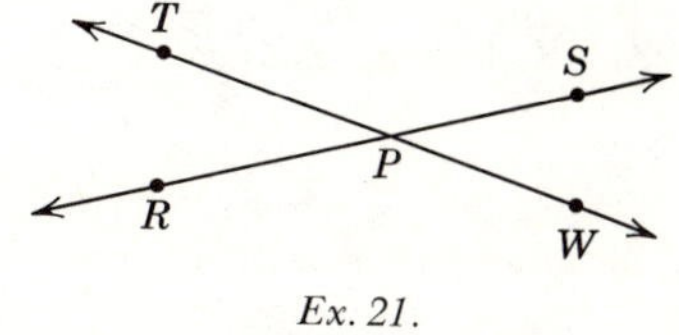

Ex. 21.

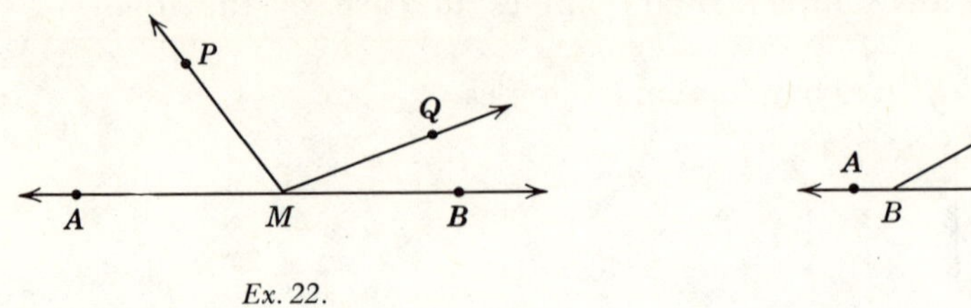

Ex. 22.

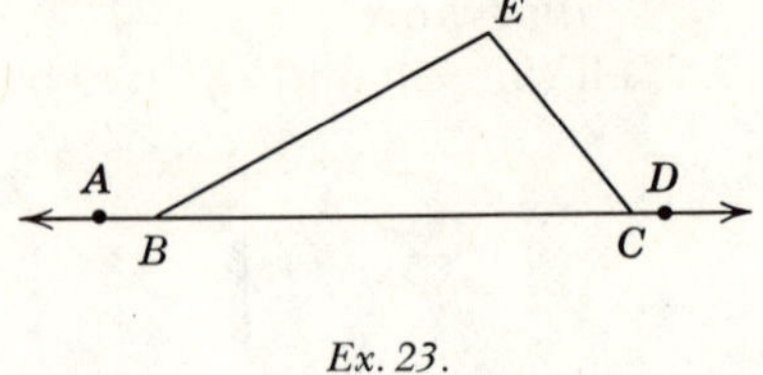

Ex. 23.

24. Find the measure of the complement of each angle whose measure is (a) 30, (b) 45, (c) 80, (d) α.
25. Find the measure of the supplement of each angle whose measure is (a) 30, (b) 45, (c) 90, (d) α.

In exercises 26–31, what conclusions about congruence can be drawn from the data given?

26. M is the midpoint of $\overline{AC}$.

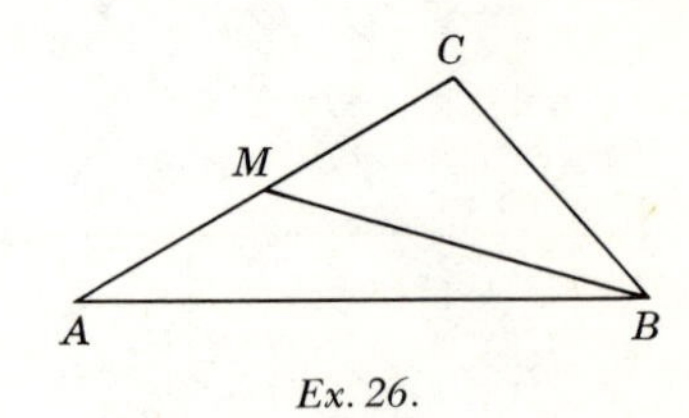

Ex. 26.

27. $\overrightarrow{BD}$ bisects $\angle ABC$.

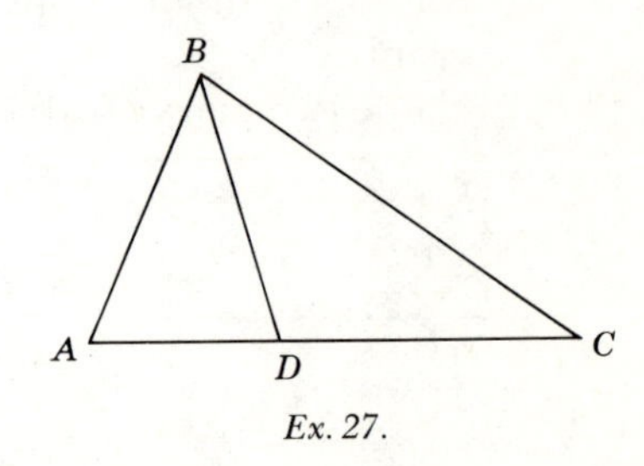

Ex. 27.

28. $\overline{OC}$ bisects $\angle ACB$.

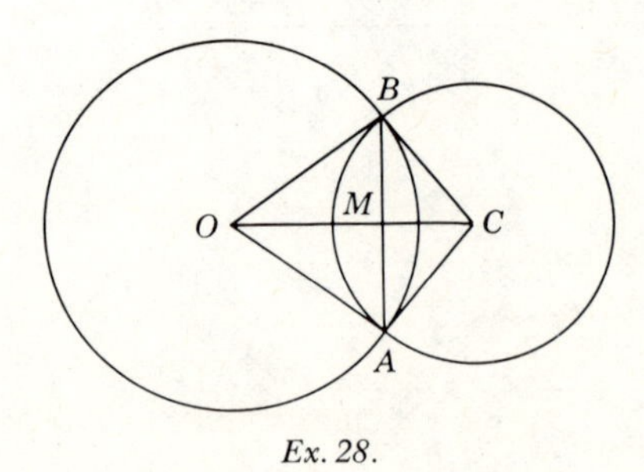

Ex. 28.

29. $\overline{AC}$ and $\overline{BD}$ bisect each other.

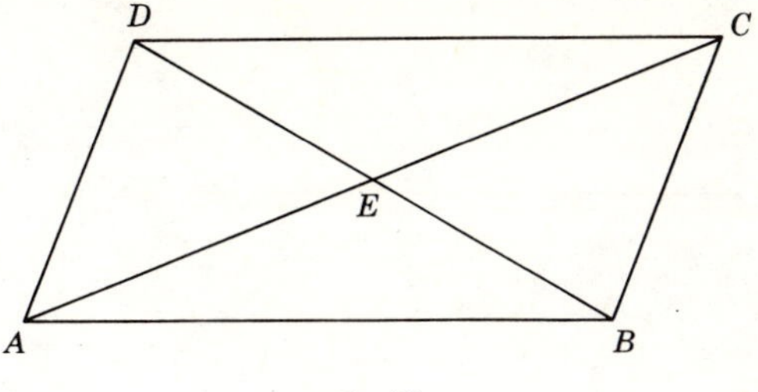

Ex. 29.

30. $\overrightarrow{DE}$ bisects $\angle ADB$.

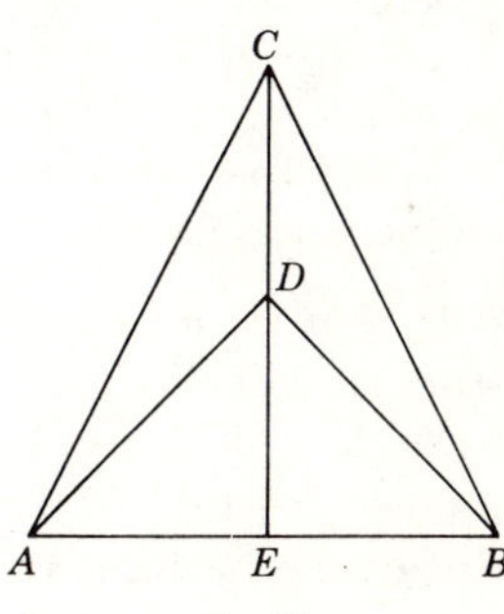

Ex. 30.

31. D is the midpoint of $\overline{BC}$.

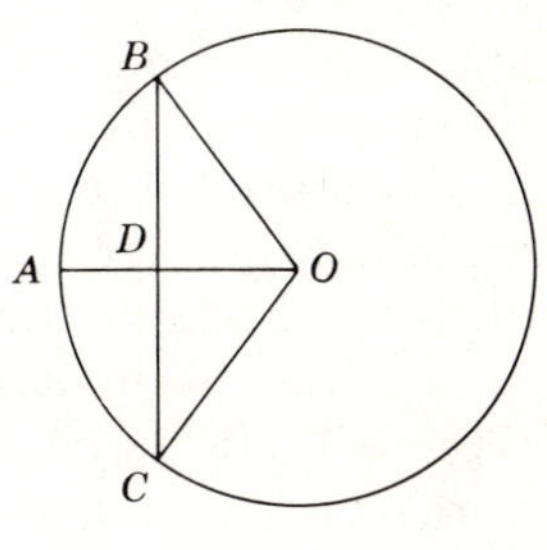

Ex. 31.

1.24. Basing conclusions on observations or measurements. Ancient mathematicians often tested the truth or falsity of a statement by direct observation or measurement. Although this is an important method of acquiring knowledge, it is not always a reliable one. Let us in the following examples attempt to form certain conclusions by the method of observation or measurement.

1. Draw several triangles. By using a protractor, determine the measure of each angle of the triangles. Find the sum of the measures of the three angles of each triangle. What conclusion do you think you might draw about the sum of the measures of the three angles of any given triangle?

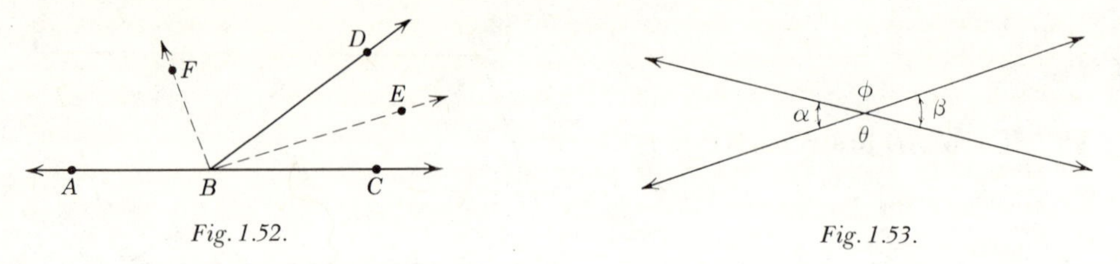

Fig. 1.52. *Fig. 1.53.*

2. In Fig. 1.52, $\angle ABD$ and $\angle CBD$ are supplementary adjacent angles. Draw $\overrightarrow{BF}$ bisecting $\angle ABD$ and $\overrightarrow{BE}$ bisecting $\angle CBD$. Determine the measure of $\angle EBF$. What conclusion might you draw from this experiment?
3. Draw two intersecting lines as in Fig. 1.53. Measure $\angle\alpha$ and $\angle\beta$. Also measure $\angle\theta$ and $\angle\phi$. Give possible conclusions about vertical angles.

If the angles of the triangles of Example 1 were measured carefully, the student will discover that the sum of the measures of the three angles of any triangle will always be near 180. Is the student, as a result of such measurements, justified in stating unequivocally that the sum of the measures of the three angles of any triangle is 180?

Let us consider the implications of making such a conclusion. First the triangles had to be drawn in order to measure the angles. The width of the lines representing the sides of these triangles will vary depending upon the fineness of the drawing instrument. The protractor with which the angles are measured is roughly divided into degrees only. Thus the protractor could not show a difference of $\frac{1}{10}$ of a degree that might exist between the sum of the measures of the angles of two triangles. No matter how fine the sides of the triangle may be drawn or how accurate the measuring instrument, there will always be a possibility that, if the accuracy of the measurements were increased, a slight error in the angle sum might be detected.

A second fallacy in stating as an absolute truth the sum of the measures of the angles of any triangle is 180 is the assumption that what may be true for a limited number of cases must be true for all cases. This is an unreliable practice. We would be safer in stating that the results of our experience lead us to believe that *probably* the angle sum of any triangle equals 180.

In like manner we would be justified in stating in Example 2 that it *appears* that the angle bisectors of two adjacent supplementary angles are perpendicular to each other. In Example 3, we could state that it *appears* that the pairs of the nonadjacent angles formed, when two lines intersect, are congruent.

In subsequent study in this text we will *prove* that each of the above apparent conclusions are truths in fact, but, until we do prove them, we can only state what *seems* to be true.

Unreliable conclusions based upon limited or inaccurate observations or measurements are common also in nonmathematical situations. For example, consider the tendency to associate sadism with the people of a whole nation because their leaders are guilty of sadistic characteristics or, on the other hand, to attribute glamour to the women of a given nation because of a limited number of celebrated beautiful women in the history of that nation.

Frequently, the athletic prowess of a whole nation is judged by the record of a very small group of athletes belonging to that nation.

The student can add many more examples to this list.

1.25. The inductive method of reasoning. In the past examples the reasoning which was used in arriving at conclusions is known as *inductive reasoning*. A general conclusion is drawn by investigating a number of particular cases. It is the method of research. Inductive reasoning has made a large contribution to civilization. In it one observes, measures, studies relations, computes, and draws conclusions. These tentative conclusions are called *hypotheses*. We will use many hypotheses in this text. The hypothesis indicates a statement that is possibly true based on observation of a limited number of cases. The finer the measuring instruments and the more careful the observations and measurements, the greater the possibility that the hypothesis is correct. National pre-election polls are conducted by observing a good representative cross section of the various regions of the nation. Experts have been able to make very accurate predictions by observing less than $\frac{1}{10}$ percent of all the eligible voters in a national election.

1.26. The deductive method of reasoning. Inductive reasoning proceeds by observing a specific common property in a limited number of cases and concluding that this property is general for all cases. Thus, it proceeds from the specific to the general. However, a theory may hold for several thousand cases and then fail on the very next one. We can never be absolutely certain that conclusions based upon inductive reasoning are always true.

A more convincing and powerful method of drawing conclusions is called *deductive reasoning*. When reasoning deductively, one proceeds from the general to the specific. One starts with a limited number of generally accepted basic *assumptions* and by a building process of logical steps proves other facts. Thus, we may build upon these accepted assumptions and derived facts in a manner that will enable us eventually to prove the desired conclusion. These proved facts are termed *theorems*.

All deductive reasoning involves acceptance of the truth of a certain statement (or statements), called an assumption. This assumption need not be obvious to the reader nor need it be a generally accepted fact, but it must be accepted for the purpose of proving a particular argument. Changing the basic assumptions will generally alter the resultant conclusions. In attempting to prove a particular argument, the originally accepted assumption may lead to a contradiction of other accepted assumptions or of other proved facts. In this event, the truth of the original assumption must be questioned; or possibly, the truth of the accepted assumptions may then be doubted.

When a certain assumption is accepted, certain conclusions inevitably follow. These conclusions may be false if the assumptions on which they are based are false. It is imperative, then, that we distinguish between *validity* and *truth*. Consider the following statements: (1) All men are brave. (2) Francis Jones is a man. (3) Francis Jones is brave. Statement 3 is a valid conclusion of assumptions 1 and 2, but it need not be true. If either statement 1 or statement 2 is false, it is possible that statement 3 is also not true. It is necessary in seeking the truth of conclusions that the truth of the basic premises upon which they are based be considered carefully.

Both induction and deduction are valuable methods of reasoning in the study of geometry. New geometric truths can be discovered by inductive reasoning. Deductive reasoning can then be used in proving that such discoveries are true.

After trying, in the next exercise, our skill at deductive reasoning, we will study at greater length in Chapter 2 what constitutes "logical" reasoning. We will then be better prepared to recognize when we have proved our theorems. The student should not be too concerned at this stage if he fails to give correct answers to the following exercise.

Exercises (A)

In the following exercises supply a valid conclusion if one can be supplied. If no conclusion is evident, explain why.

1. Mrs. Jones' dog barks whenever a stranger enters her yard. Mrs. Jones' dog is barking.
2. Water in the fish pond freezes whenever the temperature is below 32° Fahrenheit. The temperature by the fish pond is 30° Fahrenheit.
3. All college freshman students must take an orientation class. Mary Smith is a freshman college student.
4. All members of the basketball team are more than 6 feet tall. Pat Black is more than 6 feet tall.
5. College students will be admitted to the baseball game free. Henry Brown was admitted to the baseball game free.
6. Tim's dad always buys candy when he goes to the drug store. Today Tim's dad bought some candy.
7. Any person born in the United States is a citizen of the United States.

Mr. Smith was born in the city of Carpenteria. Carpenteria is in the United States.

8. All quadrilaterals have four sides. A rhombus has four sides.
9. Only students who study regularly will pass geometry. Bill Smith does not study regularly.
10. Mary is in an English class. All freshmen in college are enrolled in some English class.
11. Baseball players eat Zeppo cereal and are alert on the diamond. I eat Zeppo cereal.
12. The first- and third-period geometry classes were given the same test. Students in the first-period class did better than those in the third-period class. Dick was enrolled in the first-period class and Stan was in the third-period class.

13–22. Answer the following questions to check your reading and reasoning ability.

13. Why can't a man, living in Winston-Salem, be buried west of the Mississippi River?
14. Some months have 30 days, some have 31 days. How many have 28 days?
15. I have in my hand two U.S. coins which total 55 cents. One is not a nickel. Place that in mind. What are the two coins?
16. A farmer had 17 sheep. All but 9 died. How many did he have left?
17. Two men play checkers. They play five games and each man wins the same number of games. How do you figure that out?
18. If you had only one match and entered a room where there was a lamp, an oil heater, and some kindling wood, which would you light first?
19. Take two apples from three apples and what do you have?
20. Is it legal in North Carolina for a man to marry his widow's sister?
21. The archaeologist who said he found a gold coin marked 46 B.C. was either lying or kidding. Why?
22. A woman gives a beggar 50 cents. The woman is the beggar's sister, but the beggar is not the woman's brother. How is this possible?

Exercises (B)

Each of the following exercises include a false assumption. Disregard the falsity of the assumption and write the conclusion which you are then forced to accept.

1. Given two men, the taller man is the heavier. Bob is taller than Jack.
2. Barking dogs do not bite. My dog barks.
3. When a person walks under a ladder, misfortune will befall him. Mr. Grimes walked under a ladder yesterday.

4. All women are poor drivers. Jerry Wallace is a woman.
5. Anyone handling a toad will get warts on his hand. I handled a toad today.
6. Of two packages, the more expensive is the smaller. Mary's Christmas present was larger than Ruth's.

In the following exercise, indicate which of the following conclusions logically follow from the given assumptions.

7. *Assumption:* All members of the Ooga tribe are dark-skinned. No dark-skinned person has blue eyes.

Conclusion:

(*a*) No Ooga tribesman has blue eyes.
(*b*) Some dark-skinned tribesmen are members of the Ooga tribe.
(*c*) Some people with blue eyes are not dark-skinned.
(*d*) Some Ooga tribesmen have blue eyes.

8. *Assumption:* Only outstanding students get scholarships. All outstanding students get publicity.

Conclusion:

(*a*) All students who get publicity get scholarships.
(*b*) All students who get scholarships get publicity.
(*c*) Only students with publicity get scholarships.
(*d*) Some students who do not get publicity get scholarships.

9. *Assumption:* Some cooked vegetables are tasty. All cooked vegetables are nourishing.

Conclusion:

(*a*) Some vegetables are tasty.
(*b*) If a vegetable is not nourishing, it is not a cooked vegetable.
(*c*) Some tasty vegetables are not cooked.
(*d*) If a vegetable is not a cooked vegetable, it is not nourishing.

Summary Tests

Test 1

Indicate the one word or number that will make the following statements true.

1. The sides of a right angle are ______ to each other.
2. The pairs of nonadjacent angles formed when two lines intersect are called ______.
3. An ______ angle is larger than its supplement.
4. The side opposite the right angle of a triangle is called the ______.
5. A triangle with no two sides congruent is called a ______ triangle.
6. If the sum of the measures of two angles is 180, the angles are ______.
7. A triangle with two congruent sides is called a ______ triangle.
8. $\sqrt{3}$ is a(n) ______ real number.
9. The ______ of an angle divides the angle into two angles with equal measures.
10. Complementary angles are two angles the sum of whose measures is equal to ______.
11. The difference between the measures of the complement and the supplement of an angle is always ______.
12. ______ is the non-negative integer that is not a counting number.
13. The sum of the measures of the angles about a point is equal to ______.
14. The angle whose measure equals that of its supplement is a ______ angle.
15. An angle with a measure less than 90 is ______.
16. Angles with the same measures are ______.
17. The intersection of two distinct planes is either a null set or a ______.
18. An angle is the ______ of two rays which have a common end point.

19. The only point of a line equally distant from two of its points is the ______ of the segment with these points as its end points.
20. For each three noncollinear points A, B, and C and for each $D \in \overline{AC}$, $m\angle ADB + m\angle BDC =$ ______.

21–30. *Given:* $A = \{1, 2, 3, 4\}$, $B = \{3, 4, 5\}$, and $C = \{5, 6, 7, 8, 9\}$.

21. $A \cup B = \{ \quad \}$.
22. $A \cap B = \{ \quad \}$.
23. $A \cup C = \{ \quad \}$.
24. $A \cap C = \{ \quad \}$.
25. $(A \cup B) \cup C = \{ \quad \}$.
26. $(A \cap B) \cap C = \{ \quad \}$.
27. $(A \cup B) \cap C = \{ \quad \}$.
28. $(A \cap B) \cup C = \{ \quad \}$.
29. $A \cup (B \cup C) = \{ \quad \}$.
30. $A \cap (B \cap C) = \{ \quad \}$.

31–36. If x is a place holder for a real number, replace the "?" by the correct symbol $>$, $<$ or $=$ to make the statement true.

31. $x-3$ ___?___ $x-4$
32. $x+2$ ___?___ $x-2$
33. $3(x+2)$ ___?___ $6+3x$
34. x ___?___ $x+1$
35. $x-1$ ___?___ x
36. $2x-3$ ___?___ $2x+1$
37. $|-7| + |+3| =$
38. $|-4| + |-6| =$
39. $|-6| - |+4| =$
40. $|-5| - |-7| =$

Test 2

In each of the following indicate whether the statement is *always true* (mark T) or *not always true* (mark F).

1. If $\overline{RS} \cong \overline{AB}$, then $m\overline{RS} = m\overline{AB}$.
2. If $\overline{RS} \cong \overline{AB}$, then $\overline{RS} = \overline{AB}$.
3. An obtuse angle has a greater measure than a right angle.
4. A straight line has a fixed length.
5. If an obtuse angle is bisected, two acute angles will be formed.
6. The measure of an angle depends on the length of its sides.
7. Complementary angles are angles the sum of whose measures is equal to a right angle.
8. A definition should be a reversible statement.
9. A straight angle is a straight line.
10. It is possible to define any word in terms of other definable terms.
11. The early Greeks studied the truths of geometry in order to obtain practical applications from them.
12. All isosceles ⚠ are equilateral.
13. When the measure of an acute angle is doubled, an obtuse angle is formed.
14. Inductive reasoning can be relied upon to give conclusive results.
15. Valid conclusions can result from false (untrue) basic assumptions.
16. For any real number x, $|x| = |-x|$.

17. If $a+b < 0$, then $|a+b| < 0$.
18. The union of two sets can never be an empty set.
19. The supplement of an angle is always obtuse.
20. Adjacent angles are always supplementary.
21. If $\overline{AB} \cap l = \phi$ and $\overline{BC} \cap l = \phi$, then $\overline{AC} \cap l = \phi$.
22. If $\overline{AB} \cap l = \phi$ and $\overline{BC} \cap l \neq \phi$, then $\overline{AC} \cap l \neq \phi$.
23. If $\overline{AB} \cap l = \phi$, then A and B must be on opposite sides of l.
24. If $\overset{\circ\!-\!\circ}{AB} \cap l = \phi$, then A and B must be on opposite sides of l.
25. If $P \in$ half-plane h_2 and $Q \in$ half-plane h_1, then $\overline{PQ} \subset h_1$.
26. If $A \in l$ and $B \in h_1$, then $\overset{\circ\!-\!\circ}{AB} \subset h_1$.
27. If $C \in \overset{\circ\!-\!\circ}{AB}$, then $m(\overline{AC})$ is less than $m(\overline{AB})$.
28. If $m\angle KLN = 32$, $m\angle NLM = 28$, then $m\angle KLM = 60$.
29. If $m\overline{RS} + m\overline{RT} = m\overline{ST}$, then S is between R and T.
30. If $\overline{AB} = \overline{CD}$, then $A = C$ and $B = D$.
31. A ray has two endpoints.
32. If $m\angle ABC = m\angle RST$, then $\angle ABC \cong \angle RST$.
33. The union of two half-planes is a whole plane.
34. $FG = JK$ and $\overline{FG} \cong \overline{JK}$ are equivalent statements.
35. The intersection of two sets is the set of all elements that belong to one or both of them.
36. The sides of an angle are rays.
37. If S is not between R and T, then T is between R and S.
38. A collinear set of points is a line.
39. If G is a point in the interior of $\angle DEF$, then $m\angle DEG + m\angle FEG = m\angle DEF$.
40. If two lines intersect to form vertical angles that are supplementary, the vertical angles are right angles.
41. $4 \in \{3, 7, 4, 5, 9\}$.
42. $\{1, 3, 5, 7\} \cap \{1, 2, 3, 4\} = \{1, 3\}$.
43. $\{1, 3, 5, 7\} \cup \{1, 2, 3, 4, 7\} = \{1, 2, 3, 4, 5, 7\}$.
44. $\{2, 5, 6\} \cap \{6, 5, 3\} = \{2, 3\}$.
45. $\{2, 4, 6\} \cap \{1, 3, 5\} = \emptyset$.
46. There is a number n such that $n \in \{1, 2, 3\} \cap \{4, 5, 6\}$.
47. There is a number n such that $n \in \{1, 2, 3\} \cup \{4, 5, 6\}$.
48. $\{3, 4\} \subset \{1, 2, 3, 4, 5\}$.
49. $\{1, 3, 4\} \subset \{1, 3, 5, 7\}$.
50. $\{3, 5\} \subset \{5, 3\}$.

Test 3

PROBLEMS

1–8. Given the scaled line with points and their corresponding coordinates as indicated. Complete the following.

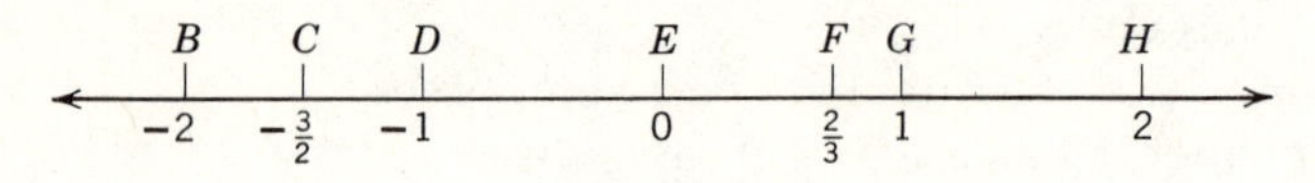

Probs. 1–8.

1. $m\overline{DH} =$ 2. $m\overline{GB} =$ 3. $m\overline{DF} =$
4. $m\overline{HB} =$ 5. $m\overline{CG} =$ 6. $m\overline{CF} =$
7. The coordinate of the midpoint of $\overline{EH}$ is ______.
8. The coordinate of the midpoint of $\overline{DH}$ is ______.
9. If $m\angle A$ is 40, what is the measure of the complement of $\angle A$?
10. If $m\angle B$ is 110, what is the measure of the supplement of $\angle B$?

11–16. Given: $m\angle AOC = 40$; $m\angle COE = 70$; $\overrightarrow{OB}$ bisects $\angle AOC$ and $\overrightarrow{OD}$ bisects $\angle COE$. Complete the following.

11. $m\angle AOB =$ 12. $m\angle COD =$
13. $m\angle BOD =$ 14. $m\angle EOB =$
15. $m\angle AOE =$ 16. $m\angle BOE =$

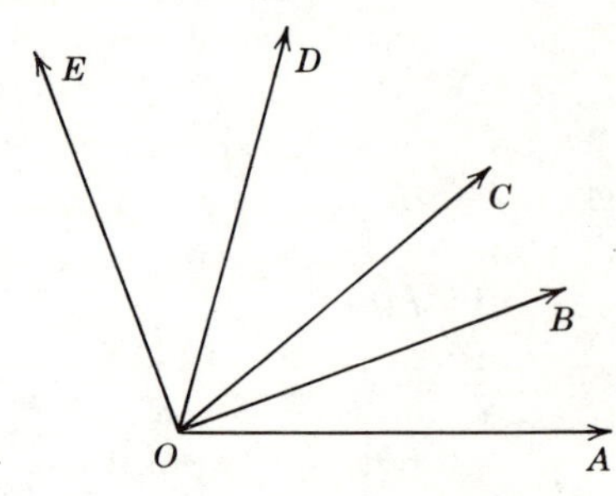

Probs. 11–16.

17. $\overleftrightarrow{AB}$ and $\overleftrightarrow{CD}$ are straight lines intersecting at E. What must the measures of angles α and ϕ be?
18. The measure of angle θ is three times the measure of $\angle\phi$. What is the measure of $\angle\phi$?

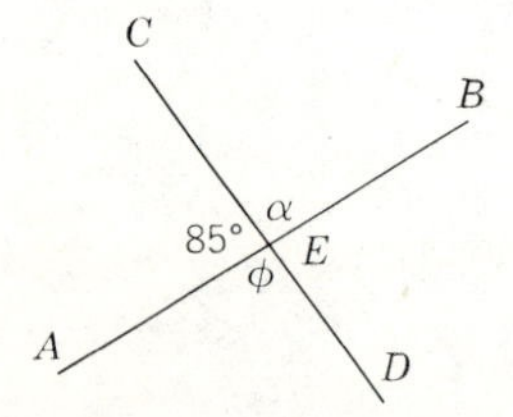

Prob. 17.

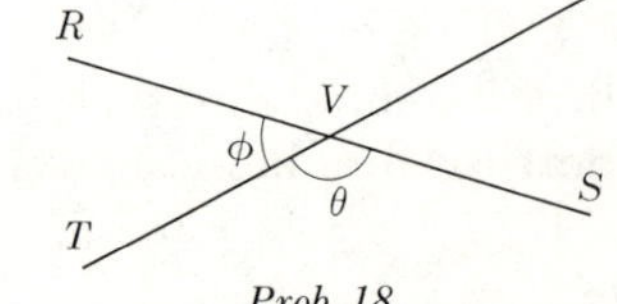

Prob. 18.

19. $\overline{CM}$ bisects $\angle ACB$; $m\angle ACB = 110$. *Then* $m\angle BCM =$ ______.
20. Angle α is the complement of an angle whose measure is 38; $\angle\beta$ is the supplement of $\angle\alpha$. Then $m\angle\beta =$ ______.

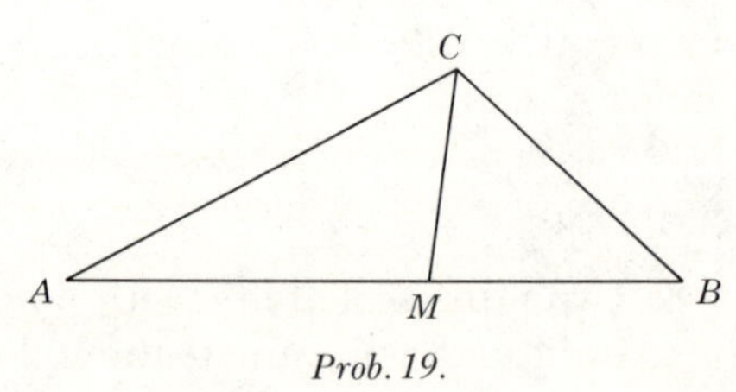

Prob. 19.

2

Elementary Logic

2.1. Logical reasoning. We have all heard the words "logic" and "logical" used. We speak of a person's action as being "logical," or of a "logical" solution to a problem A "logical" behavior is a "reasonable" behavior. The "illogical" conclusion is an "unreasonable" conclusion. When a person engages in "clear thinking" or "rigorous thinking," he is employing the discipline of logical reasoning.

In this chapter we will discuss the meanings of a few words and symbols used in present-day logic and mathematics. We will then introduce some of the methods and principles used in distinguishing correct from incorrect argument. We will systematize some of the simpler principles of valid reasoning.

Although the method of deductive logic permeates all fields of human knowledge, it is probably found in its sharpest and clearest form in the study of mathematics.

2.2. Statements. A discourse is carried on by using sentences. Some of these sentences are in the form of statements.

Definition: A *statement* is a sentence which is either true or false, but not both.

It should be noted here that the words "true" and "false" are undefined elements. Every statement is a sentence; but not every sentence is a statement. A statement is said to have a *truth value* T if it is true and F if it is false. Such things as affirmations, denials, reports, opinions, remarks, comments, and judgments are statements. Every statement is an assertion.

The sentence "San Francisco is in California" is a statement with a truth value T. The sentence "Every number is odd" is a statement with a truth value F.

All statements in the field of logic are either *simple sentences* or *compound sentences.* The simple sentence contains one grammatically independent statement. It does not contain connecting words such as *and, or,* and *but.* A compound sentence is formed by two or more clauses that act as independent sentences and are joined by connectives such as *and, or, but, if . . . then, if and only if, either . . . or,* and *neither . . . nor.*

Examples

1. Every natural number is odd or even.
2. I am going to cash this check and buy myself a new suit.
3. The wind is blowing and I am cold.
4. I will go to the show if John asks me.
5. People who do not work should not eat.

It is customary in logic to represent simple statements by letters as p, q, r, etc. Hence if we let p indicate the statement, "The wind is blowing" and q indicate, "I am cold," we can abbreviate Statement 3 above as p and q.

Exercises (A)

Consider the following sentences. Which are statements?

1. How many are there?
2. 3 plus 2 equals 5.
3. 3×2 equals 5.
4. Give me the text.
5. Tom is older than Bill.
6. All right angles have the same measure.
7. She is hungry.
8. Mrs. Jones is ill.
9. He is the most popular boy in school.
10. If I do not study, I will fail this course.
11. If I live in Los Angeles, I live in California.
12. x plus 3 equals 5.
13. Go away!
14. The window is not closed.
15. 3×2 does not equal 5.
16. How much do you weigh?

Exercises (B)

In each of the following exercises there is a compound statement or one that can be interpreted as one. State the simple components of each sentence.

1. It is hot and I am tired.
2. Baseball players eat Zeppo cereal and are alert on the diamond.
3. His action was either deliberate or careless.
4. The composer was either Chopin or Brahms.
5. The figure is neither a square nor a rectangle.
6. Either Jones is innocent or he is lying.
7. He is clever and I am not.
8. Sue and Kay are pretty.
9. Sue and Kay dislike each other.
10. That animal is either dead or alive.
11. Two lines either intersect or they are parallel.
12. If this object is neither a male nor a female, it is not an animal.
13. Every animal is either a male or a female.
14. The cost is neither cheap nor expensive.
15. I would buy the car, but it costs too much.
16. A square is a rectangle.

2.3. Conjunction. We have seen how two statements can be connected to make another statement. Some of these forms occur repeatedly in logical discourse and are indispensable for purposes of analysis. We will define and discuss some of the more common ones in this chapter.

Definition: If p and q are statements, the statement of the form p and q is called the *conjunction* of p and q. The symbol for p and q is "$p \wedge q$."

There are many other words in ordinary speech besides "and," that are used as conjunctives; e.g., "but," "although," "however," "nevertheless."

Examples

1. It is daytime; however, I cannot see the sun.
2. I am starved, but he is well fed.
3. Mary is going with George and Ruth is going with Bill.
4. Some roses are red and some roses are blue.
5. Some roses are red and today is Tuesday.

Although the definition for conjunction seems simple enough, we should not accept it blindly. You will note that our definition *takes for granted* that "p and q" will always be a statement. Remember a sentence is not a statement unless it is either true or false, but not both. It becomes necessary, then, to formulate some *rule* which we can use to determine when "p and q" is true and when it is false. Without such a rule, our definition will have no meaning.

Each of the following statements is in the form of "p and q." Check which ones are true and which ones are false, and then try to formulate a general rule for deciding upon the truths of a conjunction.

1. $2+3=5$ and $2 \times 3=5$.
2. 2 is an even number and 3 is an odd number.
3. 2 is an even number and 4 is an even number.
4. 2 is an odd number and 4 is an odd number.
5. A circle has ten sides and a triangle has three sides.
6. $\overset{\circ}{\overrightarrow{AB}} \cup \{A\} = \overrightarrow{AB}$ and $\overset{\circ\ \ \circ}{\overline{AB}} \cup \{A, B\} = \overline{AB}$.

In studying the foregoing examples, you should have discovered that "p and q" is considered true only when both p and q are true. If either p is false or q is false (or both are false), then "p and q" is false. This is sometimes shown most clearly by the truth table below.

p	q	$p \wedge q$
T	T	T
T	F	F
F	T	F
F	F	F

"p and q" is true (T) in only one case and false (F) in all others. It should be emphasized that truth tables cannot be proved. They represent agreements in truth values of statements that have proved useful to mathematicians and logicians.

2.4. Disjunction. Another way to combine statements is by using the connective "or" between them. Consider the following sentences:

1. I plan to go to the game or to the show.
2. I expect to see John or Tom at the party.
3. The music teacher told my son that he could do well as a student of the piano or the flute.

In the first sentence it is clear that the speaker will go either to the game or to the show but that *he will not do both.* It is not clear in the second sentence if the speaker will see only John or only Tom at the party. It might mean that he will see both. In the third sentence, it should be clear that the son should do well with either or both instruments.

Thus we see that the common use of the word "or" often leads to ambiguity and not uniform meaning. Sometimes it indicates only one of the statements which make up the disjunction is true. Sometimes it is used to mean at least one of the statements and possibly both are true. In logic we cannot tolerate such varied meanings. We must agree on precisely what we mean when we say "p or q." Mathematicians have agreed that, unless it is explicitly stated to the contrary, the connective "or" should be used in the inclusive sense. Thus statements of the form "p or q" are true in all cases except when p and

q are both false. It will be recalled that we interpreted "or" in the inclusive sense in our definition of the union of sets.

Definition: The *disjunction* of two statements p and q is the compound sentence "p or q." It is false when both p and q are false and true in all other cases. The symbol for the inclusive p or q is "$p \vee q$."

The truth table for the disjunction "p or q" follows:

p	q	$p \vee q$
T	T	T
T	F	T
F	T	T
F	F	F

Exercises

In each of the following exercises there are two statements. Join these statements first to form a conjunction and then to form a disjunction. Determine the truth or falsity of each of the compound sentences.

1. The diamond is hard. Putty is soft.
2. The statement is true. The statement is false.
3. The two lines intersect. The lines are parallel.
4. A ray is a half-line. A ray contains a vertex.
5. There are 30 days in February. Five is less than 4.
6. No triangle has four sides. A square has four sides.
7. Three plus zero equals 3. Three times zero equals 3.
8. Some animals are dogs. Some dogs bark.
9. A is in the interior of $\angle ABC$. C is on side AB of $\angle ABC$.
10. All women are poor drivers. My name is Mudd.
11. The sun is hot. Dogs can fly.
12. -5 is less than 2. 4 is more than 3.
13. An angle is formed by two rays. An interval includes its endpoints.
14. $\triangle ABC \cap \angle ABC = \angle ABC$. $\triangle ABC \cup \angle ABC = \triangle ABC$.
15. The sides of an angle is not a subset of the interior of the angle. Christmas occurs in December.
16. The supplement of an angle is larger than the complement of the angle. The measure of an acute angle is greater than the measure of an obtuse angle.

2.5. Negation. Statements can be made about other statements. One of the simplest and most useful statement of this type has the form "p is false." Everyone has probably made a statement that he believed true only to have someone else show his disagreement by saying, "That is not true."

Definition: The *negation* of a statement "p" is the statement "not-p." It means "p is false"; or "it is not true that p." The symbol for not-p is "$\sim p$."

The negation of a statement, however, is not usually formed by placing a "not" in front of it. This usually would make the sentence sound awkward. Thus where p symbolizes the statement, "All misers are selfish," the various statements, "It is false that all misers are selfish," "Not all misers are selfish," "Some misers are not selfish," "It is not true that all misers are selfish" are symbolized "not-p." The negation of any true statement is false, and the negation of any false statement is true. This fact can be expressed by the truth table.

p	$\sim p$
T	F
F	T

In developing logical proofs, it is frequently necessary to state the negation of statements like "All fat people are happy" and "Some fat people are happy." It should be clear that, if we can find one unhappy fat person, we will have proved the first statement to be false. Thus we could form the negation by stating "Some fat people are not happy" or "There is at least one fat person who is not happy." But we could not form the negation by the statement "No fat person is happy." This is a common error made by the loose thinker. The negation of "all are" is "some are not" or "not all are ."

The word "some" in common usage means "more than one." However, in logic it will be more convenient if we agree it to mean "one or more." This we will do in this text. Thus the negation of the second statement above would be "No fat person is happy" or "Every fat person is unhappy." The negation of "some are" is "none are" or "it is not true that some are."

Exercises

In each of the following form the negation of the statement.

1. Gold is not heavy.
2. Fido never barks.
3. Anyone who wants a good grade in this course must study hard.
4. Aspirin relieves pain.
5. A hexagon has seven sides.
6. It is false that a triangle has four sides.
7. Not every banker is rich.

8. It is not true that 2 plus four equals 6.
9. Two plus 4 equals 8.
10. Perpendicular lines form right angles.
11. All equilateral triangles are equiangular.
12. All blind men cannot see.
13. Some blind men carry white canes.
14. All squares are rectangles.
15. All these cookies are delicious.
16. Some of the students are smarter than others.
17. Every European lives in Europe.
18. For every question there is an answer.
19. There is at least one girl in the class.
20. Every player is 6 feet tall.
21. Some questions cannot be answered.
22. Some dogs are green.
23. Every ZEP is a ZOP.
24. Some pillows are soft.
25. A null set is a subset of itself.
26. Not every angle is acute.

2.6. Negations of conjunctions and disjunctions. In determining the truth of the negation of a conjunction or a disjunction we should first recall under what conditions the compound sentences are true. To form the negation of "A chicken is a fowl and a cat is a feline," we must say the statement is false. We can do this by stating that at least one of the simple statements is false. We can do this by stating "A chicken is not a fowl or a cat is not a feline." The negation of "I will study both Spanish and French" could be "I will not study Spanish or I will not study French."

It should be clear that the negation of "p and q" is the statement "not-p or not-q." In truth table form:

p	q	$p \wedge q$	$\sim(p \wedge q)$	$\sim p$	$\sim q$	$\sim p \vee \sim q$
T	T	T	F	F	F	F
T	F	F	T	F	T	T
F	T	F	T	T	F	T
F	F	F	T	T	T	T

To form the negation of the disjunction, "We are going to win or my information is incorrect" we write, "We are not going to win and my information is correct." Thus the negation of "p or q" is "not-(p or q)," and this means "not p and not q." In truth table form:

p	q	$p \vee q$	$\sim(p \vee q)$	$\sim p$	$\sim q$	$\sim p \wedge \sim q$
T	T	T	F	F	F	F
T	F	T	F	F	T	F
F	T	T	F	T	F	F
F	F	F	T	T	T	T

Exercises

Give the negation of each statement below and determine if it is true or false.

1. An apricot is a fruit and a carrot is a vegetable.
2. Lincoln was assassinated or Douglass was assassinated.
3. Some men like to hunt, others like to fish.
4. Some numbers are odd and some are even.
5. No numbers are odd and all numbers are even.
6. All lines are sets of points or all angles are right angles.
7. The sides of a right angle are perpendicular and all right angles are congruent.
8. The intersection of two parallel lines is a null set or each pair of straight lines has a point common to the two lines.
9. Every triangle has a right angle and an acute angle.
10. Every triangle has a right angle and an obtuse angle.
11. Every triangle has a right angle or an obtuse angle.
12. No triangle has two obtuse angles or two right angles.
13. Some triangles have three acute angles and some have only two acute angles.
14. $\overset{\circ\!\!\leftrightarrow\!\!\circ}{AB}$ designates a line and $\overrightarrow{AB}$ designates a ray.
15. A ray has one end-point or a segment has two end-points.

2.7. Logical implication. The most common connective in logical deduction is "if-then." All mathematical proofs employ conditional statements of this type. The *if* clause, called *hypothesis* or *premise* or *given* is a set of one or more statements which will form the basis for a conclusion. The *then* clause which follows necessarily from the premises is called the *conclusion.* The statement immediately following the "if" is also called the *antecedent,* and the statement immediately following the "then" is the *consequent.*

Here are some simple examples of such conditional sentences:

1. If $5x = 20$, then $x = 4$.
2. If this figure is a rectangle, then it is a parallelogram.

A hypothetical statement asserts that its antecedent *implies* its consequent. The statement does not assert that the antecedent is true, but only that the consequent is true *if* the antecedent is true.

It is customary in logic to represent statements by letters. Thus, we might let p represent the statement, "The figure is a rectangle" and q the statement, "The figure is a parallelogram." We could then state, "If p, then q" or "p implies q." We shall find it useful to use an arrow for "implies." We then can write "$p \rightarrow q$." Such a statement is called an *implication*.

The "if" statement does not have to come at the beginning of the compound statement. It may come last. In other cases, the premise will not start with the word "if." For example:

1. A good scout is trustworthy.
2. Apples are not vegetables.
3. The student in this class who does not study may expect to fail.

Each of the above can be arranged to the "if-then" form as follows:

1. If he is a good scout, then he is trustworthy.
2. If this is an apple, then it is not a vegetable.
3. If the student in this class does not study, then he will fail.

Other idioms that have the same meaning as "if p, then q" are: "p only if q," "p is a sufficient condition for q," "q, if p," "q, is a necessary condition for p," "whenever p, then q," "suppose p, then q."

Suppose your instructor made the statement, "If you hand in all your homework, you will pass this course." Here we can let p represent the statement, "You hand in all your homework," and q the statement, "You will pass the course." If both p and q are true, then $p \rightarrow q$ is certainly true. Suppose p is true and q is false; i.e., you hand in all your homework but still fail the course. Obviously, then, $p \rightarrow q$ is false.

Next suppose p is false. How shall we complete the truth table? If p is false and q is true, you do not hand in all your homework but you still pass the course. If p is false and q is false, you do not hand in all your homework and you do not pass the course. At first thought one might feel that no truth value should be given to such compound statement under those conditions. If we did so, we would violate the property that a statement must be either true or false.

Logicians have made the completely arbitrary decision that $p \rightarrow q$ is true when p is false, regardless of the truth value of q. Thus, $p \rightarrow q$ is considered false only if p is true and q is false. The truth table for $p \rightarrow q$ is:

p	q	$p \rightarrow q$
T	T	T
T	F	F
F	T	T
F	F	T

Exercises

In each of the following compound sentences indicate the premise and the conclusion.

1. The train will be late if it snows.
2. A person lives in California if he lives in San Francisco.
3. Only citizens over 21 have the right to vote.
4. Four is larger than three.
5. All students must take a physical examination.
6. I know he was there because I saw him.
7. Two lines which are not parallel intersect.
8. All right angles are congruent.
9. Natural numbers are either even or odd.
10. He will be punished if he is caught.
11. Every parallelogram is a quadrilateral.
12. Good scouts obey the laws.
13. Birds do not have four feet.
14. Diamonds are expensive.
15. Those who study will pass this course.
16. The sides of an equilateral triangle are congruent to each other.
17. The person who steals will surely be caught.
18. To be successful, one must work.
19. The worker will be a success.
20. You must be satisfied or your money will be refunded.
21. With your looks, I'd be a movie star.

2.8. Modus ponens. An implication by itself is of little value. However, if we know "p implies q" *and* that p is also true, we must accept q as true. This is known as the *Fundamental Rule of Inference.* This rule of reasoning is called *modus ponens.* For example, consider the implication: (*a*) "If it is raining, it is cloudy." Also, with the implication consider the statement (*b*) "It is raining." If we accept (*a*) and (*b*) together, we *must* conclude that (*c*) "It is cloudy."

In applying the Rule of Inference, it does not matter what the content of the statements p and q are. So long as "p implies q" is true and p is true, we logically must conclude that q is true. This is shown by forming the general structure:

$$\begin{array}{l} 1.\ p \rightarrow q \\ \underline{2.\ p \qquad} \\ 3.\ \therefore q \end{array} \qquad \text{or} \qquad \begin{array}{l} \underline{1.\ p \rightarrow q. \quad 2.\ p} \\ 3.\ \therefore q \end{array}$$

The symbol ∴ means "then" or "therefore." The three-step form is called a *syllogism.* Steps 1 and 2 are called the *assumptions* or *premises,* and step 3 is called the *conclusion.* The order of the steps 1 and 2 can be reversed and not change the validity of the syllogism. Thus the syllogism could also be written:

$$\begin{array}{l} 1.\ p \\ \underline{2.\ p \rightarrow q} \\ 3.\ \therefore q \end{array} \qquad \text{or} \qquad \begin{array}{l} \underline{1.\ p. \quad 2.\ p \rightarrow q} \\ 3.\ \therefore q \end{array}$$

A common type of invalid reasoning is that of *affirming the consequent.* Its structure follows:

$$\begin{array}{l} 1.\ p \rightarrow q \\ \underline{2.\ q \qquad} \\ 3.\ \therefore p \end{array}$$

INVALID

2.9. Modus tollens. A second syllogism denies the consequent of an inference and then concludes the antecedent of the conditional sentence must be denied. This mode of reasoning is called *modus tollens. Modus tollens* reasoning can be structured:

$$\begin{array}{l} 1.\ p \rightarrow q \\ \underline{2.\ \text{not-}q} \\ 3.\ \therefore \text{not-}p \end{array}$$

Consider the conditional sentence (*a*) "If it is raining, it is cloudy." Then consider with the inference the statement (*b*) "It is not cloudy." If premises (*a*) and (*b*) hold, we must conclude by *modus tollens* reasoning that (*c*) "It is not raining."

The method of *modus tollens* is a logical result of the interpretation that $p \rightarrow q$ means "q is a necessary condition for p." Thus, if we don't have q, we can't have p.

Another common type of invalid reasoning is that of *denying the antecedent.* Its structure follows:

1. $p \rightarrow q$
2. not-p
3. $\therefore$ not-q

Two other principles of logic should be mentioned here. The *Law of the Excluded Middle* asserts "p or not p" as a logical statement. The "or" in this instance is used in the limited or exclusive sense. For example, "A number is either an odd number or it is not an odd number." Another example, "Silver is heavier than gold or silver is not heavier than gold. It cannot be both."

The symbol for the "exclusive or" is "$\veebar$." The truth table for the "exclusive or" follows.

p	q	$p \veebar q$
T	*T*	*F*
T	*F*	*T*
F	*T*	*T*
F	*F*	*F*

The *Rule for Denying the Alternative* is expressed schematically by:

1. p or q	1. p or q
2. not-q	2. not-p
3. $\therefore p$	3. $\therefore q$

As an example, if we accept the statements (*a*) "The number k is odd or the number k is even," and (*b*) "The number k is not even," we must then conclude that (*c*) "The number k is odd."

We will use these two principles in developing proofs for theorems later in this book.

Exercises

In the following exercises supply a valid conclusion, if one can be supplied by the method of *modus ponens* or *modus tollens.* Assume the "or" in the following exercises to be the exclusive or. (*Note.* You are not asked to determine whether the premises or conclusions are true.)

1. The taller of two men is always the heavier. Bob is taller than Jack.
2. All quadrilaterals have four sides. A rhombus has four sides.
3. Barking dogs do not bite. My dog barks.
4. Triangle ABC is equilateral. Equilateral triangles are isosceles.
5. Every parallelogram is a quadrilateral. Figure $ABCD$ is a parallelogram.
6. If $B \in \overline{AC}$, then $m\overline{AB} + m\overline{CB} = m\overline{AC}$. $B \in \overline{AC}$.
7. If $a = b$, then $a + c = b + c$. $a = b$.
8. If $a = b$, then $c = d$. $c = d$.
9. Parallel lines do not meet. Lines l and m do not meet.
10. All women are poor drivers or I am mistaken. I am not mistaken.
11. Anyone handling a toad will get warts on his hand. I handled a toad today.
12. All goons are loons. This is a loon.
13. Jones lives in Dallas or he lives in Houston. Jones does not live in Dallas.
14. All squares are rectangles. This is not a rectangle.
15. If $a = b$, then $ac = bc$. $ac \neq bc$.
16. If $R \in \overleftrightarrow{ST}$, then $R \in \overset{\circ\!-\!\circ}{ST}$. $R \notin \overset{\circ\!-\!\circ}{ST}$.
17. If $B \in \overleftrightarrow{AC}$, then $B \in \overline{AC}$. $B \in \overrightarrow{AC}$.

Each of the following gives the pattern for arriving at a conclusion. Write the statements which complete the pattern.

18. (1) If $B \in \overleftrightarrow{AC}$, then $B \in \overset{\circ\!-\!\circ}{AC}$. (2)
    (3) Then $B \notin \overleftrightarrow{AC}$.
19. (1) If $x = 4$, then $y = 4$. (2) $x = 4$.
    (3) Then
20. (1) If $x = y$, then $x \neq z$. (2) $x = y$.
    (3) Then
21. (1) (2) If $a \neq b$, then $a = c$.
    (3) Then $a = c$.
22. (1) This is an acute or an obtuse triangle. (2)
    (3) Then this is an obtuse triangle.
23. (1) $S \in \overrightarrow{RT}$ or $S \in \overset{\circ\!\rightarrow}{RT}$. (2)
    (3) Then $S \in \overset{\circ\!\rightarrow}{RT}$
24. (1) (2) l is parallel to m.
    (3) Then $l \cap m = \emptyset$.
25. (1) l is not parallel to m. (2)
    (3) $l \cap m \neq \emptyset$.
26. (1) $\overleftrightarrow{AB} \perp \overleftrightarrow{BC}$ or $m\angle ABC \neq 90$. (2) $\overleftrightarrow{AB}$ is not $\perp \overleftrightarrow{BC}$.
    (3) Then

2.10. Converse of an implication. Many statements can be expressed in converse form. This is done by interchanging the "if" and the "then" of the statement.

Definition: The *converse* of $p \rightarrow q$ is $q \rightarrow p$.

Frequently we are prone to accept a statement and, then without realizing it, infer the converse of the statement. The converse of a statement does not always have the same truth value as the statement. An obvious example is the true statement "All horses are animals," and the false converse "All animals are horses." Broken into parts, the "if" of the statement is, "This is a horse," whereas the conclusion is, "This is an animal."

The converse of the statement "All Huftons are good radios" is "If a radio is a good one, it is a Hufton." In geometry, the converse of the statement "Perpendicular lines form right angles" is "If lines form right angles, they are perpendicular." In this case, both the statement and its converse are true. However, note the following syllogism.

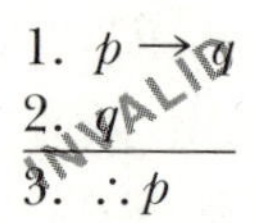

Exercises

In the following exercises determine, if possible, the truth or falsity of the given statement. Then write the converse of each statement and determine (if possible) the truth or falsity of the converse.

1. Carrots are vegetables.
2. Every U.S. citizen over 21 years of age has the right to vote.
3. Fords are cars.
4. Half-lines are rays.
5. No journalists are poor spellers.
6. If two angles are each a right angle, they are congruent.
7. Only a moron would accept your offer.
8. Only parallel lines do not meet.
9. To succeed in school one must study.
10. Only perpendicular lines form right angles.
11. Diamonds are hard.
12. A geometric figure is a set of points.
13. An equilateral triangle has three congruent sides.
14. If a is less than b, then b is larger than a.
15. If $x-y=1$, then x is larger than y.
16. Equilateral triangles are isosceles.

17. If a man lives in Los Angeles, he lives in California.
18. Parallel lines in a plane do not intersect.
19. If $x = 5$, then $x^2 = 25$.
20. If B is between A and C, then $m\overline{AC} = m\overline{AB} + m\overline{BC}$.

2.11. Logical equivalence. We have seen that the converse of a true implication does not have to have the same truth value as that of the statement but, of course, it may. If two statements mutually imply each other, they are said to be logically equivalent. Logically equivalent statements present the same information.

Definition: The statements p and q are *equivalent* if p and q have the same truth values and may be substituted for each other.

If p and q are equivalent statements, we indicate this by writing $p \leftrightarrow q$. This means $p \rightarrow q$ *and* $q \rightarrow p$. The truth table for equivalence can be developed as follows:

p	q	$p \rightarrow q$	$q \rightarrow p$	$p \leftrightarrow q$
T	T	T	T	T
T	F	F	T	F
F	T	T	F	F
F	F	T	T	T

The following are equivalent statements.

p: Line l is parallel to line m.
q: Line m is parallel to line l.

Logically equivalent sentences are often put in the form "if and only if." Thus we have, "l is parallel to m if and only if m is parallel to l."

Another obvious equivalence is the double negation, since a double negation is equivalent to the corresponding positive statement. Thus, for every statement p, we have

$$[\text{not } (\text{not-}p)] \leftrightarrow p.$$

As an example, if p means "Three is a prime number," then the double negation of p is stated "It is false that three is not a prime number." The two statements are equivalent.

Exercises

In the following exercises determine which pairs are equivalent. Note that in some exercises p and q are simple statements; in others, p and q are implications.

1. p: 5 is greater than 3.
   q: 3 is less than 5.
2. p: $a+2b=4$.
   q: $2a+4b=8$.
3. p: Line l is perpendicular to line m.
   q: Line m is perpendicular to line l.
4. p: Lines l and m are not parallel.
   q: Lines l and m intersect.
5. p: If it is a dog, it has four legs.
   q: If it does not have four legs, it is not a dog.
6. p: Perpendicular lines form right angles.
   q: Right angles form perpendicular lines.
7. p: A diameter is a chord.
   q: A chord is a diameter.
8. p: $x=y$.
   q: $y=x$.
9. p: For numbers a, b, c, $a=b$.
   q: For numbers a, b, c, $a+c=b+c$.
10. p: The present was expensive.
    q: It is not true that the present was expensive.
11. p: If he is a native of Spain, he is a native of Europe.
    q: If he is not a native of Europe, he is not a native of Spain.
12. p: If two lines meet to form right angles, they are perpendicular.
    q: If two lines are not perpendicular, they do not meet to form right angles.
13. p: Points R and S are on opposite sides of line l.
    q: Line segment RS intersects line l.
14. p: B is between A and C.
    q: $B \in \overline{AC}, B \neq A, B \neq C$.
15. p: l and m are two lines and $A \in l \cap m$.
    q: Line l and line m intersect at point A.
16. p: $R \notin \overleftrightarrow{ST}$.
    q: R lies on one side of $\overleftrightarrow{ST}$.
17. p: $\angle RST$ is an acute angle and $\angle ABC$ is an obtuse angle.
    q: $m\angle ABC > m\angle RST$.
18. p: Vertical angles are congruent.
    q: If the angles are not vertical angles, then they are not congruent.

19. *p:* If today is Saturday, then tomorrow is Sunday.
    q: Tomorrow is not Sunday; hence today is not Saturday.
20. *p:* If $a < b$, then $a - b$ is negative.
    q: If $a - b$ is positive, then $a > b$.
21. *p:* l and m are two lines and $l \cap m = \phi$.
    q: Lines l and m are parallel to each other.
22. *p:* If r, then not-s.
    q: If s, then not-r.
23. *p:* If not-r, then s.
    q: If not-s, then r.
24. *p:* The figure is a triangle.
    q: The figure is that formed by the union of three line segments.

2.12. Four rules of contraposition. Logically equivalent statements may be substituted for each other whenever they occur in a discourse. One particular type of equivalence has great value in the study of logic, namely, contraposition.

Definition: The statement not-$q \rightarrow$ not-p is called the *contrapositive* of the statement of $p \rightarrow q$.

There are four common types of contraposition. A study of the following four equivalences will reveal that the contrapositive is the negation of the clauses of the converse, as well as the converse of the negation of the clauses of the original implication.

1. $\dfrac{\text{If } p, \text{ then } q}{\text{If not-}q, \text{ then not-}p}$; $(p \rightarrow q) \leftrightarrow (\text{not-}q \rightarrow \text{not-}p)$.

2. $\dfrac{\text{If not-}p, \text{ then not-}q}{\text{If } q, \text{ then } p}$; $(\text{not-}p \rightarrow \text{not-}q) \leftrightarrow (q \rightarrow p)$.

3. $\dfrac{\text{If } p, \text{ then not-}q}{\text{If } q, \text{ then not-}p}$; $(p \rightarrow \text{not-}q) \leftrightarrow (q \rightarrow \text{not-}p)$.

4. $\dfrac{\text{If not-}p, \text{ then } q}{\text{If not-}q, \text{ then } p}$; $(\text{not-}p \rightarrow q) \leftrightarrow (\text{not-}q \rightarrow p)$.

The student should study the four types until he is satisfied that if you accept either one of a pair of contrapositives as true, you must accept the other as true also. The following examples illustrate the applications of the four types.

1. If he can vote, then he is over 21 years of age.
   If he is not over 21 years of age, then he cannot vote.
2. If l and m are not perpendicular, they do not intersect at right angles.

If l and m intersect at right angles, they are perpendicular.

3. If he drives, he should not drink.
   If he drinks, he should not drive.
4. If the natural number is not even, it is odd.
   If the natural number is not odd, it is even.

The equivalence of contrapositive statements is shown by the following truth table. The numbers under each column indicates the order of each step.

$(p$	$\rightarrow$	$q)$	$\leftrightarrow$	$(\sim q$	$\rightarrow$	$\sim p)$
T	T	T	T	F	T	F
T	F	F	T	T	F	F
F	T	T	T	F	T	T
F	T	F	T	T	T	T
1	3	1	4	2	3	2

Exercises

Each exercise contains a conditional statement. Form (*a*) its converse, (*b*) its contrapositive, and (*c*) the converse of its contrapositive.

1. If $T \in \overset{\circ}{\overrightarrow{RX}}$, then $T \in \overrightarrow{RX}$.
2. If $T \in \overrightarrow{RX}$, then $T \in \overset{\circ}{\overrightarrow{RX}}$.
3. If $C \in \overleftrightarrow{AB}$, then $C \in \overline{AB}$.
4. If $a+c=b+c$, then $a=b$.
5. If $a+b=0$, then $a=-b$.
6. If $a+b=c$, then c is greater than a.
7. I will pass this course if I study.
8. If he is an alien, he is not a citizen.
9. Parallel lines will not meet.
10. If this is not a Zap, it is a Zop.
11. If the figure is not a rectangle, it is not a square.
12. If he is not a European, he is not a native Italian.
13. If the triangle is equilateral, it is equiangular.
14. Good citizens do not create disturbances.

In the following exercises determine which of the conclusions are valid.

15. Good citizens do not create disturbances. I do not create disturbances.

    I am a good citizen.

16. If I study, I will pass this course. I study.

    I will pass this course.

17. $$\frac{\text{If } x = y, \text{ then } x^2 = y^2}{\text{If } x^2 = y^2, \text{ then } x = y}.$$

18. $$\frac{\text{If I do not study, I will not pass this course.}}{\text{If I study, I will pass this course.}}$$

19. $$\frac{\text{If this is rhombus, it is not a trapezoid.}}{\text{If this is a trapezoid, it is not a rhombus.}}$$

20. $$\frac{\text{If } a \neq b, \text{ then } c \neq d; \quad c \neq d}{a \neq b}.$$

21. $$\frac{\text{If } c \neq d, \text{ then } a \neq b; \; c \neq d}{a \neq b}.$$

22. $$\frac{\text{If } C \notin \overline{AB}, \text{ then } C \notin \overset{\circ\!-\!\circ}{AB}}{\text{If } C \in \overline{AB}, \text{ then } C \in \overset{\circ\!-\!\circ}{AB}}.$$

23. $$\frac{\text{If } l \text{ is not } \| m, \text{ then } l \cap m = \text{a point}}{\text{If } l \cap m \text{ is not a point, then } l \parallel m}.$$

24. $$\frac{\text{If } l \parallel m, \; l \cap m = \emptyset}{\text{If } l \cap m = \emptyset, \; l \parallel m}.$$

25. $$\frac{\text{If it is Thanksgiving Day, the month is November.} \quad \text{It is not December}}{\text{It is Thanksgiving Day.}}.$$

Summary Test

In each of the following indicate whether the statement is *always true* (mark T) or *not always true* (mark F).

1. Valid conclusions can result from false (untrue) basic assumptions.
2. The converse of "In triangle RST, if $m(\overline{RT}) > m(\overline{RS})$, then $m\angle S > m\angle T$" is "In triangle RST, if $m\angle S > m\angle T$, then $m(\overline{RS}) > m(\overline{RT})$."
3. The converse of "If you eat toadstools, you'll get sick" is "You will get sick if you eat toadstools."
4. "Close the door!" is a statement.
5. "It is cold and I am freezing" is a statement.
6. Given p is true, q is false. Then "p and q" is false.
7. Given p is false, q is true. Then "p and q" is false.
8. Given p is true, q is false. Then "p or q" is false.
9. Given p is false, q is true. Then "p or q" is false.
10. "p or q" is called a conjunction of p and q.
11. If p is false, then not-p is true.
12. A negation of the statement "Not every student is smart" is "Not every student is stupid."
13. A negation of the statement "a equals 2 and b equals 3" is "a does not equal 2 and b does not equal 3."
14. A negation of "Some blind men can see" is "At least one blind man can see."
15. The negation "not (p or q)" has the same meaning as "not p or not q."
16. "Not (p and q)" means the same as "not (p or q)."
17. "Not p or not q" means the same as "not (p and q)."
18. If an implication is true, its converse is also true.
19. The converse of "If $a \cap t = \phi$, then $a \parallel t$" is "If $a \parallel t$, then $a \cap t \neq \phi$."

20. The converse of a true statement is always true.
21. The negation of a false statement may result in a true statement.
22. "January has 32 days or 4 is less than 5" is a true statement.
23. "not-not p" has the same meaning as "p."
24. If p is true, not-p is also true.
25. The negation of "No A is B" is "Every A is B."
26. The negation of "Every Lak is a Luk" is "Not every Luk is a Lak."
27. $(P \to Q) \leftrightarrow (Q \to P)$.
28. not (p or q) $\leftrightarrow$ (not p or not q).
29. not (p and q) $\leftrightarrow$ (not p and not q).
30. $(p \to q) \leftrightarrow (\text{not } p \to \text{not } q)$.
31. $(\text{not } p \to \text{not } q) \leftrightarrow (p \to q)$.
32. $(p \to \text{not } q) \leftrightarrow (q \to \text{not } p)$.
33. $(\text{not } p \to q) \leftrightarrow (\text{not } q \to p)$.

34.
$$\begin{array}{c} p \to q \\ q \\ \hline \therefore p \end{array}$$

35.
$$\begin{array}{c} p \to q \\ p \\ \hline \therefore q \end{array}$$

36.
$$\begin{array}{c} p \to q \\ \text{not } q \\ \hline \therefore \text{not } p \end{array}$$

37.
$$\begin{array}{c} p \to q \\ \text{not } q \\ \hline \therefore p \end{array}$$

38.
$$\begin{array}{c} p \to q \\ \text{not } p \\ \hline \therefore \text{not } q \end{array}$$

39.
$$\begin{array}{c} p \to q \\ \text{not } p \\ \hline \therefore q \end{array}$$

40.
$$\begin{array}{c} p \leftrightarrow q \\ \hline \therefore p \to q \text{ and } q \to p \end{array}$$

41.
$$\begin{array}{c} p \to q \text{ or } q \to p \\ \hline \therefore p \leftrightarrow q \end{array}$$

42.
$$\begin{array}{c} \text{not } p \to q \\ \hline \therefore \text{not } q \to p \end{array}$$

3

Deductive Reasoning

3.1. Properties of real numbers. In your first course in algebra, you learned some basic facts about the real number system. Since you will have numerous occasions to refer to the properties of the real number system, we will present them in this section. The student is advised to thoroughly review them.

In stating the following properties, we will let the letters a, b, c, and d represent real numbers. Hereafter, you can refer to these properties either by name or by repeating the property when asked to support deductions made about real numbers.

Equality Properties

E-1 (*reflexive property*). $a = a$.

E-2 (*symmetric property*). $a = b \rightarrow b = a$.

E-3 (*transitive property*). $(a = b) \wedge (b = c) \rightarrow a = c$.

E-4 (*addition property*). $(a = b) \wedge (c = d) \rightarrow (a + c) = (b + d)$.

E-5 (*subtraction property*). $(a = b) \wedge (c = d) \rightarrow (a - c) = (b - d)$.

E-6 (*multiplication property*). $(a = b) \wedge (c = d) \rightarrow ac = bd$.

E-7 (*division property*). $(a = b) \wedge (c = d \neq 0) \rightarrow \dfrac{a}{c} = \dfrac{b}{d}$.

E-8 (*substitution property*). Any expression may be replaced by an equivalent expression in an equation without changing the truth value of the equation.

The symbol for "is greater than" is ">" and for "is less than" is "<." Thus, $a > b$ is read "a is greater than b." It should be noted that $a > b$ and $b < a$ are two ways of writing the same fact. They can be used interchangeably.

Definition: A real number is *positive* iff it is greater than zero; it is *negative* iff it is less than zero.

We say that $a > b$ iff $a - b$ is a positive number. Similarly, $a < b$ iff $a - b$ is a negative number.

The symbol for "is not greater than" is "$\ngtr$" and for "is not less than" is "$\nless$."

Order Properties

O-1 (*trichotomy property*). For every pair of real numbers, a and b, exactly one of the following is true: $a < b$, $a = b$, $a > b$.

O-2 (*addition property*). $(a < b) \wedge (c \leq d) \rightarrow (a + c) < (b + d)$.

O-3 (*subtraction property*). $(a < b) \rightarrow (a - c) < (b - c)$;
$(a < b) \rightarrow (c - a) > (c - b)$.

O-4 (*multiplication property*). $(a < b) \wedge (c > 0) \rightarrow ac < bc$;
$(a < b) \wedge (c < 0) \rightarrow ac > bc$.

O-5 (*division property*). $(a < b) \wedge (c > 0) \rightarrow a/c < b/c \wedge c/a > c/b$;
$(a < b) \wedge (c < 0) \rightarrow a/c > b/c \wedge c/a < c/b$.

O-6 (*transitive property*). $(a < b) \wedge (b < c) \rightarrow a < c$.

O-7 (*substitution property*). Any expression may be substituted for an equivalent expression in an inequality without changing the truth value of the inequality.

O-8 (*partition property*). $(c = a + b) \wedge (b > 0) \rightarrow c > a$.

Properties of a field

The following additional properties of the real number system are called "field properties."

Operations of Addition

F-1 (*closure property*). $a + b$ is a unique real number.

F-2 (*associative property*). $(a + b) + c = a + (b + c)$.

F-3 (*commutative property*). $a + b = b + a$.

F-4 (*additive property of zero*). There is a unique real number 0, the *additive identity element*, such that $a + 0 = 0 + a = a$.

F-5 (*additive inverse property*). For every real number a, there exists a real number $(-a)$, the *additive inverse of* a, such that $a + (-a) = (-a) + a = 0$.

Operations of Multiplication

F-6 (*closure property*). $a \cdot b$ is a unique real number.

F-7 (*associative property*). $(a \cdot b) \cdot c = a \cdot (b \cdot c)$.

F-8 (*commutative property*). $a \cdot b = b \cdot a$.

F-9 (*multiplicative property of 1*). There is a unique real number 1, the *multiplicative identity element,* such that $a \cdot 1 = 1 \cdot a = a$.

F-10 (*multiplicative inverse property*). For every real number $a (a \neq 0)$ there is a unique real number $1/a$, the *multiplicative inverse of a*, such that $a \cdot (1/a) = (1/a) \cdot a = 1$.

F-11 (*distributive property*). $a(b+c) = a \cdot b + a \cdot c$.

Exercises

1–6. What property of the real number system is illustrated by each of the following?

1. $4+3=3+4$.
2. $5+(-5)=0$.
3. $6+0=6$.
4. $7 \cdot 1=7$.
5. $2(5+4)=2 \cdot 5+2 \cdot 4$.
6. $5 \cdot 2=2 \cdot 5$.

7–24. Name the property of the real number system which will support the indicated conclusion.

7. If $x-2=5$, then $x=7$.
8. If $3x=12$, then $x=4$.
9. If $7=5-x$, then $5-x=7$.
10. If $a+3=7$, then $a=4$.
11. If $2a+5=9$, then $2a=4$.
12. If $a+b=10$, and $b=3$, then $a+3=10$.
13. If $\frac{1}{2}x=7$, then $x=14$.
14. $5 \cdot (\frac{1}{5})=1$.
15. If $a+3<8$, then $a<5$.
16. If $x=y$ and $y=6$, then $x=6$.
17. If $x>y$ and $z>x$, then $z>y$.
18. If $a-2>10$, then $a>12$.
19. If $-3x<15$, then $x>-5$.
20. $1/2+\sqrt{4}$ is a real number.
21. $(5 \cdot \frac{3}{4}) \cdot 12=5 \cdot (\frac{3}{4} \cdot 12)$.
22. $(17+18)+12=17+(18+12)$.
23. If $\frac{1}{3}x>-4$, then $x>-12$.
24. $3(y+5)=3y+15$.

25–30. Name the property of real numbers which justify each of the numbered steps in the following problems.

Illustrative Problem. $8-3x=2(x+6)$.

Solution

EQUATIONS	REASONS
1. $8-3x=2(x-6)$.	1. Given.
2. $8-3x=2x-12$.	2. Distributive property of equality.
3. $-3x=2x-20$.	3. Subtractive property of equality.
4. $-5x=-20$.	4. Subtraction property of equality.
5. $x=4$.	5. Division property of equality.

25. 1. $5x-7=2x+8$.
    2. $5x=2x+15$.
    3. $3x=15$.
    4. $x=5$.

26. 1. $8=2(x-3)$.
    2. $8=2x-6$.
    3. $14=2x$.
    4. $2x=14$.
    5. $x=7$.

27. 1. $3(x-5)=4(x-2)$.
    2. $3x-15=4x-8$.
    3. $3x=4x+7$.
    4. $-x=7$.
    5. $x=-7$.

28. 1. $5x-7>3x+9$.
    2. $5x>3x+16$.
    3. $2x>16$.
    4. $x>8$.

29. 1. $3x-9<7x+15$.
    2. $3x<7x+24$.
    3. $-4x<24$.
    4. $x>-6$

30. 1. $2(x-3)>5(x+7)$.
    2. $2x-6>5x+35$.
    3. $2x>5x+41$.
    4. $-3x>41$.
    5. $x<-\frac{41}{3}$.

3.2. Initial postulates. In this course, we are interested in determining and proving geometric facts. We have, with the aid of the undefined geometric concepts, defined as clearly and as exactly as we could other geometric concepts and terms. We will next *agree on or assume* certain properties that can be assigned to these geometric figures. These agreed-upon properties we will call *postulates.* They should seem almost obvious, even though they may be difficult, if not impossible, to prove. The postulates are not made up at random, but have been carefully chosen to develop the geometry we intend to develop. With definitions, properties of the real number system, and postulates as a foundation, we will establish many new geometric facts by giving logical proofs. When statements are to be logically proved, we will call them *theorems.*

Once a theorem has been proved, it can be used with definitions and postulates in proving other theorems.

It should be clear that the theorems which we can prove will, to a great extent, depend upon the postulates we agree to enumerate. Altering two or three postulates can completely change the theorems that can be proved in a given geometry course. Hence, we should recognize the importance of the selection of postulates to be used.

The postulates we will agree on will in great part reflect the world about us.

Definition: A statement that is accepted as being true without proof is called a *postulate*.

Postulate 1. *A line contains at least two points; a plane contains at least three points not all collinear; and space contains at least four points not all coplanar.*

Postulate 2. *For every two distinct points, there is exactly one line that contains both points.*

Notice that this postulate states *two* things, sometimes called *existence* and *uniqueness*:

1. There exists one line that contains the two given points.
2. This line is unique; that is, it is the only one that contains the two points.

Postulate 3. *For every three distinct noncollinear points, there is exactly one plane that contains the three points.*

Postulate 4. *If a plane contains two points of a straight line, then all points of the line are points of the plane.*

Postulate 5. *If two distinct planes intersect, their intersection is one and only one line* (see Fig. 3.1).

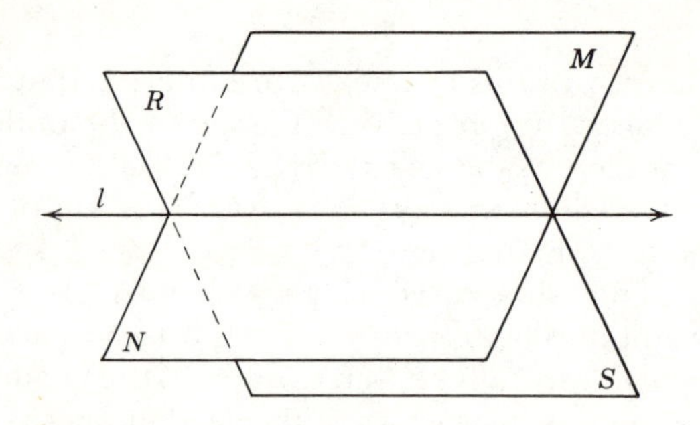

Fig. 3.1.

With the above postulates we can start proving some theorems. These first theorems will state what to most of us will seem intuitively obvious. Unfortunately, their formal proofs get tricky and not too meaningful to the geometry student beginning the study of proofs. Consequently, we will give informal proofs of the theorems. You will not be required to reproduce them. However, you should understand clearly the statements of the theorems, since you will be using them later in proving other theorems.

Theorem 3.1

3.3. If two distinct lines in a plane intersect, then their intersection is at most, one point.

Supporting argument. Let l and m be two distinct lines that intersect at S. Using the law of the excluded middle, we know that either lines l and m intersect in more than one point or they do not intersect in more than one point. If they intersect in more than one point, such as at R and S, then line l and line m must be the same line (applying Postulate 2). This contradicts the given conditions that l and m are distinct lines. Therefore, applying the rule for denying the alternative, lines l and m intersect in, at most, one point.

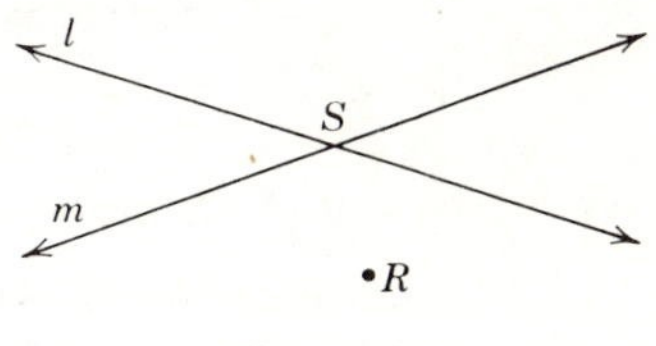

Theorem 3.1.

Theorem 3.2

3.4. If a point *P* lies outside a line *l*, exactly one plane contains the line and the point.

Supporting argument. By Postulate 1, line l contains at least two different points, say A and B. Since P is a point not on l, we have three distinct noncollinear points A, B, and P. Postulate 3, then, assures the existence and uniqueness of a plane M through line l and point P.

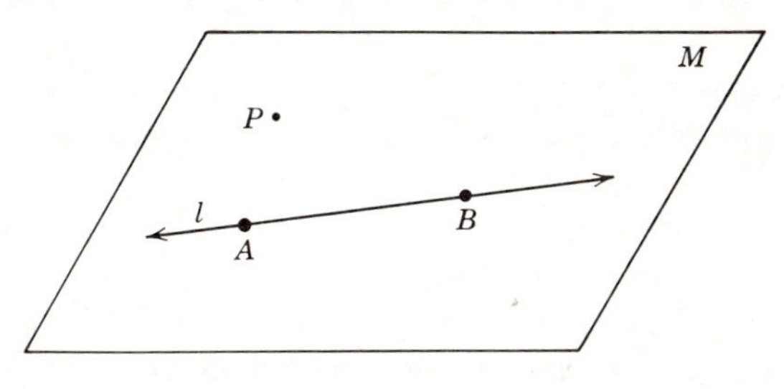

Theorem 3.2.

Theorem 3.3

3.5. If two distinct lines intersect, exactly one plane contains both lines.

Supporting argument. Let Q be the point where lines l and m intersect. Postulate 1 guarantees that a line must contain at least two points; hence, there must be another point on l and another point on m. Let these points be lettered R and P, respectively. Postulate 3 tells us that there is exactly one plane that contains points Q, R, and P. We also know that both l and m must lie in this plane by postulate 4.

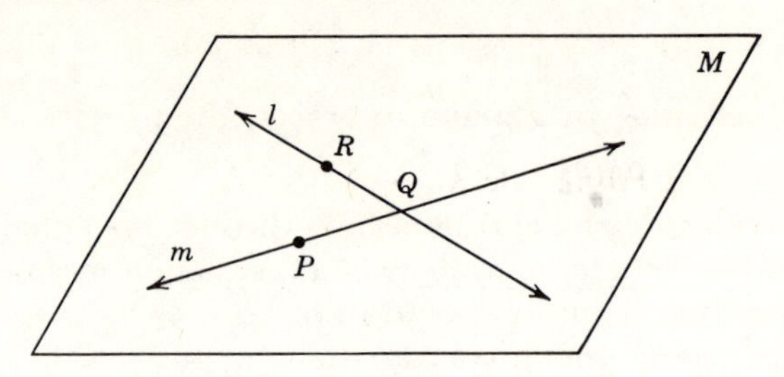

Theorem 3.3.

Summarizing. A plane is determined by

1. Three noncollinear points.
2. A straight line and a point not on the line.
3. Two intersecting straight lines.

Exercises

1. How many planes can be passed (*a*) through two points? through three points not in a straight line?
2. What figure is formed at the intersection of the front wall and the floor of a classroom?
3. Hold a pencil so that it will cast a shadow on a piece of paper. Will the shadow be parallel to the pencil?
4. How many planes, in general, can contain a given straight line and a point not on the line?
5. How many planes can contain a given straight line and a point not on the line?
6. Why is a tripod (three legs) used for mounting cameras and surveying instruments?
7. How many planes are fixed by four points not all lying in the same plane?
8. Why will a four-legged table sometimes rock when placed on a level floor?
9. Two points A and B lie in plane RS. What can be said about line AB?
10. If two points of a straight ruler touch a plane surface, how many other points of the ruler touch the surface?
11. Can a straight line be perpendicular to a line in a plane without being perpendicular to the plane?
12. Can two straight lines in space not be parallel and yet not meet? Explain.
13. On a piece of paper draw a line AB. Place a point P on $\overleftrightarrow{AB}$. In how many positions can you hold a pencil and make the pencil appear perpendicular to $\overleftrightarrow{AB}$ at P?
14. Are all triangles plane figures? Give reasons for your answer.
15. How many different planes are determined by pairs of the four different lines $\overleftrightarrow{AP}$, $\overleftrightarrow{BP}$, $\overleftrightarrow{CP}$, and $\overleftrightarrow{DP}$ no three of which are coplanar?

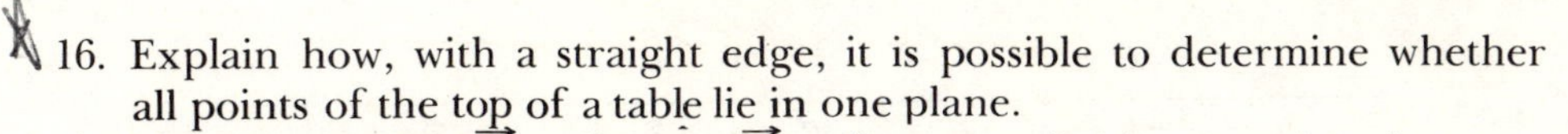

16. Explain how, with a straight edge, it is possible to determine whether all points of the top of a table lie in one plane.
17. If in plane MN, $\overrightarrow{AB} \perp$ line m, $\overrightarrow{AC} \perp$ line m, and A is on m, does it necessarily follow that $\overleftrightarrow{AB} = \overleftrightarrow{AC}$?
18. Is it possible for the intersection of two planes to be a line segment? Explain your answer.
19. Using the accompanying diagram (a 3-dimensional figure), indicate which sets of points are (1) collinear, (2) coplanar but not collinear, (3) not coplanar.

   (*a*) $\{A, C, D\}$

   (*b*) $\{D, A, F\}$

   (*c*) $\{F, G, A\}$

   (*d*) $\{F, D, G\}$

   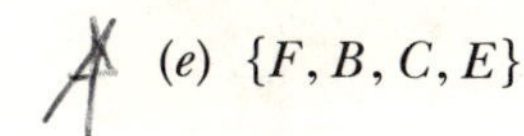

   (*e*) $\{F, B, C, E\}$

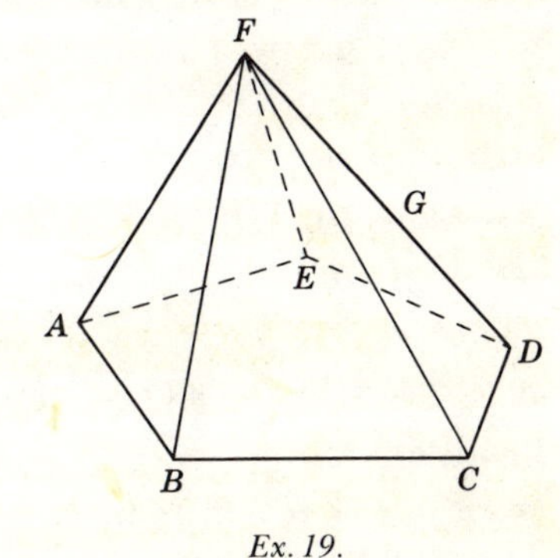

Ex. 19.

20. Which of the following choices correctly completes the statement:
    Three distinct planes cannot have in common (a) exactly one point, (b) exactly two points, (c) exactly one line, (d) more than two points.

3.6. Additional postulates. In Chapter 1 we discussed the real number line. We showed the correspondence between points on the number line and the real numbers. In order that we may use in subsequent deductive proofs the conclusions we arrived at, we will now restate them as postulates.

Postulate 6. (the ruler postulate). *The points on a line can be placed in a one-to-one correspondence with real numbers in such a way that:*

1. *For every point of the line there corresponds exactly one real number*;
2. *for every real number, there corresponds exactly one point of the line*; and
3. *the distance between two points on a line is the absolute value of the difference between the corresponding numbers*.

Postulate 7. *To each pair of distinct points there corresponds a unique positive number, which is called the distance between the points*.

The correspondence between points on a line and real numbers is called the *coordinate system* for the line. The number corresponding to a given point is called the *coordinate* of the point. In Fig. 3.2, the coordinate of A is -4, of B is -3, of C is 0, of E is 2, and so on.

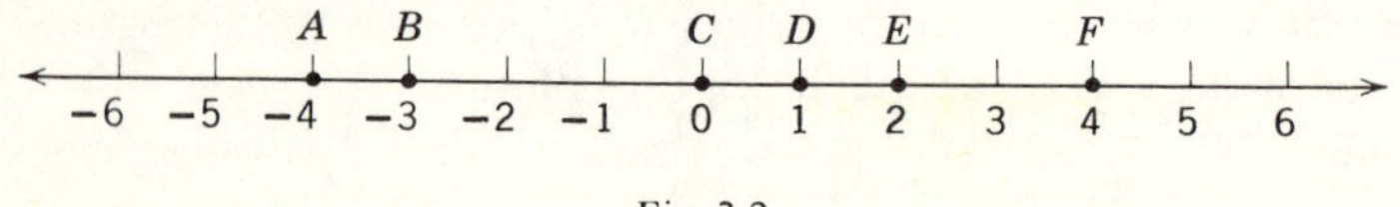

Fig. 3.2.

Postulate 8. *For every three collinear points, one and only one is between the other two.* That is, if A, B, and C are (distinct) collinear points, then one and only one of the following statements is true: (a) A lies between B and C; (b) B lies between A and C; (c) C lies between A and B.

Postulate 9. *If A and B are two distinct points, then there is at least one point C such that $C \in \overline{AB}$.* This is, in effect, saying that every line segment has at least three points.

Postulate 10. *If A and B are two distinct points, there is at least one point D such that $\overline{AB} \subset \overline{AD}$.*

Postulate 11. *For every $\overrightarrow{AB}$ and every positive number n there is one and only one point P of $\overrightarrow{AB}$ such that $m\overline{AP} = n$.* This is called the *point plotting postulate*.

Postulate 12. *If $\overrightarrow{AB}$ is a ray on the edge of the half-plane h, then for every n between 0 and 180 there is exactly one ray AP, with P in h, such that $m\angle PAB = n$.* This is called the *angle construction postulate*.

Postulate 13. (segment addition postulate). *A set of points lying between the endpoints of a line segment divides the segment into a set of consecutive segments the sum of whose lengths equals the length of the given segment.*

Thus, in Fig. 3.3, if A, B, C, D are collinear, then $m\overline{AB} + m\overline{BC} + m\overline{CD} =$

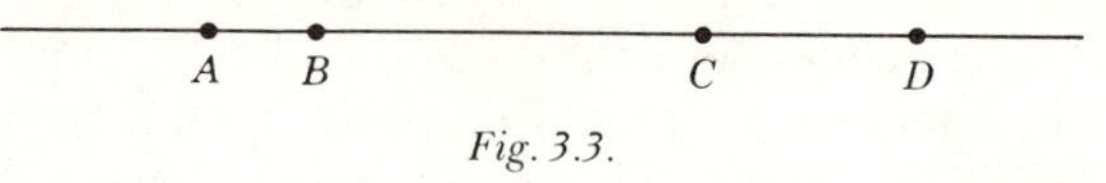

Fig. 3.3.

$m\overline{AD}$. Using the symmetric property of equality we could also write $m\overline{AD} = m\overline{AB} + m\overline{BC} + m\overline{CD}$. This postulate is often stated as "the whole equals the sum of its parts."

Postulate 14. (angle addition postulate). *In a given plane, rays from the vertex of an angle through a set of points in the interior of the angle divide the angle into consecutive angles the sum of whose measures equals the measure of the given angle.*

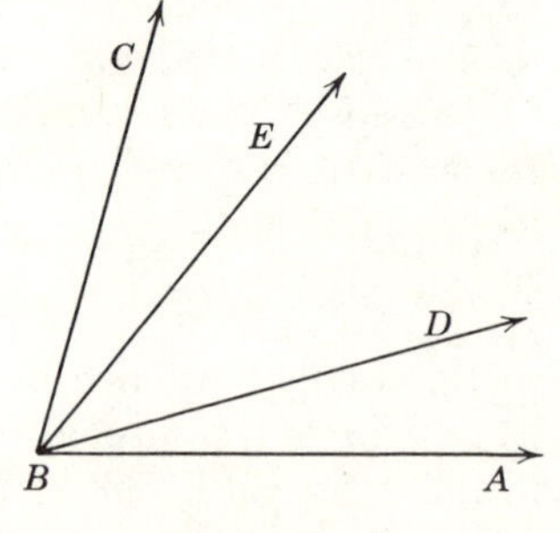

Fig. 3.4.

Thus, in Fig. 3.4, if D and E lie in the interior of $\angle ABC$, then $m\angle ABD + m\angle DBE + m\angle EBC = m\angle ABC$. Using the symmetric property of equality, we could also write $m\angle ABC = m\angle ABD + m\angle DBE + m\angle EBC$. This, too, is referred to as "the measure of the whole is equal to the sum of the measures of its parts."

Postulate 15. *A segment has one and only one midpoint.*

Postulate 16. *An angle has one and only one bisector.*

Exercises

1. What point has a coordinate of -3?
2. What is the distance from B to E?
3. What is the $m\overline{BD}$?
4. What is the $m\overline{DA}$?
5. What is the coordinate of the midpoint of $\overline{BE}$?

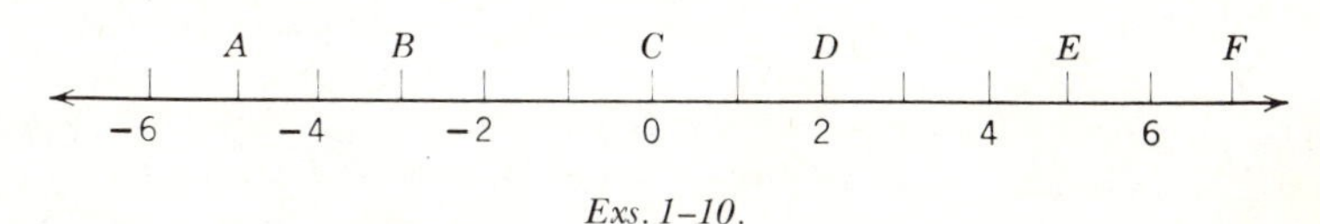

Exs. 1–10.

6. What is the coordinate of the endpoint of $\overrightarrow{FA}$?
7. What is the $m\overline{BC} + m\overline{CD} + m\overline{DE}$? Does this equal the $m\overline{BE}$?
8. What is the $m\overline{BA}$?
9. Is the coordinate of point A greater than the coordinate of point D?
10. Does $m\overline{BD} = m\overline{DB}$?
11. $a, b, c,$ are the coordinates of the corresponding points A, B, C. If $a > c$ and $c > b$, which point lies between the other two?
12. If T is a point on $\overline{RS}$, complete the following:
    (a) $m\overline{RT} + m\overline{TS} =$ (b) $m\overline{RS} - m\overline{TS} =$
13. A, B, and C are three collinear points, $m\overline{BC} = 15$, $m\overline{AB} = 11$. Which point cannot lie between the other two?
14. R, S, T are three collinear points. If $m\overline{RS} < m\overline{ST}$, which point cannot lie between the other two?

15–22. *Given:* $m\angle AEB = 44$, $m\angle BED = 34$, $m\angle AEF = 120$ $\overrightarrow{EC}$ bisects $\angle BED$. Complete the following:

15. $m\angle AEB + m\angle BEC = m\angle$
16. $m\angle BED - m\angle CED = m\angle$
17. $m\angle DEC + m\angle CEB + m\angle BEA = m\angle$
18. $m\angle BEC =$
19. $m\angle AED =$
20. $m\angle AEC =$
21. $m\angle DEF =$
22. $m\angle BEF =$

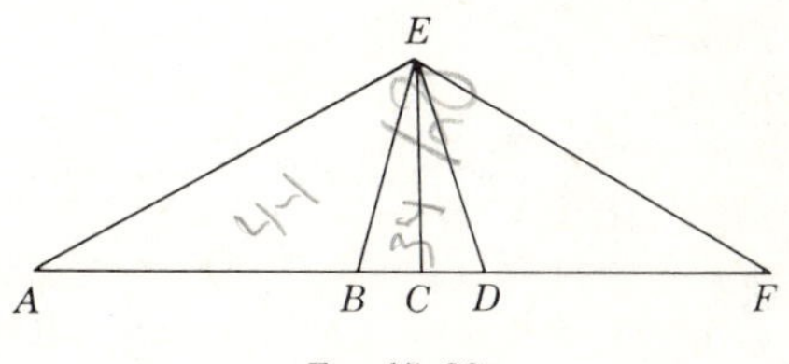

Exs. 15–22.

3.7. Formal proofs of theorems. A *theorem* is a statement or principle that is accepted only after it has been proved by reasoning. Every theorem in geometry consists of two parts: a part which states what is given or known, called the "*given*" or "*hypothesis,*" and a part which is to be proved, called the "*conclusion*" or "*prove.*"

Theorems can be written in either of two forms: (1) As a complex sentence. In this form the given is a clause beginning with "if" or "when" and the conclusion is a clause beginning with "then." For example, in the theorem, "If two angles are right angles, then the angles are congruent," "Two angles are right angles" is the given, and "The angles are congruent" is the conclusion. (2) As a declarative sentence. In this form the given and the conclusion are not so readily evident. For example, the above theorem could be written, "Two right angles are congruent." Frequently the simplest way to determine the given and conclusion of a declarative sentence is to rewrite it in the if-then form.

The formal proof of a theorem consists of five parts: (1) a statement of the theorem; (2) a general figure illustrating the theorem; (3) a statement of what is given in terms of the figure, (4) a statement of what is to be proved in terms of the figure, and (5) a logical series of statements substantiated by accepted definitions, postulates, and previously proved theorems.

Of course, it is not necessary to present proofs in formal form as we will do. The proofs could be given just as conclusively in paragraph form. However, the beginning geometry student will likely find that by putting statements of the proof in one column and reasons justifying the statements in a neighboring column, it will be easier for others, as well as himself, to follow his line of reasoning.

Most of the theorems in this text hereafter will be proved formally. The student will be expected to give the same type of proofs in the exercises that follow.

Theorem 3.4

3.8. For any real numbers, a, b, and c, if $a = c$, and $b = c$, then $a = b$.

Given: a, b, and c are real numbers. $a = c$; $b = c$.
Prove: $a = b$.
Proof:

STATEMENTS	REASONS
1. $a = c$; $b = c$.	1. Given.
2. $c = b$.	2. Symmetric property of equality.
3. $a = b$.	3. Transitive property of equality (from Statements 1 and 2).

Theorem 3.5

3.9. For any real numbers a, b, and c, if $c = a$, $c = b$, then $a = b$.
Hypothesis: a, b, and c are real numbers. $c = a$; $c = b$.
Conclusion: $a = b$.
Proof:

STATEMENTS	REASONS
1. $c = a$; $c = b$.	1. Given
2. $a = c$; $b = c$.	2. Symmetric property of equality.
3. $a = b$.	3. Theorem 3.4.

Theorem 3.6

3.10. For any real numbers a, b, c, and d if $c = a$, $d = b$, and $c = d$, then $a = b$.
Hypothesis: a, b, c, and d are real numbers; $c = a$, $d = b$, $c = d$.
Conclusion: $a = b$.
Proof:

STATEMENTS	REASONS
1. $c = a$; $d = b$; $c = d$.	1. Given.
2. $c = b$.	2. Transitive property of equality.
3. $a = b$.	3. Theorem 3.5 ($c = a \land c = b \rightarrow a = b$).

Theorem 3.7

3.11. All right angles are congruent.
Given: $\angle\alpha$ and $\angle\beta$ are right angles.
Conclusion: $\angle\alpha \cong \angle\beta$
Proof:

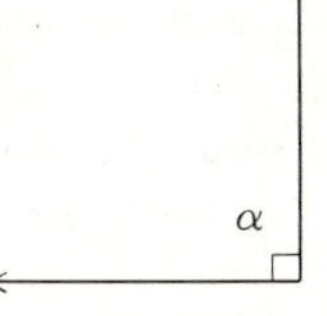

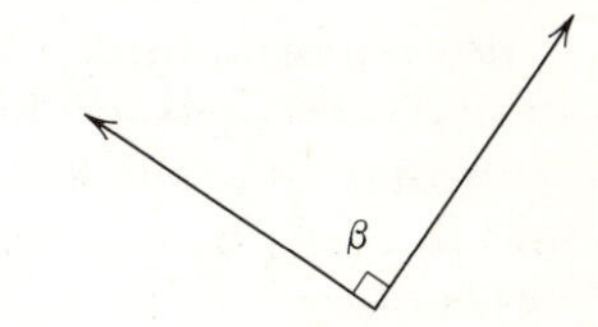

Theorem 3.7.

STATEMENTS	REASONS
1. $\angle\alpha$ is a right angle. $\angle\beta$ is a right angle.	1. Given.
2. $m\angle\alpha = 90$; $m\angle\beta = 90$.	2. The measure of a right angle is 90.
3. $m\angle\alpha = m\angle\beta$.	3. If $a = c$, $b = c$, then $a = b$.
4. $\angle\alpha \cong \angle\beta$.	4. $\angle a \cong \angle b \leftrightarrow m\angle a = m\angle b$.

Theorem 3.8

3.12. Complements of the same angle are congruent.

Given: $\angle x$ and $\angle\theta$ are complementary angles.
$\angle y$ and $\angle\theta$ are complementary angles.

Prove: $\angle x \cong \angle y$.

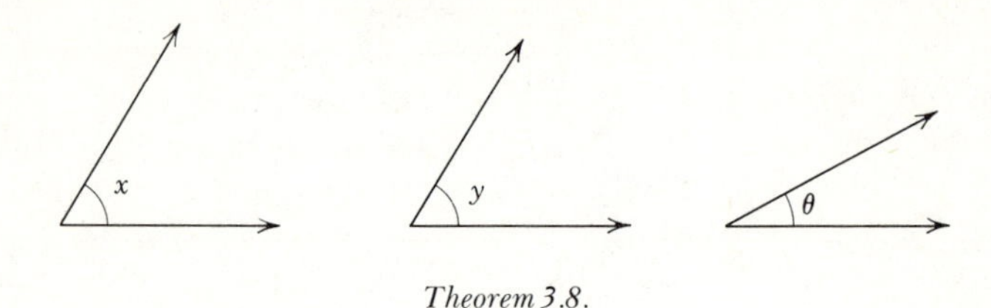

Theorem 3.8.

Proof:

STATEMENTS	REASONS
1. $\angle x$ and $\angle\theta$ are complementary $\angle$s. $\angle y$ and $\angle\theta$ are complementary $\angle$s.	1. Given.
2. $m\angle x + m\angle\theta = 90$. $m\angle y + m\angle\theta = 90$.	2. If two $\angle$s are complementary, the sum of their measures equals 90.
3. $m\angle x + m\angle\theta = m\angle y + m\angle\theta$.	3. If $a = c$, $b = c$, then $a = b$.
4. $m\angle x = m\angle y$.	4. Subtractive property of equality.
5. $\angle x \cong \angle y$.	5. $\angle x \cong \angle y \leftrightarrow m\angle x = m\angle y$.

It is important that each statement in the proof be substantiated by a reason for its correctness. These reasons must be written in full, and only abbreviations that are clear and commonly accepted may be used. The reader will find in the appendix a list of the common abbreviations which we will use in this book.

The student can easily prove the following theorems:

Theorem 3.9

3.13. All straight angles are congruent.

Theorem 3.10

3.14. Supplements of the same angle are congruent.

These theorems will subsequently be used in proving new theorems. A *corollary* of a geometric theorem is another theorem which is easily derived from the given theorem. Consider the following:

3.15. Corollary to Theorem 3.8. Complements of congruent angles are congruent.

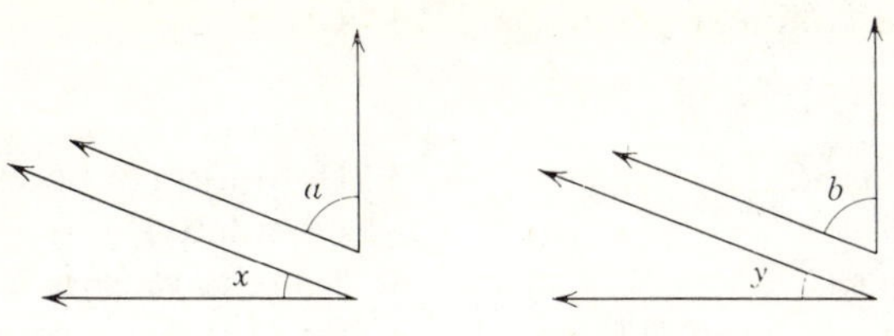

Corollary to Theorem 3.8.

Given: $\angle x$ is the complement of $\angle a$; $\angle y$ is the complement of $\angle b$; $\angle a \cong \angle b$.
Conclusion: $\angle x \cong \angle y$.
Proof:

STATEMENTS	REASONS
1. $\angle x$ is the complement of $\angle a$. $\angle y$ is the complement of $\angle b$.	1. Given.
2. $\angle a \cong \angle b$	2. Given.
3. $m\angle a = m\angle b$.	3. $\angle a \cong \angle b \leftrightarrow m\angle a = m\angle b$.
4. $m\angle x + m\angle a = 90$.	4. If two angles are complementary, the sum of their measures is 90.
5. $m\angle x + m\angle b = 90$.	5. A quantity may be substituted for its equal in an equation.
6. $\angle x$ is the complement of $\angle b$.	6. If the sum of the measures of two angles = 90, they are complementary.
7. $\angle x \cong \angle y$.	7. Complements of the same angle are congruent.

In like manner the student can prove:

3.16. Corollary to Theorem 3.10. Supplements of congruent angles are congruent.

3.17. Illustrative Example 1:
Given: Collinear points A, B, C, D as shown; $m\overline{AC} = m\overline{BD}$.
Prove: $m\overline{AB} = m\overline{CD}$.
Proof:

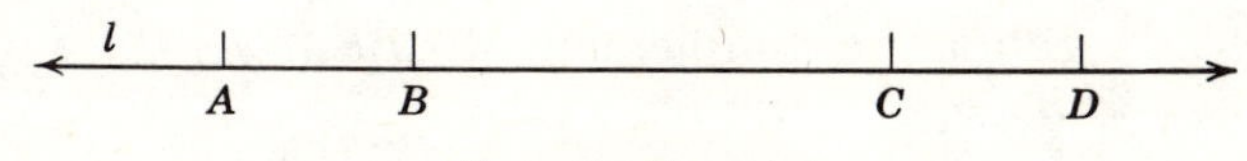

Illustrative Example 1.

STATEMENTS	REASONS
1. A, B, C, D are collinear points as shown.	1. Given
2. $m\overline{AC} = m\overline{BD}$	2. Given.
3. $m\overline{AC} = m\overline{AB} + m\overline{BC}$.	3. Definition of betweenness (also by postulate 13).
4. $m\overline{BD} = m\overline{BC} + m\overline{CD}$.	4. Same as reason 3.
5. $m\overline{AB} + m\overline{BC} = m\overline{BC} + mCD$.	5. Substitution property (statements 3 and 4 in statement 2).
6. $m\overline{AB} = m\overline{CD}$.	6. Subtractive property of equality.

3.18. Illustrative Example 2:

Given: $\angle ABC$ with $\overrightarrow{BE}$ and $\overrightarrow{BD}$ as shown. A, B, C, D, E are coplanar points. $m\angle ABD > m\angle EBC$.
Prove: $m\angle ABE > m\angle DBC$.
Proof:

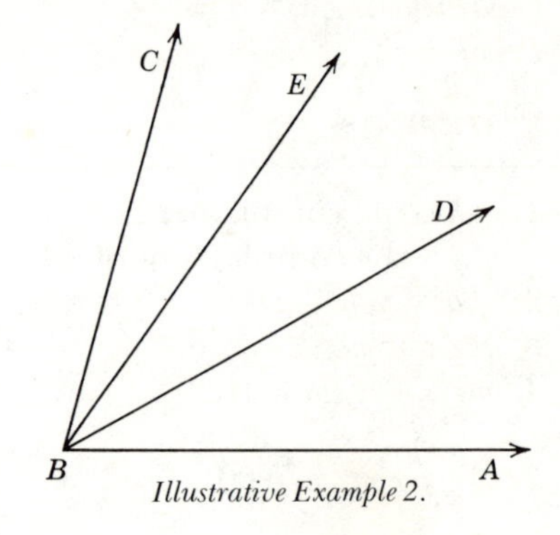

Illustrative Example 2.

STATEMENTS	REASONS
1. $\overrightarrow{BE}$ and $\overrightarrow{BD}$ are rays drawn from from the vertex of $\angle ABC$ as shown. A, B, C, D, E are coplanar.	1. Given.
2. $m\angle ABD > m\angle EBC$.	2. Given.
3. $m\angle ABD + m\angle DBE > m\angle EBC + m\angle DBE$.	3. Additive property of order.
4. $m\angle ABD + m\angle DBE = m\angle ABE$; $m\angle EBC + m\angle DBE = m\angle DBC$.	4. Angle addition postulate.
5. $m\angle ABE > m\angle DBC$.	5. Substitution property of order.

3.19. Illustrative Example 3:

Given: $\overleftrightarrow{BD} \perp \overleftrightarrow{AC}$. $\angle EBC$ is the complement of $\angle EBA$.
Prove: $\angle ABC \cong \angle AEB$.
Proof:

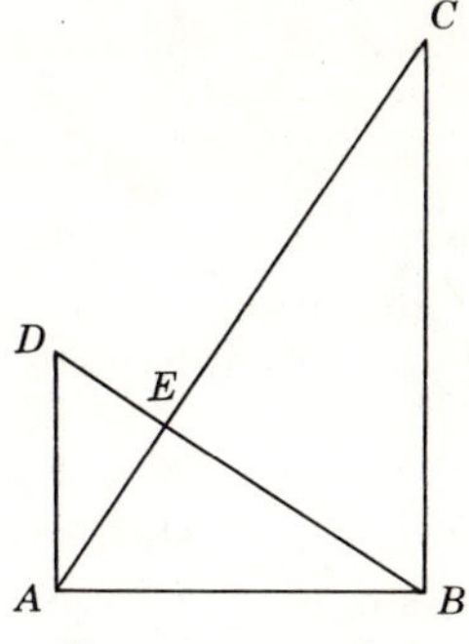

Illustrative Example 3.

STATEMENTS	REASONS
1. $\overleftrightarrow{BD} \perp \overleftrightarrow{AC}$.	1. Given.
2. $\angle AEB$ is a right angle.	2. Perpendicular lines form right angles.
3. $m\angle AEB = 90$.	3. Definition of right angle.
4. $\angle EBC$ is the complement of $\angle EBA$.	4. Given.
5. $m\angle EBC + m\angle EBA = 90$.	5. Two angles are complementary iff the sum of their measures is 90.
6. $m\angle EBC + m\angle EBA = m\angle ABC$.	6. Angle addition postulate.
7. $m\angle ABC = 90$.	7. Substitution property of equality (or Theorem 3.5).
8. $m\angle ABC = m\angle AEB$.	8. $(a = c) \wedge (b = c) \rightarrow a = b$.
9. $\angle ABC \cong \angle AEB$.	9. Angles with the same measure are congruent.

Exercises

In the following exercises complete the proofs, using for reasons only the given, definitions, properties of the real number system, postulates, theorems, and corollaries we have proved thus far.

1. Prove Theorem 3.9.
2. Prove the corollary to Theorem 3.10.
3. *Given:* Collinear points A, B, C, D as shown; $m\overline{AC} = m\overline{BD}$.
   Prove: $m\overline{AB} = m\overline{CD}$.

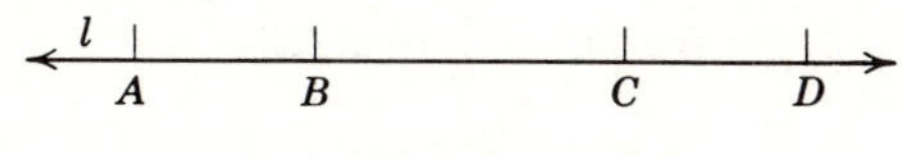

Exs. 3–4.

4. *Given:* Collinear points A, B, C, D as shown; $m\overline{AC} > m\overline{BD}$.
   Prove: $m\overline{AB} > m\overline{CD}$.

5. *Given:* $\angle ABC$ with $\overrightarrow{BD}$ and $\overrightarrow{BE}$ as shown. A, B, C, D, E are coplanar; $m\angle ABD = m\angle EBC$.
   Prove: $m\angle ABE = m\angle DBC$.

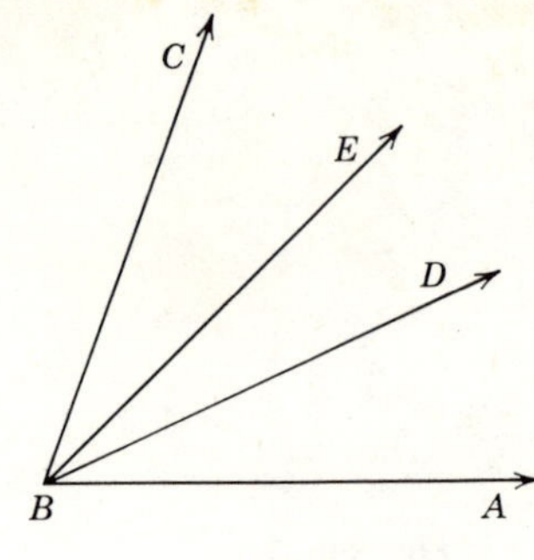

6. *Given:* $\angle ABC$ with $\overrightarrow{BD}$ and $\overrightarrow{BE}$ as shown. A, B, C, D, E are coplanar; $m\angle ABE = m\angle DBC$.
   Prove: $m\angle ABD = m\angle CBE$.

Exs. 5–6.

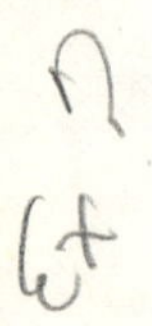

7. *Given:* $m\overline{PS} = m\overline{PR}$; $m\overline{TS} = m\overline{QR}$.
   Prove: $m\overline{PT} = m\overline{PQ}$.

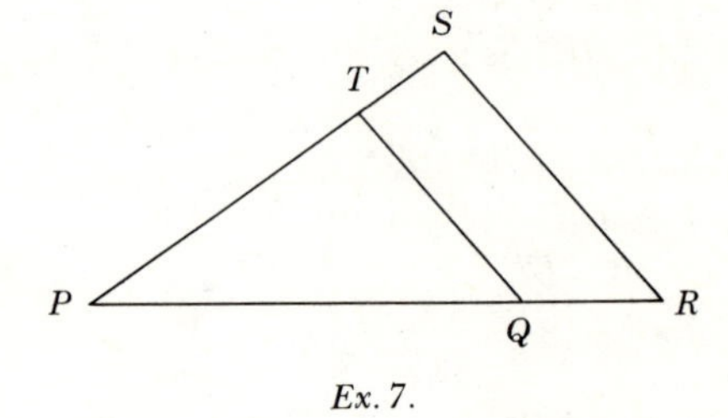

Ex. 7.

8. *Given:* $m\angle ABC = m\angle RST$ and $m\angle ABD = m\angle RSP$.
   Prove: $m\angle DBC = m\angle PST$.

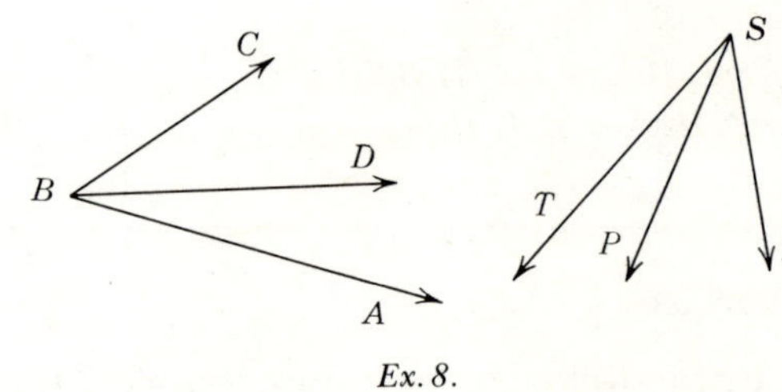

Ex. 8.

9. *Given:* $\angle ABC \cong \angle RST$; $\angle\phi \cong \angle\theta$; $\angle\alpha \cong \angle\beta$.
   Prove: $\angle ABD \cong \angle RSP$.

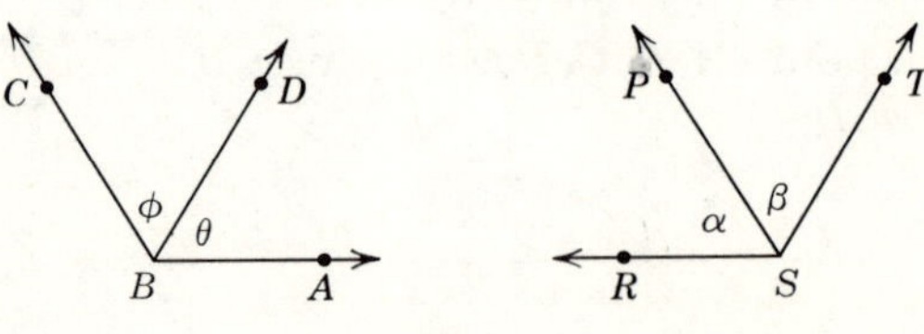

Ex. 9.

10. *Given:* Points D and E lie on sides AC and BC of $\triangle ABC$ as shown; $m\overline{AD} = m\overline{BE}$; $m\overline{DC} = m\overline{EC}$.
Prove: $\triangle ABC$ is isosceles.

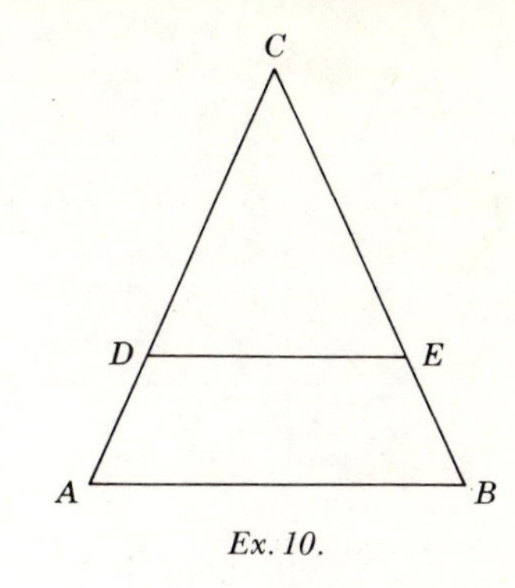

Ex. 10.

11. *Given:* A, B, C, D are collinear points; $m\angle EBC = m\angle ECB$.
Prove: $m\angle ABE = m\angle DCE$.

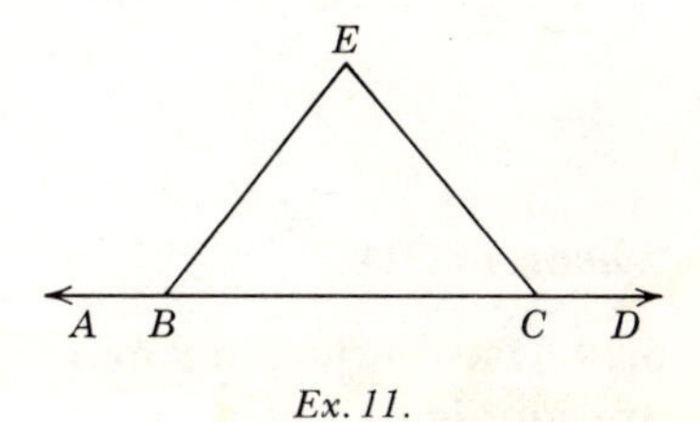

Ex. 11.

12. *Given:* Points D and E on sides AC and BC of $\triangle ABC$ as shown; $m\angle BAC = m\angle ABC$; $m\angle CAE = m\angle CBD$.
Prove: $m\angle EAB = m\angle DBA$.

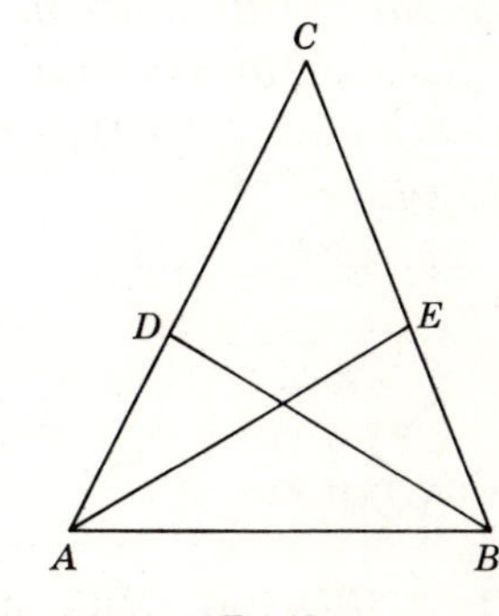

Ex. 12.

13. *Given:* $\overleftrightarrow{BC} \perp \overleftrightarrow{AB}$; $\angle BAC$ and $\angle CAD$ are complementary.
Prove: $m\angle DAB = m\angle ABC$.

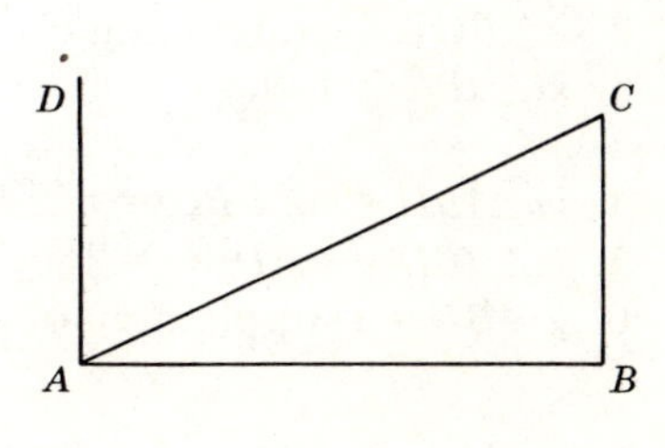

Ex. 13.

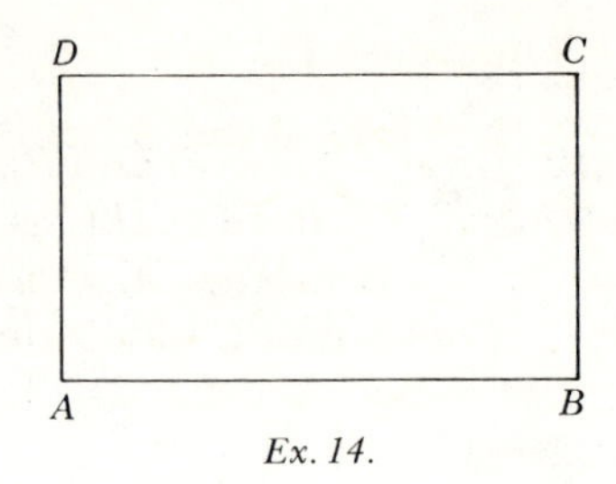

14. *Given:* $\overline{AD} \perp \overline{AB}$; $\overline{BC} \perp \overline{AB}$; $\angle B \cong \angle C$.
Prove: $\angle A \cong \angle C$.

Ex. 14.

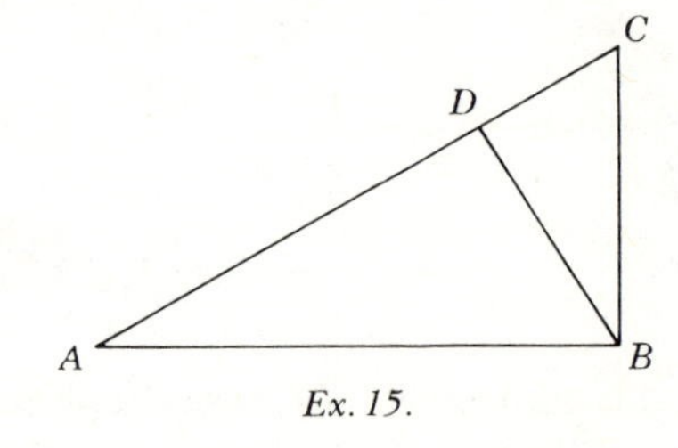

15. *Given:* $\overline{BC} \perp \overline{AB}$; $\angle C$ is the complement of $\angle ABD$.
Prove: $m\angle C = m\angle DBC$.

Ex. 15.

Theorem 3.11

3.20. Two adjacent angles whose noncommon sides form a straight angle are supplementary.

Given: $\angle ABD$ and $\angle DBC$ are adjacent angles.
$\angle ABC$ is a straight angle.
Conclusion: $\angle ABD$ is a supplement of $\angle DBC$.
Proof:

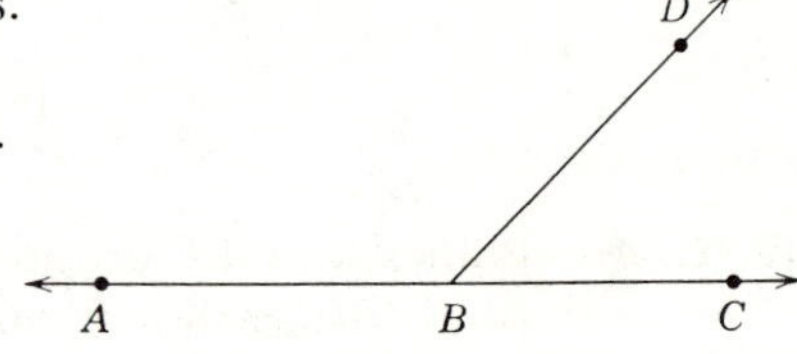

Theorem 3.11.

STATEMENTS	REASONS
1. $\angle ABD$ and $\angle DBC$ are adjacent angles.	1. Given.
2. $\angle ABC$ is a straight angle.	2. Given.
3. $m\angle ABC = 180$.	3. The measure of a straight angle is 180.
4. $m\angle ABD + m\angle DBC = m\angle ABC$.	4. Angle addition postulate.
5. $m\angle ABD + m\angle DBC = 180$.	5. $a = b \wedge b = c \rightarrow a = c$.
6. $\angle ABD$ is a supplement of $\angle DBC$.	6. If the sum of the measures of two angles is 180, the angles are supplementary.

Theorem 3.12

3.21. Vertical angles are congruent.

Given: AB and CD are straight lines intersecting at E, forming vertical angles $\angle x$ and $\angle y$.
Conclusion: $\angle x \cong \angle y$.
Proof:

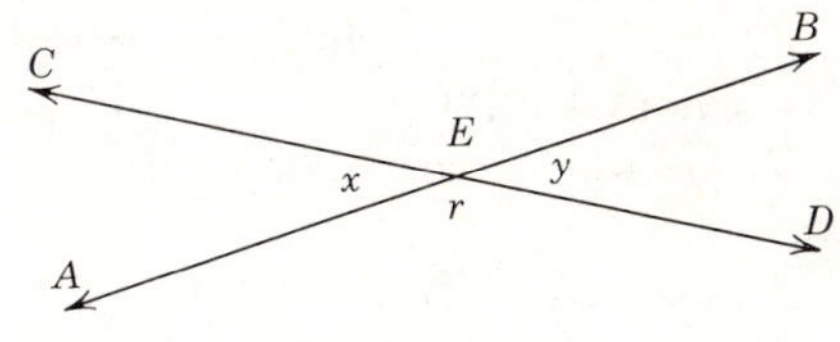

Theorem 3 12

STATEMENTS	REASONS
1. AB and CD are straight lines.	1. Given.
2. $\angle CED$ and $\angle AEB$ are straight angles.	2. Definition of straight angle.
3. $\angle x$ and $\angle r$ are supplementary angles.	3. Theorem 3.11.
4. $\angle y$ and $\angle r$ are supplementary angles.	4. Theorem 3.11.
5. $\angle x \cong \angle y$.	5. Supplements of the same angle are congruent.

Theorem 3.13

3.22. Perpendicular lines form four right angles.

Given: $\overleftrightarrow{CD} \perp \overleftrightarrow{AB}$ at O.
Conclusion: $\angle AOC$, $\angle BOC$, $\angle BOD$, and $\angle AOD$ are right angles.
Proof:

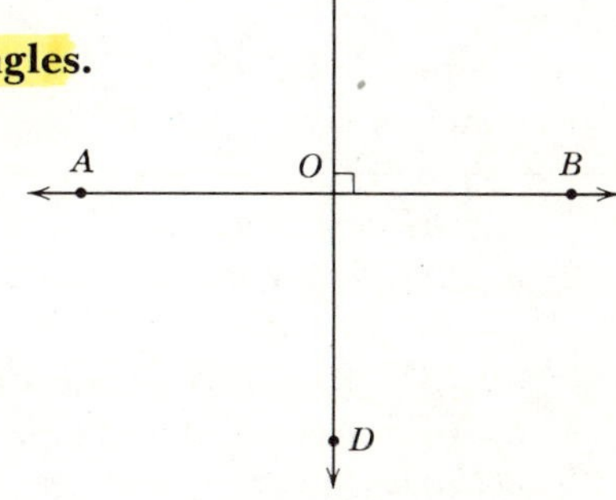

Theorem 3.13.

STATEMENTS	REASONS
1. $\overleftrightarrow{CD} \perp \overleftrightarrow{AB}$ at O.	1. Given.
2. $\angle BOC$ is a right angle.	2. $\perp$ lines form a right angle.
3. $m\angle BOC = 90$.	3. The measure of a right angle is 90.

4. $\angle AOB$ is a straight angle.	4. Definition of a straight angle.
5. $\angle BOC$ and $\angle AOC$ are adjacent angles.	5. Definition of adjacent angles.
6. $\angle BOC$ and $\angle AOC$ are supplementary.	6. Theorem 3.11.
7. $m\angle BOC + m\angle AOC = 180$.	7. The sum of the measures of two supplementary angles is 180.
8. $m\angle AOC = 90$.	8. Subtraction property of equality.
9. $m\angle AOD = m\angle BOC$; $m\angle BOD = m\angle AOC$.	9. Vertical angles are congruent.
10. $m\angle AOD = 90$; $m\angle BOD = 90$.	10. Substitution property of equality.
11. $\angle AOC$, $\angle BOC$, $\angle BOD$, $\angle AOD$ are right angles.	11. Statements 2, 8, 9, and definition of right angle.

Theorem 3.14

3.23. If two lines meet to form congruent adjacent angles, they are perpendicular.

Given: $\overleftrightarrow{CD}$ and $\overleftrightarrow{AB}$ intersect at O;
$\angle AOC \cong \angle BOC$.
Prove: $\overleftrightarrow{CD} \perp \overleftrightarrow{AB}$.
Proof:

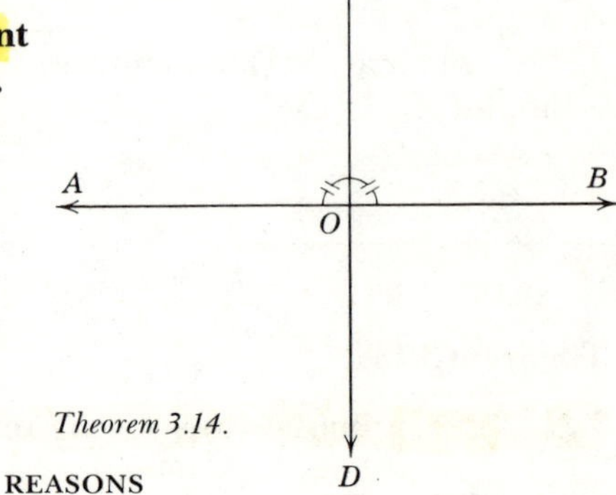

Theorem 3.14.

STATEMENTS	REASONS
1. $\angle AOC \cong \angle BOC$.	1. Given.
2. $m\angle AOC = m\angle BOC$.	2. Congruent $\measuredangle$s have equal measures.
3. $m\angle AOC + m\angle BOC = m\angle AOB$.	3. Angle addition postulate.
4. $\angle AOB$ is a straight angle.	4. Definition of straight angle.
5. $m\angle AOB = 180$.	5. The measure of a straight angle is 180.
6. $m\angle AOC + m\angle BOC = 180$.	6. $a = b \wedge b = c \rightarrow a = c$.
7. $m\angle BOC + m\angle BOC = 180$.	7. Substitution property of equality.
8. $m\angle BOC = 90$.	8. Division property of equality.
9. $\angle BOC$ is a right angle.	9. Definition of right angle.
10. $\overleftrightarrow{CD} \perp \overleftrightarrow{AB}$.	10. Definition of perpendicular lines.

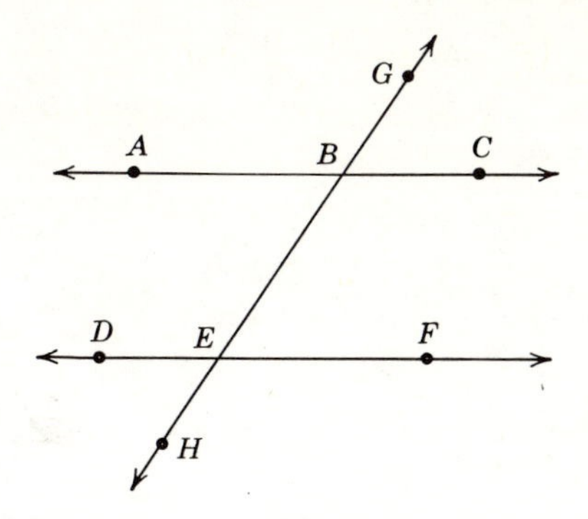

Illustrative Example.

3.24. Illustrative Example:
Given: AC, DF and GH are straight lines. $\angle GBC \cong \angle BEF$.
Prove: $\angle ABG \cong \angle DEB$.
Proof:

STATEMENTS	REASONS
1. AC is a straight line.	1. Given.
2. $\angle ABC$ is a straight angle.	2. Definition of straight angle.
3. $\angle ABG$ is the supplement of $\angle GBC$.	3. Theorem 3.11.
4. DF is a straight line.	4. Given.
5. $\angle DEF$ is a straight angle.	5. Same as reason 2.
6. $\angle DEB$ is the supplement of $\angle BEF$.	6. Theorem 3.11.
7. $\angle GBC \cong \angle BEF$.	7. Given.
8. $\angle ABG \cong \angle DEB$.	8. Supplements of congruent angles are congruent.

Exercises

In the following exercises give formal proofs, using for reasons only the given statements, definitions, postulates, theorems, and corollaries.

1. *Given:* CD and CE are straight lines. $\angle CAB \cong \angle CBA$.
   Prove: $\angle BAD \cong \angle ABE$.
2. *Given:* $\overleftrightarrow{AC}$, $\overleftrightarrow{DE}$, and $\overleftrightarrow{DF}$ as shown in the figure. $\angle ABE$ is the supplement of $\angle BCF$.
   Prove: $\angle DCB \cong \angle ABE$.

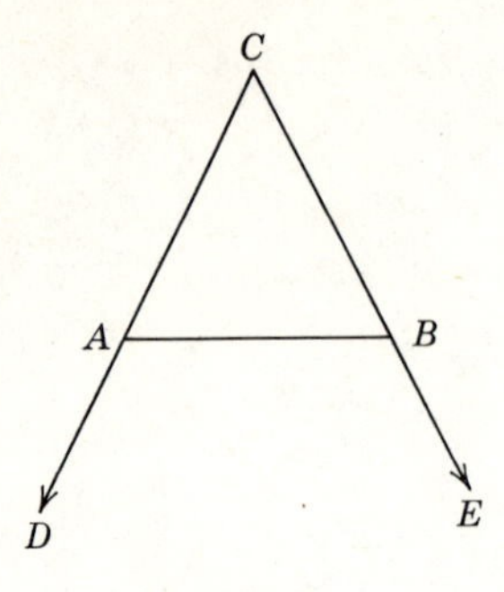

Ex. 1.

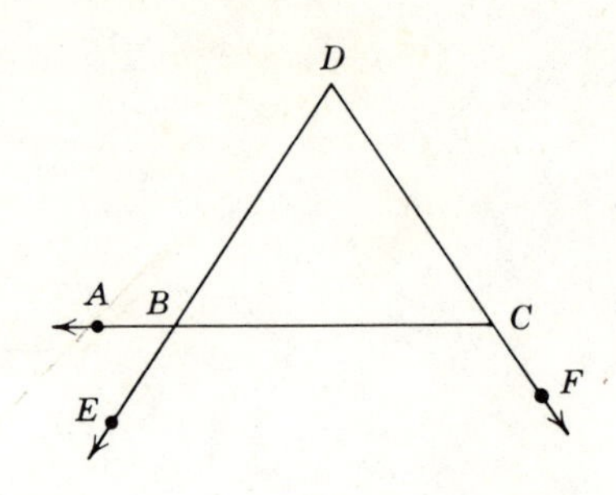

Ex. 2.

3. *Given:* $\overleftrightarrow{AB}$, $\overleftrightarrow{CD}$, and $\overleftrightarrow{EF}$ are straight lines; $\angle b \cong \angle c$.
   Prove: $\angle a \cong \angle c$.

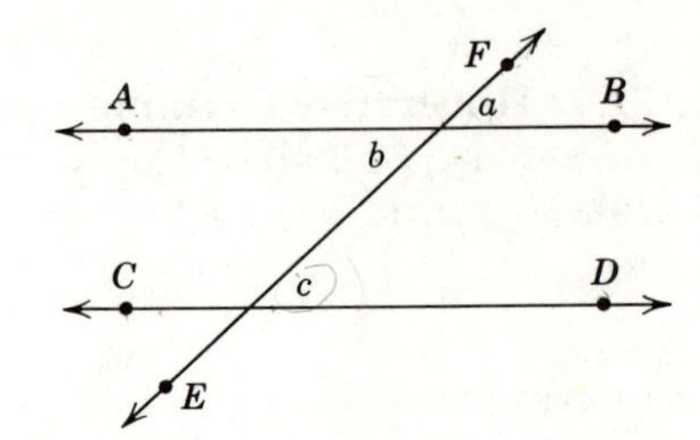

Ex. 3.

4. *Given:* $\overleftrightarrow{AB}$, $\overleftrightarrow{EC}$, and $\overleftrightarrow{CD}$ are straight lines; $\angle a \cong \angle d$.
   Prove: $\angle b \cong \angle c$.

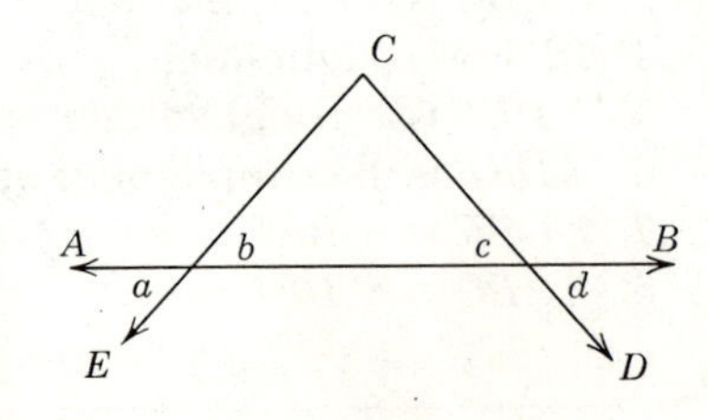

Ex. 4.

5. *Given:* $\overleftrightarrow{AE}$ and $\overleftrightarrow{BD}$ intersecting at C; $\angle A$ is the complement of $\angle ACB$; $\angle E$ is the complement of $\angle DCE$.
   Prove: $\angle A \cong \angle E$.

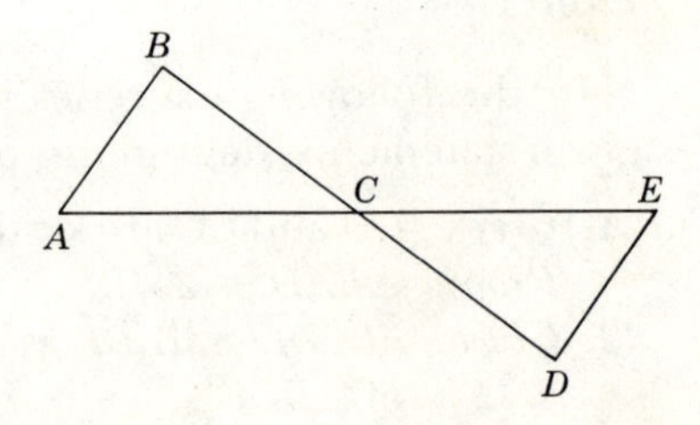

Ex. 5.

6. *Given:* $\overleftrightarrow{AB}$, $\overleftrightarrow{CD}$, and $\overleftrightarrow{EF}$ are straight lines;
$\angle a \cong \angle b$
Prove: $\overleftrightarrow{EF}$ bisects $\angle BOD$.

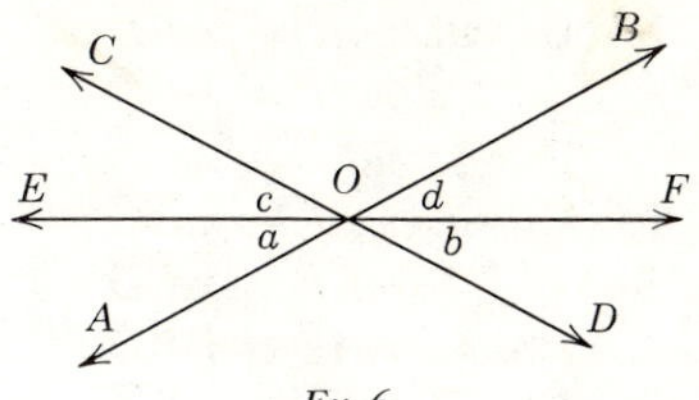

Ex. 6.

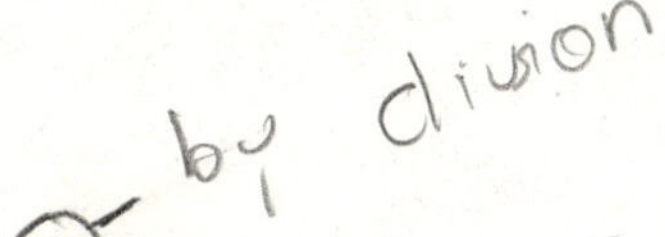

7. *Given:* $\angle a \cong \angle b$;
$\angle c \cong \angle d$;
$\angle A \cong \angle B$.
Prove: $\angle a \cong \angle c$.

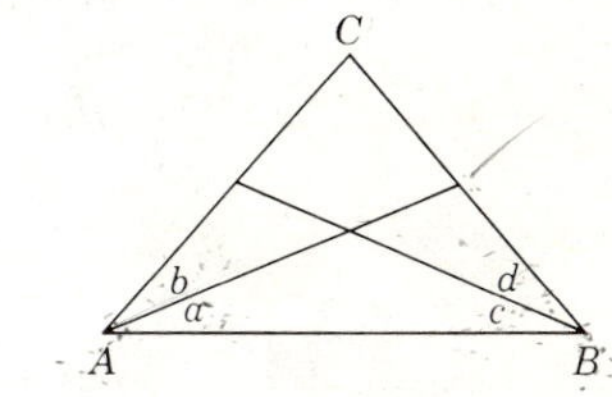

Ex. 7.

8. *Given:* $\overline{AD} \perp \overline{AB}$; $\overline{BC} \perp \overline{AB}$;
$\angle BAC \cong \angle ABD$.
Prove: $\angle DAC \cong \angle CBD$.

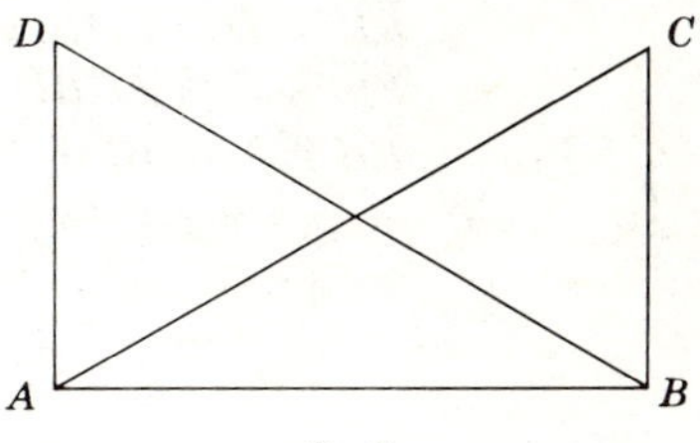

Ex. 8.

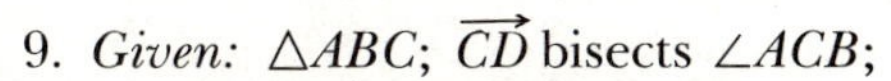

9. *Given:* $\triangle ABC$; $\overrightarrow{CD}$ bisects $\angle ACB$;
$\angle A$ is the complement of $\angle ACD$;
$\angle B$ is the complement of $\angle BCD$.
Prove: $\angle A \cong \angle B$.

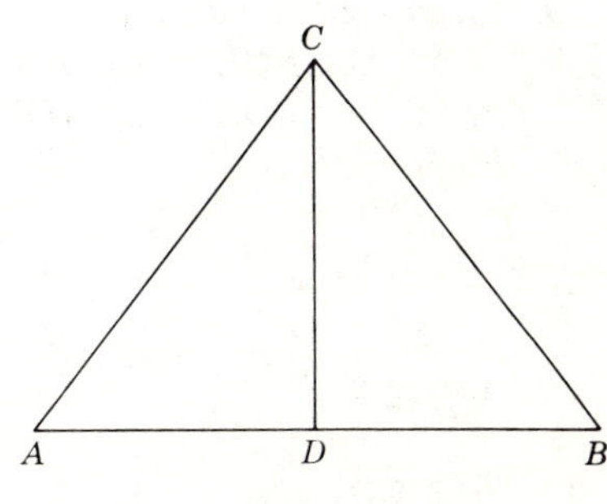

Ex. 9.

10. *Given:* $m\overline{AD} = m\overline{BD}$;
$m\overline{AE} = m\overline{ED}$;
$m\overline{BC} = m\overline{CD}$.
Prove: $m\overline{ED} = m\overline{CD}$.

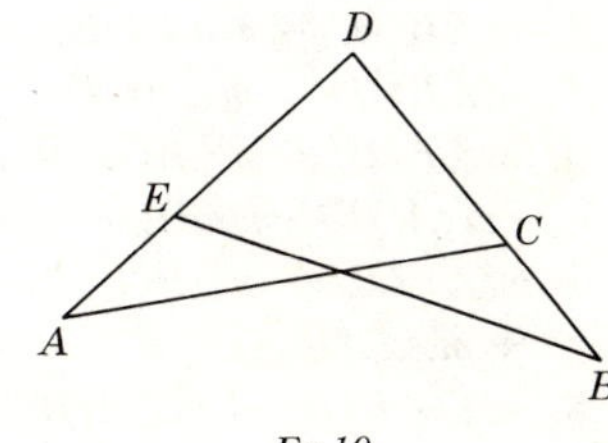

Ex 10.

11. *Given:* $m\overline{AD} = m\overline{BD}$;
$m\overline{AE} = m\overline{BC}$.
Prove: $m\overline{ED} = m\overline{CD}$.

12. *Given:* $m\overline{AE} = m\overline{BC}$;
$m\overline{ED} = m\overline{CD}$.
Prove: $m\overline{AD} = m\overline{BD}$.

13. *Given:* $m\overline{ED} = m\overline{CD}$;
$m\overline{AE} = m\overline{ED}$;
$m\overline{BC} = m\overline{CD}$.
Prove: $m\overline{AE} = m\overline{BC}$.

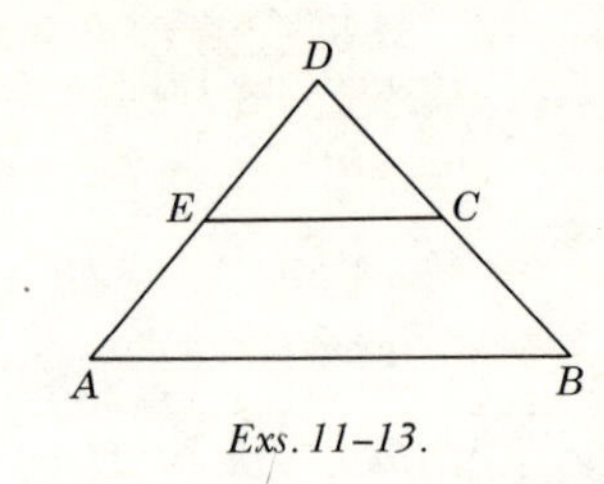

Exs. 11–13.

14. *Given:* $\angle ABC$ is a straight $\angle$;
$\overleftrightarrow{EB}$ bisects $\angle ABD$;
$\overleftrightarrow{FB}$ bisects $\angle CBD$.
Prove: $\overleftrightarrow{EB}$ is $\perp$ to $\overleftrightarrow{BF}$.
(*Hint:* $m\angle x + m\angle y + m\angle r + m\angle s = ?$;
$m\angle x = m\angle ?$; $m\angle r = m\angle ?$).

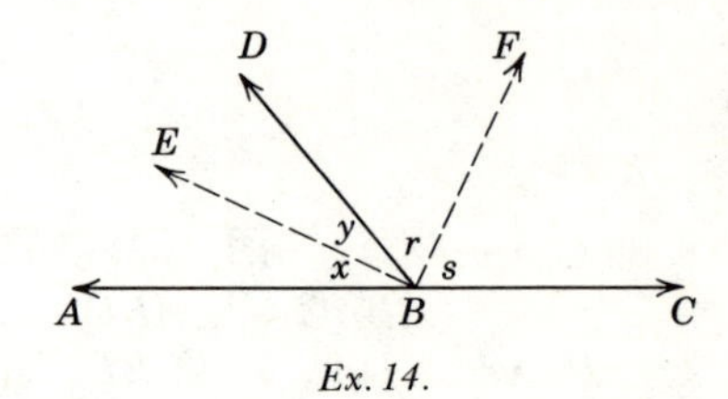

Ex. 14.

15. *Given:* $\overleftrightarrow{AB} \perp \overleftrightarrow{CD}$ at O;
$m\angle BOE = m\angle DOF$.
Prove: $m\angle EOD = m\angle AOF$.
Proof:

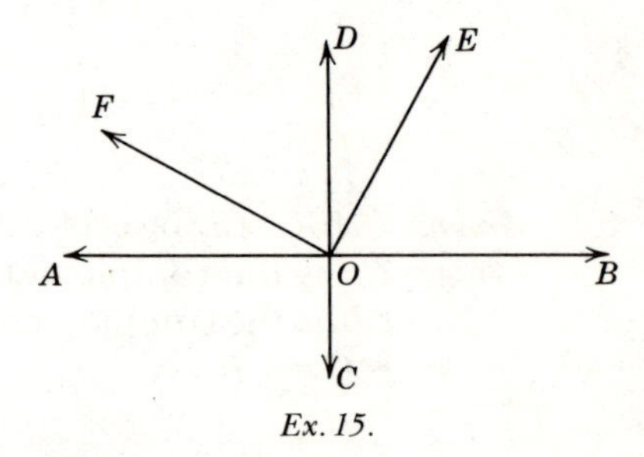

Ex. 15.

STATEMENTS	REASONS
1. $\overleftrightarrow{AB} \perp \overleftrightarrow{CD}$.	1. Why?
2. $\angle BOD$ and $\angle AOD$ are right $\angle$s.	2. Why?
3. $m\angle BOE = m\angle DOF$.	3. Why?
4. $m\angle BOD = m\angle AOD$.	4. Why?
5. $m\angle EOD + m\angle BOE = m\angle BOD$.	5. Why?
6. $m\angle AOF + m\angle DOF = m\angle AOD$.	6. Why?
7. $m\angle EOD + m\angle BOE = m\angle AOF + m\angle DOF$.	7. Why?
8. $\therefore m\angle EOD = m\angle AOF$.	8. Why?

16. *Given:* $\overleftrightarrow{AB} \perp \overleftrightarrow{CD}$ at O;
$m\angle BOE = m\angle DOF$.
Prove: $\overleftrightarrow{OF} \perp \overleftrightarrow{OE}$.
Proof:

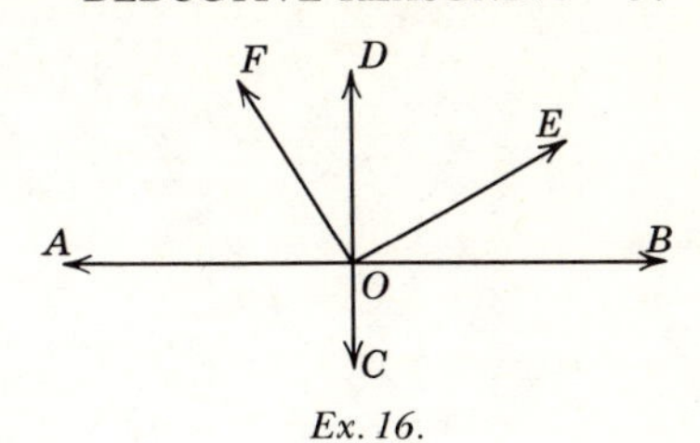

Ex. 16.

STATEMENTS	REASONS
1. $\overleftrightarrow{AB} \perp \overleftrightarrow{CD}$.	1. Why?
2. $\angle BOD$ is a right $\angle$.	2. Why?
3. $m\angle BOD = 90$.	3. Why?
4. $m\angle BOE + m\angle EOD = m\angle BOD$.	4. Why?
5. $m\angle BOE + m\angle EOD = 90$.	5. Why?
6. $m\angle BOE = m\angle DOF$.	6. Why?
7. $m\angle DOF + m\angle EOD = 90$.	7. Why?
8. $m\angle DOF + m\angle EOD = m\angle EOF$.	8. Why?
9. $m\angle EOF = 90$.	9. Why?
10. $\therefore \overleftrightarrow{OF} \perp \overleftrightarrow{OE}$.	10. Why?

17. *Given:* $\overleftrightarrow{AB} \perp \overleftrightarrow{CD}$; $\overleftrightarrow{OE} \perp \overleftrightarrow{OF}$.
Prove: $\angle BOE \cong \angle FOD$.
Proof:

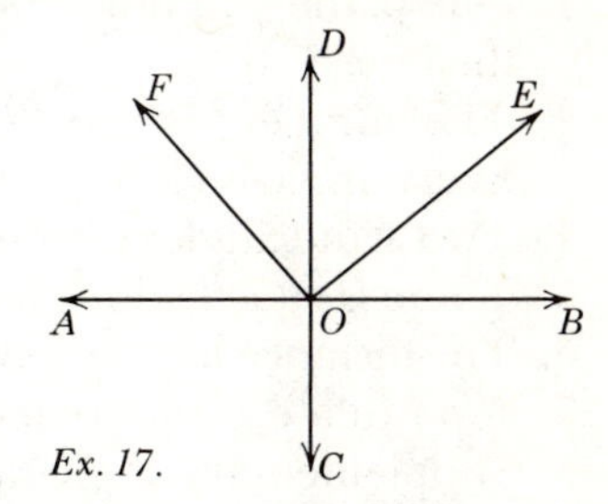

Ex. 17.

STATEMENTS	REASONS
1. $\overleftrightarrow{AB} \perp \ldots$	1. Why?
2. $\overleftrightarrow{OE} \perp \ldots.$	2. Why?
3. $\angle BOD$ is a . . . ; $\angle FOE$ is a	3. Why?
4. $m\angle BOD = \ldots$; $m\angle FOE = \ldots.$	4. Why?
5. $m\angle ? + m\angle ? = m\angle BOD$.	5. Why?
6. $m\angle BOE + m\angle DOE = 90$.	6. Why?
7. $m\angle ? + m\angle ? = m\angle FOE$.	7. Why?
8. $m\angle ? + m\angle ? = 90$.	8. Why?
9. $\angle BOE$ is the . . . of $\angle DOE$.	9. Why?
10. $\angle FOD$ is the . . . of $\angle DOE$.	10. Why?
11. $\angle BOE \cong \angle FOD$.	11. Why?

Summary Tests

Test 1

TRUE-FALSE STATEMENTS

1. One plane and only one plane can contain a given line and a point not on the line.
2. The number $\frac{2}{3}$ is a real number that is not rational.
3. Every angle is congruent to itself.
4. Two acute angles cannot be supplementary.
5. There is exactly one plane containing a given line.
6. The distance between two points is a positive number.
7. A postulate is a statement that has been proved.
8. Supplementary angles are congruent.
9. The bisectors of two adjacent supplementary angles are perpendicular to each other.
10. Vertical angles have equal measures.
11. The absolute value of every nonzero real number is positive.
12. If two lines intersect, there are two and only two points that are contained by both lines.
13. If two lines intersect to form vertical angles that are supplementary, the vertical angles are right angles.
14. A corollary is a theorem.
15. If an obtuse angle is bisected, two acute angles will be formed.
16. Vertical angles cannot be supplementary.
17. Adjacent angles are supplementary.
18. A perpendicular to a line bisects the line.
19. Two adjacent angles are either complementary or supplementary.

20. It is not possible for vertical angles to be adjacent angles.
21. It is possible for three lines to be mutually perpendicular.
22. A perpendicular is a line running up and down.
23. If two angles are complementary, then each of them is acute.
24. If two angles are supplementary, then one of them is acute and the other is obtuse.
25. Two angles are vertical angles if their union is the union of two intersecting lines.

Test 2

COMPLETION STATEMENTS

1. A statement considered true without proof is called a(n) ______ .
2. If two angles are either complements or supplements of the same angle they are ______ .
3. The sides of a right angle are ______ to each other.
4. The pairs of nonadjacent angles formed when two lines intersect are called ______ .
5. A(n) ______ angle has a larger measure than its supplement.
6. The bisectors of two complementary adjacent angles forms an angle whose measure is ______ .
7. Angle A is the complement of an angle whose measure is 42. Angle B is the supplement of $\angle A$. Then the measure of $\angle B$ is ______ .
8. Point B lies on line RS. Line AB is perpendicular to line RS. Then $m\angle ABR =$ ______ .
9. Two angles complementary to the same angle are ______ .
10. The difference between the measures of the supplement and complement of an angle is ______ .
11. The bisectors of a pair of vertical angles form a ______ angle.
12. The measure of an angle that is congruent to its complement is ______ .
13. The measure of an angle that has half the measure of its supplement is ______ .
14. The sum of the measures of two adjacent angles formed by two intersecting lines is ______ .
15. For every three distinct noncollinear points, there is exactly one ______ that contains the three points.
16. If $m\angle A < m\angle B$, then the measure of the supplement of $\angle A$ is ______ the measure of the supplement of $\angle B$.
17. If the noncommon sides of two adjacent angles are perpendicular to each other, then the angles are ______ .
18. The correspondence between points on a line and the real numbers is called the ______ ______ for the line.

19. If two planes intersect, their intersection is a ______.
20. If two distinct lines intersect, how many planes can contain both lines?

Test 3

PROBLEMS

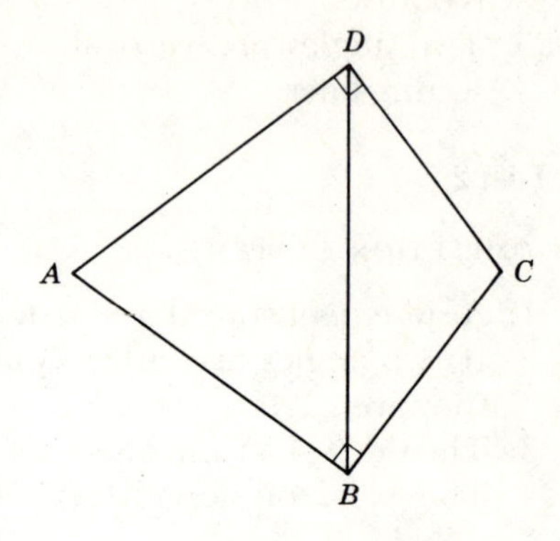

Ex. 1.

1. *Given:* $\overline{AD} \perp \overline{DC}$; $\overline{AB} \perp \overline{BC}$;
   $\angle CDB \cong \angle CBD$.
   Prove: $\angle ADB \cong \angle ABD$.

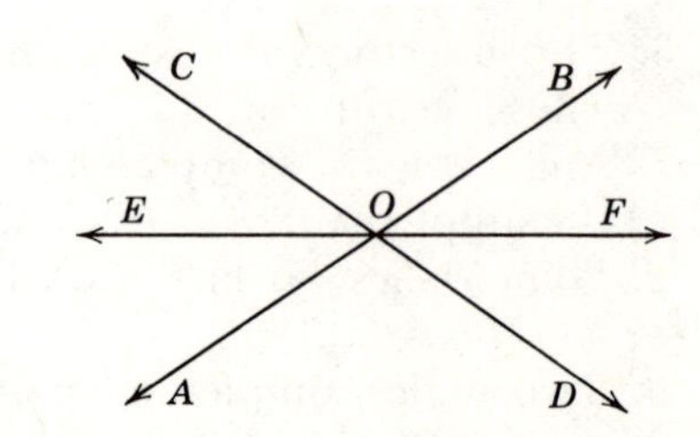

Ex. 2.

2. *Given:* AB, CD, and EF are straight lines;
   $\overleftrightarrow{EF}$ bisects $\angle AOC$.
   Prove: $\overleftrightarrow{EF}$ bisects $\angle BOD$.

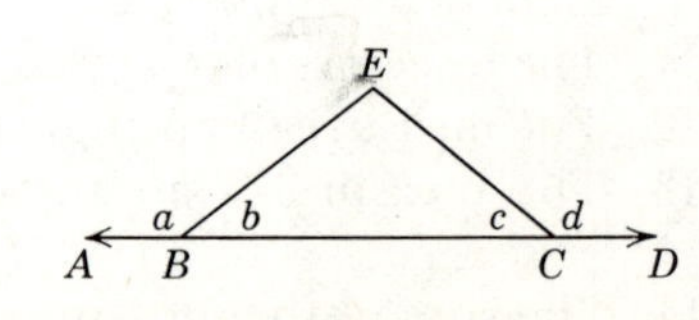

Ex. 3.

3. *Given:* $\overleftrightarrow{ABCD}$ is a straight line;
   $m\angle a + m\angle c = 180$.
   Prove: $\angle b \cong \angle c$.

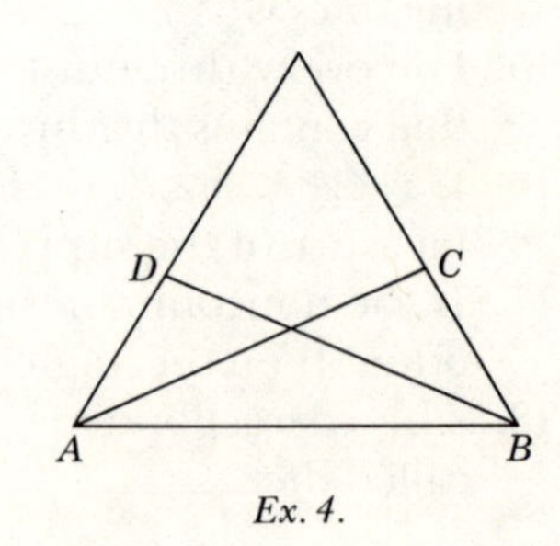

Ex. 4.

4. *Given:* $\angle BAD \cong \angle ABC$;
   $\angle DAC \cong \angle CBD$.
   Prove: $\angle CAB \cong \angle DBA$.

|4|

Congruence – Congruent Triangles

4.1. Congruent figures. Industry today relies a great deal on mass production and assembly line manufacture. Often each part of a machine or household article is made by precision manufacture to have exactly the same shape and size. These parts are then sent to an assembly plant where the parts can be fitted together to form a complete unit.

The mass production and repair of automobiles, airplanes, television sets, automatic washers, refrigerators, and the many other products of modern industry depend on the manufacture of thousands of parts having exactly the same shape and size. It is especially important in repairing a complex machine that the necessary replacement parts match exactly the original parts.

In this chapter we will study the geometry of figures that have the same shape and size.

Definition: Two figures are *congruent* when they have the same shape and size. (see §1.19)

The word congruent is derived from the Latin words *con* meaning "with" and *gruere,* meaning "to agree." Congruent figures can be made to coincide, part by part. The coincident parts are called *corresponding parts.* The symbol for congruence is $\cong$. This symbol is a combination of the two symbols $=$, meaning have the same size, and $\sim$, meaning having the same shape. Thus, $\triangle ABC \cong \triangle DEF$ means $\triangle ABC$ is congruent to $\triangle DEF$.

4.2. Congruence relations. The following theorems are a direct consequence of the properties of the real number system. They can be used to shorten many proofs of other theorems. Proofs of some of the theorems will be given; others will be left as exercises.

Congruence Theorems for Segments

Theorem 4.1

4.3. Reflexive theorem. Every segment is congruent to itself.

Given: $\overline{AB}$.
Conclusion: $\overline{AB} \cong \overline{AB}$.
Proof:

A B

Theorem 4.1.

STATEMENTS	REASONS
1. $m\overline{AB} = m\overline{AB}$.	1. Reflexive axiom (E-1).
2. $\overline{AB} \cong \overline{AB}$.	2. Definition of congruent segments.

Theorem 4.2

4.4. Symmetric theorem. If $\overline{AB} \cong \overline{CD}$, then $\overline{CD} \cong \overline{AB}$.

Theorem 4.3

4.5. Transitive theorem. If $\overline{AB} \cong \overline{CD}$ and $\overline{CD} \cong \overline{EF}$, then $\overline{AB} \cong \overline{EF}$.

Theorem 4.4

4.6. Addition theorem. If B is between A and C, E between D and F and if $\overline{AB} \cong \overline{DE}$ and $\overline{BC} \cong \overline{EF}$, then $\overline{AC} \cong \overline{DF}$.

Given: $\overline{AB} \cong \overline{DE}$; $\overline{BC} \cong \overline{EF}$; B is between A and C; E is between D and F.
Conclusion: $\overline{AC} \cong \overline{DF}$.
Proof:

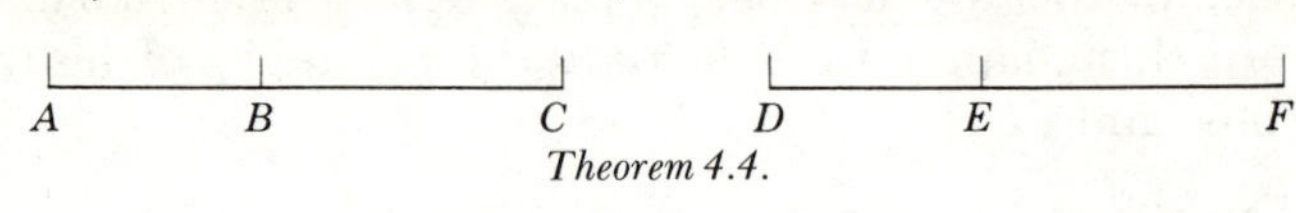

Theorem 4.4.

STATEMENTS	REASONS
1. $\overline{AB} \cong \overline{DE}$; $\overline{BC} \cong \overline{EF}$.	1. Given.
2. $m\overline{AB} = m\overline{DE}$; $m\overline{BC} = m\overline{EF}$.	2. Definition of congruent segments.
3. $m\overline{AB} + m\overline{BC} = m\overline{DE} + m\overline{EF}$.	3. Addition property of real numbers.
4. B is between A and C; E is between D and F.	4. Given.
5. $m\overline{AB} + m\overline{BC} = m\overline{AC}$. $m\overline{DE} + m\overline{EF} = m\overline{DF}$.	5. Definition of betweenness.
6. $m\overline{AC} = m\overline{DF}$.	6. Substitution property of equality.
7. $\overline{AC} \cong \overline{DF}$.	7. Definition of congruent segments.

Theorem 4.5

4.7. Subtractive theorem. **If B is between A and C, E is between D and F, $\overline{AC} \cong \overline{DF}$ and $\overline{BC} \cong \overline{EF}$, then $\overline{AB} \cong \overline{DE}$.**

Congruence Theorems for Angles

Theorem 4.6

4.8. Reflexive theorem. **Every angle is congruent to itself. $\angle A \cong \angle A$.**

Theorem 4.7

4.9. Symmetric theorem. **If $\angle A \cong \angle B$, then $\angle B \cong \angle A$.**

Theorem 4.8

4.10. Transitive theorem. **If $\angle A \cong \angle B$ and $\angle B \cong \angle C$, then $\angle A \cong \angle C$.**

Theorem 4.9

4.11. Angle addition theorem. **If D is in the interior of $\angle ABC$, P is in the interior of $\angle RST$, $\angle ABD \cong \angle RSP$, and $\angle DBC \cong \angle PST$, then $\angle ABC \cong \angle RST$.** (See figure for Theorem 4.10.)

Theorem 4.10

4.12. Angle subtraction theorem. **If D is in the interior of $\angle ABC$, P is in the interior of $\angle RST$, $\angle ABC \cong \angle RST$, and $\angle ABD \cong \angle RSP$, then $\angle DBC \cong \angle PST$.**

Given: D is in interior of $\angle ABC$; P is in interior of $\angle RST$;
$\angle ABC \cong \angle RST$; $\angle ABD \cong \angle RSP$.
Conclusion: $\angle DBC \cong \angle PST$.
Proof:

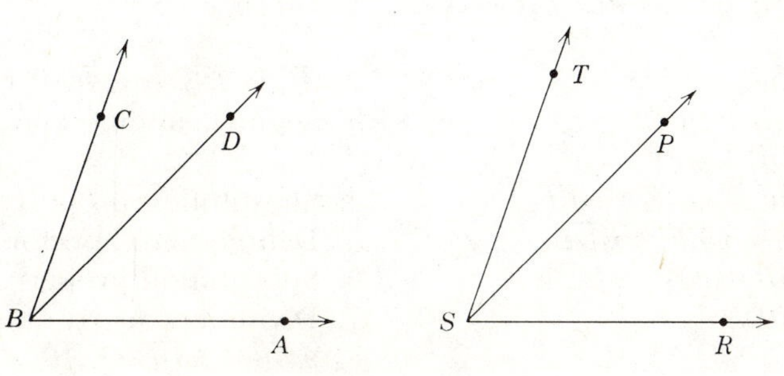

Theorem 4.10.

STATEMENTS	REASONS
1. D in interior of $\angle ABC$; P is in interior of $\angle RST$; $\angle ABC \cong \angle RST$; $\angle ABD \cong \angle RSP$.	1. Given.
2. (a) $m\angle ABC = m\angle RST$. (b) $m\angle ABD = m\angle RSP$.	2. Definition of angle congruence.
3. $m\angle ABC = m\angle ABD + m\angle DBC$; $m\angle RST = m\angle RSP + m\angle PST$.	3. Angle addition postulate.
4. $m\angle ABD + m\angle DBC = m\angle RSP + m\angle PST$.	4. Substitution property of equality (statement 3 in statement 2a).
5. $m\angle DBC = m\angle PST$.	5. Subtraction property of equality (statement 2b from statement 4).
6. $\angle DBC \cong \angle PST$.	6. Definition of angle congruence.

Theorems on Bisectors

Theorem 4.11

4.13. Segment bisector theorem. If $\overline{AC} \cong \overline{DF}$, B bisects $\overline{AC}$, E bisects $\overline{DF}$, then $\overline{AB} \cong \overline{DE}$.

Given: $\overline{AC} \cong \overline{DF}$; B bisects $\overline{AC}$; E bisects $\overline{DF}$.
Conclusion: $\overline{AB} \cong \overline{DE}$.
Proof:

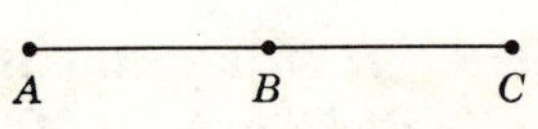

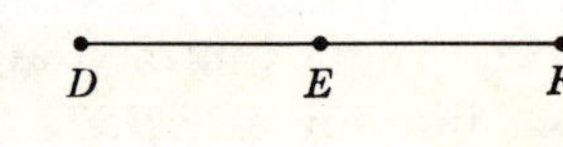

Theorem 4.11.

STATEMENTS	REASONS
1. $\overline{AC} \cong \overline{DF}$; B bisects $\overline{AC}$; E bisects $\overline{DF}$.	1. Given.
2. $m\overline{AC} = m\overline{DF}$.	2. $\overline{AC} \cong \overline{DF} \leftrightarrow m\overline{AC} = m\overline{DF}$.
3. $m\overline{AC} = m\overline{AB} + m\overline{BC}$; $m\overline{DF} = m\overline{DE} + m\overline{EF}$.	3. Segment addition postulate.
4. $m\overline{AB} + m\overline{BC} = m\overline{DE} + m\overline{EF}$.	4. Substitution property of equality.
5. $m\overline{BC} = m\overline{AB}$; $m\overline{EF} = m\overline{DE}$.	5. Definition of bisector of segment.
6. $m\overline{AB} + m\overline{AB} = m\overline{DE} + m\overline{DE}$.	6. Substitution property of equality.
7. $m\overline{AB} = m\overline{DE}$.	7. Division property of equality.
8. $\overline{AB} \cong \overline{DE}$.	8. $m\overline{AB} = m\overline{DE} \leftrightarrow \overline{AB} \cong \overline{DE}$.

Theorem 4.12

4.14. Angle bisector theorem. **If $\angle ABC \cong \angle RST$, $\overrightarrow{BD}$ bisects $\angle ABC$, $\overrightarrow{SP}$ bisects $\angle RST$, then $\angle ABD \cong \angle RSP$.**

(The proof is similar to that of Theorem 4.11.)

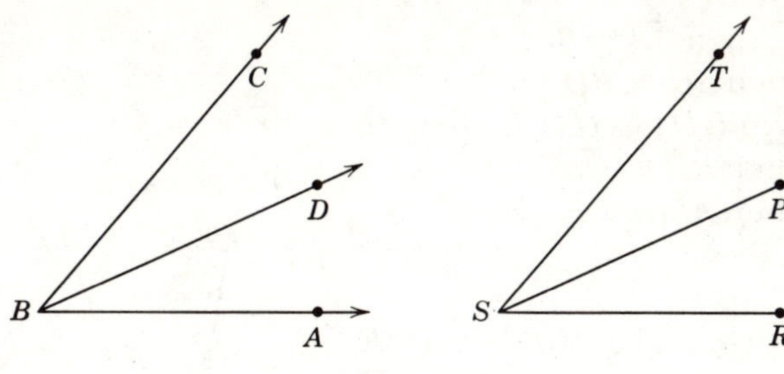

Theorem 4.12.

4.15. Illustrative Example:

Given: $\angle ABE \cong \angle DBC$.

Prove: $\angle ABD \cong \angle EBC$.

Proof:

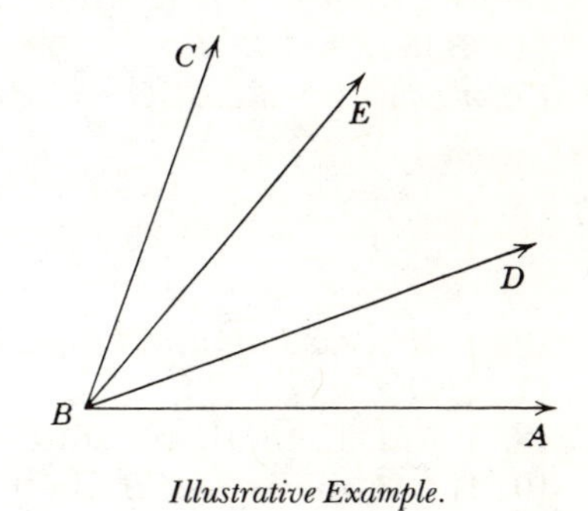

Illustrative Example.

STATEMENTS	REASONS
1. $\angle ABE \cong \angle DBC$.	1. Given.
2. $\angle DBE \cong \angle DBE$.	2. Reflexive theorem of $\cong$ $\angle$s.
3. $\angle ABD \cong \angle EBC$.	3. Subtractive theorem of $\cong$ $\angle$s.

Exercises (A)

In the following indicate which statements are *always true* and which are *not always true*.

1. $\overline{AB} \cong \overline{CD} \rightarrow \overline{AB} = \overline{CD}$.
2. $m\overline{AB} = m\overline{CD} \rightarrow \overline{AB} \cong \overline{CD}$.
3. A ray has one, and only one, midpoint.
4. If $\overline{AB} \cong \overline{BC}$, then B bisects $\overline{AC}$.

5. $\overline{AB} \cong \overline{BA}$.
6. If $\overline{AB} \cong \overline{RS}$ and $\overline{RS} \cong \overline{CD}$, then $\overline{AB} \cong \overline{CD}$.

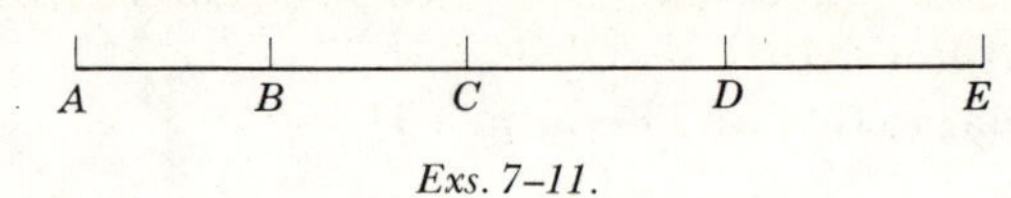

Exs. 7–11.

7. If B bisects $\overline{AC}$, then $AB = BC$.
8. If $\overline{AB} \cong \overline{BC}$, then $\overline{AC} \cong \overline{BD}$.
9. If B bisects $\overline{AC}$ and D bisects $\overline{CE}$, then $\overline{AB} \cong \overline{DE}$.
10. If $\overline{BC} \cong \overline{DE}$, then $\overline{BD} \cong \overline{CE}$.
11. If $\overline{AD} \cong \overline{BE}$, then $\overline{AB} \cong \overline{DE}$.

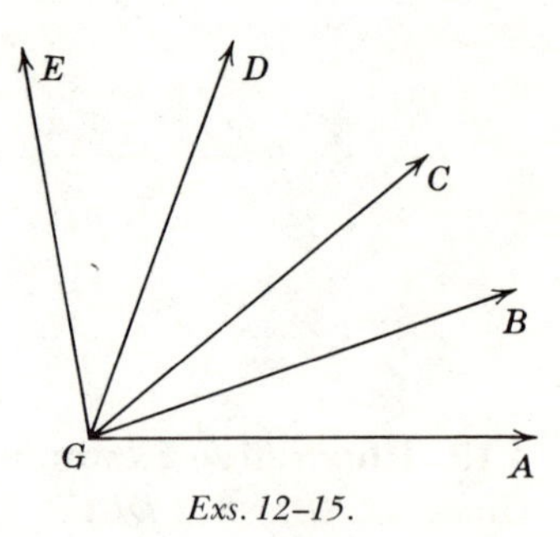

Exs. 12–15.

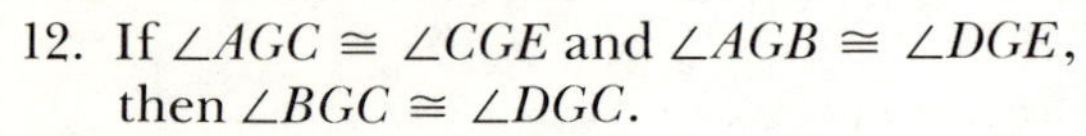

12. If $\angle AGC \cong \angle CGE$ and $\angle AGB \cong \angle DGE$, then $\angle BGC \cong \angle DGC$.
13. If $\angle DGA \cong \angle BGE$, then $\angle AGB \cong \angle EGD$.
14. $\angle AGD \cong \angle BGE$.
15. $m\angle DGA = m\angle BGC + m\angle DGC + m\angle AGB$.

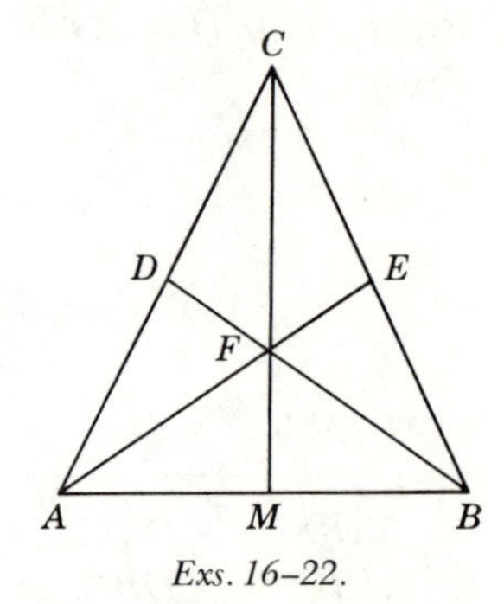

Exs. 16–22.

16. If D bisects $\overline{AC}$ and E bisects $\overline{BC}$, then $\overline{AD} \cong \overline{BE}$.
17. $\angle CFE \cong \angle MFA$.
18. If $\overline{CM} \perp \overline{AB}$, then $\angle AMC \cong \angle BMC$.
19. If $\overline{CM}$ bisects $\angle ACB$, then $\overline{AM} \cong \overline{BM}$.
20. If F bisects $\overline{AE}$, then F bisects $\overline{BD}$.
21. If $\angle AFM \cong \angle BFM$. Then $\angle CFA \cong \angle CFB$.
22. $\angle AFC$ and $\angle BFC$ are vertical angles.

Exercises (B)

By using the theorems on congruence, what conclusions can be drawn in each of the following exercises? Write your conclusions and reasons in the same manner shown in the following example.

Illustrative example:
Given: $\angle AED \cong \angle CDE$; $\angle DEB \cong \angle EDB$.
Conclusion: $\angle AEB \cong \angle CDB$.
Reason: Angle subtraction theorem.

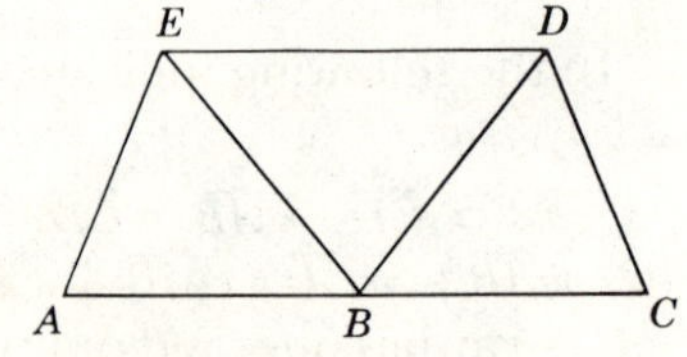

Illustrative Example.

1. *Given:* $\overline{AB} \cong \overline{FE}$; $\overline{BC} \cong \overline{ED}$.
2. *Given:* B bisects $\overline{AC}$; E bisects $\overline{FD}$; $\overline{AC} \cong \overline{FD}$.
3. *Given:* $\overline{FB}$ and $\overline{AE}$ bisect each other at G; $\overline{AE} \cong \overline{FB}$.
4. *Given:* $\overline{AE} \perp \overline{FB}$.

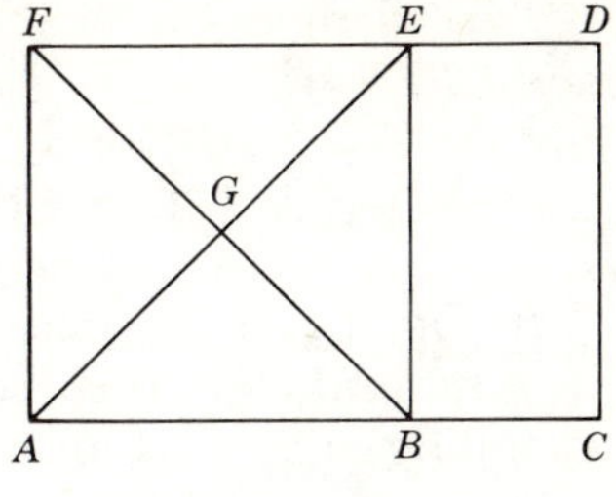

Exs. 1–4.

5. *Given:* $\angle DAB \cong \angle CBA$; $\angle BAG \cong \angle ABE$.
6. *Given:* $\overline{HF}$ bisects both $\angle EJG$ and $\angle AJB$; $\angle EJG \cong \angle AJB$.
7. *Given:* $\overline{AE} \cong \overline{ED}$; $\overline{ED} \cong \overline{BG}$.
8. *Given:* $\overline{AG} \cong \overline{BE}$; $\overline{HF}$ bisects both $\overline{AG}$ and $\overline{BE}$ at J.

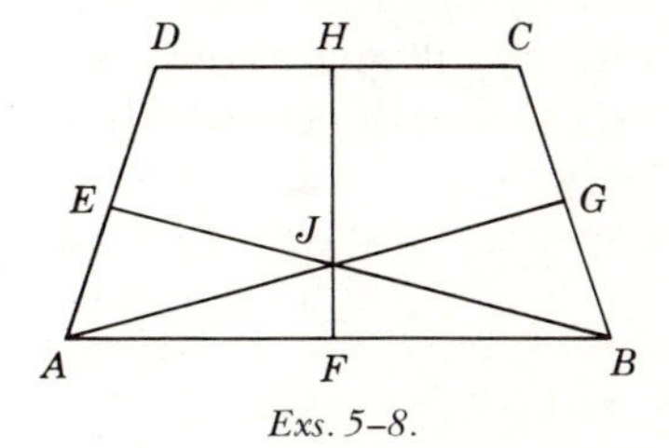

Exs. 5–8.

4.16. Corresponding parts of geometric figures. Congruent figures can be made to coincide, part by part. For example, in $\triangle ABC$ and $\triangle DEF$ of Fig. 4.1, if it is possible to move the triangles so that the three vertices and the three sides of $\triangle ABC$ fit exactly the three vertices and the three sides of $\triangle DEF$, the triangles are congruent to each other. We write this fact as $\triangle ABC \cong \triangle DEF$. (It should be understood that the triangles need not actually be moved, but the movement is done abstractly in the mind.)

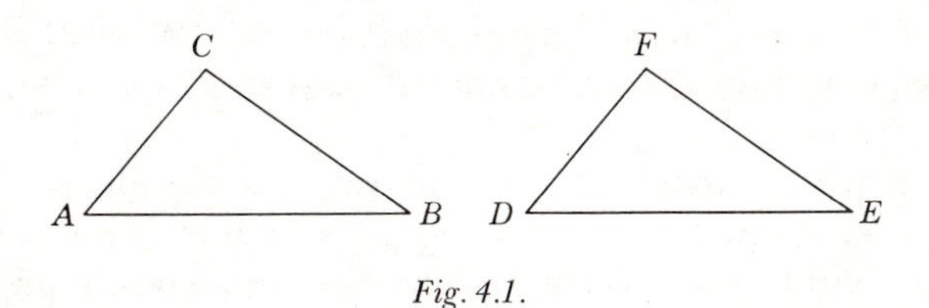

Fig. 4.1.

This matching of vertices and sides of a geometric figure is called a *one-to-one correspondence.* The matched parts are called *corresponding parts.* Thus we speak of *corresponding sides* and *corresponding angles.* The matching-up scheme of corresponding vertices can be shown by the symbolism: $A \leftrightarrow D$, $B \leftrightarrow E$, $C \leftrightarrow F$. We can also show this matching by writing $ABC \leftrightarrow DEF$.

Thus given a correspondence $ABC \leftrightarrow DEF$ between the vertices of two triangles, if each pair of corresponding sides are congruent, and if each pair of corresponding angles are congruent, then the correspondence $ABC \leftrightarrow DEF$ is a *congruence between the two triangles.*

Two triangles can be matched-up six ways. Other ways of matching $\triangle ABC$ and $\triangle DEF$ are:

$$ABC \leftrightarrow FED \qquad ABC \leftrightarrow EFD \qquad ABC \leftrightarrow DFE$$
$$ABC \leftrightarrow FDE \qquad ABC \leftrightarrow EDF$$

In Fig. 4.1, if the matching $ABC \leftrightarrow DEF$ gives a congruence, we can state that $\overline{AC}$ and $\overline{DF}$ are corresponding sides, and $\angle BCA$ and $\angle EFD$ are corresponding angles. Can you find the other pairs of corresponding sides and corresponding angles?

Two scalene triangles can have only a single one-to-one correspondence which will give a congruence. Two isosceles triangles can have two one-to-one correspondences which will give congruence.

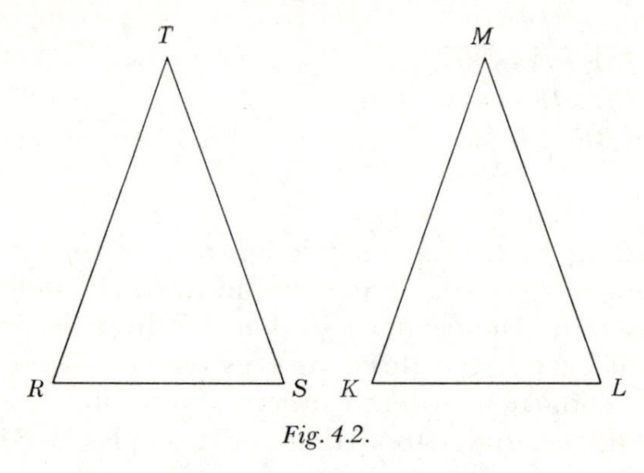

Fig. 4.2.

In Fig. 4.2, if $\overline{RT} \cong \overline{ST}$ and $\overline{KM} \cong \overline{LM}$, the two correspondences $RST \leftrightarrow KLM$ and $RST \leftrightarrow LKM$ might give congruences. We will determine later what additional conditions must be known before the triangles can be proved congruent to each other.

The order in which matching pairs of vertices are given is not important in expressing a congruence and the vertex you start with is not important. In Fig. 4.3, we could describe the one-to-one correspondence in one line as $DEFG \leftrightarrow HKJI$ or $EFGD \leftrightarrow KJIH$. There are two others. Can you find them? All that matters is that corresponding points be matched.

It should be evident that a triangle can be made to coincide with itself. The

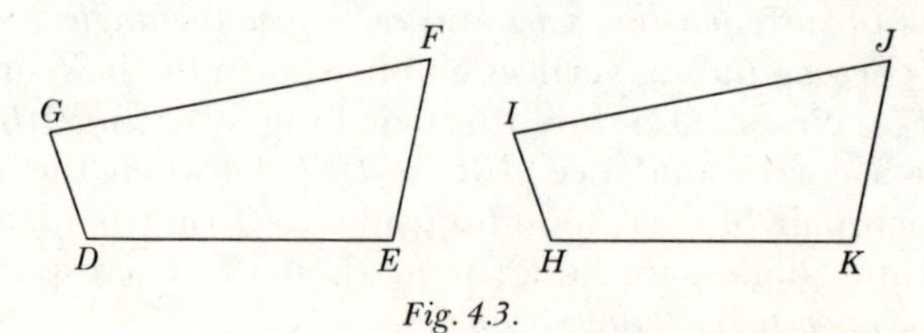

Fig. 4.3.

one-to-one correspondence in which every vertex is matched with itself is called the *identity* congruence. Thus

$$ABC \leftrightarrow ABC$$

is an identity congruence.

For the isosceles triangle RST (Fig. 4.4), where $\overline{RT} \cong \overline{ST}$, it can be shown that, under the one-to-one correspondence $RST \leftrightarrow SRT$, the figure can be made to coincide with itself.

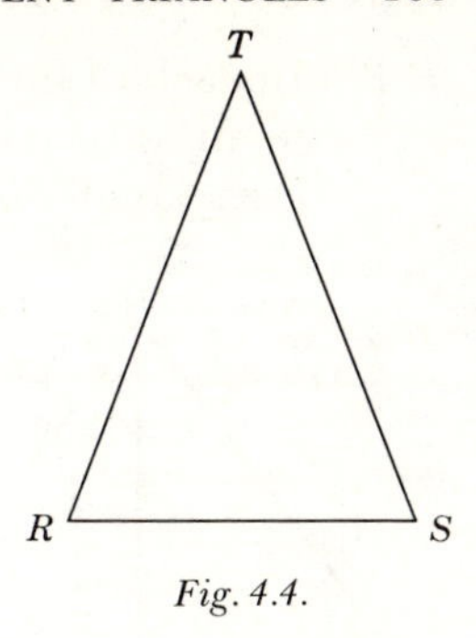

Fig. 4.4.

Exercises

1. Draw a $\triangle GHJ$ and a $\triangle KLM$. List all the possible matchings of the second triangle with the ordered sequence GHJ of the first triangle.
2. If the matching $RST \leftrightarrow LMK$ gives a congruence between $\triangle RST$ and $\triangle LMK$, list all the pairs of corresponding sides and corresponding angles of the two triangles.
3. Write down the six matchings of equilateral $\triangle ABC$ with itself, beginning with the identity congruence $ABC \leftrightarrow ABC$.
4. Write down the four matchings of rectangle $ABCD$ with itself.
5. In matching $\triangle ABC$ with $\triangle RST$, $\overline{AC}$ and $\overline{RT}$ were matched as corresponding sides. Does it then follow that (1) $\angle B$ and $\angle S$ are corresponding angles? (2) BC and ST are corresponding sides?
6. Which of the following figures form matched pairs that are congruent to each other?

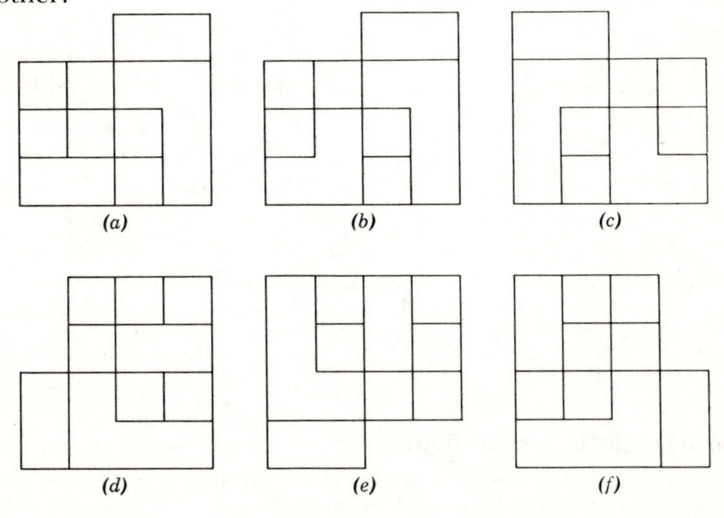

Prob. 6.

7–12. In each of the following use ruler and protractor to find which triangles seem to be congruent. Then indicate the pairs of sides and angles in the triangles which seem to match in a congruence.

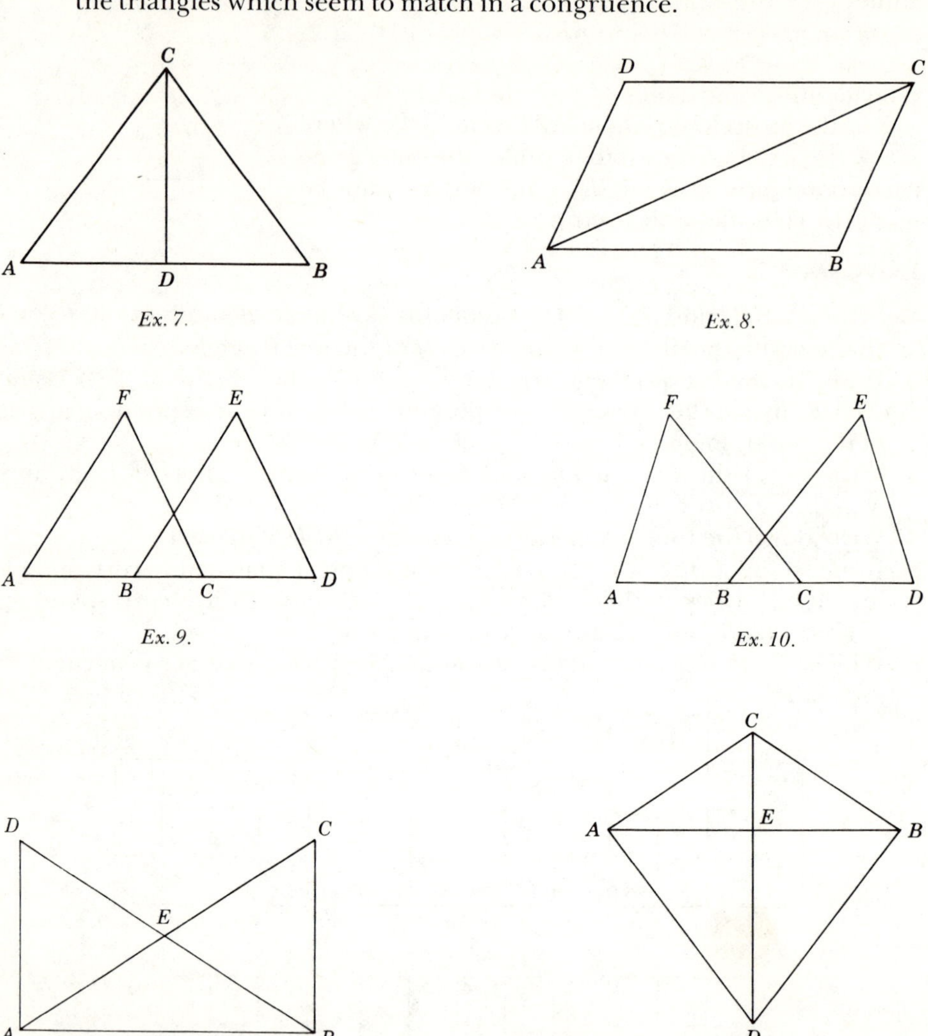

Ex. 7. *Ex. 8.* *Ex. 9.* *Ex. 10.* *Ex. 11.* *Ex. 12.*

4.17. The triangle is a rigid figure. Much of our study of congruence of geometric figures deals with triangles. The triangle is the most widely used of all the geometric figures formed by straight lines. The *triangle is rigid* in structural design. If three boards are bolted together at A, B, and C, as shown in Fig. 4.5, the shape of the triangle is fixed. It cannot be changed without bending or breaking the pieces of wood. However, if we bolt together four (or more) boards, forming a four-sided figure as shown in Fig. 4.6, the shape of the frame can be changed by exerting a force on one of the bolts. The measures of the angles formed by the boards can be changed in size even though the lengths of the sides of the figure remain the same. The frame of Fig. 4.6 can be made rigid by bolting a board across D and F (or E and G), thus forming two rigid triangles.

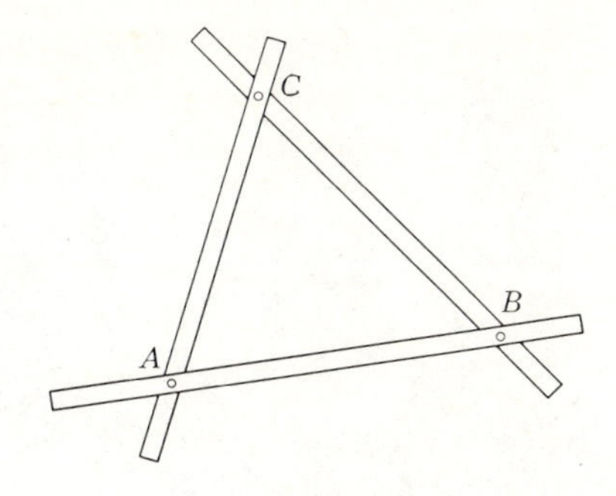

Fig. 4.5.

The rigidity of triangles is illustrated in the practical applications of this property in the construction of many types of structures, such as bridges, towers, and gates (Fig. 4.7).

4.10. Congruence of triangles. The engineer and the draftsman are continually using congruence of triangles in their work. By applying their knowledge of congruent triangles, they are able to study measures of the three sides and the three angles of a given triangle and to compute areas of triangles. Often they apply this knowledge in constructing triangular structures which will be exact duplicates of an original structure.

Definition: If there exists some correspondence $ABC \leftrightarrow DEF$ of the vertices of $\triangle ABC$ with those of $\triangle DEF$ such that each pair of corresponding sides are congruent and each pair of corresponding angles are congruent, the correspondence $ABC \leftrightarrow DEF$ is called a *congruence* between the triangles. The triangles are congruent triangles. Or we may state that $\triangle ABC$ is congruent to $\triangle DEF$, written $\triangle ABC \cong \triangle DEF$.

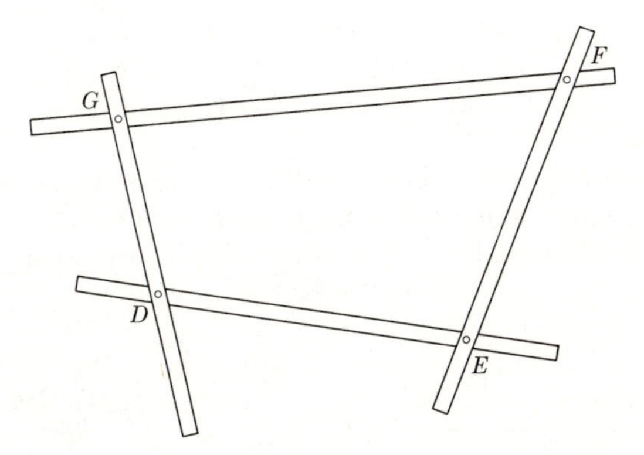

Fig. 4.6.

Fig. 4.7.

Thus, if $\triangle ABC \cong \triangle DEF$ (Fig. 4.8), we know six relationships between the sides and angles of the two triangles, namely

$$
\begin{array}{ll}
m\overline{AB} = m\overline{DE} & \overline{AB} \cong \overline{DE} \\
m\overline{BC} = m\overline{EF} & \overline{BC} \cong \overline{EF} \\
m\overline{AC} = m\overline{DF} & \overline{AC} \cong \overline{DF} \\
m\angle A = m\angle D & \angle A \cong \angle D \\
m\angle B = m\angle E & \angle B \cong \angle E \\
m\angle C = m\angle F & \angle C \cong \angle F
\end{array}
$$

The equations in the left column and the congruences in the right column mean the same thing. They can be used interchangeably.
In Section 9.2 we will introduce a third way to indicate congruency of segments.

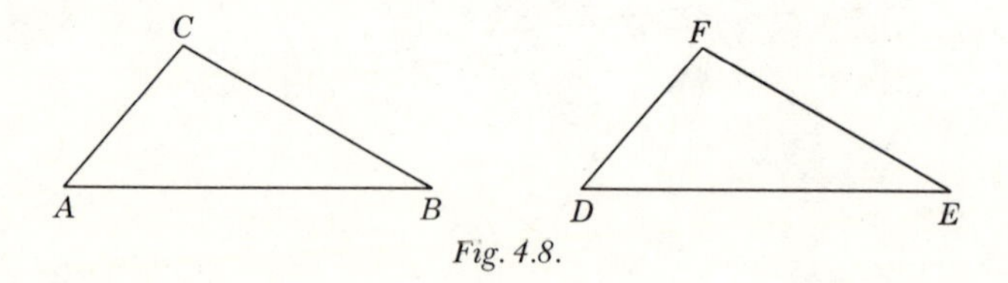

Fig. 4.8.

4.19. Basic congruence postulate. Although we *defined* two triangles as congruent if three pairs of sides and three pairs of angles are congruent, triangles can be *proved* congruent if fewer pairs of corresponding parts are known to be congruent. We must first accept a new postulate.

Postulate 17 (the S.A.S. postulate). *Two triangles are congruent if two sides and the included angle of one are, respectively, congruent to the two sides and the included angle of the other.*

This postulate states that, in Fig. 4.9, if $\overline{AB} \cong \overline{ED}$, $\overline{AC} \cong \overline{EF}$, and $\angle A \cong \angle E$, then $\triangle ABC \cong \triangle DEF$.

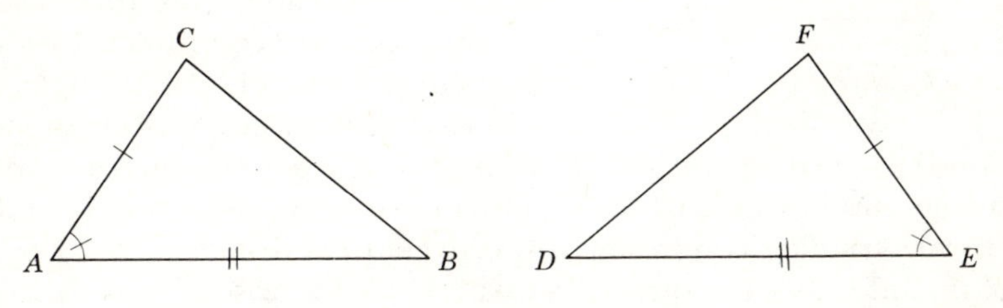

Fig. 4.9.

The student often will find that he is aided in making a quick selection of the congruent sides and congruent angles in the two triangles by designating them with similar check marks for the congruent pairs of congruent sides and congruent angles. In this text we will frequently use hash marks to indicate "given" congruences. Thus, in Fig. 4.10, if it is given that $\overline{AC} \cong \overline{DE}$, $\overline{AB} \cong \overline{DB}$, $\overline{AC} \perp \overline{AD}$ and $\overline{DE} \perp \overline{AD}$, the student can readily see which are the congruent pairs.

It will also be helpful if, in proving a congruence for two triangles, the student names the triangles in such a way as to indicate the matching vertices. For example, in Fig. 4.10, since $ABC \leftrightarrow DBE$ can be proved a congruence, it would be more explicit to refer to these triangles as "$\triangle ABC$ and $\triangle DBE$" rather than, say, "$\triangle ABC$ and $\triangle DEB$." Although the sentence "$\triangle ABC \cong \triangle DEB$" can be proved correct, the sentence "$\triangle ABC \cong \triangle DBE$" will prove more helpful since it aids in picking out the corresponding parts of the two figures.

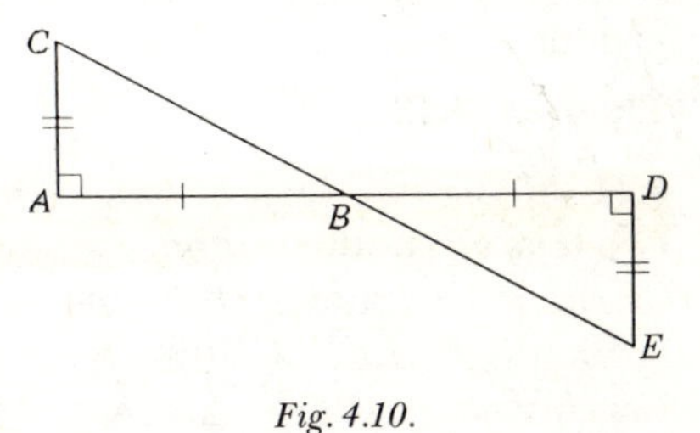

Fig. 4.10.

It is important that the student recognize, in using Postulate 17 to prove triangles congruent, that the congruent angles must be *between* (formed by)

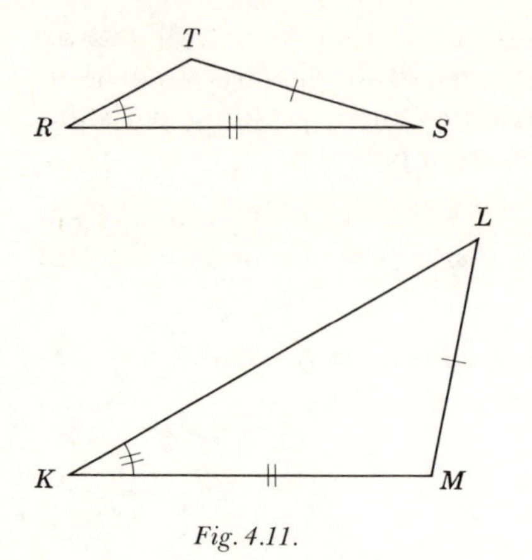

Fig. 4.11.

the corresponding congruent sides. If the congruent angles are not between the two known congruent sides, it does not necessarily follow that the correspondence will give a congruence. In $\triangle RST$ and $\triangle KLM$ (Fig. 4.11) note that, though $\overline{RS} \cong \overline{KM}$, $\overline{ST} \cong \overline{ML}$, and $\angle R \cong \angle K$, the triangles certainly are not congruent.

4.20. Application of Postulate 17. In Postulate 17 we have stated that two triangles, each made up of three sides and three angles, are congruent if only three particular parts of one triangle can be shown congruent respectively to the three corresponding parts of the second triangle. Hereafter, when we are given *any* two triangles in which we know, or can prove, two sides and the included angle of one triangle congruent respectively to two sides and the included angle of the other, we can quote Postulate 17 as the reason for stating that the two triangles are congruent.

It is essential that the student memorize, or can state the equivalent in his own words, the statement of Postulate 17 because he will be required frequently in subsequent proofs to give it as a reason for statements in these proofs. After the student has shown competence in stating the postulate, the instructor may permit him to refer briefly to it by the abbreviation S.A.S. (side-angle-side). This abbreviation will be used hereafter in this text.

Once Postulate 17 is accepted as true, it becomes possible to prove various congruence theorems for triangles. We will next consider a theorem and two other examples of how this postulate can be used in proving other congruences.

Theorem 4.13

4.21. If the two legs of one right triangle are congruent respectively to the two legs of another right triangle, the triangles are congruent.

Given: $\triangle ABC$ and $\triangle DEF$ with $\overline{AC} \cong \overline{DF}$, $\overline{BC} \cong \overline{EF}$; $\angle C$ and $\angle F$ are right $\angle$s.

Conclusion: $\triangle ABC \cong \triangle DEF$.

Proof:

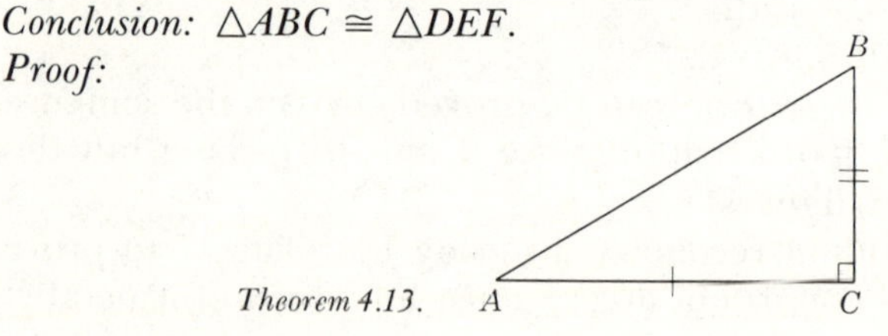

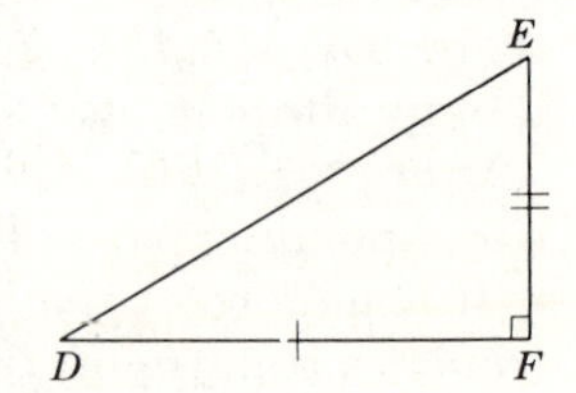

Theorem 4.13.

STATEMENTS	REASONS
1. $\overline{AC} \cong \overline{DF}$; $\overline{BC} \cong \overline{EF}$.	1. Given
2. $\angle C$ and $\angle F$ are right $\measuredangle$s.	2. Given.
3. $\angle C \cong \angle F$.	3. Right angles are congruent.
4. $\triangle ABC \cong \triangle DEF$.	4. S.A.S.

4.22. Illustrative Example 1: The bisector of the vertex of an isosceles triangle divides it into two congruent triangles.

Given: Isosceles triangle ABC with $\overline{AC} \cong \overline{BC}$; $\overline{CD}$ bisects $\angle ACB$.

Conclusion: $\triangle ADC \cong \triangle BDC$.

Proof:

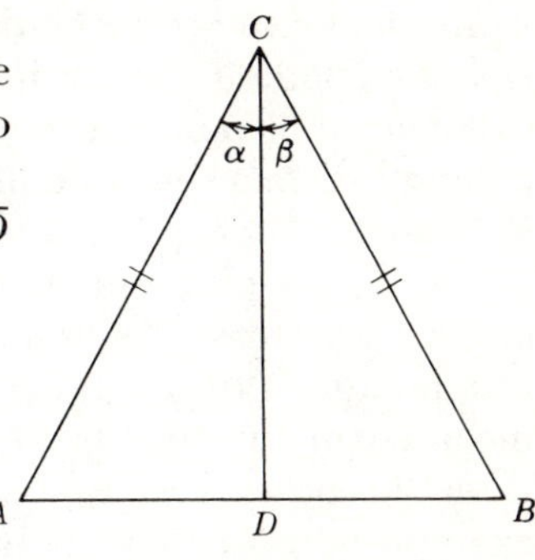

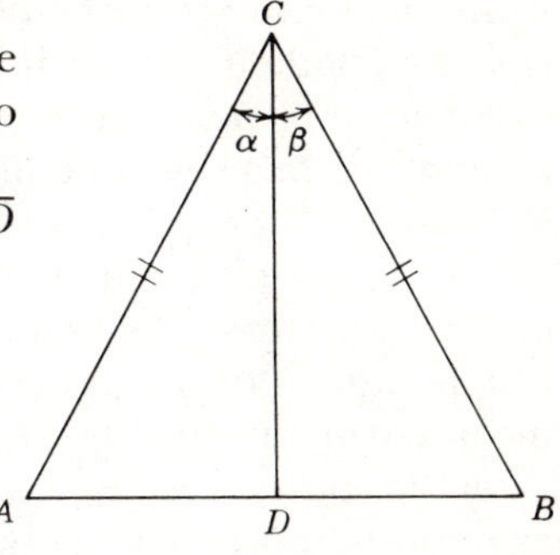

Illustrative Example 1.

STATEMENTS	REASONS
1. $\overline{AC} \cong \overline{BC}$.	1. Given.
2. $\overline{CD} \cong \overline{CD}$.	2. Reflexive theorem of segments.
3. $\overline{CD}$ bisects $\angle ACB$.	3. Given.
4. $\angle \alpha \cong \angle \beta$.	4. A bisector divides an angle into two congruent angles.
5. $\triangle ADC \cong \triangle BDC$.	5. S.A.S.

4.23. Illustrative Example 2:

Given: The adjacent figures with $\overline{AD}$ and $\overline{CE}$ bisecting each other at B.

Conclusion: $\triangle ABC \cong \triangle DBE$.

Proof:

Illustrative Example 2.

STATEMENTS	REASONS
1. $\overline{AD}$ and $\overline{CE}$ bisect each other at B.	1. Given.
2. $\overline{BA} \cong \overline{BD}$.	2. Definition of bisector.
3. $\overline{BC} \cong \overline{BE}$.	3. Reason 2.
4. $\angle ABC \cong \angle DBE$.	4. Vertical angles are congruent.
5. $\triangle ABC \cong \triangle DBE$.	5. S.A.S.

4.24. Use of figures in geometric proofs. Every valid geometric proof should be independent of the figure used to illustrate the problem. Figures are used merely as a matter of convenience. Strictly speaking, before Example 2 could be proved, it should be stated that: (1) $A, B, C, D,$ and E are five points lying in the same plane; (2) B is between A and D; and (3) B is between C and E.

To include such information, which can be inferred from the figure, would make the proof tedious and repetitious. In this text it *will* be permissible to use the figure to infer (without stating it) such things as betweenness, collinearity of points, the location of a point in the interior or the exterior of an angle or in a certain half-plane, and the general relative position of points, lines, and planes.

The student should be careful *not* to infer congruence of segments and angles, bisectors of segments and angles, perpendicular and parallel lines just because "they appear that way" in the figure. Such things must be included in the hypotheses or in the developed proofs. It would not, for example, be correct to assume $\angle A$ and $\angle D$ are right angles in the second example because they might look like it.

Exercises (A)

The triangles of each of the twelve following problems are marked to show congruent sides and angles. Indicate the pairs of triangles which can be proved congruent by Postulate 17 or Theorem 4.13.

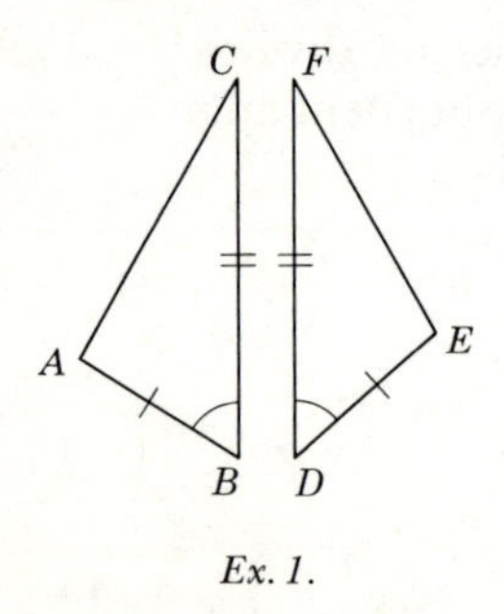

Ex. 1.

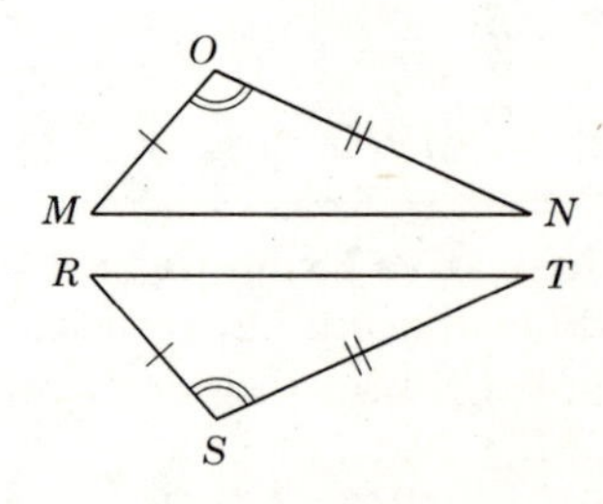

Ex. 2.

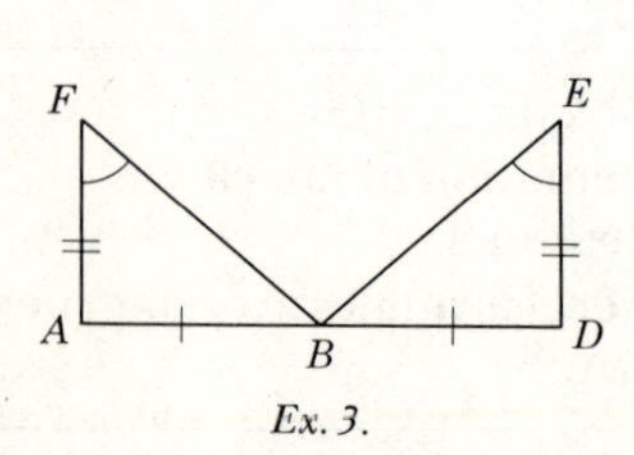

Ex. 3.

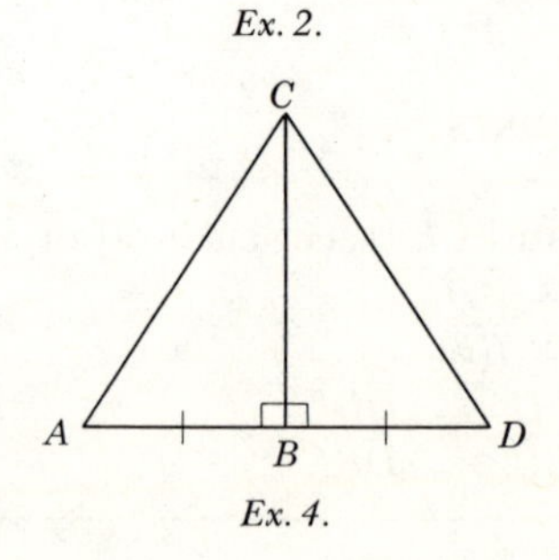

Ex. 4.

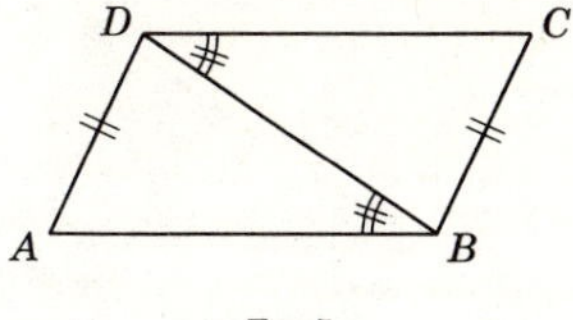

Ex. 5.

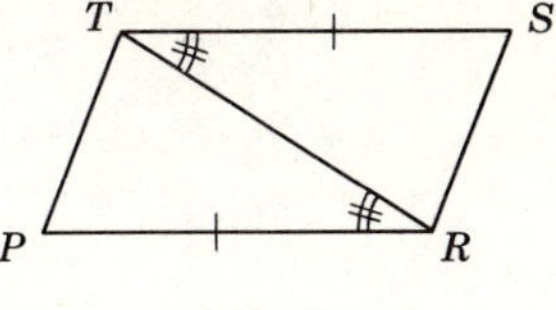

Ex. 6.

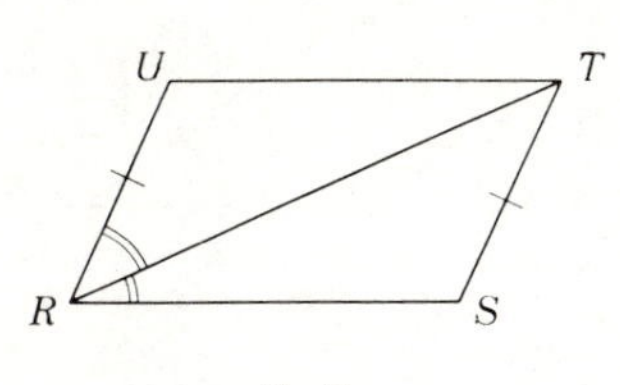

Ex. 7.

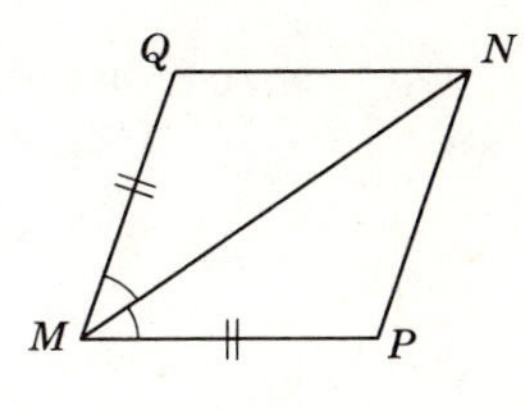

Ex. 8.

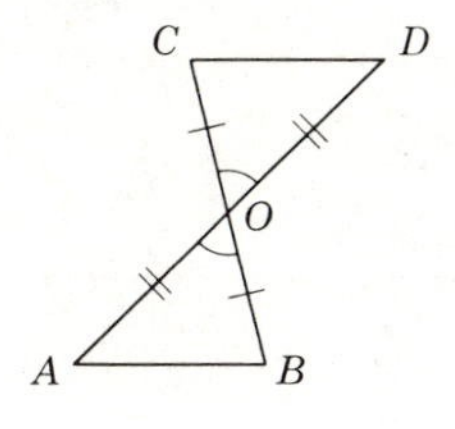

Ex. 9.

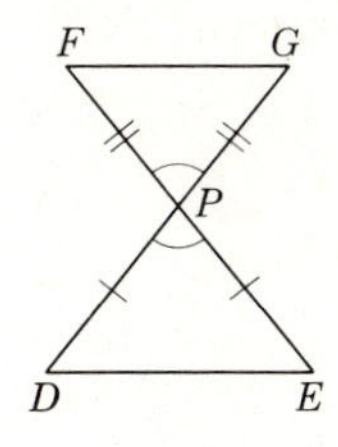

Ex. 10.

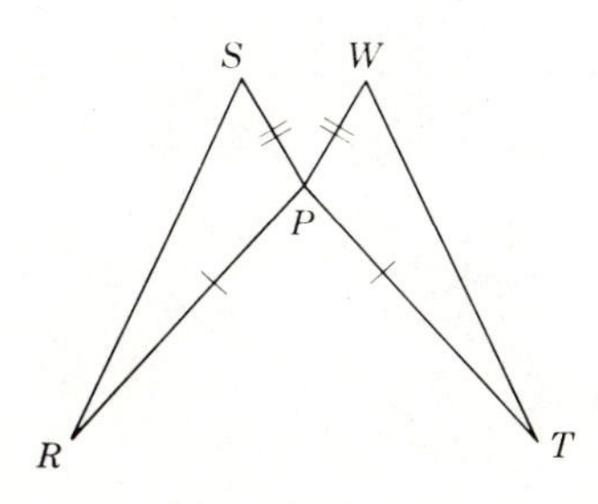

Ex. 11.

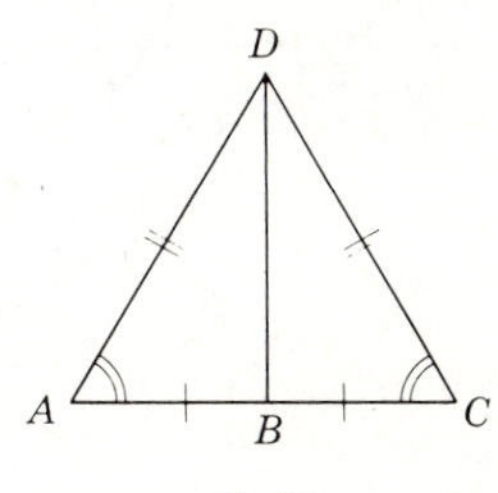

Ex. 12.

Exercises (B)

Prove the following exercises:

13. *Given:* $\overline{AC} \perp \overline{AB}$; $\overline{DE} \perp \overline{BD}$; $\overline{AC} \cong \overline{DE}$; B bisects $\overline{AD}$.
Conclusion: $\triangle ABC \cong \triangle DBE$.

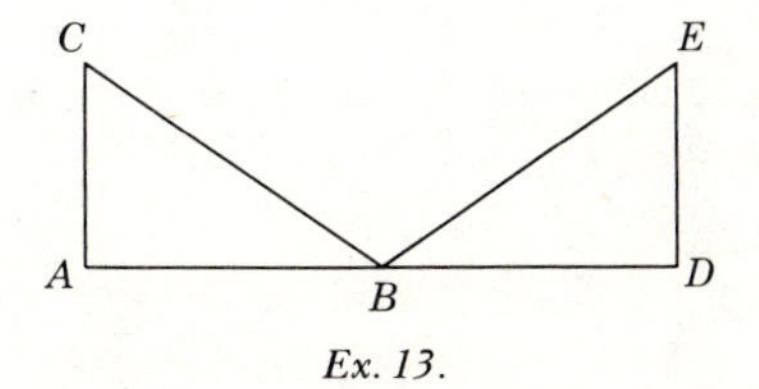

Ex. 13.

14. *Given:* $\overline{AD}$ and $\overline{BE}$ intersecting at C;
$\overline{CE} \cong \overline{CB}$; $\overline{AC} \cong \overline{DC}$.
Conclusion: $\triangle ABC \cong \triangle DEC$.

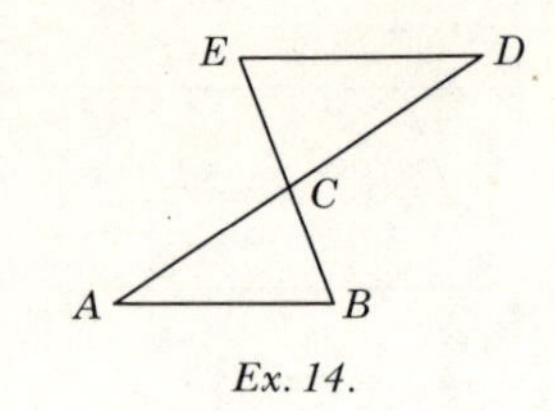

Ex. 14.

15. *Given:* $\overline{QS} \perp \overline{RT}$; S bisects $\overline{RT}$.
Conclusion: $\triangle RSQ \cong \triangle TSQ$.

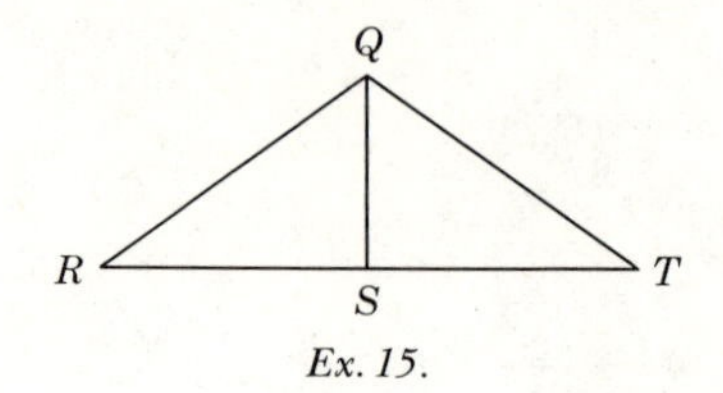

Ex. 15.

16. *Given:* $\angle DAB \cong \angle CBA$; $\overline{EA} \cong \overline{BF}$.
Conclusion: $\triangle ABE \cong \triangle BAF$.

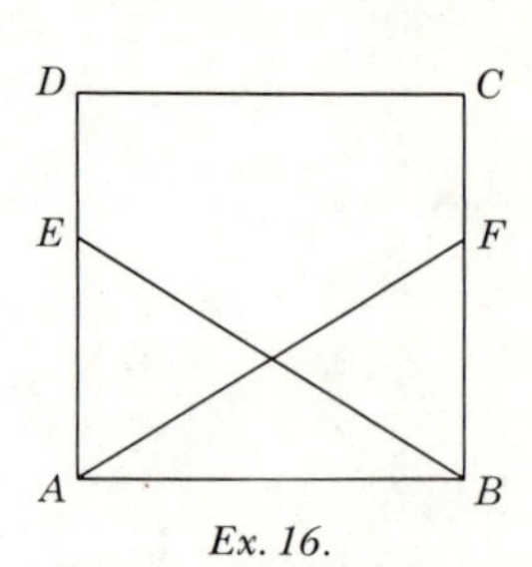

Ex. 16.

17. *Given:* $\overline{RS} \cong \overline{QT}$, $\overline{PS} \cong \overline{PT}$;
$\angle RTP \cong \angle QSP$.
Conclusion: $\triangle RTP \cong \triangle QSP$.

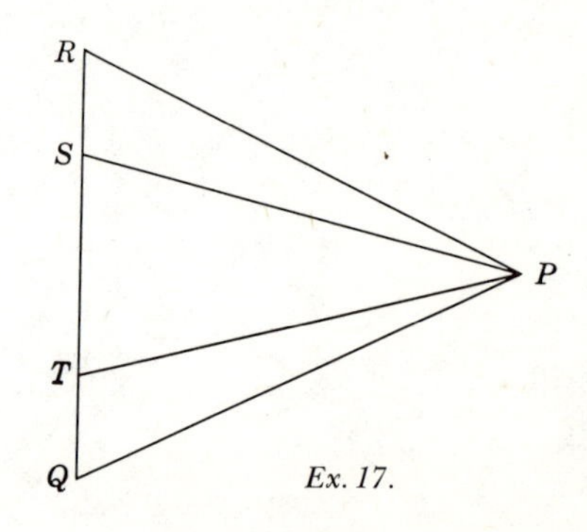

Ex. 17.

18. *Given:* $\overline{AC} \cong \overline{AD}$; $\overline{BC} \cong \overline{BD}$;
$\angle\alpha \cong \angle\theta$; $\angle\beta \cong \angle\gamma$.
Conclusion: $\triangle ABC \cong \triangle ABD$.

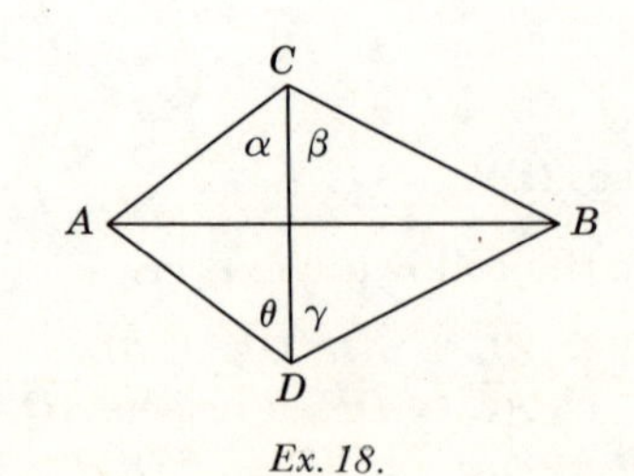

Ex. 18.

19. *Given:* Isosceles $\triangle ABC$ with $\overline{AC} \cong \overline{BC}$; D the midpoint of $\overline{AC}$; E the midpoint of $\overline{BC}$.
Conclusion: $\triangle ACE \cong \triangle BCD$.

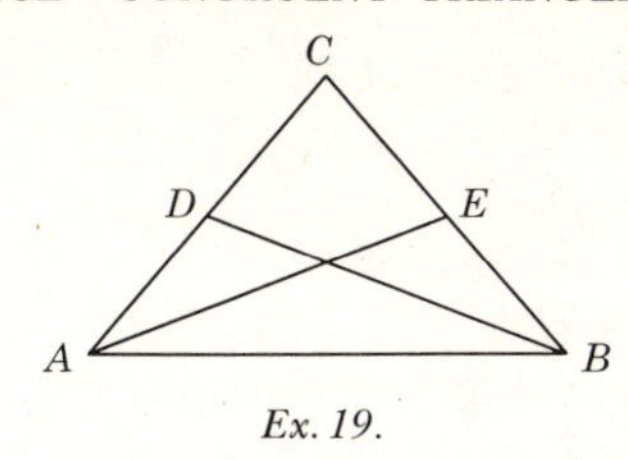

Ex. 19.

20. *Given:* $\triangle QRS$ with $\angle SQR \cong \angle SRQ$; T the midpoint of $\overline{QS}$; W the midpoint of $\overline{RS}$; $\overline{QS} \cong \overline{RS}$.
Conclusion: $\triangle TQR \cong \triangle WRQ$.

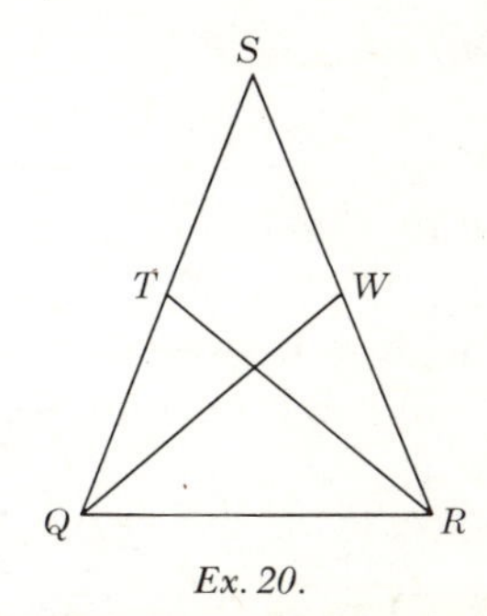

Ex. 20.

Theorem 4.14 ASA

4.25. If two triangles have two angles and the included side of one congruent to the corresponding two angles and the included side of the other, the triangles are congruent.

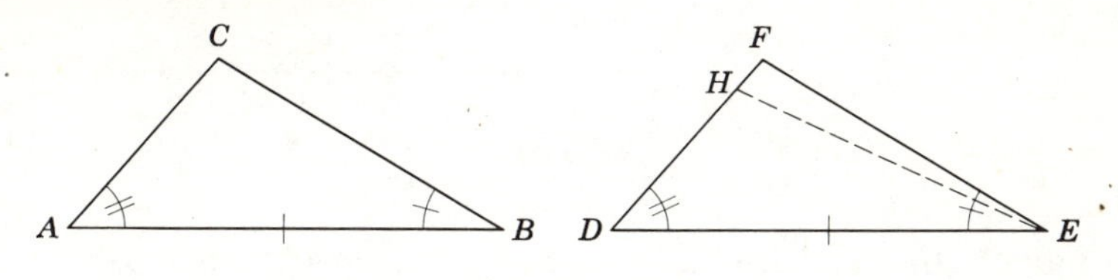

Theorem 4.14.

Given: $\triangle ABC$ and $\triangle DEF$ with $\angle A \cong \angle D$, $\angle B \cong \angle E$, $\overline{AB} \cong \overline{DE}$.
Conclusion: $\triangle ABC \cong \triangle DEF$.
Proof:

STATEMENTS	REASONS
1. $\overline{AB} \cong \overline{DE}$, $\angle A \cong \angle D$.	1. Given.
2. On $\overrightarrow{DF}$ there is a point H such that $m\overline{DH} = m\overline{AC}$.	2. Point plotting postulate.
3. Draw $\overline{HE}$.	3. Two points determine a line.

4. $\triangle ABC \cong \triangle DEH$.	4. S.A.S.
5. $\angle DEH \cong \angle B$.	5. Corresponding $\angle$s of congruent $\triangle$s are $\cong$ to each other.
6. $\angle B \cong \angle E$.	6. Given.
7. $\angle DEH \cong \angle E$.	7. Congruence of $\angle$s is transitive.
8. $\overrightarrow{EH}$ and $\overrightarrow{EF}$ are the same ray.	8. Angle construction postulate.
9. $H = F$.	9. Two lines intersect in at most one point.
10. $\triangle ABC \cong \triangle DEF$.	10. Replacing H of Statement 4 by F (from Statement 9).

It will be noted that in drawing the figure for the proof of Theorem 4.14, the point H is shown between D and F. The point could just as well be drawn with F between H and D. This would not alter the validity of the proof. The abbreviation for the statement of this theorem is A.S.A.

Theorem 4.15

4.26. If a leg and the adjacent acute angle of one right triangle are congruent respectively to a leg and the adjacent acute angle of another, the right triangles are congruent.

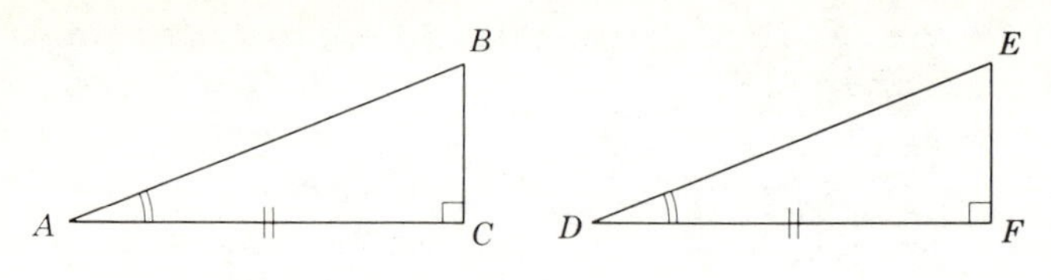

Theorem 4.15.

Given: Right $\triangle$s ABC and DEF with $\angle A \cong \angle D$, leg $AC \cong$ leg DF, $\angle C$ and $\angle F$ are right angles.
Conclusion: $\triangle ABC \cong \triangle DEF$.
Proof:

STATEMENTS	REASONS
1. $\angle A \cong \angle D$.	1. Given.
2. $\angle C$ and $\angle F$ are right $\angle$s.	2. Given.
3. $\angle C \cong \angle F$.	3. Right angles are congruent.
4. $\overline{AC} \cong \overline{DF}$.	4. Given.
5. $\therefore \triangle ABC \cong \triangle DEF$.	5. A.S.A.

4.27. Illustrative Example 1:

Given: $\overline{CD}$ bisects $\angle ACB$; $\overline{CD} \perp \overline{AB}$.
Conclusion: $\triangle ADC \cong \triangle BDC$.
Proof:

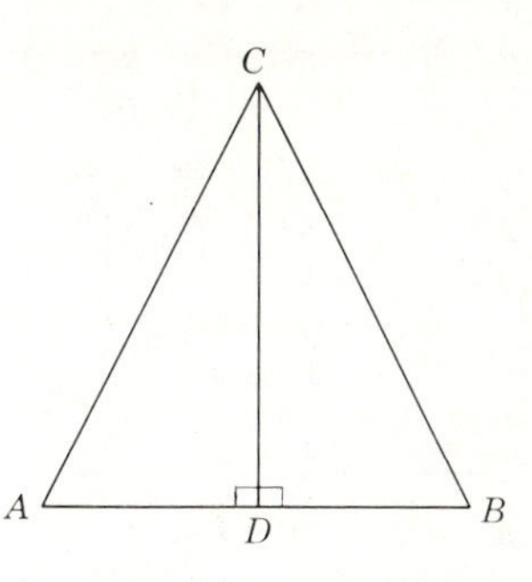

Illustrative Example 1.

STATEMENTS	REASONS
1. $\overline{CD}$ bisects $\angle ACB$.	1. Given.
2. $\angle ACD \cong \angle BCD$.	2. A bisector divides an angle into two congruent angles.
3. $\overline{CD} \perp \overline{AB}$.	3. Given.
4. $\angle ADC$ and $\angle BDC$ are right angles.	4. Two perpendicular lines form right angles.
5. $\angle ADC \cong \angle BDC$.	5. Right angles are congruent.
6. $\overline{CD} \cong \overline{CD}$.	6. Congruence of segments is reflexive.
7. $\therefore \triangle ADC \cong \triangle BDC$.	7. A.S.A.

The student will note how the method of *modus ponens* has been applied in the above proof. The logic used could be written:

(a) 1. A bisector divides an angle into two congruent angles.
2. $\overline{CD}$ bisects $\angle ACB$.
3. $\angle ACD \cong \angle BCD$.

(b) 1. Two perpendicular lines form right angles.
2. $\overline{CD}$ is perpendicular to $\overline{AB}$.
3. $\angle ADC$ and $\angle BDC$ are right angles.

(c) 1. All right angles are congruent.
2. $\angle ADC$ and $\angle BDC$ are right angles.
3. $\angle ADC \cong \angle BDC$.

(d) 1. If two triangles have two angles and the included side of one congruent to the corresponding two angles and the included side of the other, the triangles are congruent.
2. $\angle ACD \cong \angle BCD$; $\overline{CD} \cong \overline{CD}$; $\angle ADC \cong \angle BDC$.
3. $\triangle ADC \cong \triangle BDC$.

4.28. Illustrative Example 2:

Given: $\overline{AB} \perp \overline{BC}, \overline{DC} \perp \overline{BC}$,
$\angle ABD \cong \angle DCA$.

Prove: $\triangle ABC \cong \triangle DCB$.

Proof:

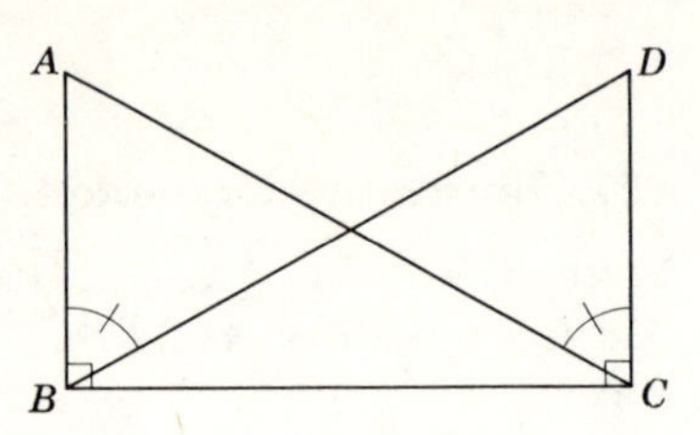

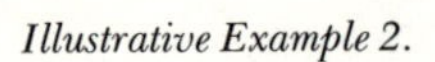
Illustrative Example 2.

STATEMENTS	REASONS
1. $\overline{AB} \perp \overline{BC}$; $\overline{DC} \perp \overline{BC}$.	1. Given.
2. $\angle ABC$ is a right angle; $\angle DCB$ is a right angle.	2. Perpendicular lines meet to form right angles.
3. $\angle ABC \cong \angle DCB$.	3. Right angles are congruent.
4. $\angle ABD \cong \angle DCA$.	4. Given.
5. $\angle DBC \cong \angle ACB$.	5. Subtraction of angles theorem.
6. $\angle ACB \cong \angle DBC$.	6. Symmetric theorem of $\cong$ angles.
7. $\overline{BC} \cong \overline{BC}$.	7. Reflexive theorem of congruent segments.
8. $\therefore \triangle ABC \cong \triangle DCB$.	8. A.S.A.

Exercises (A)

The triangles of each of the following ten problems are marked to show congruent sides and angles. Indicate the pairs of triangles that can be proved congruent by Theorem 4.14 or Theorem 4.15. (See figures for Exercises 1 through 10.)

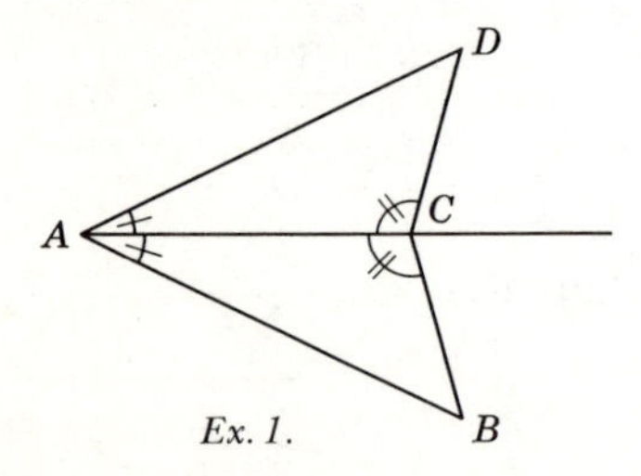

Ex. 1.

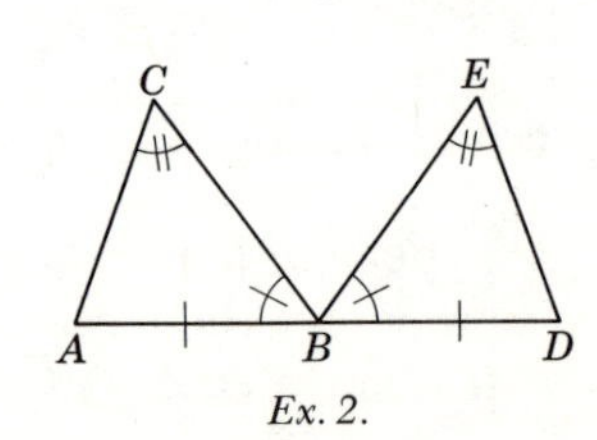

Ex. 2.

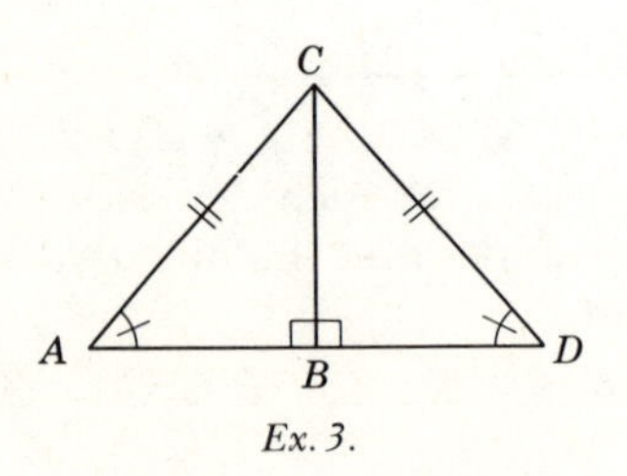

Ex. 3.

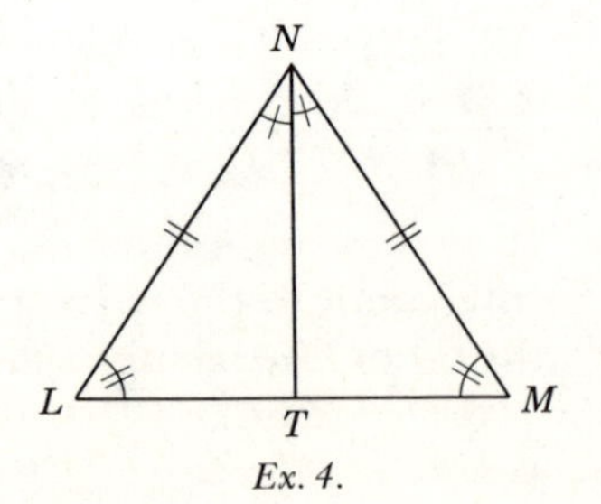

Ex. 4.

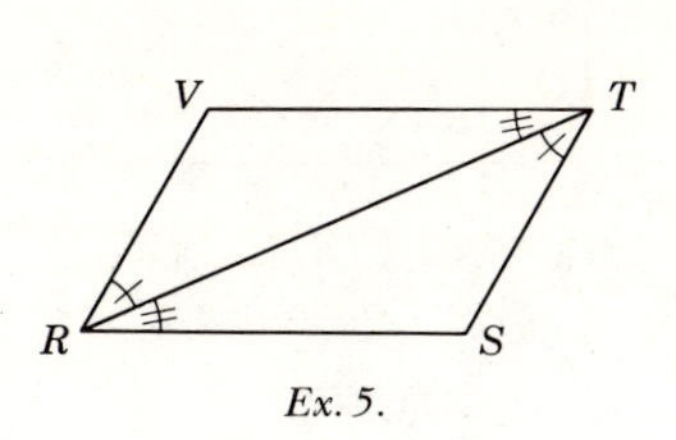

Ex. 5.

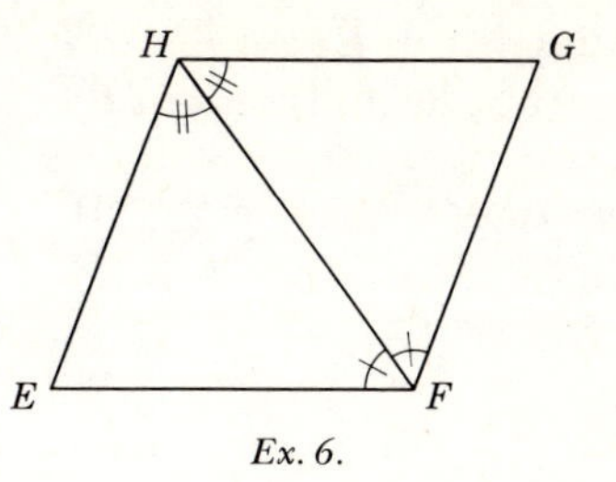

Ex. 6.

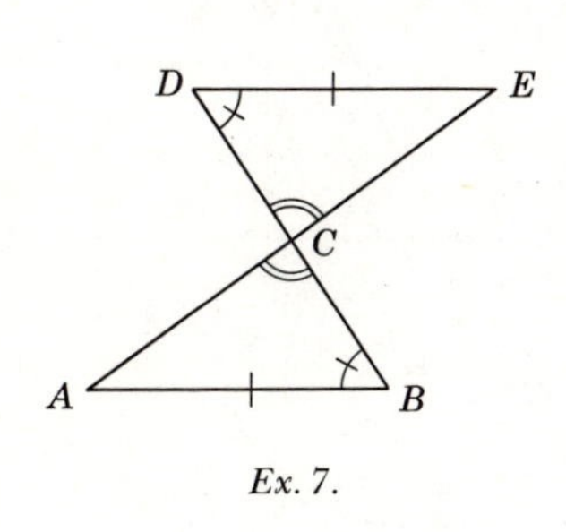

Ex. 7.

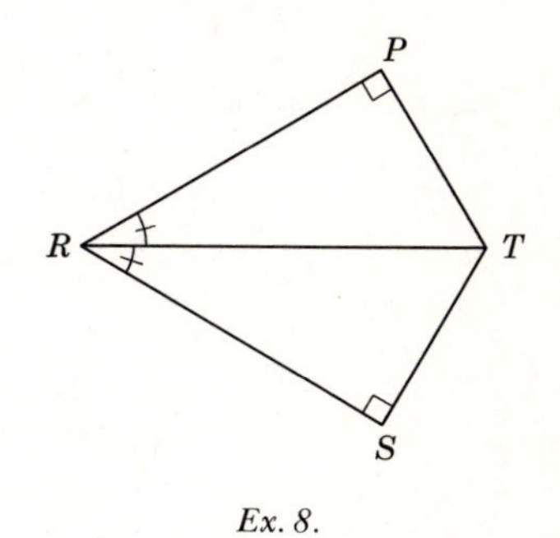

Ex. 8.

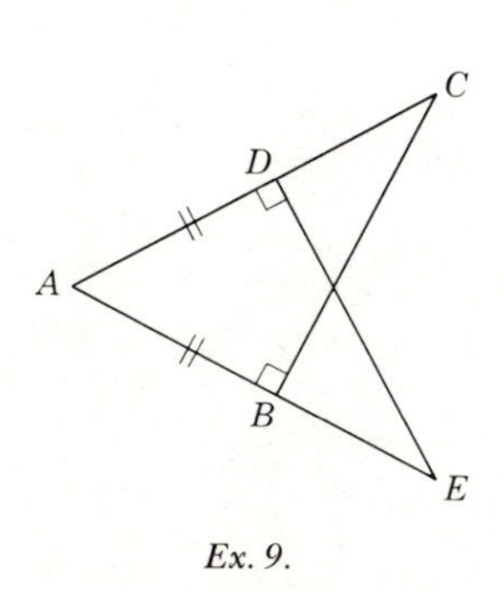

Ex. 9.

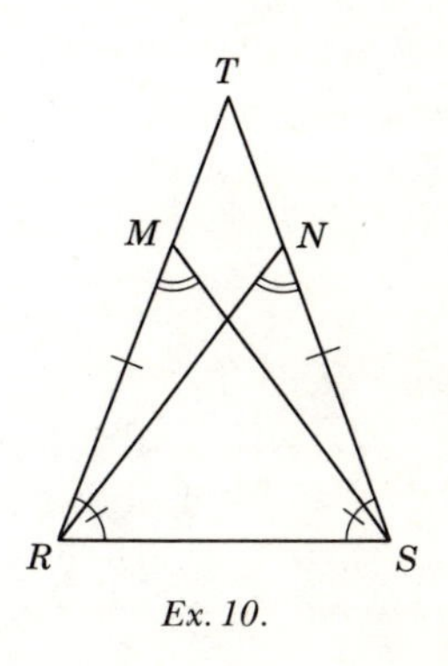

Ex. 10.

Exercises (B)

Prove formally the following exercises:

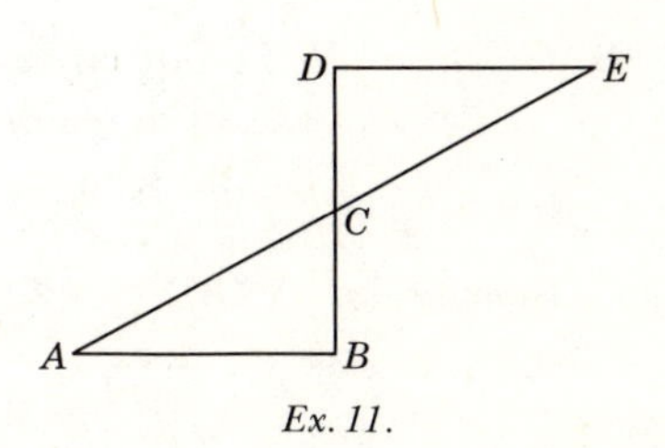

Ex. 11.

11. *Given:* $\overline{AE}$ and $\overline{BD}$ bisect each other at C; $\overline{DE} \perp \overline{BD}$; $\overline{AB} \perp \overline{BD}$.
 Conclusion: $\triangle ABC \cong \triangle EDC$.

12. *Given:* $\overline{VR} \perp \overline{RT}$; $\overline{WT} \perp \overline{RT}$;
S the midpoint of $\overline{RT}$;
$\angle RSV \cong \angle TSW$.
Conclusion: $\triangle RSV \cong \triangle TSW$.

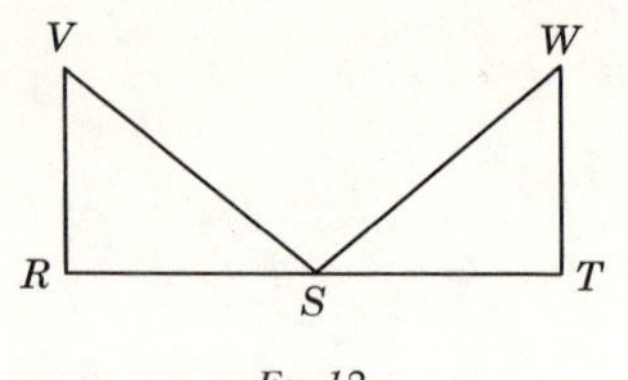

Ex. 12.

13. *Given:* $\overline{PQ}$ bisects $\angle MPN$;
$\overline{PQ} \perp \overline{MN}$.
Conclusion: $\triangle MQP \cong \triangle NQP$.

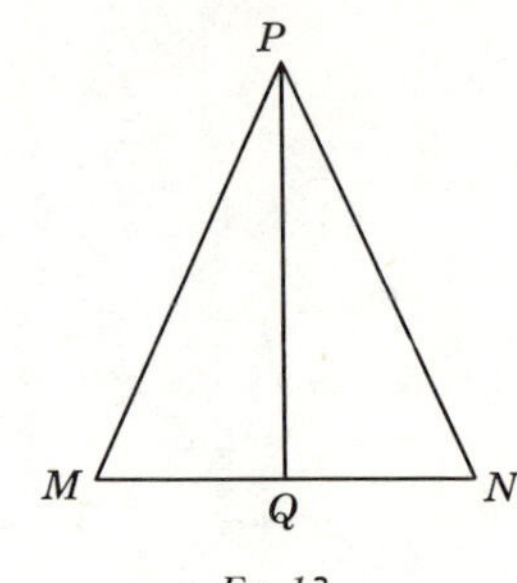

Ex. 13.

14. *Given:* R, S, T are collinear;
$m\angle RSW = m\angle RSV$;
$m\angle RTW = m\angle RTV$.
Conclusion: $\triangle STW \cong \triangle STV$.

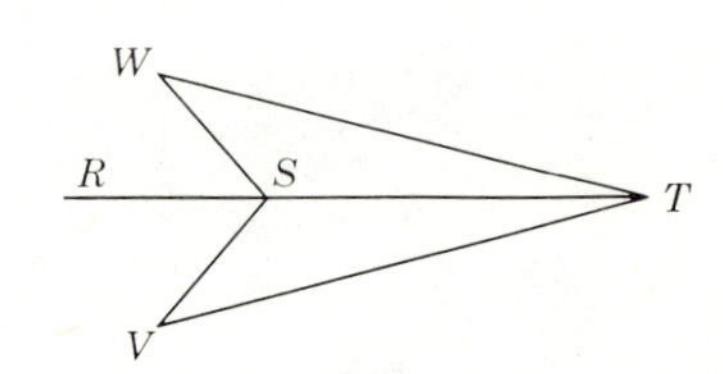

Ex. 14.

15. *Given:* $\overline{AC} \cong \overline{BC}$;
D is the midpoint of $\overline{AC}$;
E is the midpoint of $\overline{BC}$;
$\angle AEC \cong \angle BDC$.
Conclusion: $\triangle AEC \cong \triangle BDC$.

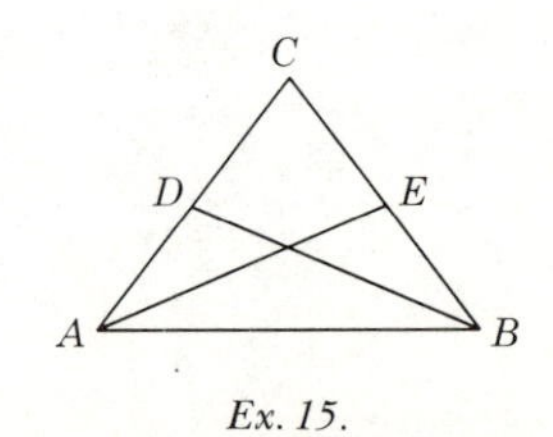

Ex. 15.

16. *Given:* $\triangle ABC$ is equiangular;
D is the midpoint of $\overline{AC}$;
E is the midpoint of $\overline{BC}$;
$\angle ABD \cong \angle BAE$.
Conclusion: $\triangle ABD \cong \triangle BAE$.

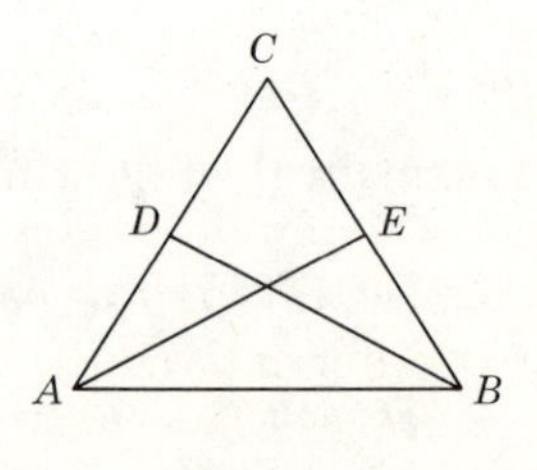

Ex. 16.

17. *Given:* $\overline{MS} \perp \overline{SN}$; $\overline{NG} \perp \overline{MG}$;
$\overline{ST} \cong \overline{TG}$.
Conclusion: $\triangle STM \cong \triangle GTN$.

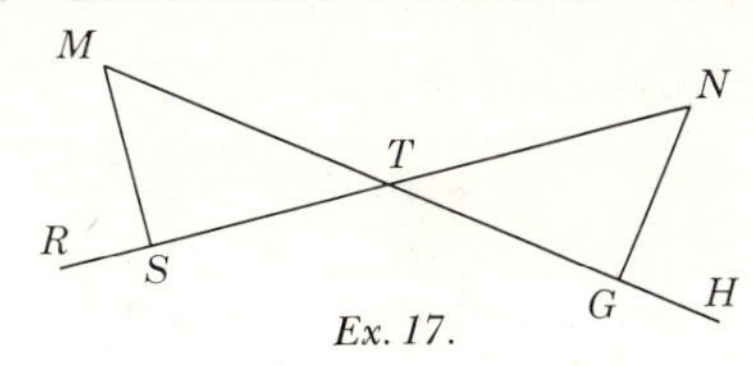

Ex. 17.

18. *Given:* $\overline{CD} \cong \overline{CB}$;
$\overline{AB} \perp \overline{CE}$; $\overline{ED} \perp \overline{AC}$.
Conclusion: $\triangle ABC \cong \triangle EDC$.

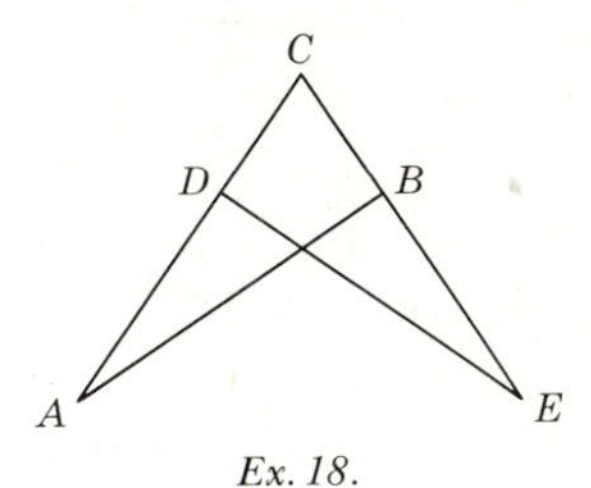

Ex. 18.

19. *Given:* A, D, E, B are collinear;
$\overline{AC} \cong \overline{BC}$;
$\angle A \cong \angle B$;
$\angle ACD \cong \angle BCE$.
Conclusion: $\triangle AEC \cong \triangle BDC$.

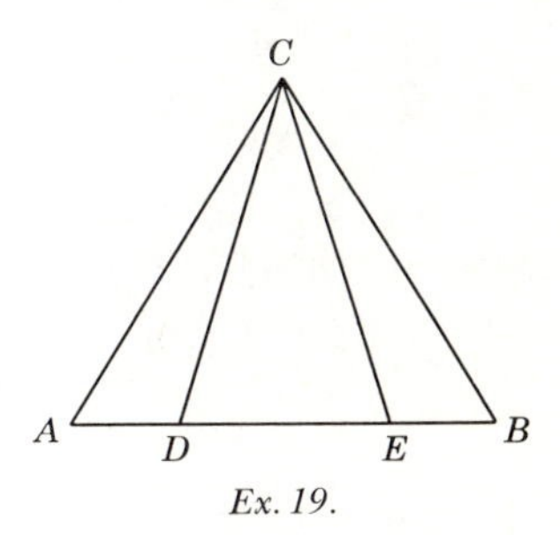

Ex. 19.

20. *Given:* R, S, T, W, are collinear;
$\overline{PS} \cong \overline{PT}$;
$m\angle\alpha = m\angle\beta$;
$m\angle\gamma = m\angle\delta$.
Conclusion: $\triangle RSP \cong \triangle WTP$.

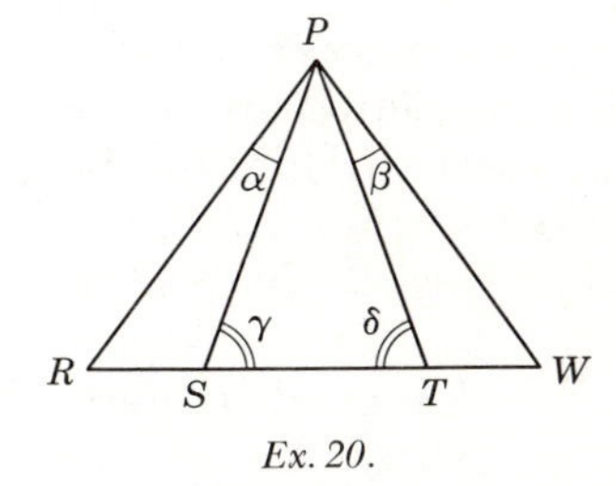

Ex. 20.

4.29. Corresponding parts of congruent triangles. The chief value in proving triangles congruent lies in the fact that *when two triangles are congruent we know that the corresponding sides and angles of the triangles are congruent.* In two congruent triangles a pair of corresponding sides is found opposite a pair of corresponding angles. Conversely, corresponding angles are formed opposite corresponding sides.

We have been using hypotheses, definitions, postulates, and theorems thus far to prove line segments and angles congruent. Now, if we can prove two figures congruent, we have still another means of determining congruent lines and angles.

Theorem 4.16

4.30. The base angles of an isosceles triangle are congruent.

Given: Isosceles $\triangle ABC$ with $\overline{AC} \cong \overline{BC}$.
Conclusion: $\angle A \cong \angle B$.

Proof:

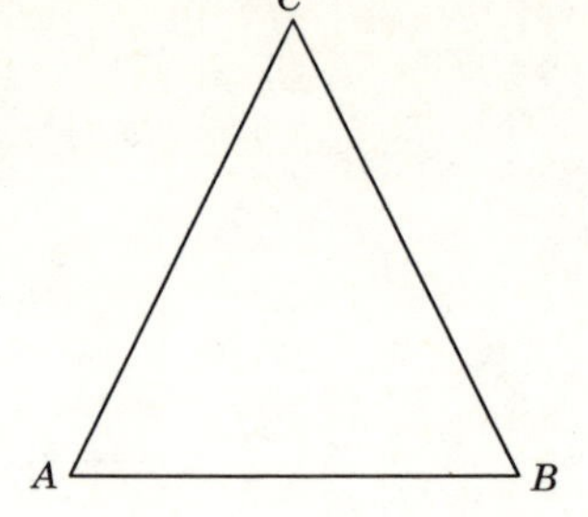

Theorem 4.16.

STATEMENTS	REASONS
Consider the correspondence $ABC \leftrightarrow BAC$.	
1. $\overline{AC} \cong \overline{BC}$.	1. Given.
2. $\overline{BC} \cong \overline{AC}$.	2. Symmetric property of congruence.
3. $\angle C \cong \angle C$.	3. Reflexive property of congruence.
4. $\triangle ABC \cong \triangle BAC$.	4. S.A.S.
5. $\angle A \cong \angle B$.	5. Corresponding parts of congruent figures are congruent.

4.31. Corollary: An equilateral triangle is also equiangular.

4.32. Illustrative Example. The segments joining the midpoints of the congruent sides of an isosceles triangle to the midpoint of the base are congruent.

Given: Isosceles triangle ABC with $\overline{AC} \cong \overline{BC}$, M the midpoint of $\overline{AC}$, N the midpoint of $\overline{BC}$, and Q the midpoint of $\overline{AB}$.
Prove: $\overline{MQ} \cong \overline{NQ}$.

Proof:

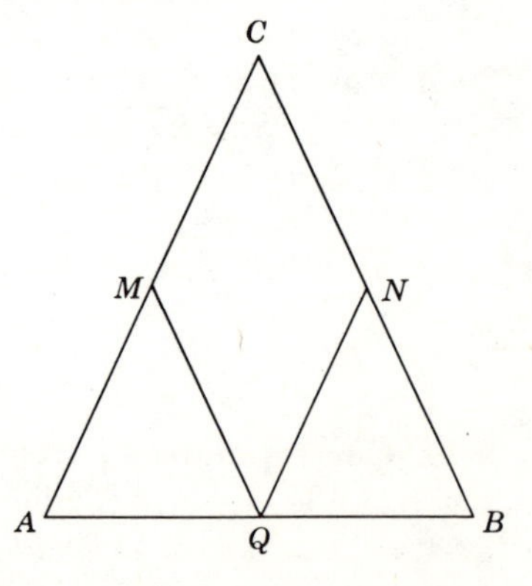

Illustrative Example.

STATEMENTS	REASONS
1. $\overline{AC} \cong \overline{BC}$.	1. Given.
2. M is the midpoint of $\overline{AC}$, N is the midpoint of $\overline{BC}$.	2. Given.
3. $\overline{AM} \cong \overline{BN}$.	3. Theorem 4.11.
4. $\angle A \cong \angle B$.	4. Theorem 4.16.

5. Q is the midpoint of $\overline{AB}$. — 5. Given.
6. $\overline{AQ} \cong \overline{BQ}$. — 6. Definition of midpoint.
7. $\triangle AQM \cong \triangle BQN$. — 7. S.A.S.
8. $\therefore \overline{MQ} \cong \overline{NQ}$. — 8. Corresponding parts of congruent triangles are congruent.

Exercises (A)

Prove the following exercises:

1. *Given:* $\overrightarrow{RP}$ bisects $\angle TRS$, $\overline{PT} \perp \overrightarrow{RT}$, $\overline{PS} \perp \overrightarrow{RS}$, $\overline{RT} \cong \overline{RS}$.
   Conclusion: $\overline{PT} \cong \overline{PS}$.

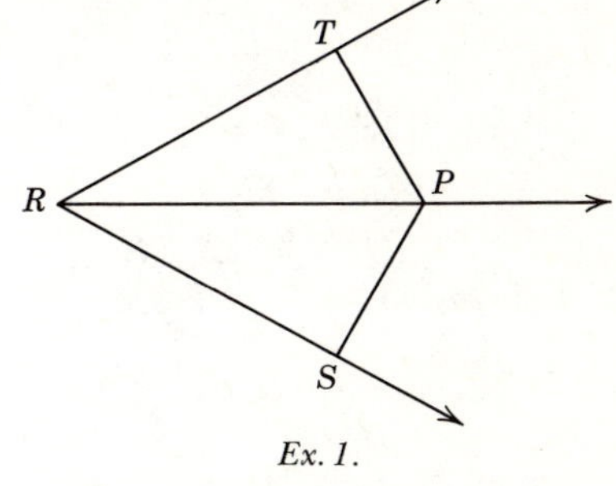

Ex. 1.

2. *Given:* C is the midpoint of $\overline{AD}$ and $\overline{BE}$.
   Conclusion: $\angle E \cong \angle B$.

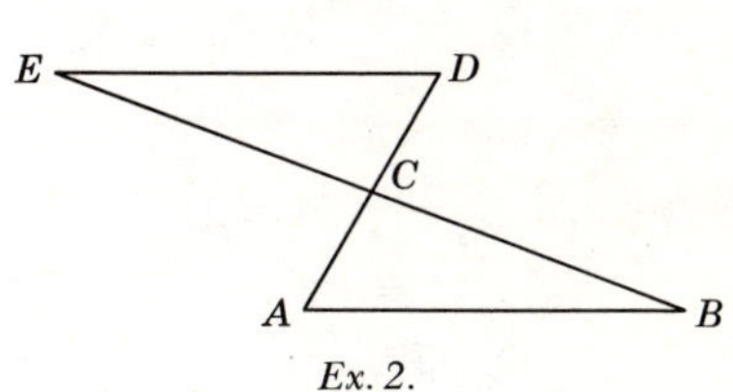

Ex. 2.

3. *Given:* $\overline{AC} \cong \overline{BC}$, $\overline{AD} \cong \overline{BD}$.
   Conclusion: $\angle CAD \cong \angle CBD$.
   (*Hint:* Draw $\overline{AB}$.)

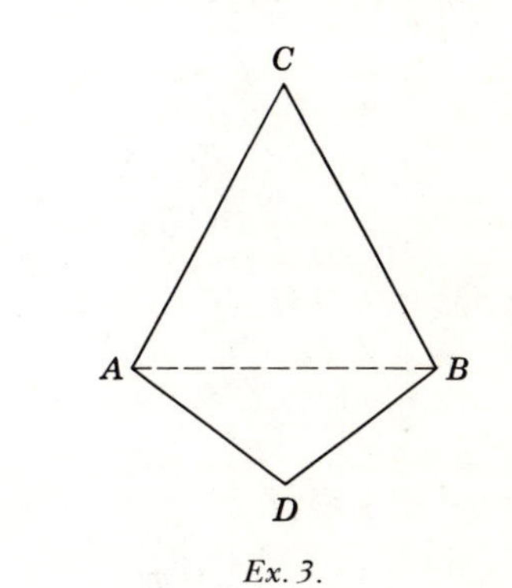

Ex. 3.

4. *Given:* $\overrightarrow{AC}$ bisects $\angle PAT$; $\overline{AP} \cong \overline{AT}$.
   Conclusion: $\angle PBC \cong \angle TBC$.

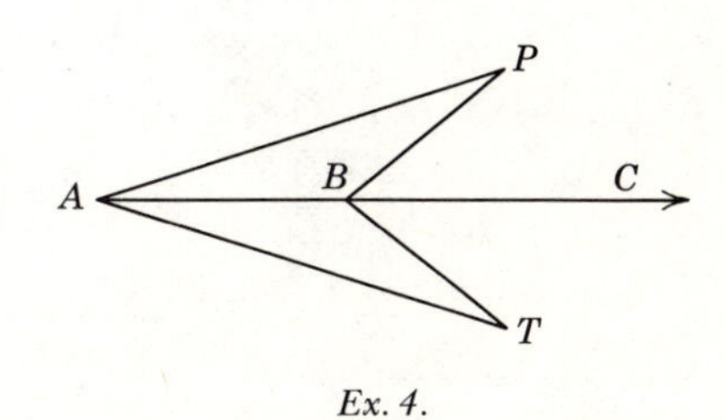

Ex. 4.

5. *Given:* $\overline{AD} \perp \overline{AB}$;
$\overline{BC} \perp \overline{AB}$;
$\overline{AD} \cong \overline{BC}$.
Conclusion: $\angle D \cong \angle C$.

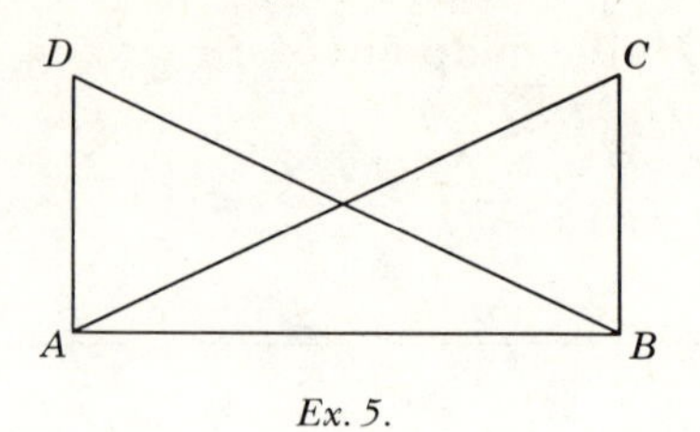

Ex. 5.

6. *Given:* $\overline{PQ} \cong \overline{TQ}$;
$\angle QPS \cong \angle QTR$.
Conclusion: $\angle R \cong \angle S$.

Ex. 6.

7. *Given:* R, S, T, P are collinear;
$\overline{RQ} \cong \overline{PQ}$;
$\angle\alpha \cong \angle\alpha'$.
Conclusion: $\angle QST \cong \angle QTS$.

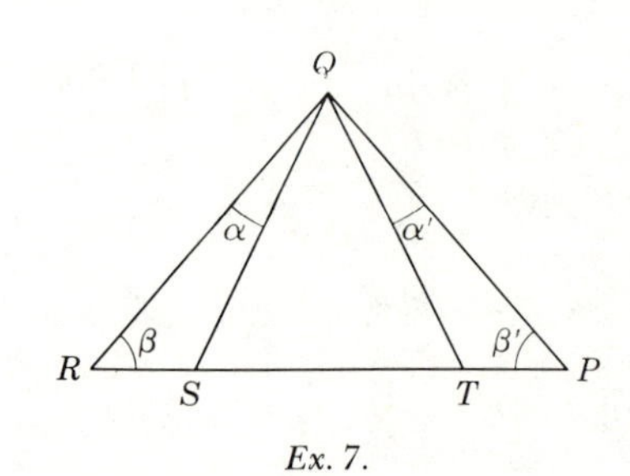

Ex. 7.

8. *Given:* $\angle DAB \cong \angle CBA$;
$\angle DBA \cong \angle CAB$.
Conclusion: $\overline{AD} \cong \overline{BC}$.

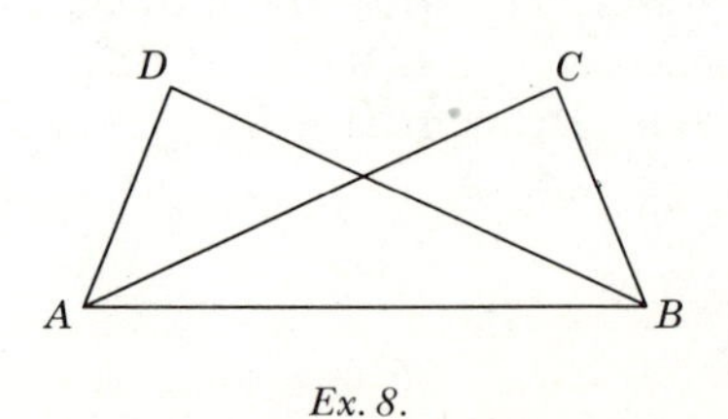

Ex. 8.

9. *Given:* A, B, C, D are collinear;
$\overline{AB} \cong \overline{CD}$;
$\overline{BE} \cong \overline{CE}$;
Conclusion: $\angle A \cong \angle D$.

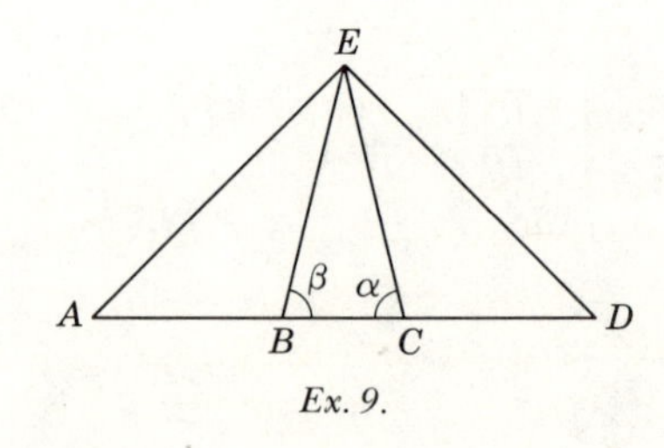

Ex. 9.

10. *Given:* $\overline{RT} \cong \overline{ST}, \overline{RP} \cong \overline{SP}$.
Conclusion: $\angle RSP \cong \angle SRP$.

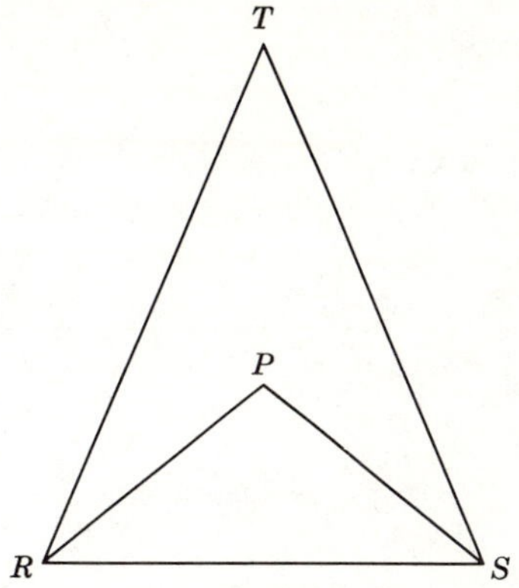

Ex. 10.

Exercises (B)

In the following exercises draw the figures to illustrate the problem, determine the given, the conclusion, and give a formal proof of the problem.

11. If the two legs of a right triangle are congruent respectively to the two legs of a second right triangle, the hypotenuses of the two triangles are congruent.
12. The line joining the vertex and the midpoint of the base of an isosceles triangle is perpendicular to the base.
13. The bisector of the vertex angle of an isosceles triangle is perpendicular to and bisects the base.
14. If the line joining the vertex B of triangle ABC to the midpoint of the opposite side AC is extended its own length to E, the distance from E to C will equal $m\overline{AB}$.
15. The lines joining the midpoints of the sides of an equilateral triangle form another equilateral triangle.

4.33. Lines and angles connected with triangles. Any one of the three sides may be designated as the base of a given triangle. The angle opposite the base of a triangle is called the *vertex angle.* A triangle has three bases and three vertex angles. The angles adjacent to the base are termed *base angles.*

Definition: A segment is an *altitude* of a triangle iff it is the perpendicular segment from a vertex to the line containing the opposite side. Every triangle has three altitudes.

The dotted line segments of Fig. 4.12 illustrate the three altitudes of an acute and an obtuse triangle.

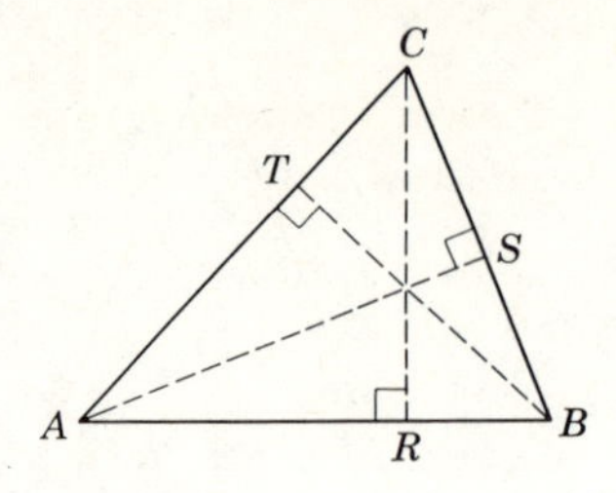

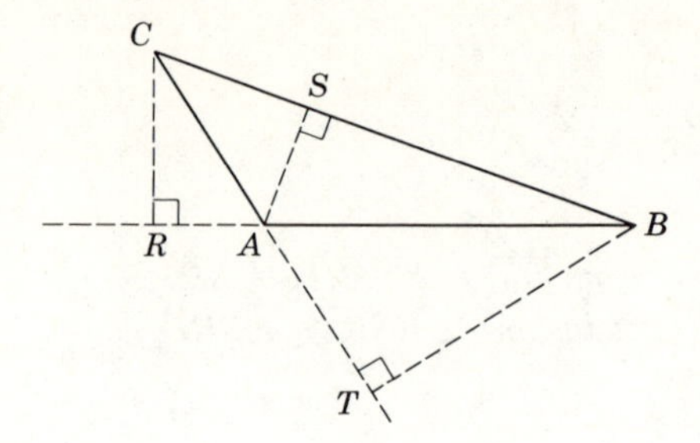

Fig. 4.12.

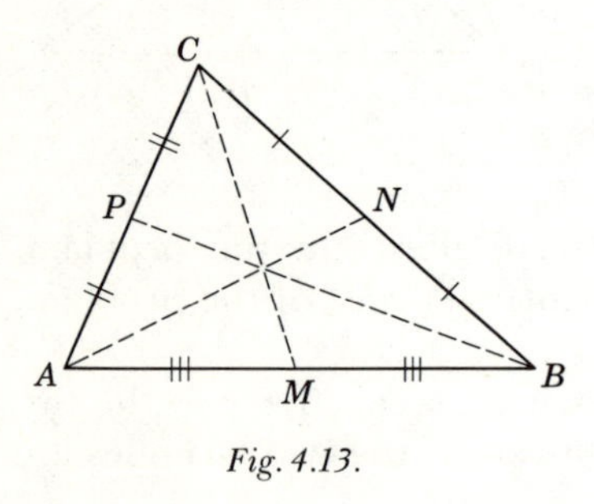

Fig. 4.13.

Definition: A segment is a *median* of a triangle iff its endpoints are a vertex of the triangle and the midpoint of the opposite side. Every triangle has three medians. The dotted line segments of Fig. 4.13 illustrate medians of a triangle. It can be shown that the three medians of a triangle pass through a common point.

Definition: An *angle bisector* of a triangle is a segment which divides an angle of the triangle into two congruent angles and has its endpoints on a vertex and the side opposite the angle. $\overline{BD}$ is the bisector of $\angle B$ of $\triangle ABC$ in Fig. 4.14. Every triangle has three angle bisectors. It can be shown that the three angle bisectors meet in a common point which is equidistant from the three sides of the triangle.

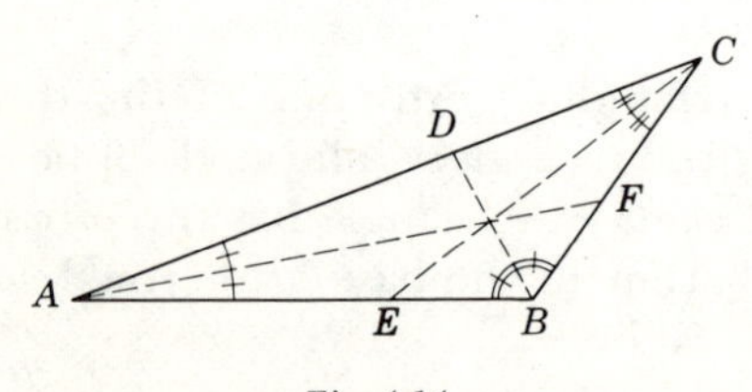

Fig. 4.14.

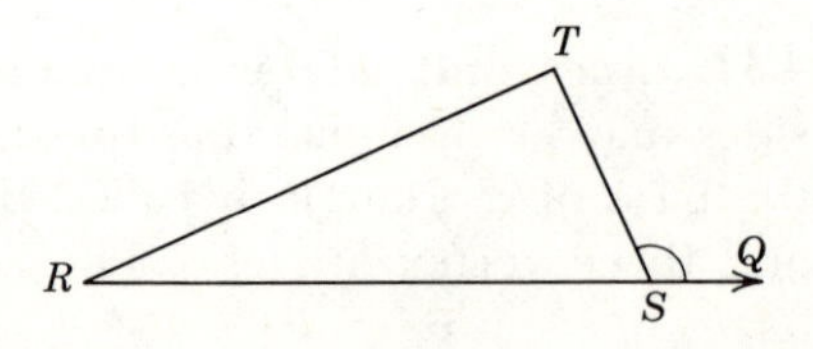

Fig. 4.15.

Definitions: If S is between R and Q, then $\angle QST$ is an *exterior angle* of $\triangle RST$ (Fig. 4.15). Every triangle has six exterior angles. These exterior angles form three pairs of vertical angles. $\angle R$ and $\angle T$ are called *nonadjacent interior angles* of $\angle QST$.

Theorem 4.17

4.34. The measure of an exterior angle of a triangle is greater than the measure of either of the two nonadjacent interior angles.

Given: $\triangle ABC$ with exterior $\angle CBE$.
Conclusion: $m\angle CBE > m\angle C$;
$m\angle CBE > m\angle A$.

Proof:

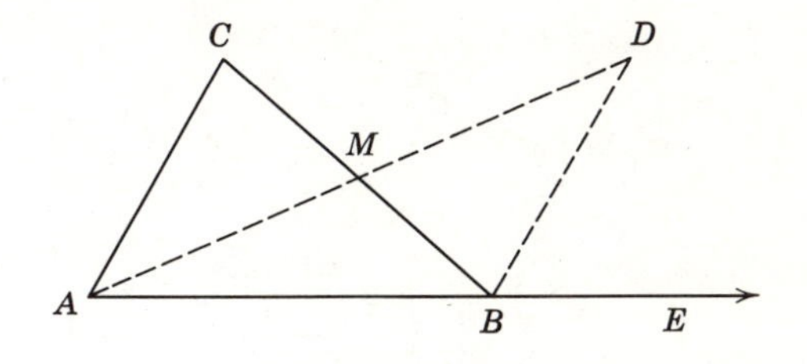

Theorem 4.17.

STATEMENTS	REASONS
1. Let M be the midpoint of $\overline{BC}$.	1. Every segment has one and only one midpoint.
2. $\overline{BM} \cong \overline{CM}$.	2. Definition of midpoint.
3. Let D be a point on the ray opposite $\overrightarrow{MA}$, such that $\overline{MD} \cong \overline{MA}$.	3. Segment construction postulate.
4. Draw $\overline{BD}$.	4. Postulate 2.
5. $\angle BMD \cong \angle CMA$.	5. Vertical angles are congruent.
6. $\triangle BMD \cong \triangle CMA$.	6. S.A.S.
7. $m\angle MBD = m\angle C$.	7. Corresponding $\angle$s of $\cong$ $\triangle$s are $\cong$.
8. $m\angle CBE = m\angle MBD + m\angle DBE$.	8. Postulate 14.
9. $m\angle CBE = m\angle C + m\angle DBE$.	9. Substitution property.
10. $m\angle CBE > m\angle C$.	10. $c = a + b \wedge b > 0 \rightarrow c > a$.

$m\angle CBE$ can be proved greater than $m\angle A$, in like manner, by taking M as the midpoint of $\overline{AB}$ and drawing $\overrightarrow{CM}$.

Theorem 4.18

4.35. If two triangles have the three sides of one congruent respectively to the three sides of the other, the triangles are congruent to each other.

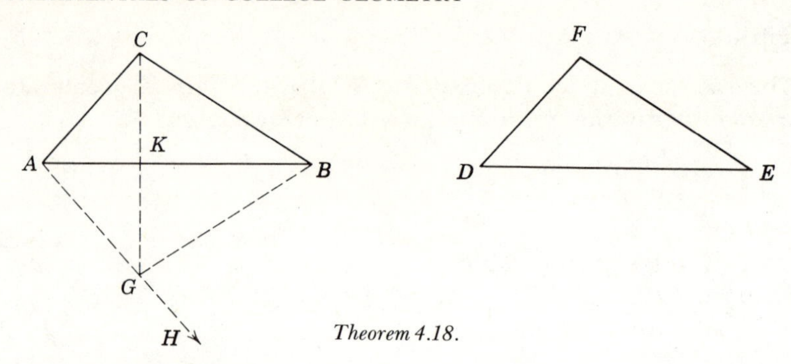

Theorem 4.18.

Given: $\triangle ABC$ and $\triangle DEF$ with
$\overline{AB} \cong \overline{DE}, \overline{BC} \cong \overline{EF}, \overline{AC} \cong \overline{DF}$.
Conclusion: $\triangle ABC \cong \triangle DEF$.

Proof:

STATEMENTS	REASONS
1. $\overline{AB} \cong \overline{DE}$.	1. Given.
2. There is a ray AH such that $\angle BAH \cong \angle EDF$, and such that C and G are on opposite sides of $\overleftrightarrow{AB}$.	2. Angle construction postulate.
3. There is a point G on $\overrightarrow{AH}$ such that $\overline{AG} \cong \overline{DF}$.	3. Point plotting postulate.
4. Draw segment BG.	4. Postulate 2.
5. $\triangle ABG \cong \triangle DEF$.	5. S.A.S.
6. $\overline{AC} \cong \overline{DF}$.	6. Given.
7. $\overline{AG} \cong \overline{AC}$.	7. Theorem 3.4.
8. $\overline{BG} \cong \overline{EF}$.	8. Corresponding parts of $\cong$ $\triangle$s are $\cong$.
9. $\overline{BC} \cong \overline{EF}$.	9. Given.
10. $\overline{BG} \cong \overline{BC}$.	10. Theorem 3.4.
11. Draw segment CG.	11. Postulate 2.
12. $\angle ACK \cong \angle AGK$.	12. Theorem 4.16.
13. $\angle BCK \cong \angle BGK$.	13. Theorem 4.16.
14. $\angle ACB \cong \angle AGB$.	14. Angle addition theorem.
15. $\angle AGB \cong \angle DFE$.	15. Reason 8.
16. $\angle ACB \cong \angle DFE$.	16. Congruence of $\angle$s is transitive.
17. $\triangle ABC \cong \triangle DEF$.	17. S.A.S.

Exercises (A)

1. *Given:* $\overline{RT} \cong \overline{RU}; \overline{TS} \cong \overline{US}$.
   Prove: (*a*) $\triangle RTS \cong \triangle RUS$.
   (*b*) $\overline{RS}$ bisects $\angle TRU$.

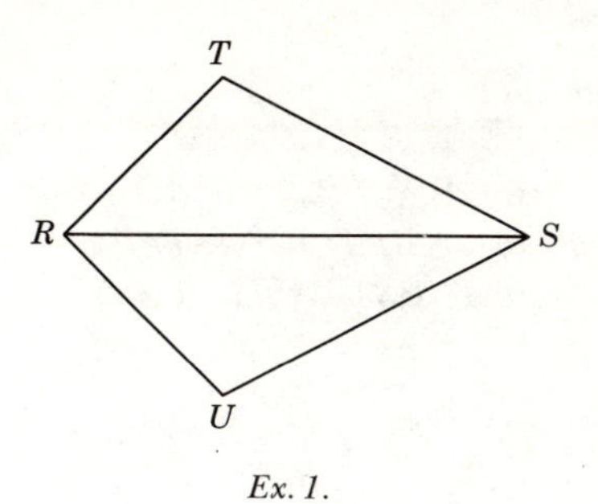

Ex. 1.

2. *Given:* Isosceles $\triangle ABC$ with $\overline{AC} \cong \overline{BC}$;
   $\overline{CD}$ bisects $\angle ACB$.
   Prove: (*a*) $\triangle ADC \cong \triangle BDC$.
   (*b*) $\overline{AD} \cong \overline{BD}$.
   (*c*) $\overline{CD} \perp \overline{AB}$.

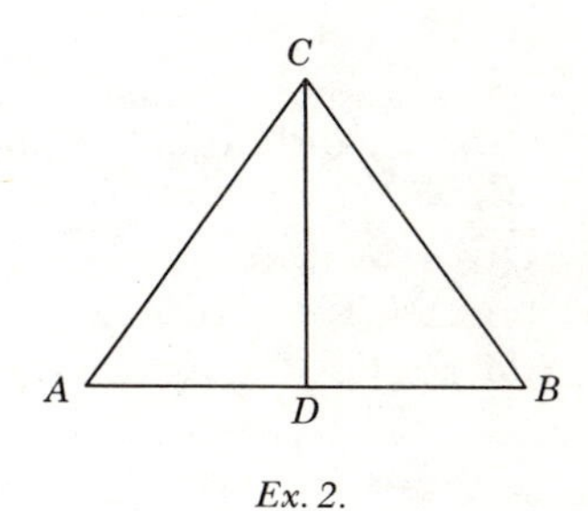

Ex. 2.

3. *Given:* $\overline{JM} \cong \overline{KL}$;
   $\overline{JL} \cong \overline{KM}$.
   Prove: (*a*) $\angle M \cong \angle L$.
   (*b*) $\angle LJK \cong \angle ?$
   (*c*) $\angle LKM \cong \angle ?$

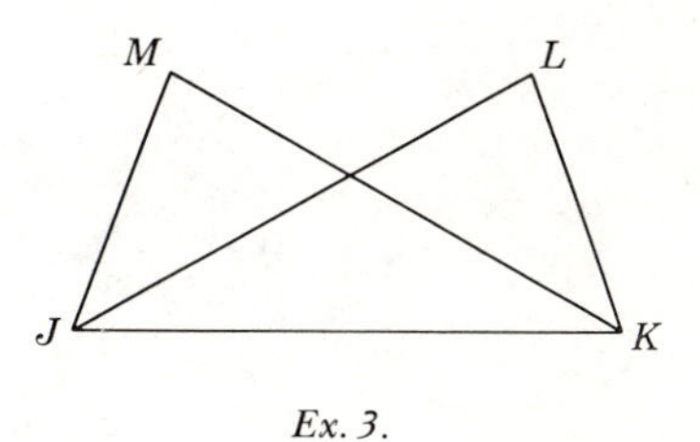

Ex. 3.

4. *Given:* $\overline{AC} \cong \overline{BC}$;
   $\overline{AD} \cong \overline{BD}$.
   Prove: $\overline{CD}$ is the perpendicular bisector of $\overline{AB}$.

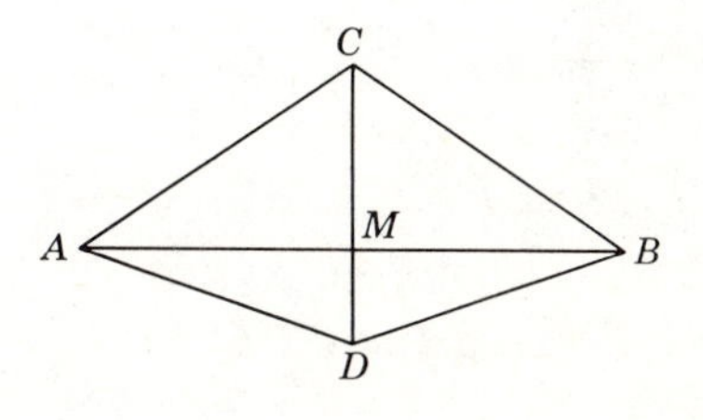

Ex. 4.

5. *Given:* $\overline{RT} \cong \overline{R'T'}$;
$\overline{RS} \cong \overline{R'S'}$;
median $TP \cong$ median $T'P'$.
Prove: (*a*) $\triangle RPT \cong \triangle R'P'T'$.
(*b*) $\triangle RST \cong \triangle R'S'T'$.

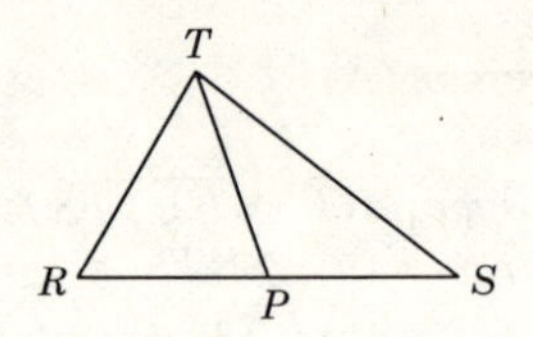

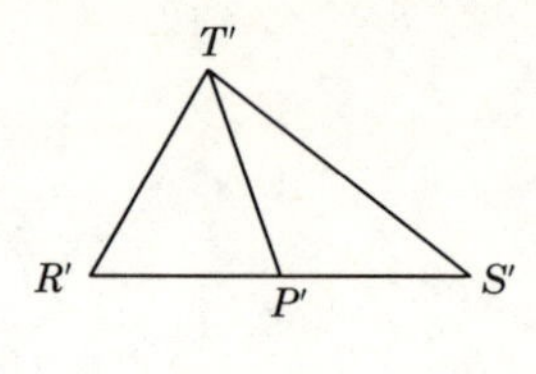

Ex. 5.

6. *Given:* AE, BD, and FG are straight lines.
$\overline{AC} \cong \overline{EC}$;
$\overline{DC} \cong \overline{BC}$.
Prove: (*a*) $\triangle ABC \cong \triangle EDC$.
(*b*) $\triangle AFC \cong \triangle EGC$.

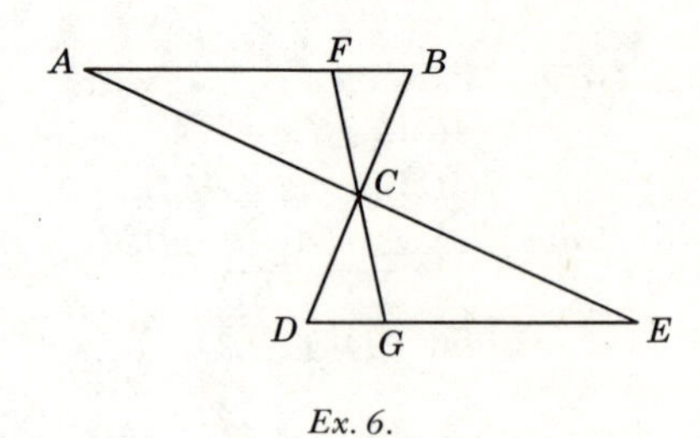

Ex. 6.

7. *Given:* Isosceles $\triangle ABC$ with $\overline{AC} \cong \overline{BC}$:
M, N, P, are midpoints of $\overline{AC}$, $\overline{BC}$, and $\overline{AB}$ respectively.
Prove: $\angle APM \cong \angle BPN$.

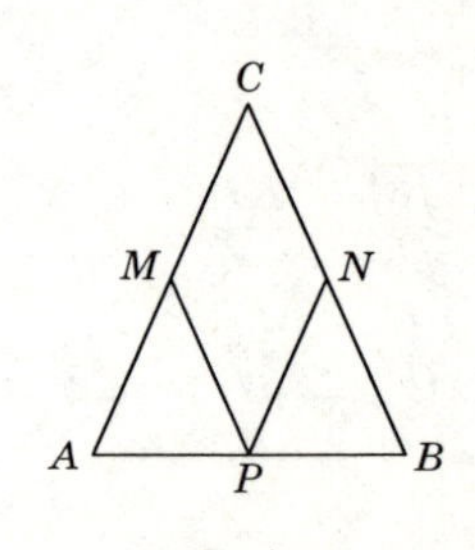

Ex. 7.

8. *Given:* $\overline{RP} \cong \overline{LP}$;
$\overline{RS} \cong \overline{LT}$;
$\overline{PS} \cong \overline{PT}$.
Prove: (*a*) $\triangle RTP \cong \triangle LSP$.
(*b*) $\angle PSR \cong \angle PTL$.

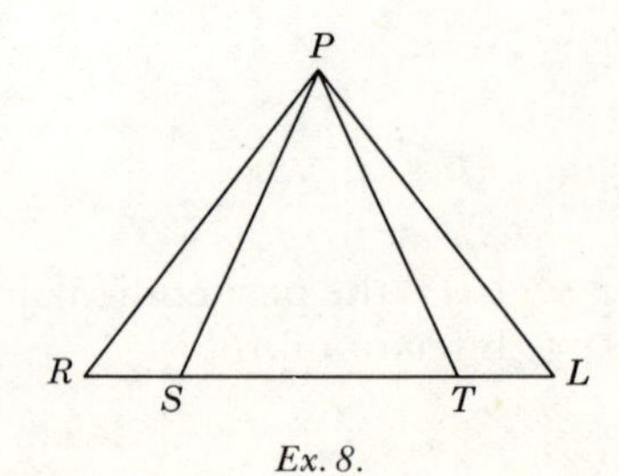

Ex. 8.

Exercises (B)

9. In the figure for Ex. 9 it is desired to determine the distance between two stations R and S on opposite sides of a building. Explain how two men with only a tape measure can accomplish the task. Prove your method.

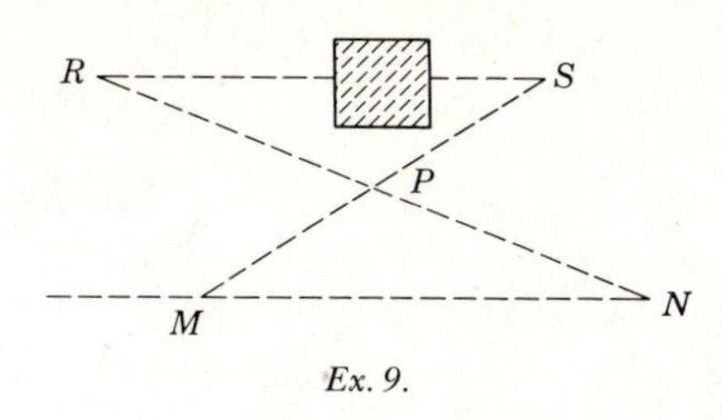

Ex. 9.

10. Describe a means of, with tape and protractor, measuring roughly the distance GP across a stream. Prove the validity of the method.

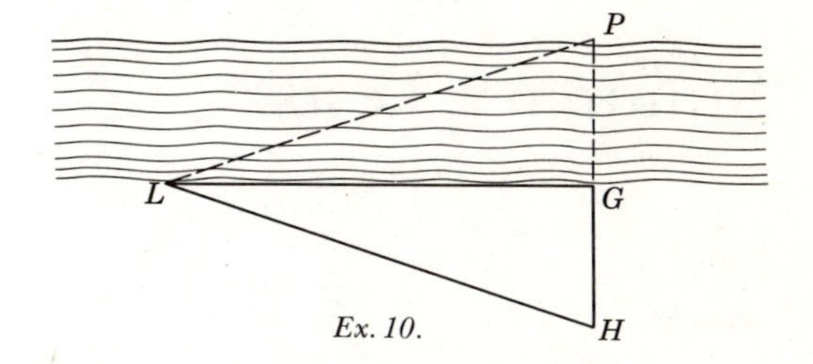

Ex. 10.

11. Prove that the median to the base of an isosceles triangle equals the altitude to that base.
12. Prove that the medians to the two congruent sides of an isosceles triangle are congruent.
13. Prove that, if the median of a triangle is also an altitude of that triangle, the triangle must be isosceles.
14. Prove that, if a point on the base of an isosceles triangle is equidistant from the midpoints of the congruent sides, the point bisects the base.
15. Prove that the intersection of the perpendicular bisectors of any two sides of a triangle are equidistant from the three vertices of the triangle.

Summary Tests

Test 1

COMPLETION STATEMENTS

1. An _______ angle of a triangle is an angle formed by one side of the triangle and the prolongation of another side through their common point.
2. A _______ of a triangle is the line segment joining a vertex and the midpoint of the opposite side of the triangle.
3. Corresponding sides of congruent triangles are found opposite the _______ angles of the triangles.
4. An _______ of a triangle is the line segment drawn from a vertex perpendicular to the opposite side.
5. _______ parts of congruent triangles are congruent.
6. The bisector of the vertex angle of an isosceles triangle is _______ to the base.
7. The _______ angles of an isosceles triangle are congruent.
8. If the median of a triangle is also an altitude, the triangle is _______ .
9. The bisectors of two supplementary adjacent angles form a _______ angle.
10. The side of a right triangle opposite the right angle is called the _______ .

Test 2

TRUE-FALSE STATEMENTS

1. Two triangles are congruent if two angles and the side of one are congruent respectively to two angles and the side of the other.
2. If two right triangles have the legs of one congruent respectively to the two legs of the other, the triangles are congruent.

3. Two triangles are congruent if two sides and an angle of one are ≅ respectively to two sides and an angle of the other.
4. Two triangles that have ≅ bases and ≅ altitudes are congruent.
5. The bisectors of two adjacent supplementary angles are perpendicular to each other.
6. The bisectors of two angles of a triangle are perpendicular to each other.
7. Two equilateral triangles are congruent if a side of one triangle is ≅ to a side of the other.
8. If the sides of one isosceles triangle are ≅ to the sides of a second isosceles triangle, the triangles are congruent.
9. The altitude of a triangle passes through the midpoint of a side.
10. The measure of the exterior angle of a triangle is greater than the measure of either of the two nonadjacent interior angles.
11. An exterior angle of a triangle is the supplement of at least one interior angle of the triangle.
12. If two triangles have their corresponding sides congruent, then the corresponding angles are congruent.
13. If two triangles have their corresponding angles congruent, then the corresponding sides are congruent.
14. No two angles of a scalene triangle are congruent.
15. The sides of triangles are lines.
16. There is possible a triangle RST in which $\angle R = \angle T$.
17. If $\triangle RST \cong \triangle STR$, then $\triangle RST$ is equilateral.
18. Adjacent angles are supplementary.
19. The supplement of an angle is always an obtuse angle.
20. A perpendicular to a line bisects the line.
21. The median to the base of an isosceles triangle is perpendicular to the base.
22. An equilateral triangle is equiangular.
23. If two angles are congruent their supplements are congruent.
24. The bisector of an angle of a triangle bisects the side opposite that angle.
25. If two isosceles triangles have the same base, the line passing through their vertices bisects the base.

Test 3

EXERCISES

1. Supply the reasons for the statements in the following proof:

Given: $\overline{AC} \cong \overline{BC}$; $\overline{AD} \cong \overline{BD}$.
Prove: $\overline{AB} \perp \overline{CD}$.
Proof:

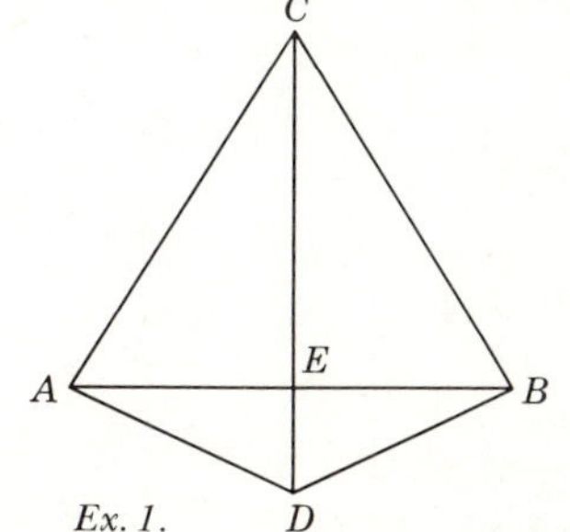

Ex. 1.

STATEMENTS	REASONS
1. $\overline{AC} \cong \overline{BC}$; $\overline{AD} \cong \overline{BD}$.	1.
2. $\angle CAE \cong \angle CBE$; $\angle DAE \cong \angle DBE$.	2.
3. $\angle DAC \cong \angle DBC$.	3.
4. $\triangle DAC \cong \triangle DBC$.	4.
5. $\angle ACE \cong \angle BCE$.	5.
6. $\overline{CE} \cong \overline{CE}$.	6.
7. $\triangle ACE \cong \triangle BCE$.	7.
8. $\angle AEC \cong \angle BEC$.	8.
9. $\therefore \overline{AB} \perp \overline{CD}$.	9.

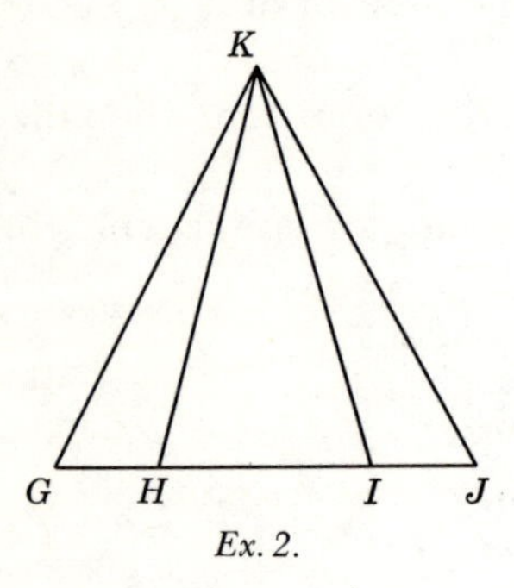

Ex. 2.

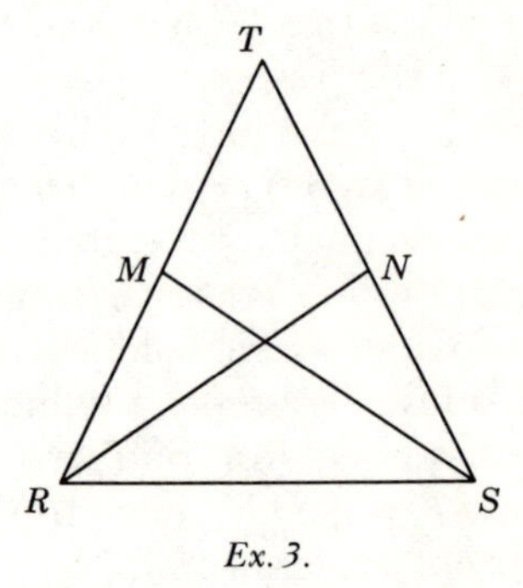

Ex. 3.

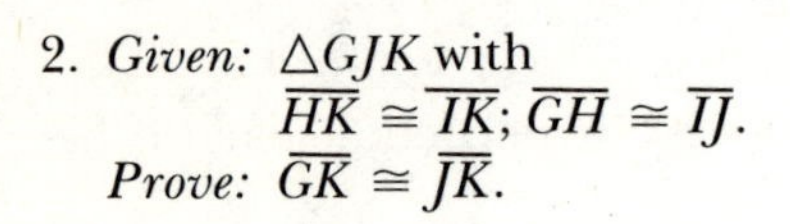

2. *Given:* $\triangle GJK$ with $\overline{HK} \cong \overline{IK}$; $\overline{GH} \cong \overline{IJ}$.
   Prove: $\overline{GK} \cong \overline{JK}$.

3. *Given:* Isosceles $\triangle RST$ with $\overline{RT} \cong \overline{ST}$; medians SM and RN.
   Prove: $\overline{SM} \cong \overline{RN}$.

5

Parallel and perpendicular lines

5.1. Parallel lines. Parallel lines are commonplace in the everyday experiences of man. Illustrations of parallel lines are the yard markers on a football field, the top and bottom edges of this page, a series of vertical fence posts, and the rails on which the trains run. (See Fig. 5.1.)

Parallel lines occur in a number of geometric figures. These lines have certain properties that produce consequences in these figures. A knowledge of these consequences is useful to the craftsman, the artisan, the architect, and the engineer.

Just as we began our study of congruent triangles with a definition of congruent triangles and with certain accepted postulates, so we will begin our study of parallel lines with a definition and a postulate. By means of this definition and postulate and the theorems already proved, we shall prove several additional theorems on parallel lines.

Definition: Two lines are *parallel* iff they lie in one plane and will not meet.

The symbol for "parallel" or "is parallel to" is "$\|$". As a matter of convenience, we will state that segments are parallel if the lines that contain them are parallel. We will similarly refer to the parallelism of two rays, a ray and a segment, a line and a segment, and so on. Thus, in Fig. 5.2, the statements $\overline{AB} \parallel \overline{DE}$, $\overrightarrow{AC} \parallel l_2$, $\overrightarrow{AC} \parallel \overline{DE}$, $l_1 \parallel \overline{DF}$, are each equivalent to the statement $l_1 \parallel l_2$.

Two straight lines in the same plane must either intersect or be parallel. However, it is possible for two straight lines not to intersect and yet not be parallel if they do not lie in the same plane. The front horizontal edge $\overleftrightarrow{DC}$ of the box of Fig. 5.3, for example, will not intersect the back vertical edge $\overleftrightarrow{HG}$ because they do not lie in the same plane. These lines are termed *skew*

Fig. 5.1. Parallel pipeways at an oil refinery

lines. Hereafter we will omit the words "in the same plane" in defining and discussing parallel lines since we are dealing essentially with plane geometry in these first chapters.

Definitions: Two planes are *parallel* if their intersection is a null set. A line and a plane are *parallel* if their intersection is a null set.

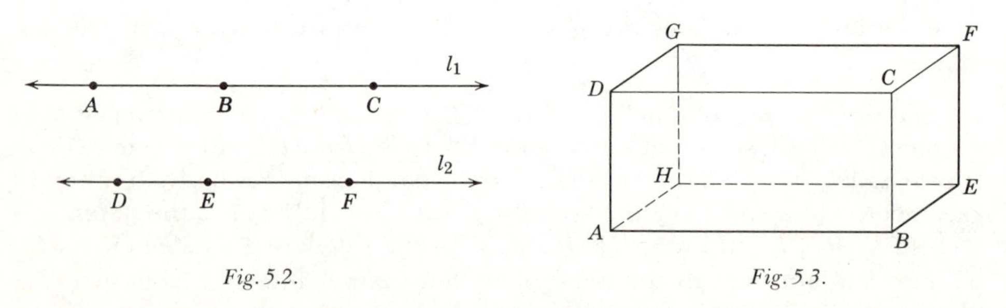

Fig. 5.2. *Fig. 5.3.*

If planes M and N (Fig. 5.4) are parallel we write $M \parallel N$. If line l_2 and plane M are parallel, we write $l_2 \parallel M$ or $M \parallel l_2$. Unless lines l_1 and l_2 of Fig. 5.4 lie in a common plane, they are called *skew lines*.

Fig. 5.4. Parallel planes.

We have already proved that two lines are perpendicular if they meet to form congruent adjacent angles. Perpendicular planes are defined in a similar way.

Definition: Two planes are *perpendicular* iff they form congruent adjacent dihedral angles. Plane M and plane N (Fig. 5.5) are perpendicular iff $\angle PQS \cong \angle PQR$.

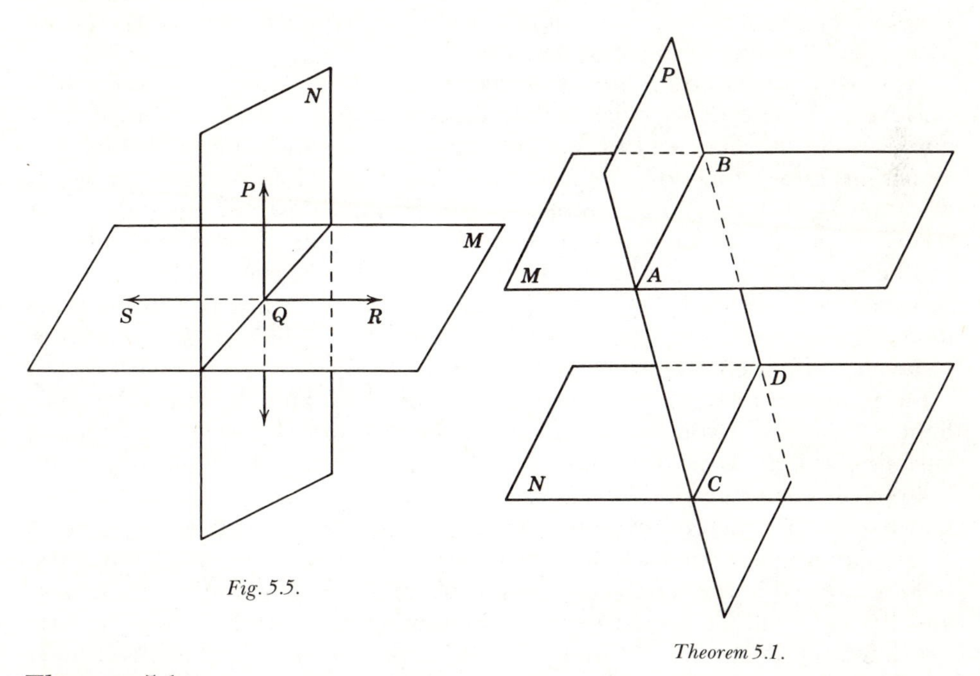

Fig. 5.5.

Theorem 5.1.

Theorem 5.1

5.2. If two parallel planes are cut by a third plane, the lines of intersection are parallel.

Given: Plane P intersecting parallel planes M and N, with $\overleftrightarrow{AB}$ and $\overleftrightarrow{CD}$ their lines of intersection.

Prove: $\overleftrightarrow{AB} \parallel \overleftrightarrow{CD}$.

Proof:

STATEMENTS	REASONS
1. Plane P intersects planes M and N in $\overleftrightarrow{AB}$ and $\overleftrightarrow{CD}$, respectively.	1. Given.
2. Plane $M \parallel$ plane N.	2. Given.
3. $\overleftrightarrow{AB}$ and $\overleftrightarrow{CD}$ lie in plane P.	3. Given.
4. $\overleftrightarrow{AB} \cap \overleftrightarrow{CD} = \emptyset$.	4. Definition of parallel planes.
5. $\overleftrightarrow{AB} \parallel \overleftrightarrow{CD}$.	5. Definition of parallel lines.

5.3. Indirect method of proof. Thus far the methods we used in proving theorems and original exercises have been direct. We have considered the information given in the problem, and, by using certain accepted truths in the form of definitions, postulates, and theorems, have developed a logical step-by-step proof of the conclusion. It has not been necessary to assume or consider one or more other conclusions.

However, not always is the information complete enough or sufficiently positive to enable us to reach a definite conclusion. Often the given facts and assumptions may lead to two or more possible conclusions. It then becomes necessary to know the exact number of possible conclusions which must be considered. Each of these conclusions must be investigated in terms of previously known facts. If all the possible conclusions but one can be shown to lead to a contradiction or violation of previously proved or accepted facts, we then can state with authority that the one remaining must be a correct conclusion. This method is called the *indirect method of proof by exclusion.* It is used extensively by all of us.

Suppose you turn on the switch to a floor lamp and the lamp does not light. How might you find the cause of the difficulty? Let us consider the various possible causes for failure. They might be: unscrewed light bulb, bulb burned out, faulty wiring in lamp, lamp unplugged in wall socket, fuse blown out, no current in your neighborhood, bad wiring in the house. Assume that in checking you find that other lights in the house will burn, the bulb is screwed in the socket, the lamp is properly plugged in the wall socket, and the bulb will light when screwed in another floor lamp. By these tests you have eliminated all but one possible cause for failure. Thus you must conclude that the failure lies in the wiring in the lamp.

A lawyer frequently uses the indirect method of proof in proving his client innocent of misconduct. Let us suppose the client is accused of armed robbery of a theater at 21st and Main street at 7:30 p.m. on a given night. It is evident that the client was either (1) at that locality at the specified time and date or (2) he was somewhere else. If the lawyer can prove that the client was at some other spot at the time of the robbery, only one conclusion can result. His client could not have been the robber.

The automobile mechanic in determining why an engine will not start must first consider the various causes for such failure. Suppose he concludes that the fault must be either (1) no gasoline reaching the cylinder or (2) no spark at the spark plug. If he can show that one of these definitely cannot be the fault, he then concludes that the other must be and acts on that basis.

The student may ask, "What if I cannot exclude all but one of the assumed possibilities?" All he can be certain of in that event is that he has no proof. It is possible that one of the alternatives he has chosen is actually true. There is no one way to determine which alternatives to select for testing in an indirect proof. Perhaps several examples will help here.

5.4. Illustrative Example 1:

Given: $m\overline{AB} = m\overline{BC}$; $m\overline{CD} \neq m\overline{AD}$.
Prove: $\overrightarrow{BD}$ does not bisect $\angle ABC$.
Proof:

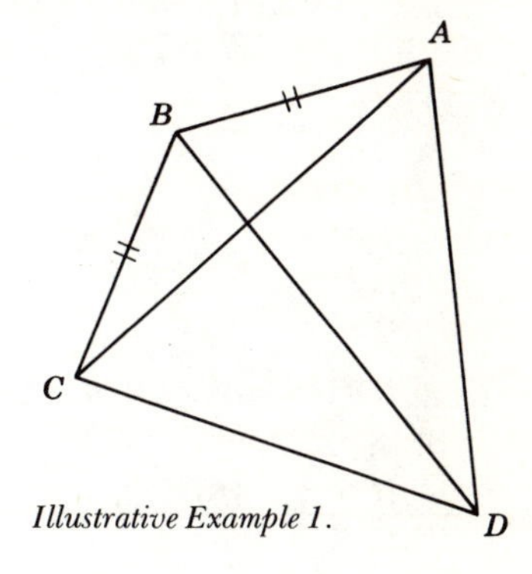

Illustrative Example 1.

STATEMENTS	REASONS
1. $m\overline{AB} = m\overline{BC}$.	1. Given.
2. $m\overline{CD} \neq m\overline{AD}$.	2. Given.
3. $\overline{AB} \cong \overline{BC}$.	3. Definition of congruent segments.
4. Either $\overrightarrow{BD}$ bisects $\angle ABC$ or $\overrightarrow{BD}$ does not bisect $\angle ABC$.	4. Law of the excluded middle.
5. Assume $\overrightarrow{BD}$ bisects $\angle ABC$.	5. Temporary assumption.
6. $\angle CBD \cong \angle ABD$.	6. Definition of angle bisector.
7. $\overline{BD} \cong \overline{BD}$.	7. Congruence of segments is reflexive.
8. $\triangle CBD \cong \triangle ABD$.	8. S.A.S.
9. $\overline{CD} \cong \overline{AD}$.	9. Corresponding parts of congruent triangles are congruent.
10. $m\overline{CD} = m\overline{AD}$.	10. Reason 3.
11. Statement 10 contradicts statement 2.	11. Statements 10 and 2.
12. Hence assumption 5 is false and $\overrightarrow{BD}$ does not bisect $\angle ABC$.	12. Rule for denying the alternative.

5.5. Illustrative Example 2:

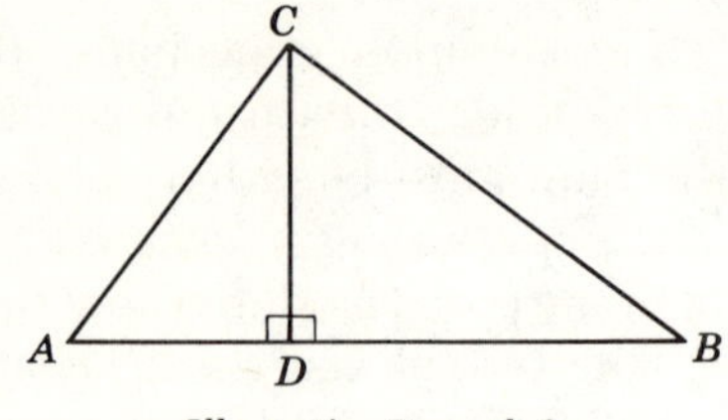

Illustrative Example 2.

Given: $\triangle ABC$ with $\overline{CD} \perp \overline{AB}$; $m\overline{AC} \neq m\overline{BC}$.
Prove: $m\overline{AD} \neq m\overline{BD}$.
Proof:

STATEMENTS	REASONS
1. $\triangle ABC$ with $\overline{CD} \perp \overline{AB}$.	1. Given.
2. $m\overline{AC} \neq m\overline{BC}$.	2. Given.
3. Either $m\overline{AD} = m\overline{BD}$ or $m\overline{AD} \neq m\overline{BD}$.	3. Law of the excluded middle.
4. Assume $m\overline{AD} = m\overline{BD}$.	4. Temporary assumption.
5. $\overline{AD} \cong \overline{BD}$.	5. Definition of congruent segments.
6. $\overline{CD} \cong \overline{CD}$.	6. Congruence of segments is reflexive.
7. $\angle ADC$ and $\angle BDC$ are right angles.	7. Perpendicular lines form right angles.
8. $\angle ADC \cong \angle BDC$.	8. Right angles are congruent.
9. $\triangle ADC \cong \triangle BDC$.	9. S.A.S.
10. $\overline{AC} \cong \overline{BC}$.	10. Corresponding parts of congruent triangles are congruent.
11. $m\overline{AC} = m\overline{BC}$.	11. Reason 5.
12. Statement 11 contradicts statement 2.	12. Statements 11 and 2.
13. Assumption 4 is false and $m\overline{AD} \neq m\overline{BD}$.	13. Rule for denying the alternative.

Exercises (A)

1. Tom, Jack, Harry, and Jim have just returned from a fishing trip in Jim's car. After Jim has taken his three friends to their homes, he discovers a bone-handled hunting knife which one of his friends has left in the car. He recalls that Tom used a fish-scaling knife to clean his fish and that Harry borrowed his knife to clean his fish. Discuss how Jim could reason whose knife was left in his car. Indicate what assumptions he would have to make to be definitely certain of his conclusion.
2. Two boys were arguing whether or not a small animal in their possession was a rat or a guinea pig. What was proved if the boys agreed that guinea pigs have no tails and the animal in question had a tail?

3. A customer returned a clock to the jeweler, claiming that the clock would not run. He offered as evidence the fact that the clock stopped at 2:17 a.m. after his butler wound it before retiring a few hours earlier. When the jeweler checked the clock he could find nothing wrong with the clock except that it was run down. Upon winding the clock it functioned properly. What conclusion would you make if you were the jeweler?
4. The story is told of Tom Jones asking permission of the local jailkeeper to see a prisoner. He was told that only relatives were permitted to see the inmate. Being a proud man, Mr. Jones did not want to admit his relationship to the prisoner. He stated, "Brothers and sisters have I none, but that man's father is my father's son." Whereupon the jailer permitted him to see the prisoner.

   Consider the following possible relationships between the prisoner and Mr. Jones: cousin, uncle, father, grandfather, grandson, son, brother. By indirect reasoning determine the true relationship between the prisoner and Mr. Jones.
5. Give an example either from your own experience or a hypothetical case in which the indirect method of proof was used.

Exercises (B)

Prove the following statements by assuming that the conclusion is not true and then show that this assumption leads to an impossible result.

6. If the measures of two angles of a triangle are unequal, the measures of sides opposite them are unequal.

7. *Given:* $m\overline{AC}$ does not equal $m\overline{BC}$; $\overline{CD}$ bisects $\angle ACB$.
   Conclusion: $\overline{CD}$ cannot be perpendicular to $\overline{AB}$.

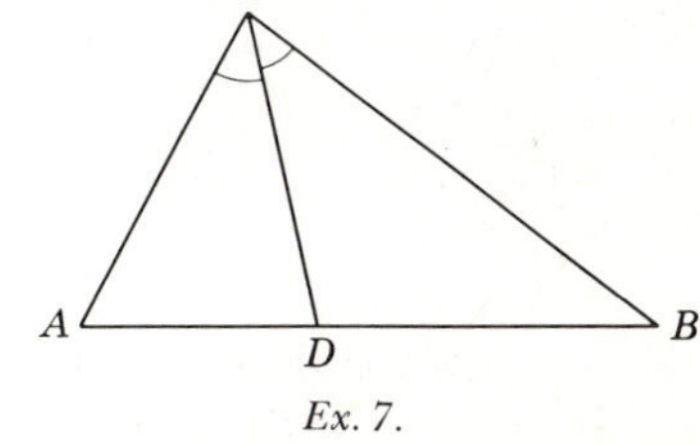

Ex. 7.

8. *Given:* $m\overline{RT}$ is not equal to $m\overline{ST}$; M bisects $\overline{RS}$.
   Conclusion: $\overline{TM}$ is not perpendicular to $\overline{RS}$.

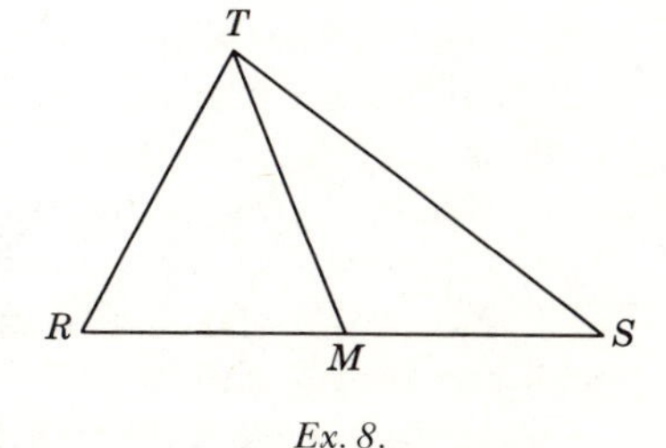

Ex. 8.

9. *Given:* $m\overline{LN} \neq m\overline{MN}$;
$\overline{PN}$ is $\perp$ $\overline{LM}$.
Conclusion: $\overline{PN}$ does not bisect $\angle LNM$.

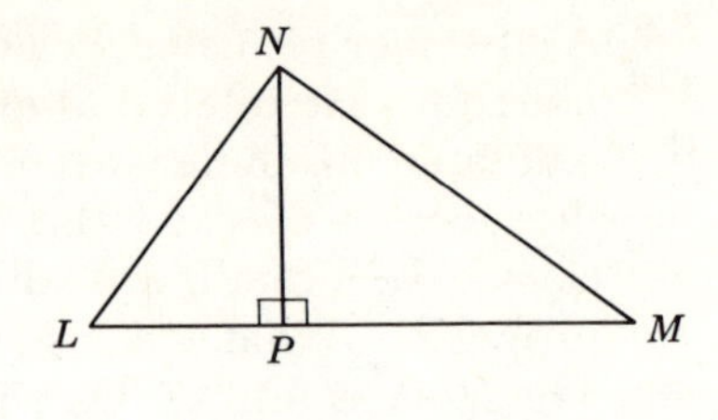

Ex. 9.

10. *Given:* $\triangle ABC$ and $A'B'C'$ with
$m\overline{AB} = m\overline{A'B'}$, $m\overline{AC} = m\overline{A'C'}$,
$m\angle A \neq m\angle A'$.
Conclusion: $m\overline{BC} \neq m\overline{B'C'}$.

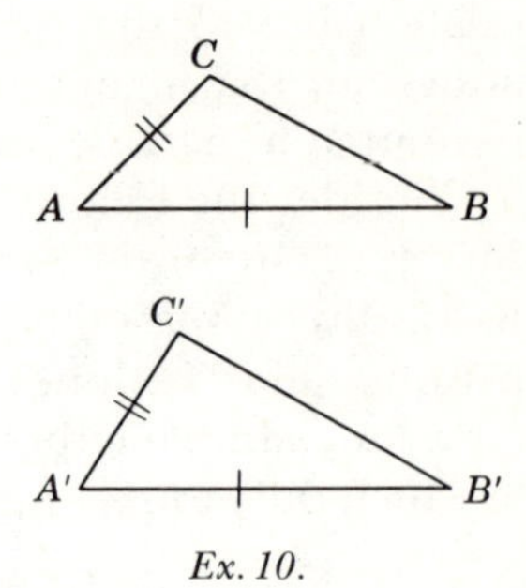

Ex. 10.

11. What conclusions can you draw if the following statements are true?
    1. $p \to q$.
    2. $q \to w$.
    3. p is true.
12. What conclusions can be drawn if the following four statements are true?
    1. $p \to q$.
    2. $w \to p$.
    3. $v \to w$.
    4. q is false.
13. Given the following three true statements.
    1. If x is α, then y is β.
    2. If x is γ, then y is δ.
    3. If y is β, then z is ψ.
       (*a*) *Complete:* x is α; then y is ____ and z is ____.
       (*b*) Can you draw any conclusions about x if you know y is δ?
14. What conclusions can be drawn from the following true statements?
    1. No one can join the bridge club unless he can play bridge.
    2. No lobster can play bridge.
    3. No one is allowed to talk at the bridge table unless he is a member of the bridge club.
    4. I always talk at the bridge table.

5.6. Properties of existence and uniqueness. The definition of parallel lines furnishes us with an impractical, if not impossible, direct method of determining whether two lines are parallel. We must resort to the indirect method. But first we must prove two basic theorems about perpendicular lines. These proofs involve the properties of *existence* and *uniqueness*.

We have asserted the idea of existence and uniqueness in several postulates and theorems in previous chapters. The student may refresh his memory on this by referring to postulates 2, 3, 5, 15, 16, and Theorems 3.1, 3.2, 3.3. The expressions "exactly one" or "one and only one" mean two things:

(1) There is *at least one* of the things being discussed.
(2) There is *at most one* of the things being discussed.

Statement (1) alone leaves open the possibility that there may be more than one such thing. Statement (2) leaves open the possibility that there are none of the things being discussed. Together, statements (1) and (2) assert there is exactly one thing having the given properties being discussed.

Theorem 5.2

5.7. In a given plane, through any point of a straight line there can pass one and only one line perpendicular to the given line.

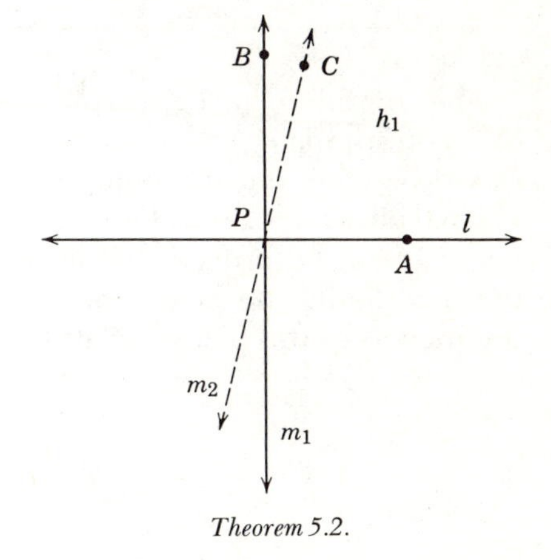

Theorem 5.2.

Given: Line l, point P of l.

Conclusion:
1. There is a line $m_1 \perp l$ such that $P \in m_1$ (existence).
2. There is at most one line $m_1 \perp l$ such that $P \in m_1$ (uniqueness).

Proof:

STATEMENTS	REASONS
Proof of Existence:	
Let A be a point on l.	
1. There is a point B in half-plane h_1 such that $m\angle APB = 90$.	1. Angle construction postulate.
2. Then $\overleftrightarrow{PB}$ (or m_1) is $\perp l$.	2. Definition of $\perp$ lines.

Proof of Uniqueness:

3. Either there is more than one line through P and $\perp$ to l or there is not more than one line through P and $\perp$ to l.	3. Law of excluded middle.
4. Suppose there is a second line $m_2 \perp$ to l at P. Let C be a point on m_2 and in the half-plane h_1.	4. Temporary assumption.
5. $m\angle APB = 90$.	5. Statement 1.
6. $m\angle APC = 90$.	6. Perpendicular lines form right ∠s, the measure of a right angle = 90.
7. This is impossible.	7. Statements 5 and 6 contradict the angle construction postulate.
8. This contradiction means that our Assumption 4 is false. Hence, there is only one line, satisfying the conditions of the theorem.	8. Rule for denying the alternative.

The condition "in a given plane" is an essential part of Theorem 5.2. If we did not stipulate "in a given plane," the existence part of the theorem would be true, but the uniqueness of the perpendicular would not be true. Fig. 5.6 illustrates several lines perpendicular to a line l through a point of the line. It can be proved that all perpendiculars to a line through a point on that line lie in one plane and that plane is perpendicular to the line. The uniqueness of this plane can also be proved.

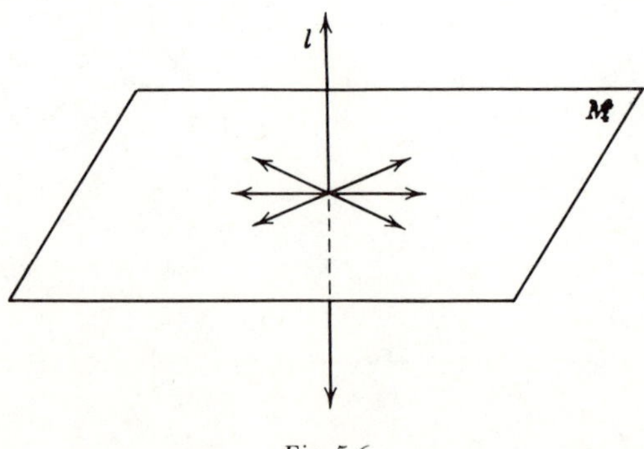

Fig. 5.6.

Definition: A line which intersects a plane in exactly one point but is not perpendicular to the plane is said to be *oblique* to the plane.

Theorem 5.3

5.8. Through a point not on a given line there is at least one line perpendicular to that given line.

Given: Line l. Point P not contained in l.

Conclusion: At least one line can contain P and be perpendicular to l.

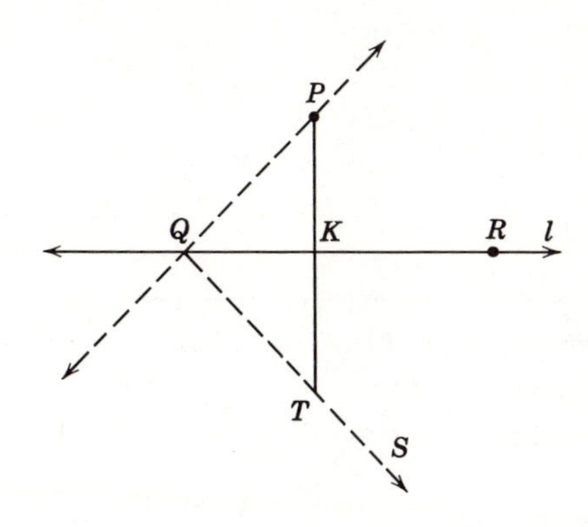

Theorem 5.3.

Proof:

STATEMENTS	REASONS
1. Let Q and R be any two points of l. Draw $\overleftrightarrow{PQ}$.	1. Postulate 2.
2. $\angle PQR$ is formed.	2. Definition of angle.
3. In the half-plane of l not containing P there is a ray, $\overrightarrow{QS}$, such that $\angle RQS \cong \angle RQP$.	3. Angle construction postulate.
4. There is a point T on $\overrightarrow{QS}$ such that $\overline{QT} \cong \overline{QP}$.	4. Point plotting postulate.
5. Draw $\overline{PT}$.	5. Postulate 2.
6. $\overline{QK} \cong \overline{QK}$.	6. Reflexive property of congruence.
7. $\triangle PKQ \cong \triangle TKQ$.	7. S.A.S.
8. $\angle PKQ \cong \angle TKQ$.	8. Corresponding ∠s of ≅ △s are ≅.
9. $\overleftrightarrow{PT} \perp l$.	9. Theorem 3.14.

Theorem 5.4

5.9. Through a given external point there is at most one perpendicular to a given line.

Given: Line l, point P not contained in l.

Conclusion: No more than one line can contain P and be $\perp$ to l.

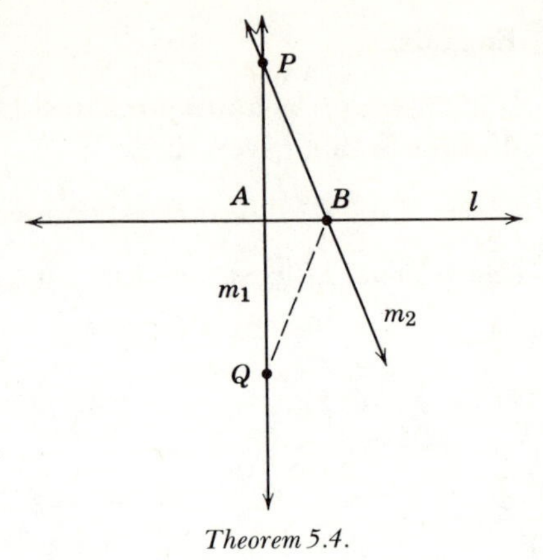

Theorem 5.4.

Proof:

STATEMENTS	REASONS
1. Either there is more than one line through $P \perp$ to l or there is not more than one line through $P \perp$ to l.	1. Law of excluded middle.
2. Assume m_1 and m_2 are two such lines and intersecting l at A and B respectively.	2. Temporary assumption.
3. On the ray opposite $\overrightarrow{AP}$ construct $\overline{AQ}$ such that $\overline{AQ} \cong \overline{AP}$.	3. Point plotting postulate.
4. Draw $\overline{QB}$.	4. Postulate 2.
5. $\overline{AB} \cong \overline{AB}$.	5. Congruence of segments is reflexive.
6. $\angle PAB$ and $\angle QAB$ are right $\angle$s.	6. Theorem 3.13.
7. $\angle PAB \cong \angle QAB$.	7. Theorem 3.7.
8. $\triangle PAB \cong \triangle QAB$.	8. S.A.S.
9. $\angle QBA \cong \angle PBA$.	9. Corresponding $\angle$s of $\cong$ $\triangle$s are $\cong$.
10. $\angle PBA$ is a right angle.	10. Perpendicular lines form right angles.
11. $\angle QBA$ is a right angle.	11. Substitution property.
12. $\overleftrightarrow{BQ} \perp l$.	12. Definition of perpendicular lines.
13. Statements 12 and 2 contradict Theorem 5.2.	13. Statements 12, 2, and Theorem 5.2.
14. Assumption 2 must be false; then there is at most one perpendicular from P to l.	14. Rule for denying the alternative.

Exercises

In each of the following indicate whether a statement is *always true* or *not always true*.

1. A triangle determines a plane.
2. Two perpendicular lines determine a plane.
3. Two planes either intersect or are parallel.
4. In space there is one and only one line through a point P on line l that is perpendicular to l.
5. More than one line in space can be drawn from a point not on line l perpendicular to l.
6. If line $l \in$ plane M, line $q \in$ plane M, $l \cap q = P$, line $r \perp$ line l at P, then line $r \perp$ plane M.
7. If line $l \in$ plane M, line $q \in$ plane M, $l \cap q = P$, line $r \perp l$, $r \perp q$, then $r \perp M$.
8. If line $l \perp$ plane M, then plane $M \perp$ line l.
9. If $P \in$ plane M, then only one line containing P can be perpendicular to M.
10. *If* $P \in$ plane M, $\angle APB$ and $\angle CPB$ are right angles, and $m\angle APC = 91$, then $PB \perp M$.

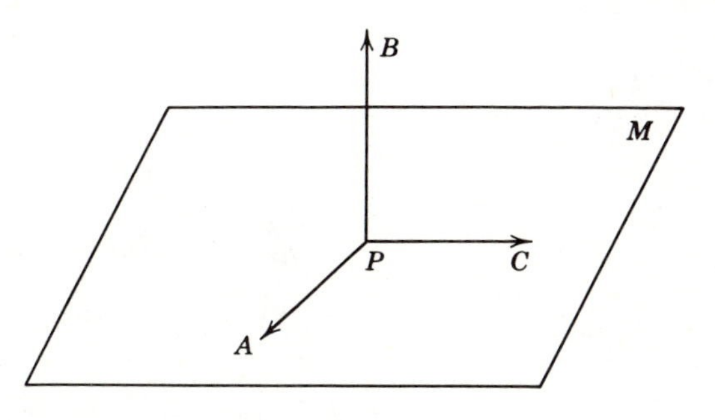

Ex. 10.

11. Any number of lines can be drawn perpendicular to a given line from a point not on the line.
12. When we prove the existence of some thing, we prove that there is exactly one object of a certain kind.

13–18. In the figure (Exs. 13–18), plane $M \perp$ plane N, plane $M \cap$ plane $N = \overleftrightarrow{RS}$, $\overleftrightarrow{AC}$ lies in M, $\overleftrightarrow{BD}$ lies in N, $\overleftrightarrow{AC} \cap \overleftrightarrow{BD} = P$.

13. $\overleftrightarrow{RS} \perp \overleftrightarrow{AC}$.
14. $\overleftrightarrow{BD} \perp \overleftrightarrow{AC}$.
15. $m\angle APB = 90$.
16. $m\angle CPD = 90$.

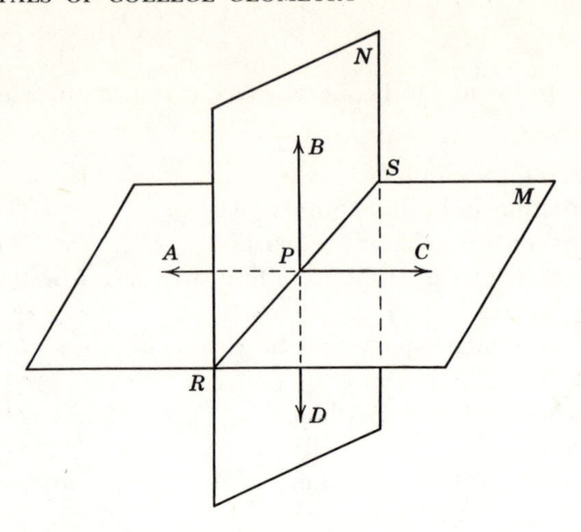

Exs. 13–18.

17. $m\angle CPD$ is the measure of a dihedral angle.
18. $\angle BPS$ and $\angle BPR$ are adjacent dihedral angles.

19–24. In the figure (Exs. 19–24), A, B, C, D are points in plane M; $\overleftrightarrow{PQ} \cap M = C$.

19. $\overleftrightarrow{PC}$ and $\overleftrightarrow{CB}$ determine a unique plane.
20. $\overleftrightarrow{PD}$ and $\overleftrightarrow{DQ}$ determine a unique plane.

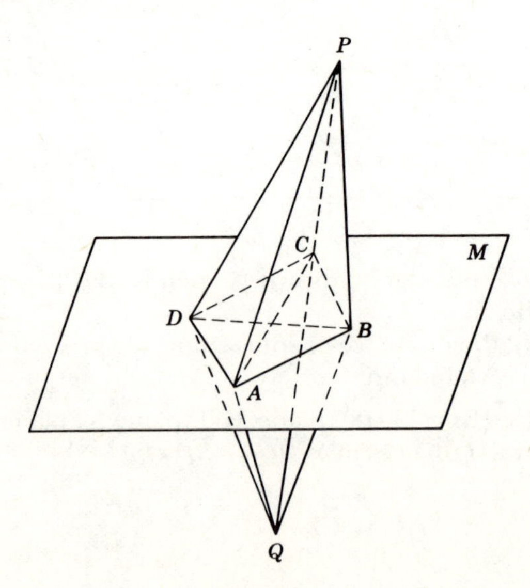

Exs. 19–24.

21. If $\overline{PA} \cong \overline{QA}$ and $\overline{PC} \cong \overline{QC}$, then $\overline{AC}$ is the perpendicular bisector of $\overline{PQ}$.
22. If $\overline{PA} \cong \overline{QA}$ and $\overline{PA} \cong \overline{BA}$, then $\overline{DA} \cong \overline{BA}$.
23. If $\triangle DBQ \cong \triangle DBP$, then $\overline{PA} \cong \overline{QA}$.
24. If $\overline{DC}$ is the perpendicular bisector of $\overline{PQ}$, then $\overline{PD} \cong \overline{QD}$.

Theorem 5.5

5.10. If two lines in a plane are perpendicular to the same line, they are parallel to each other.

Given: m and n are coplanar, $m \perp l, n \perp l$;
Conclusion: $m \parallel n$.

Proof:

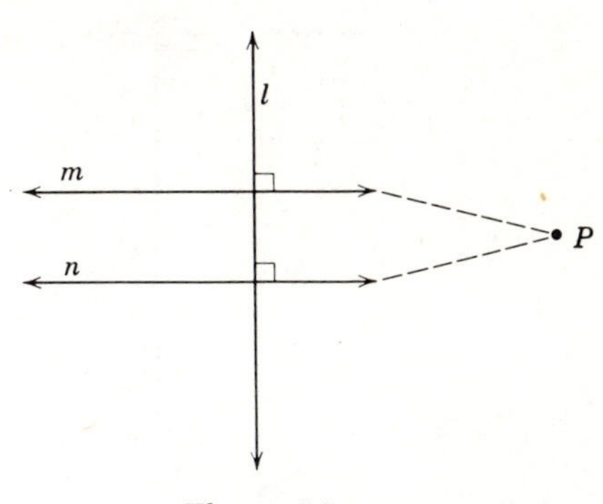

Theorem 5.5.

STATEMENTS	REASONS
1. *m and n are coplanar,* $m \perp l, n \perp l$.	1. Given.
2. Either $m \parallel n$ or $m \nparallel n$.	2. Law of excluded middle.
3. Assume $m \nparallel n$.	3. Temporary assumption.
4. m and n must meet, say at P.	4. Nonparallel coplanar lines intersect.
5. Then m and n are two lines passing through an external point and $\perp$ to the same line.	5. Statements 1 and 3.
6. Statement 5 contradicts Theorem 5.4.	6. Statements 5 and Theorem 5.4.
$m \parallel n$ is the only possible conclusion remaining.	Either p or not-p; not (not-p) $\leftrightarrow p$.

Theorem 5.6

5.11. Two planes perpendicular to the same line are parallel.
Given: Plane $M \perp$ line l; plane $N \perp$ line l.
Conclusion: Plane $M \parallel$ plane N.

(This theorem is proved by the indirect method of proof and is left as an exercise for the student.)

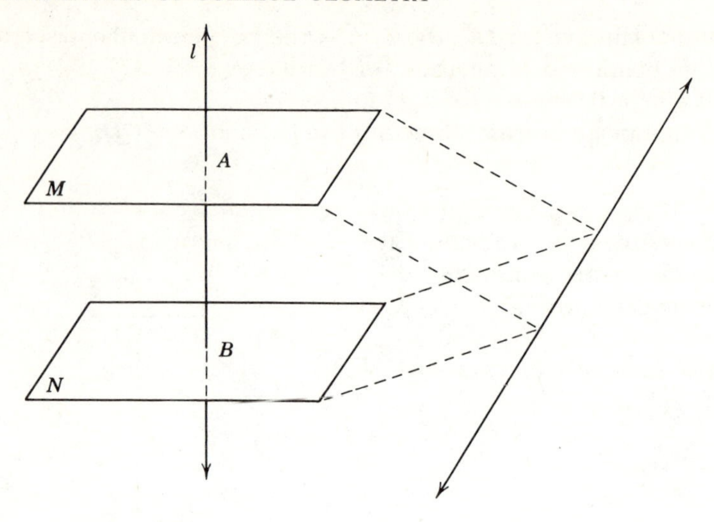

Theorem 5.6.

Theorem 5.7

5.12. In a plane containing a line and a point not on the line, there is at least one line parallel to the given line.

Given: Line l with point P not contained in l.

Conclusion: In the plane of P and l there is at least one line l_1 that can be drawn through P and parallel to l.

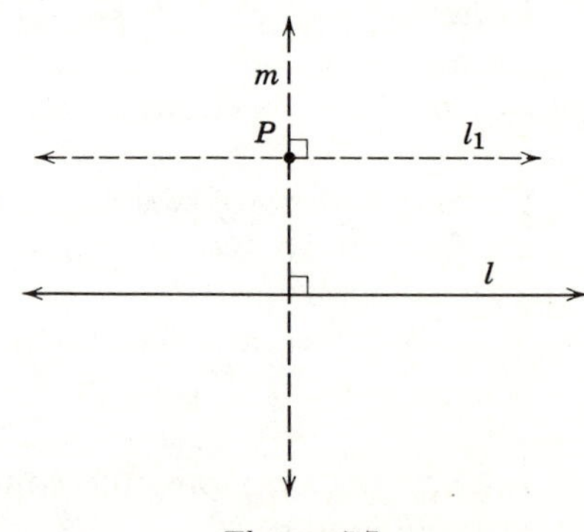

Theorem 5.7.

Proof:

STATEMENTS	REASONS
1. P is a point not on line l.	1. Given.
2. Let m be a line through P and $\perp$ to l.	2. Theorem 5.3.
3. Let l_1 be a line through P in the plane of l and P and $\perp$ to m.	3. Theorem 5.2.
4. $l_1 \parallel l$.	4. Theorem 5.5.

5.13. The parallel postulate. Having proved the existence of a line through an external point and parallel to a second line, it would seem that the next step would logically be to prove its uniqueness. Strange as it may seem at first, this cannot be done if we are to use only the postulates we have stated thus far. We must assume this uniqueness as a postulate.

Postulate 18 (the parallel postulate or Playfair's postulate).* *Through a given point not on a given line there is at most one line which can be drawn parallel to the given line.*

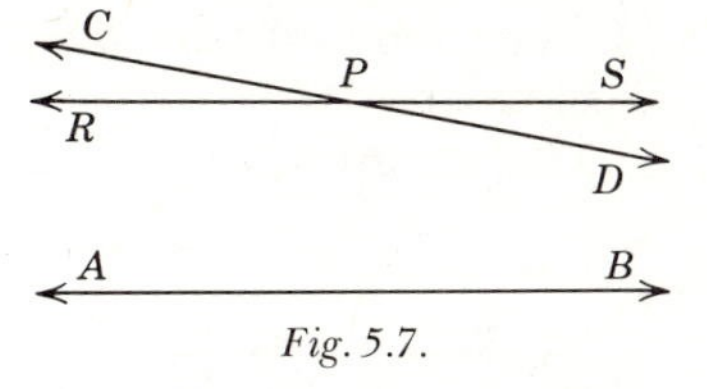

Fig. 5.7.

Thus in Fig. 5.7, if in a plane we know that $\overleftrightarrow{RS}$ is parallel to $\overleftrightarrow{AB}$ and passes through P, we must assume that, if $\overleftrightarrow{CD}$ passes through P, $\overleftrightarrow{CD}$ cannot be parallel to $\overleftrightarrow{AB}$; or, on the other hand, if $\overleftrightarrow{CD}$ is parallel to $\overleftrightarrow{AB}$, $\overleftrightarrow{CD}$ cannot pass through P.

Postulate 18 was assumed by Euclid. Since that time many mathematicians have tried to prove or disprove this postulate by means of other postulates and axioms. Each effort met with failure. As a consequence, mathematicians have considered what kind of geometry would result if this property were not assumed true, and several geometries different from the one which we are studying have been developed. Such a geometry is known as *non-Euclidean geometry.*

During the nineteenth century Nicholas Lobachevsky (1793–1856), a Russian mathematician, developed a new geometry based upon the postulate that through a given point there can be any number of lines parallel to a given line.

In 1854 a still different non-Euclidean geometry was developed by Bernhard Rieman (1826–1866), a German mathematician, who based his development on the assumption that all lines must intersect. A geometry somewhat different from any of these was used by Albert Einstein (1879–1955) in developing his Theory of Relativity.

These geometries are quite complex. Euclidean geometry is much simpler and serves adequately for solving the common problems of the surveyor, the contractor, and the structural engineer.

Theorem 5.8

5.14. Two lines parallel to the same line are parallel to each other.

Given: $l \parallel n$, $m \parallel n$.

Conclusion: $l \parallel m$.

Proof:

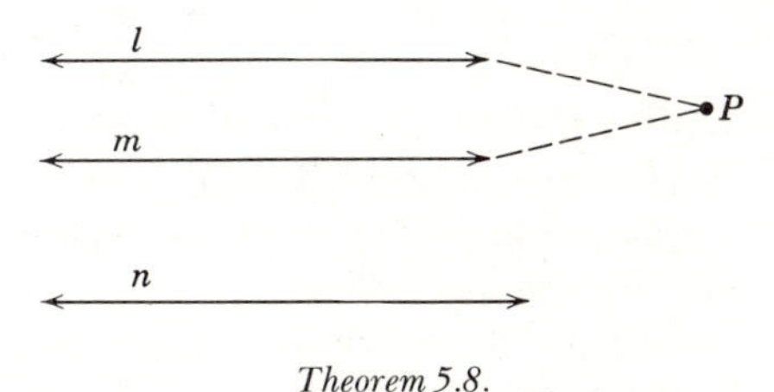

Theorem 5.8.

*This statement is attributed to John Playfair (1748–1819), brilliant Scottish physicist and mathematician.

STATEMENTS	REASONS
1. $l \parallel n, m \parallel n$.	1. Given.
2. Either $l \parallel m$ or $l \nparallel m$.	2. Law of the excluded middle.
3. Assume $l \nparallel m$.	3. Temporary assumption.
4. Then l and m meet at, say P.	4. Two nonparallel lines lying in the same plane intersect.
5. Then l and m pass through the same point and are parallel to the same line.	5. Statements 1 and 3.
6. This is impossible.	6. Postulate 18.
7. $\therefore l \parallel m$.	7. Rule for denying the alternative. Either p or not-p; not(not-p) $\leftrightarrow$ p.

Theorem 5.9

5.15. In a plane containing two parallel lines, if a line is perpendicular to one of the two parallel lines it is perpendicular to the other also.

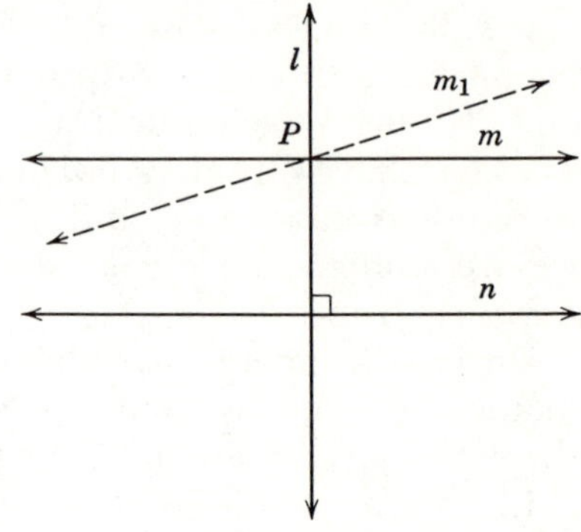

Theorem 5.9.

Given: $m \parallel n$, l in the plane of m and n, $l \perp n$.
Conclusion: $l \perp m$.

Proof:

STATEMENTS	REASONS
1. $m \parallel n$; l lies in plane of m and n.	1. Given.
2. $l \perp n$(or $n \perp l$).	2. Given. (Definitions are reversible.)
3. Either $l \perp m$ or l is not $\perp$ to m.	3. Law of excluded middle.
4. Assume l is not $\perp$ to m (or m is not $\perp$ to l).	4. Temporary assumption.
5. Then there is a line m_1 in the plane of m and n that is $\perp$ to l at the point P where m intersects l.	5. Theorem 5.2.
6. Then $m_1 \parallel n$.	6. Theorem 5.5.
7. This is impossible.	7. Postulate 18.
8. $\therefore l \perp m$.	8. Rule for denying the alternative. Either p or not-p; not(not-p) $\leftrightarrow$ p.

Theorem 5.10

5.16. A line perpendicular to one of two parallel planes is perpendicular to the other.

Given: Plane M is parallel to plane N; $\overleftrightarrow{AB}$ is perpendicular to plane M.
Conclusion: $\overleftrightarrow{AB}$ is perpendicular to plane N.

Proof:

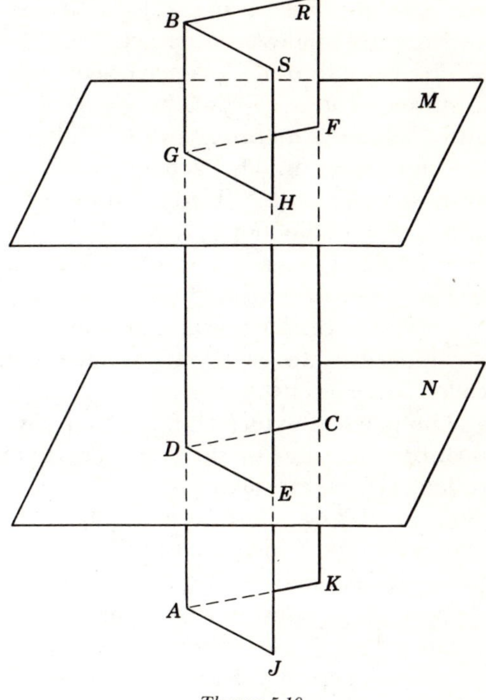

Theorem 5.10.

STATEMENTS	REASONS
1. Plane $M \parallel$ plane N; $\overleftrightarrow{AB} \perp$ plane M.	1. Given.
2. Through $\overleftrightarrow{AB}$ pass plane R intersecting planes M and N in $\overleftrightarrow{GF}$ and $\overleftrightarrow{DC}$, respectively; also through $\overleftrightarrow{AB}$ pass plane S intersecting planes M and N in $\overleftrightarrow{GH}$ and $\overleftrightarrow{DE}$, respectively.	2. Postulate 5.
3. $\overleftrightarrow{GF} \parallel \overleftrightarrow{DC}$ and $\overleftrightarrow{GH} \parallel \overleftrightarrow{DE}$.	3. Theorem 5.1.
4. $\overleftrightarrow{AB} \perp \overleftrightarrow{GF}$; $\overleftrightarrow{AB} \perp \overleftrightarrow{GH}$.	4. Definition of perpendicular to plane.
5. $\overleftrightarrow{AB} \perp \overleftrightarrow{DC}$; $\overleftrightarrow{AB} \perp \overleftrightarrow{DE}$.	5. Theorem 5.9.
6. $\overleftrightarrow{AB} \perp$ plane N.	6. Reason 4.

5.17. Transversals and special angles. A *transversal* is a line which intersects two or more straight lines. In Fig. 5.8, t is a transversal of lines l and m. When two straight lines are cut by a transversal, eight angles are formed.

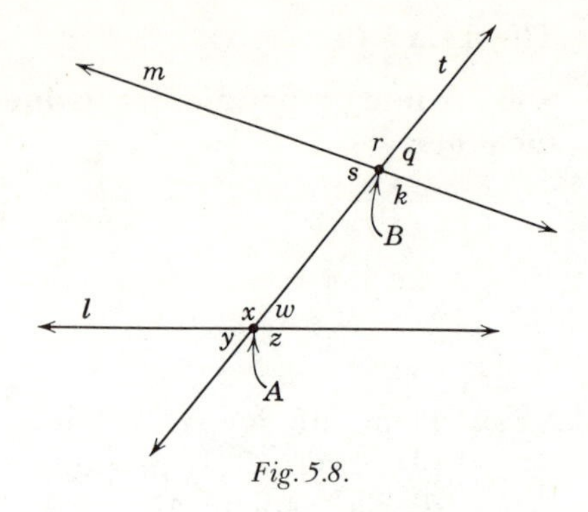

Fig. 5.8.

There are four angles each of which is a subset of $t \cup l$. Two of these ($\angle x$ and $\angle w$) contain both A and B. There are also four angles each of which is a subset of $t \cup m$. Two of these ($\angle s$ and $\angle k$) contain both A and B. The four angles ($\angle x$, $\angle w$, $\angle s$, $\angle k$) which contain both A and B are called *interior angles*. The other four ($\angle y$, $\angle z$, $\angle r$, $\angle q$) are called *exterior angles*.

The pairs of interior angles that have different vertices and contain points on opposite sides of the transversal (such as $\angle s$ and $\angle w$ or $\angle x$ and $\angle k$) are called *alternate interior angles*.

The pairs of exterior angles that have different vertices and contain points on opposite sides of the transversal (such as $\angle r$ and $\angle z$ or $\angle q$ and $\angle y$) are called *alternate exterior angles*.

Corresponding angles are a pair consisting of an interior angle and an exterior angle which have different vertices and lie in the same closed half-plane determined by the transversal. Examples of corresponding angles are $\angle q$ and $\angle w$. There are four pairs of corresponding angles in Fig. 5.8.

Since we will use the term "transversal" only when the lines lie in one plane, we will not repeat this fact in each of the following theorems.

Theorem 5.11

5.18 If two straight lines form congruent alternate interior angles when they are cut by a transversal, they are parallel.

Given: Lines l and m cut by transversal t at R and S; $\angle\alpha \cong \angle\beta$.
Conclusion: $l \parallel m$.

Proof:

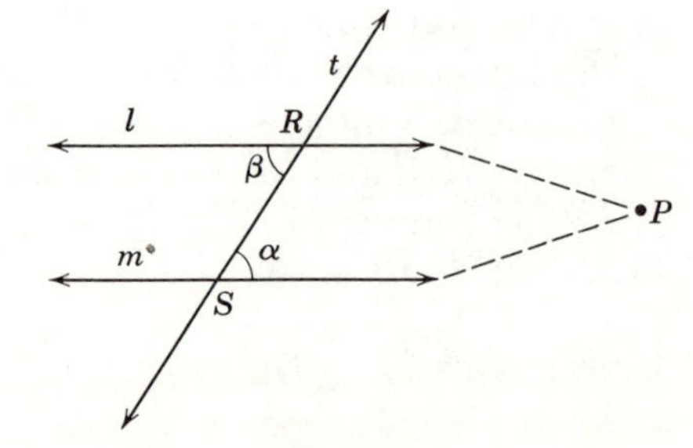

Theorem 5.11.

STATEMENTS	REASONS
1. $\angle\alpha \cong \angle\beta$.	1. Given.
2. Either $l \parallel m$ or $l \nparallel m$.	2. Law of the excluded middle.
3. Assume $l \nparallel m$.	3. Temporary assumption.
4. Then l must meet m, say, at P.	4. Nonparallel lines in a plane must intersect.
5. $\triangle RSP$ is formed.	5. Definition of a triangle.
6. $\angle\beta$ is an exterior $\angle$ of $\triangle RSP$.	6. Definition of exterior $\angle$ of a $\triangle$.
7. $\angle\beta > \angle\alpha$.	7. Theorem 4.17.
8, This is impossible.	8. Statements 1 and 7 conflict.
9. $\therefore l \parallel m$.	9. Rule for denying the alternative. not(not-p) $\leftrightarrow$ p.

Theorem 5.12

5.19. If two straight lines are cut by a transversal so as to form a pair of congruent corresponding angles, the lines are parallel.

Given: Lines l and m cut by transversal t; $\angle\alpha \cong \angle\gamma$.
Conclusion: $l \parallel m$.
Proof:

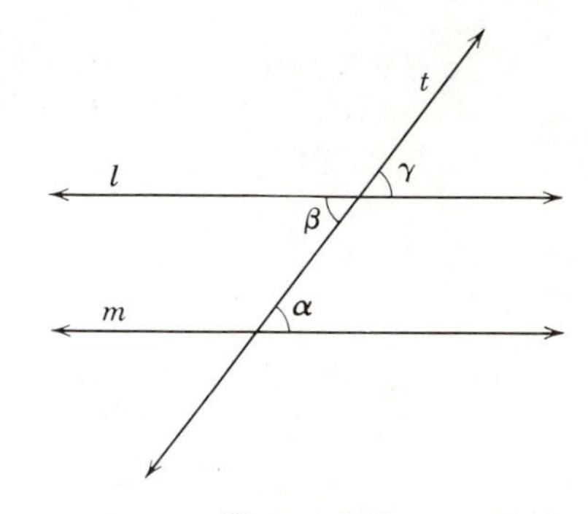

Theorem 5.12.

STATEMENTS	REASONS
1. $\angle\alpha \cong \angle\gamma$.	1. Given.
2. $\angle\gamma \cong \angle\beta$.	2. Vertical angles are $\cong$.
3. $\angle\alpha \cong \angle\beta$.	3. Congruence of angles is transitive.
4. $\therefore l \parallel m$.	4. Theorem 5.11.

5.20. Corollary: If two lines are cut by a transversal so as to form interior supplementary angles in the same closed half-plane of the transversal, the lines are parallel. (The proof of this corollary is left to the student.)

Exercises

1. In the following figure list the pairs of angles of each of the following types.
   (*a*) Alternate interior angles.
   (*b*) Alternate exterior angles.
   (*c*) Corresponding angles.
   (*d*) Vertical angles.
   (*e*) Adjacent angles.

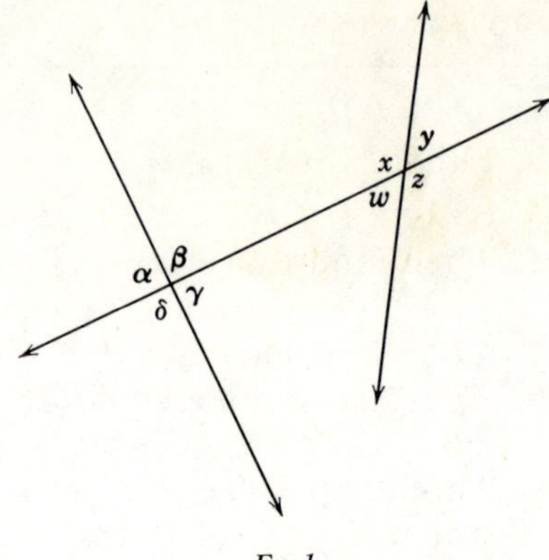

Ex. 1.

2. In the figure, if the angles are of the measures indicated, which lines would be parallel?

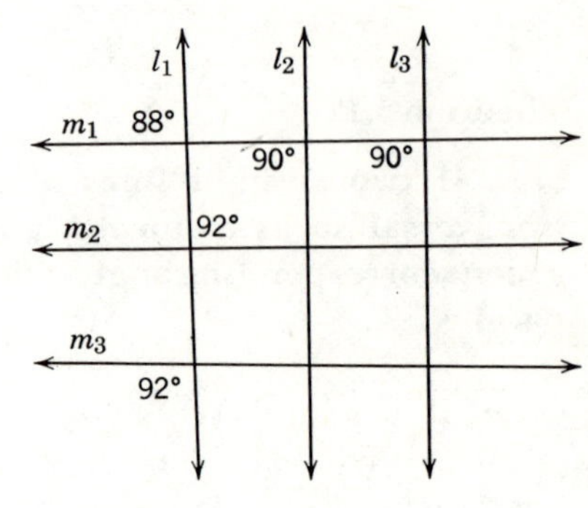

Ex. 2.

3. *Given:* B is the midpoint of $\overline{AE}$ and $\overline{CD}$.
   Prove: $\overline{AC} \parallel \overline{DE}$.

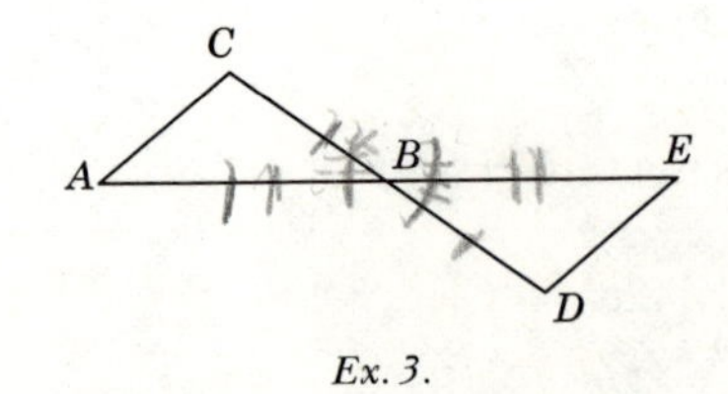

Ex. 3.

4. *Given:* $\overline{RT}$ and $\overline{PS}$ are diagonals;
   $\overline{PQ} \cong \overline{QS}$;
   $\overline{RQ} \cong \overline{QT}$.
   Prove: $\overline{PT} \parallel \overline{RS}$;
   $\overline{RP} \parallel \overline{ST}$.

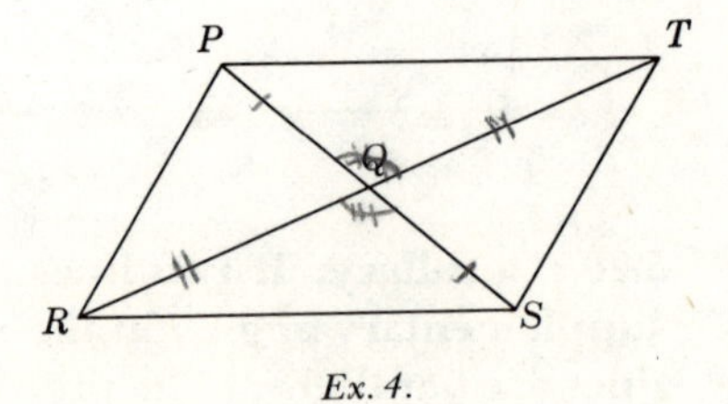

Ex. 4.

5. *Given:* $\overline{LM} \cong \overline{NT}$,
$\overline{LT} \cong \overline{NM}$.
Prove: $\overline{TN} \parallel \overline{LM}$.

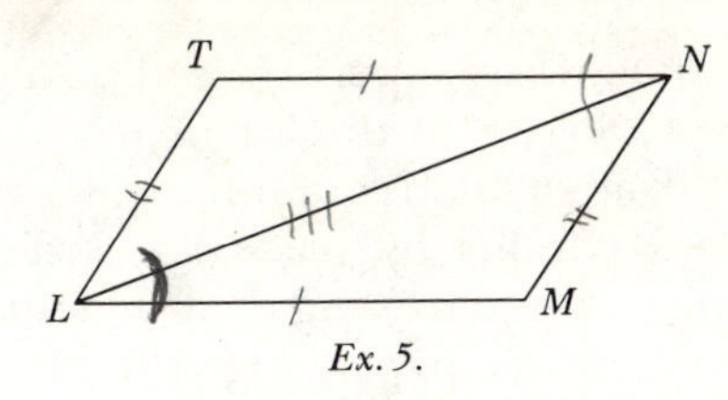

Ex. 5.

6. *Given:* A, D, B, E are collinear;
$\overline{BC} \cong \overline{EF}$; $\overline{AD} \cong \overline{BE}$;
$\overline{AC} \cong \overline{DF}$.
Prove: $\overline{BC} \parallel \overline{EF}$.

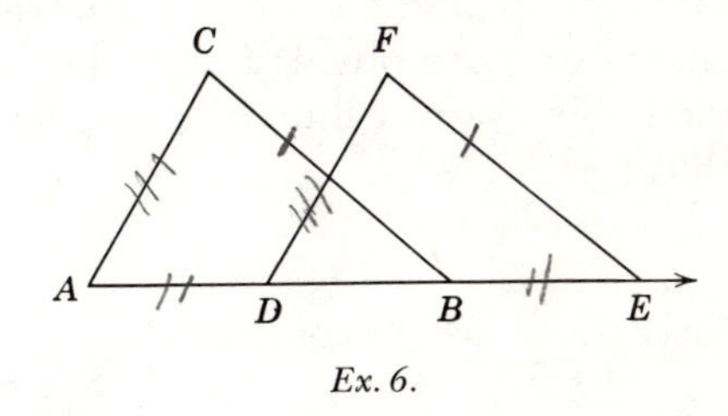

Ex. 6.

7. *Given:* $l \parallel m$; $n \perp l$; $k \perp m$.
Prove: $n \parallel k$.

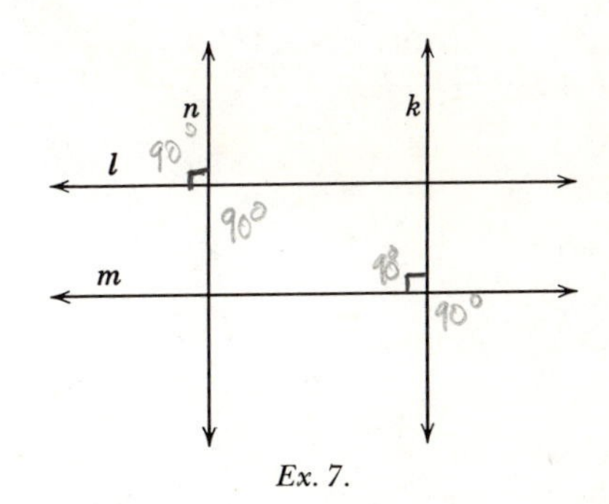

Ex. 7.

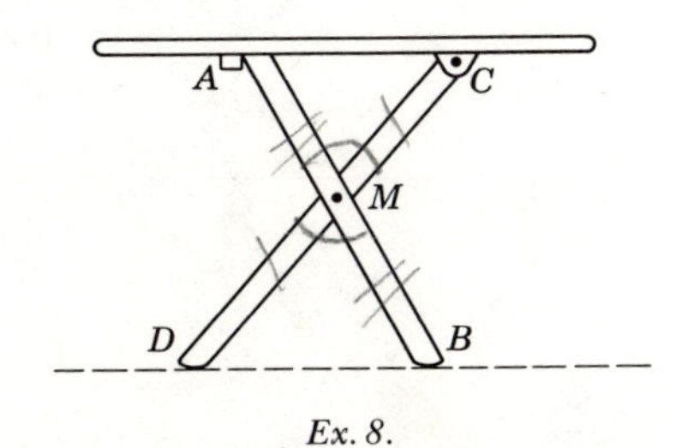

Ex. 8.

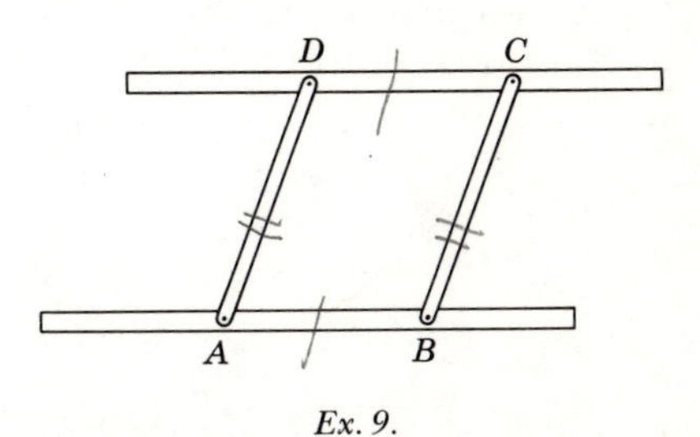

Ex. 9.

8. A collapsible ironing board is constructed to that the supports bisect each other. Show why the board will always be parallel to the floor.
9. The draftsman frequently uses a device, called a parallel ruler, to draw parallel lines. The ruler is so constructed that $\overline{AB} \cong \overline{DC}$ and $\overline{AD} \cong \overline{BC}$. The pins at the vertices permit the ruler to be opened up or collapsed.

If line AB is superimposed on a given line m, the edge DC will be parallel to m. Show why this is true.

10. A draftsman frequently draws two lines parallel by placing a straight edge (T-square) rigid at a desired point on the paper. He then slides a celluloid triangle with base flush with the straight edge. With triangles in positions I and II he is then able to draw line $AB \parallel$ line CD. Why?

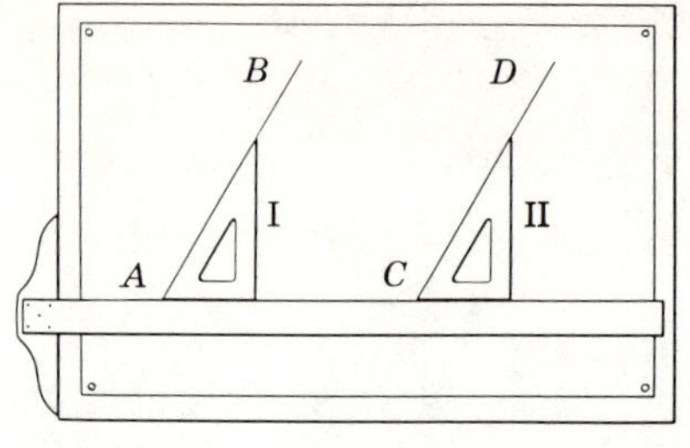

Ex. 10.

11. *Given:* K, L, M are collinear; $\overline{KL} \cong \overline{NL}$, $m\angle MLN = m\angle K + m\angle N$; $\overrightarrow{LQ}$ bisects $\angle MLN$.
Prove: $\overrightarrow{LQ} \parallel \overline{KN}$.

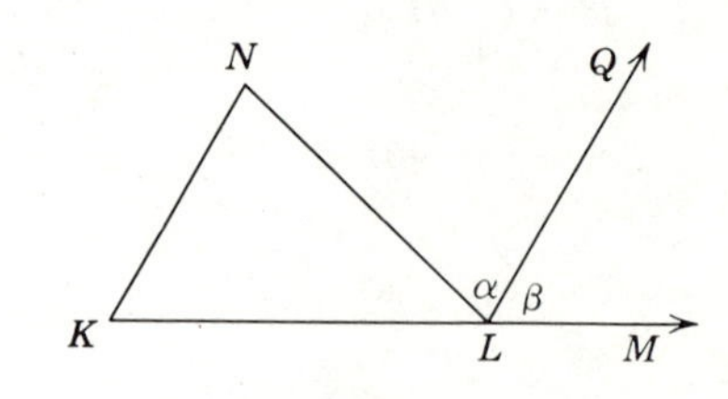

Ex. 11.

12. *Given:* $\overline{AC} \cong \overline{BC}$; $\overline{DC} \cong \overline{EC}$.
Prove: $\overline{DE} \parallel \overline{AB}$. (*Hint:* Draw $\overrightarrow{CG}$ in a manner which will help your proof.)

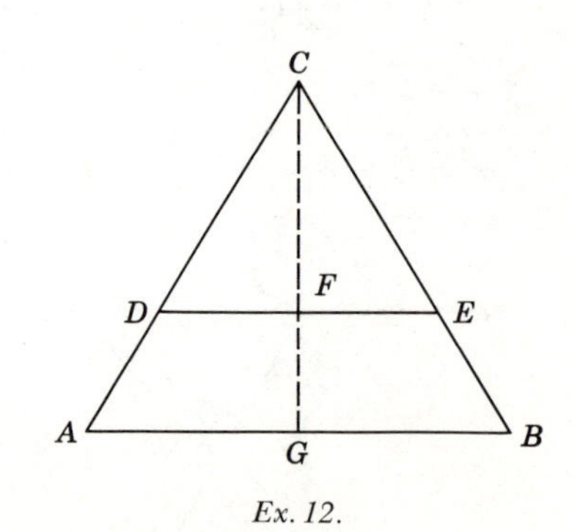

Ex. 12.

13. *Given:* $\overline{EF}$ bisects $\overline{DC}$ and $\overline{AB}$; $\angle A \cong \angle B$; $\overline{AD} \cong \overline{BC}$.
Prove: $\overline{DC} \parallel \overline{AB}$.
(*Hint:* Use Theorem 5.5)

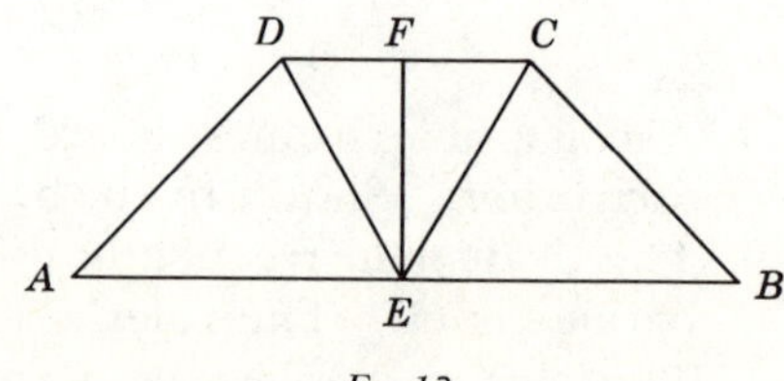

Ex. 13.

Theorem 5.13

5.21. If two parallel lines are cut by a transversal, the alternate interior angles are congruent.

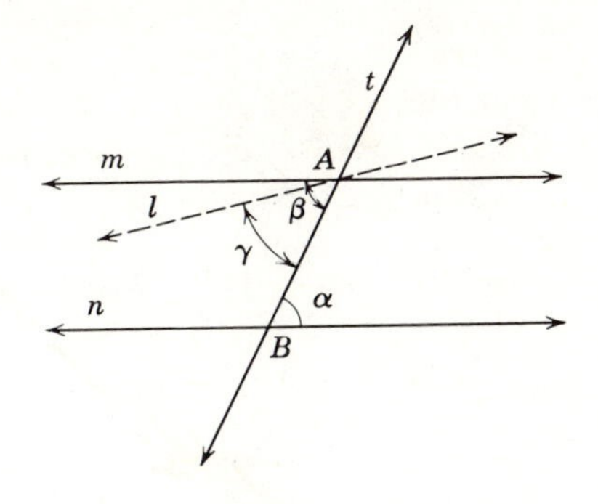

Theorem 5.13.

Given: $m \parallel n$; transversal t cutting n at B and m at A.
Conclusion: $\angle\alpha \cong \angle\beta$.

Proof:

STATEMENTS	REASONS
1. $m \parallel n$.	1. Given.
2. Either $\angle\alpha \cong \angle\beta$ or $\angle\alpha$ is not $\cong \angle\beta$.	2. Law of the excluded middle
3. Assume $\angle\alpha$ is not $\cong \angle\beta$.	3. Temporary assumption.
4. Let l be a line through A for which the alternate angles are congruent, i.e., $\angle\alpha \cong \angle\gamma$.	4. Angle construction postulate.
5. Then $l \parallel n$.	5. Theorem 5.11.
6. This is impossible.	6. Statements 1 and 5 contradict Postulate 18.
7. $\therefore \angle\alpha \cong \angle\beta$.	7. Rule for denying the alternative.

Theorem 5.14

5.22. If two parallel lines are cut by a transversal, the corresponding angles are congruent. (The proof of this theorem is left as an exercise for the student.)

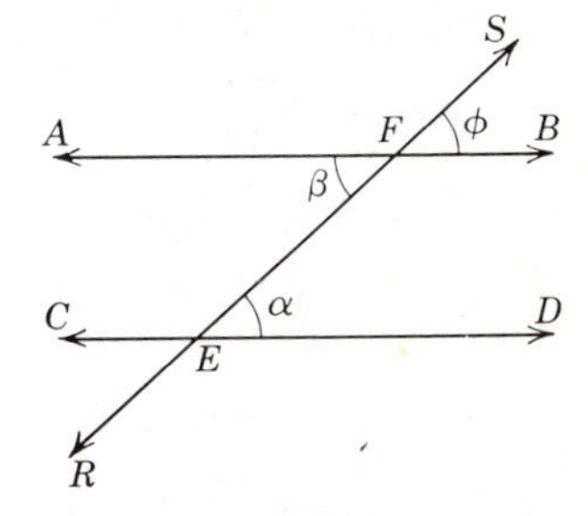

Theorem 5.14.

Theorem 5.15

5.23. If two parallel lines are cut by a transversal, the interior angles on the same side of the transversal are supplementary. (The proof is left as an exercise for the student.)

Exercises

1. *Given:* $\overrightarrow{AB} \parallel \overrightarrow{CD}$;
   $\overrightarrow{AE} \parallel \overrightarrow{CF}$.
   Prove: $\angle\alpha \cong \angle\gamma$.

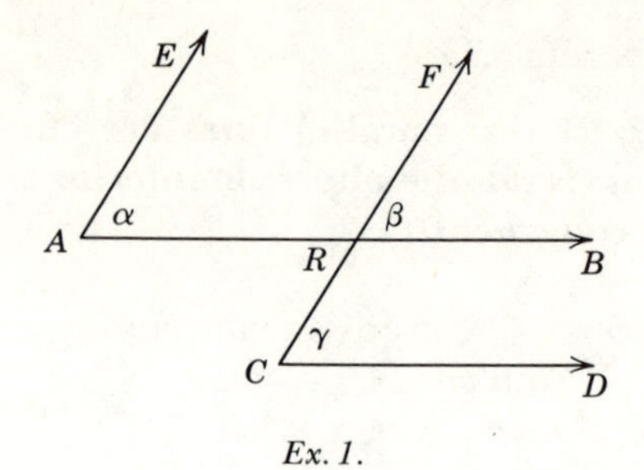

Ex. 1.

2. *Given:* Isosceles $\triangle ABC$ with
   $\overline{AC} \cong \overline{BC}$; line $l \parallel \overline{AB}$.
   Prove: $\angle CDE \cong \angle CED$.

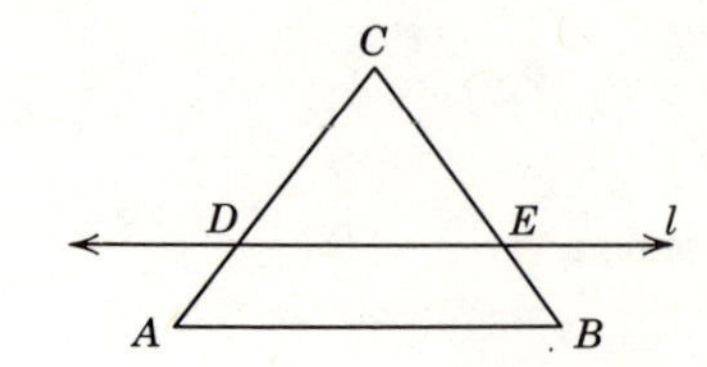

Ex. 2.

3. *Given:* $\overline{AB} \parallel \overline{DC}$;
   $\overline{AB} \cong \overline{DC}$.
   Prove: $\overline{AD} \cong \overline{CB}$.

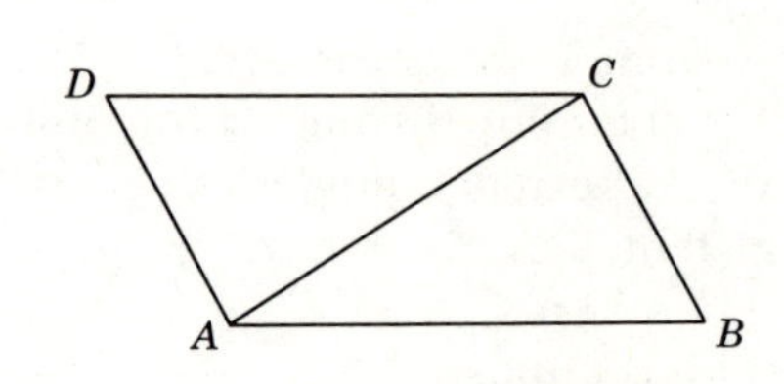

Ex. 3.

4. *Given:* $\overline{RT} \perp \overline{TP}$;
   $\overline{PQ} \perp \overline{TP}$;
   $\overline{TS} \cong \overline{PS}$.
   Prove: $\overline{RS} \cong \overline{QS}$.

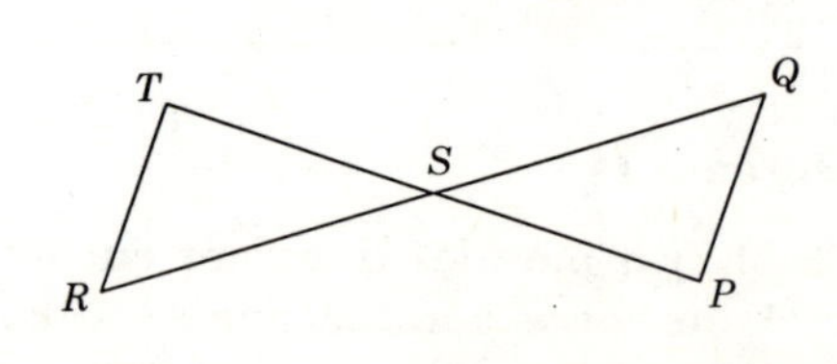

Ex. 4.

5. *Given:* $\overleftrightarrow{RS} \parallel \overleftrightarrow{PQ}$;
   O is midpoint of $\overline{AB}$.
   Prove: O is midpoint of $\overline{CD}$.

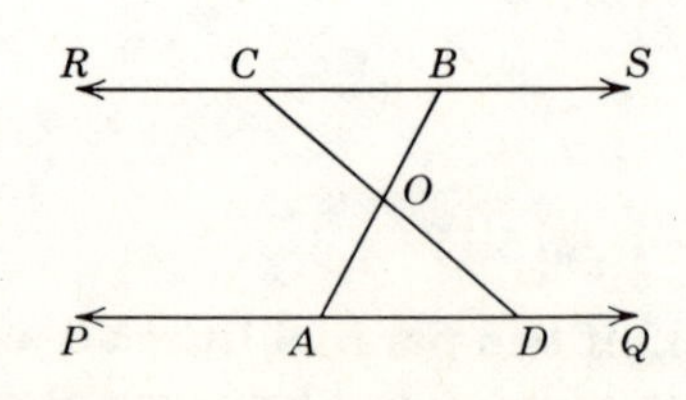

Ex. 5.

6. *Given:* $\overrightarrow{BA} \parallel \overrightarrow{DC}$;
$m\angle B = 40$;
$m\angle BPD = 70$.
Find: the number of degrees in $\angle D$. (*Hint:* Draw auxiliary line through $P \parallel$ to $\overleftrightarrow{CD}$.)

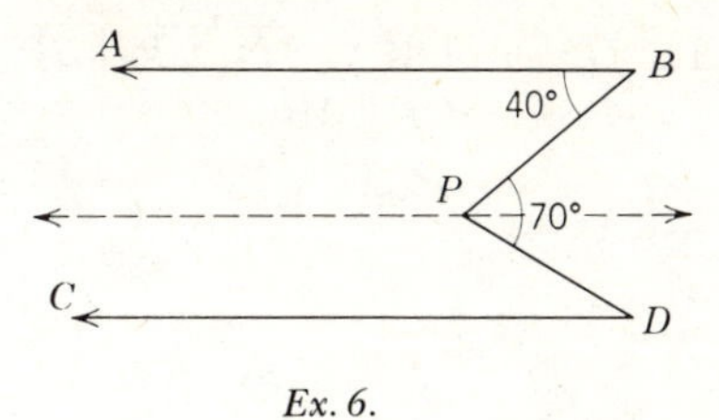

Ex. 6.

7. *Given:* $\overrightarrow{AB} \parallel \overrightarrow{CD}$;
$m\angle A = 130$;
$m\angle E = 80$.
Find: $m\angle C =$ ________.

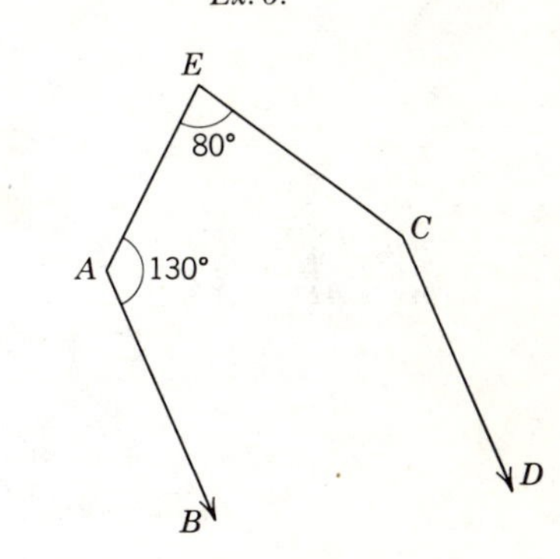

Ex. 7.

8. *Given:* $\overline{AB} \parallel \overline{DC}$;
$\overline{AD} \parallel \overline{BC}$.
Prove: $\overline{AB} \cong \overline{DC}$;
$\overline{AD} \cong \overline{BC}$.

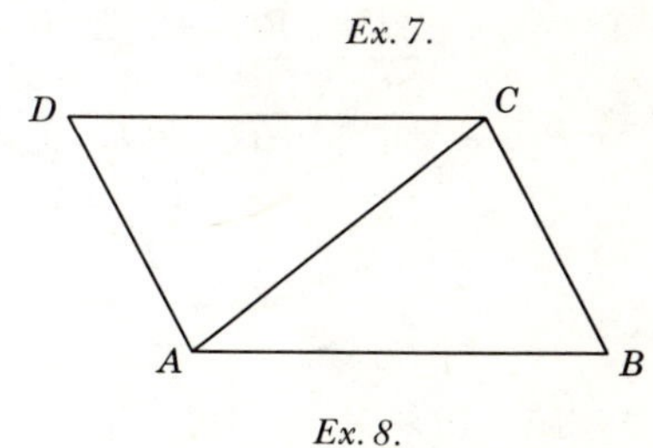

Ex. 8.

9. *Given:* $\overline{AD} \cong \overline{BC}$;
$\overline{AD} \parallel \overline{BC}$.
Prove: $\overline{AO} \cong \overline{CO}$;
$\overline{DO} \cong \overline{BO}$.

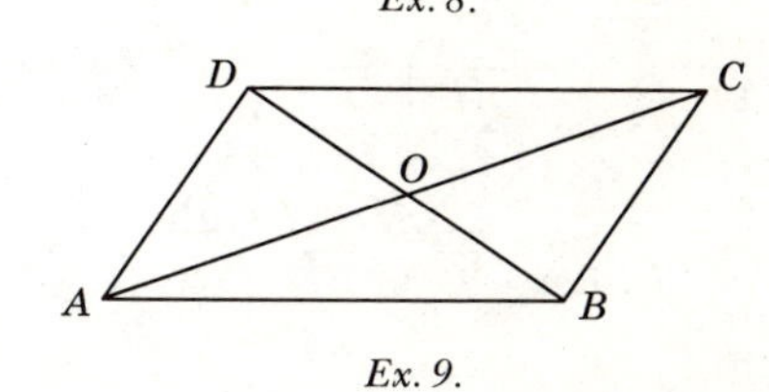

Ex. 9.

10. Prove that a line drawn parallel to the base of an isosceles triangle and through its vertex will bisect the exterior angle at the vertex.

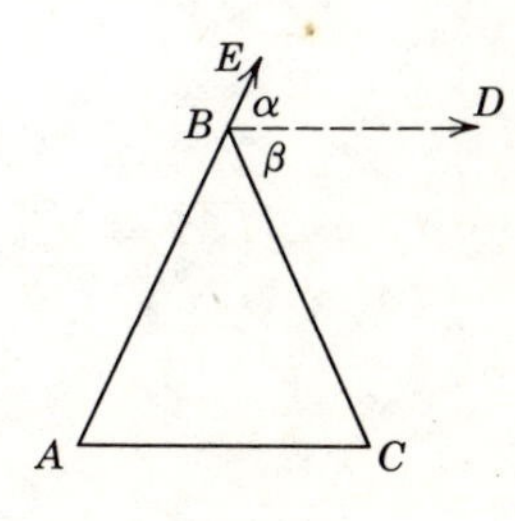

Ex. 10.

11. *Given:* $\overline{LM} \cong \overline{TN}$; $\overline{LM} \parallel \overline{TN}$.
Prove: $\overline{LT} \parallel \overline{MN}$.

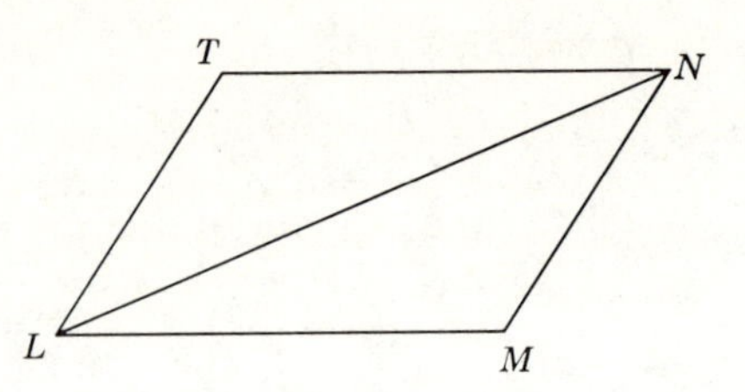

Ex. 11.

12. *Given:* $l \parallel m$; $\overline{AD} \perp m$; $\overline{BC} \perp m$.
Prove: $\overline{AD} \cong \overline{BC}$.

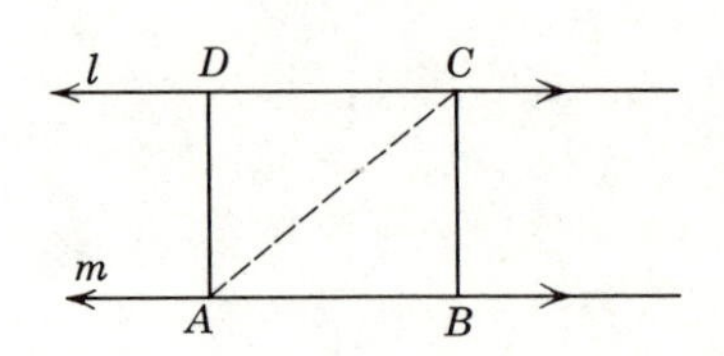

Ex. 12.

13. *Given:* $\overline{AD} \cong \overline{BC}$; $\overline{AD} \perp \overline{AB}$; $\overline{BC} \perp \overline{AB}$.
Prove: $\overline{DC} \cong \overline{AB}$ and $\overline{DC} \parallel \overline{AB}$.

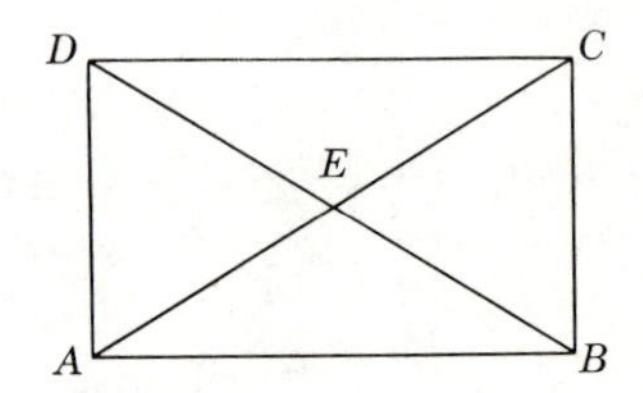

Ex. 13.

14. *Given:* $\overline{AB} \parallel \overline{DC}$;
$\overline{AD} \parallel \overline{BC}$;
$\angle BAD$ is a right $\angle$.
Prove: $\overline{AC} \cong \overline{BD}$.

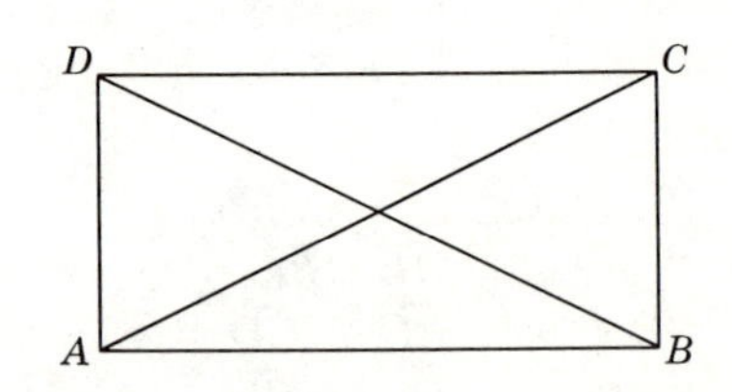

Ex. 14.

15. *Given:* $\overline{LM} \cong \overline{MN} \cong \overline{TN} \cong \overline{LT}$.
Prove: $\overline{LN} \perp \overline{TM}$.

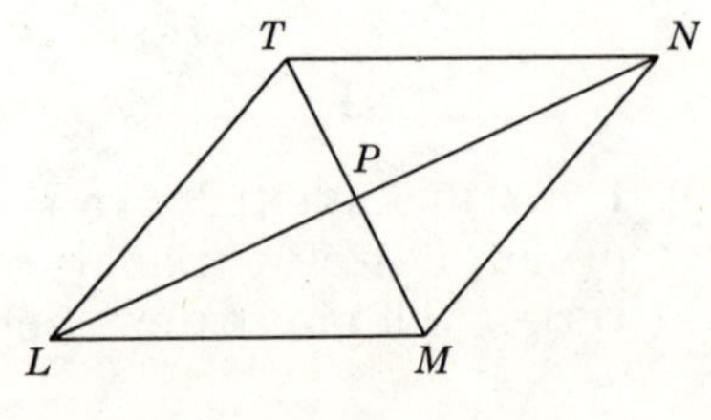

Ex. 15.

Theorem 5.16

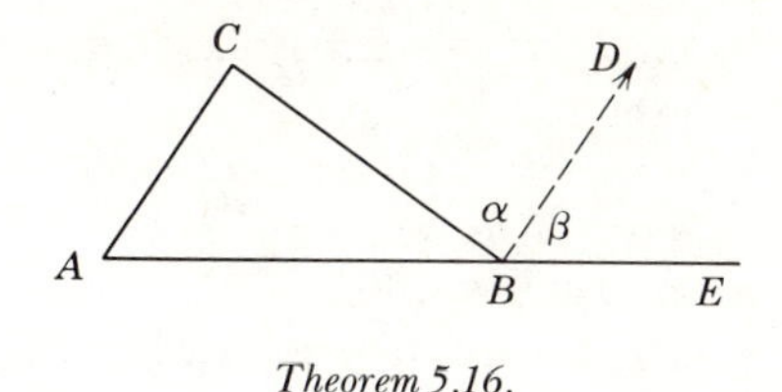

Theorem 5.16.

5.24. The measure of an exterior angle of a triangle is equal to the sum of the measures of the two nonadjacent interior angles.

Given: $\angle CBE$ is an exterior $\angle$ *of* $\triangle ABC$.
Conclusion: $m\angle CBE = m\angle C + m\angle A$.
Proof:

STATEMENTS	REASONS
1. Draw $\overrightarrow{BD} \parallel \overline{AC}$.	1. Postulate 18; Theorem 5.7.
2. $m\angle\alpha = m\angle C$.	2. Theorem 5.13.
3. $m\angle\beta = m\angle A$.	3. Theorem 5.14.
4. $m\angle CBE = m\angle\alpha + m\angle\beta$.	4. Angle addition postulate.
5. $m\angle CBE = m\angle C + m\angle A$.	5. Substitution property of equality.

5.25. Sum of the angle measures of a triangle. Acceptance of the parallel postulate makes possible the proof of the next theorem, one of the most widely used theorems in dealing with figures in a plane. You are probably familiar with it, having learned about it inductively by measuring in other mathematics courses. We now proceed to prove it deductively.

Theorem 5.17

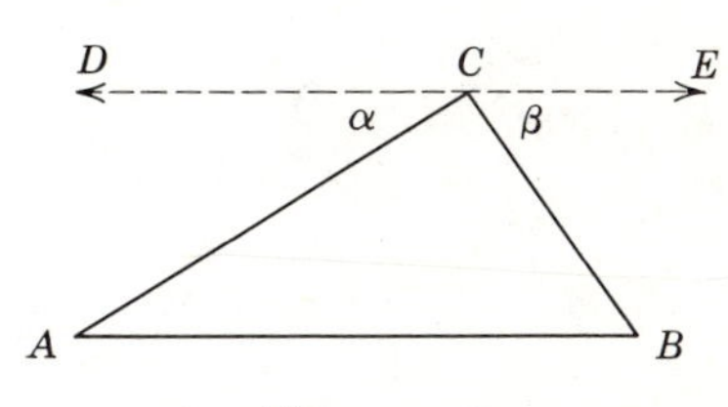

Theorem 5.17.

5.26. The sum of the measures of the angles of a triangle is 180.

Given: $\triangle ABC$ is any triangle.
Conclusion: $m\angle A + m\angle ACB + m\angle B = 180$.
Proof:

STATEMENTS	REASONS
1. $\triangle ABC$ is any triangle.	1. Given.
2. Through C draw $\overleftrightarrow{DE} \parallel \overline{AB}$.	2. Theorem 5.7; Postulate 18.
3. $m\angle\alpha = m\angle A$, $m\angle\beta = m\angle B$.	3. Theorem 5.13.
4. $m\angle DCE = m\angle\alpha + m\angle ACE$.	4. Postulate 14.
5. $m\angle ACE = m\angle ACB + m\angle\beta$.	5. Postulate 14.

6. $m\angle DCE = m\angle\alpha + m\angle ACB + m\angle\beta$.	6. Substitution property of equality.
7. $m\angle DCE = m\angle A + m\angle ACB + m\angle B$.	7. Substitution property of equality.
8. $m\angle DCE = 180$.	8. Definition of straight angle.
9. $m\angle A + m\angle ACB + m\angle B = 180$.	9. Theorem 3.5.

It should be evident that the sum of the measures of the angles of a triangle depends on our assuming true Euclid's postulate that only one line can be drawn through a point parallel to a given line. As a matter of interest, non-Euclidean geometry proves the sum of the measures of the three angles of a triangle different from 180. In this course we will agree with Euclid, since it will prove satisfactory for all our needs.

The proofs to the following are left to the student.

5.27. Corollary: Only one angle of a triangle can be a right angle or an obtuse angle.

5.28. Corollary: If two angles of one triangle are congruent respectively to two angles of another triangle, the third angles are congruent.

5.29. Corollary: The acute angles of a right triangle are complementary.

Exercises

Determine the number of degrees in the required angles in Exs. 1 through 8.

1. *Given:* $\overrightarrow{AB} \parallel \overrightarrow{CD}$;
   $m\angle BAE = 50$;
   $m\angle DCE = 40$.
   Find: $m\angle\alpha + m\angle\beta =$ _____ .

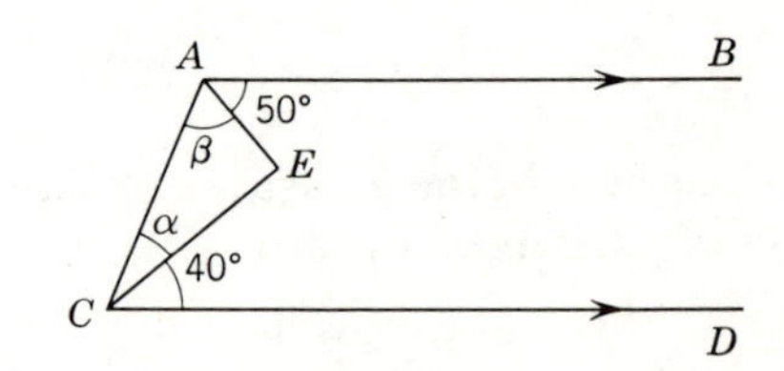

Ex. 1.

2. *Given:* $\overline{AB} \cong \overline{BC} \cong \overline{AC}$.
   Find: $m\angle A =$ _____ .

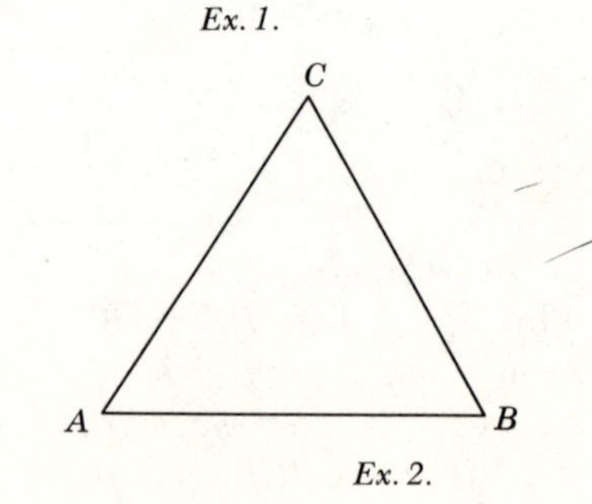

Ex. 2.

3. *Given:* $m\angle A = 70$;
$m\angle C = 80$.
Find: $m\angle x =$ _____ .

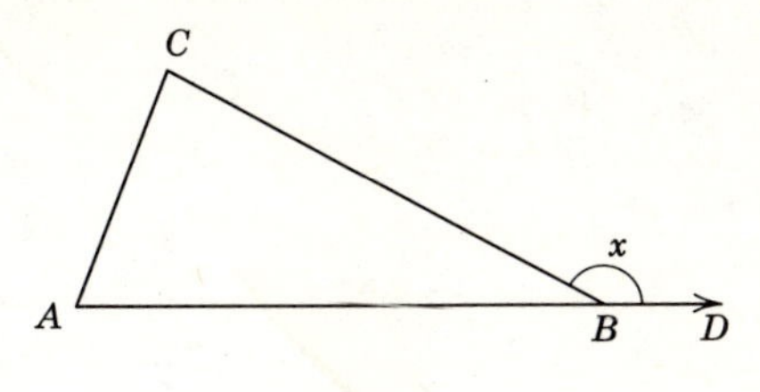

Ex. 3.

4. *Given:* $m\angle C = 110$;
$m\angle CBD = 155$.
Find: $m\angle \alpha =$ _____ .

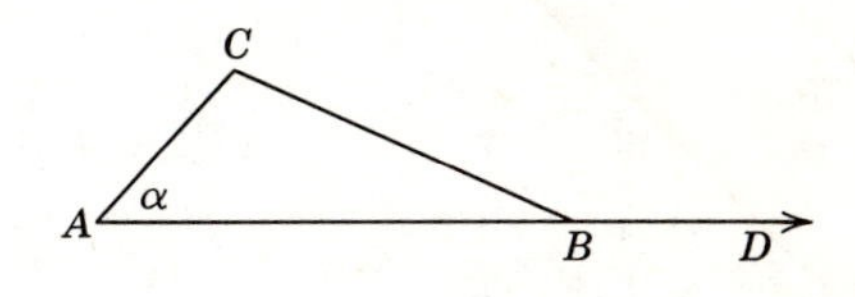

Ex. 4.

5. *Given:* $\overline{AC} \cong \overline{BC}$;
$\overline{AC} \perp \overline{BC}$.
Find: $m\angle A =$ _____ .

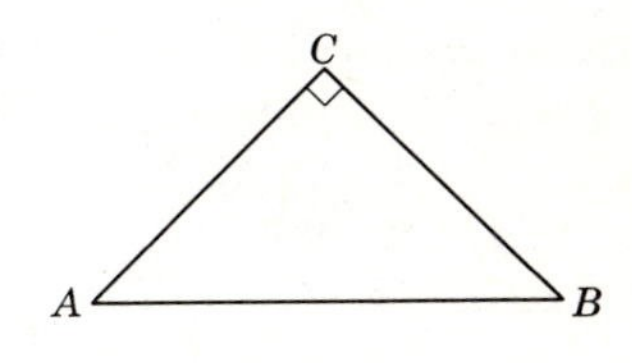

Ex. 5.

6. *Given:* $m\angle A = 50$;
$m\angle B = 60$;
$\angle\alpha \cong \angle\beta$.
Find: $m\angle \alpha =$ _______ .

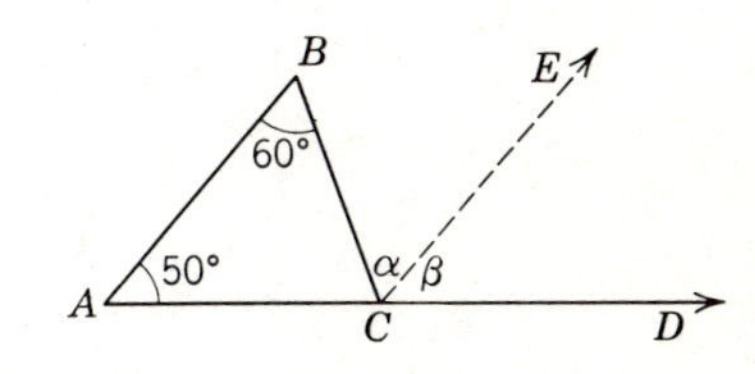

Ex. 6.

7. *Given:* $\angle\alpha \cong \angle\alpha'$;
$\angle\beta \cong \angle\beta'$;
$m\angle D = 130$.
Find: $m\angle C =$ _____ .

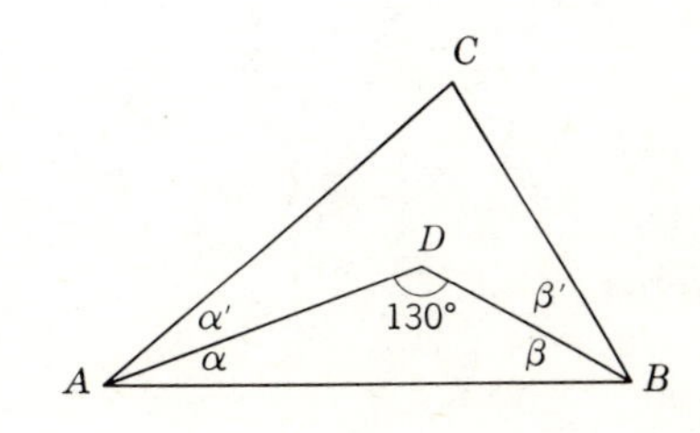

Ex. 7.

8. *Given:* $\overline{AC} \cong \overline{AB}$;
$\overline{AC} \perp \overrightarrow{AB}$.
Find: $m\angle x =$ ______ .

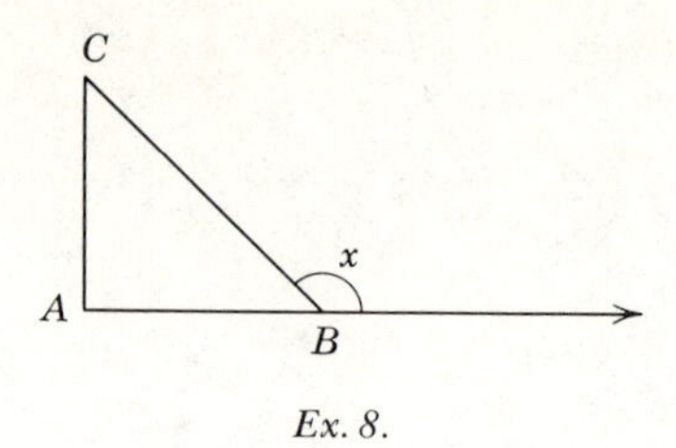

Ex. 8.

9. *Given:* $\overline{AE} \perp \overline{AC}$;
$\overline{CD} \perp \overline{AC}$;
$\angle\alpha \cong \angle\beta$.
Prove: $\angle E \cong \angle D$.

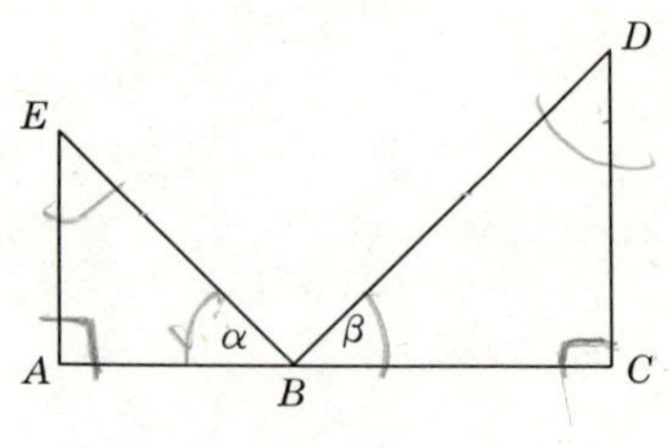

Ex. 9.

10. *Given:* $\triangle ABC$ with
$\angle CDE \cong \angle B$.
Prove: $\angle CED \cong \angle A$.

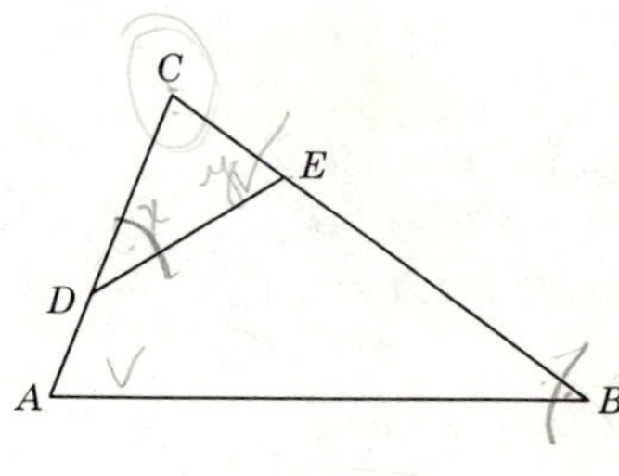

Ex. 10.

11. *Given:* $\angle A \cong \angle B$;
$\overline{DE} \perp \overline{AB}$.
Prove: $\angle\alpha \cong \angle E$.

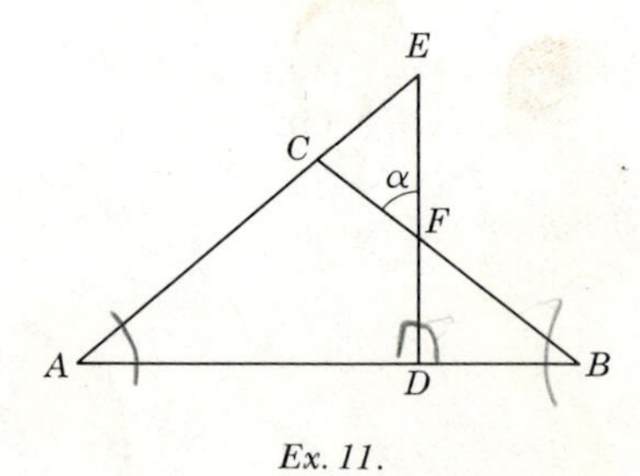

Ex. 11.

12. *Given:* $\overline{BC} \perp \overline{AC}$;
$\overline{DC} \perp \overline{AB}$.
Prove: $\angle\alpha \cong \angle\beta$.

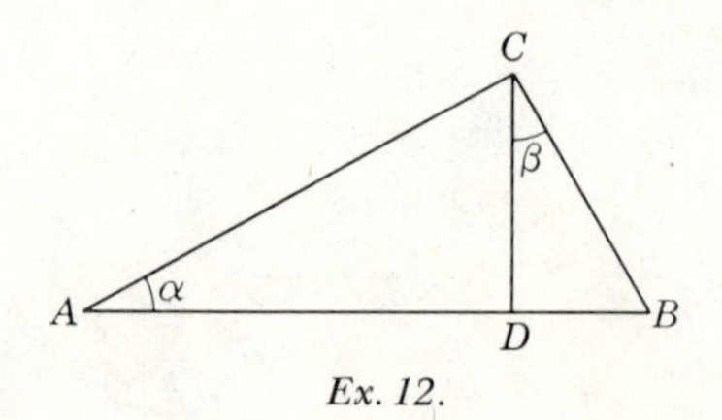

Ex. 12.

13. *Given:* $\overline{ST} \parallel \overline{AC}$;
$\overline{SR} \parallel \overline{BC}$;
$\angle\alpha \cong \angle\beta$.
Prove: $\angle A \cong \angle B$.

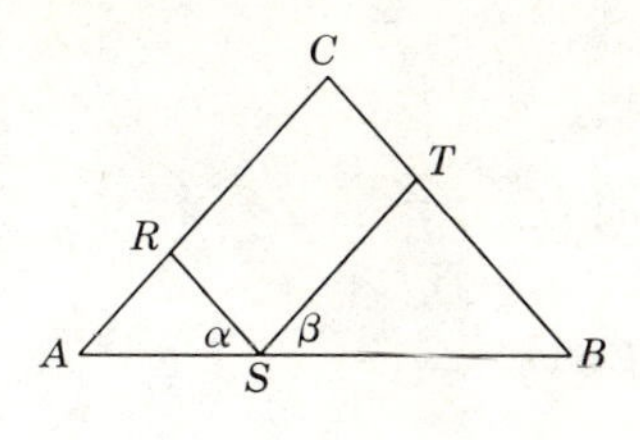

Ex. 13.

14. *Given:* $\overline{CE} \perp \overline{AB}$.
$\overline{DB} \perp \overline{AC}$.
Prove: $\angle DCE \cong \angle EBD$.

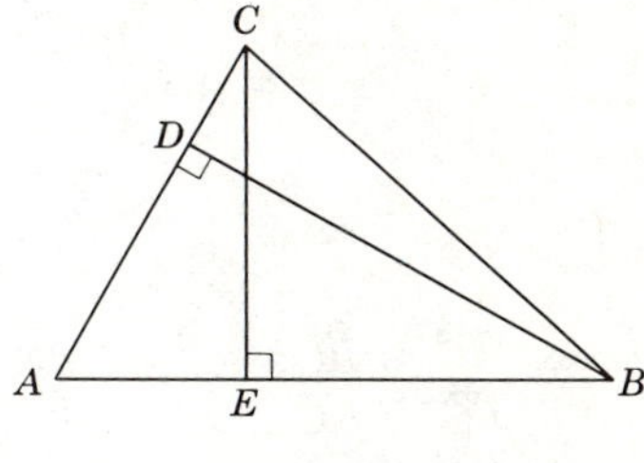

Ex. 14.

15. *Given:* $\overleftrightarrow{AB} \parallel \overleftrightarrow{CD}$;
$\overline{FG}$ bisects $\angle BFE$;
$\overline{EG}$ bisects $\angle DEF$.
Prove: $\overline{EG} \perp \overline{GF}$.

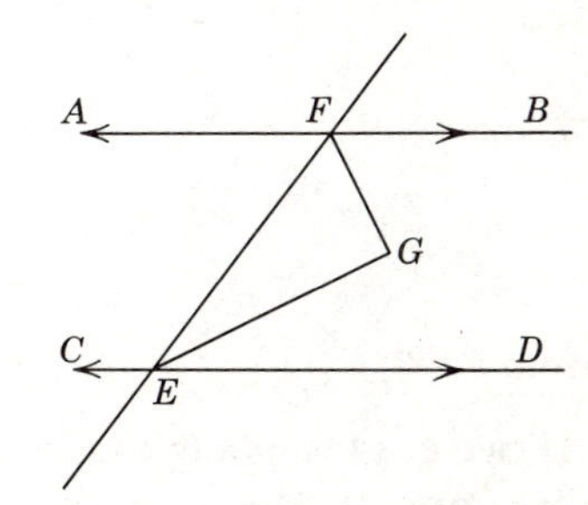

Ex. 15.

16. *Given:* $\overline{AC} \cong \overline{BC}$;
$\overline{DF} \perp \overline{AC}$;
$\overline{EF} \perp \overline{BC}$.
Prove: $\angle AFD \cong \angle BFE$.

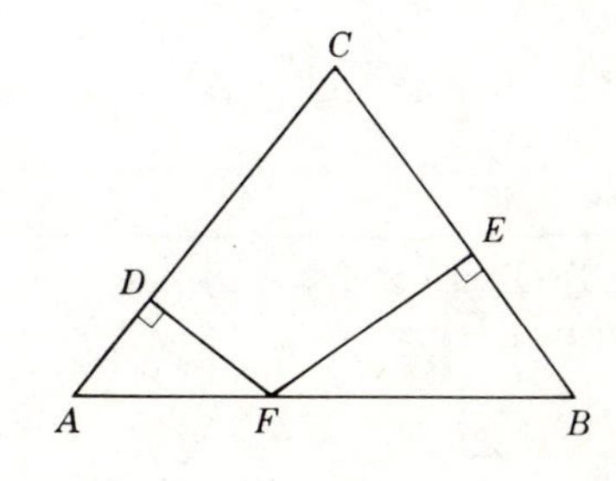

Ex. 16.

17. *Given:* $\overline{CE} \perp \overrightarrow{AC}$;
$\overline{DE} \perp \overrightarrow{AD}$.
Prove: $\angle A \cong \angle E$.

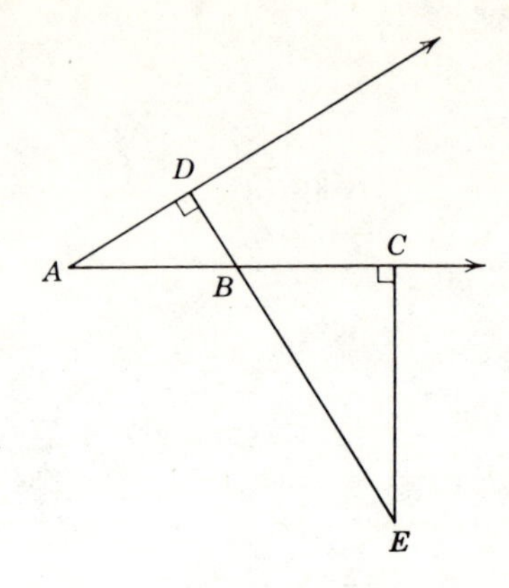

Ex. 17.

18. *Given:* $\overline{AC} \cong \overline{BC}$;
$\overline{FC} \cong \overline{EC}$.
Prove: $\overline{DE} \perp \overline{AB}$.

Ex. 18.

Theorem 5.18

5.30. If two angles of a triangle are congruent, the sides opposite them are congruent.

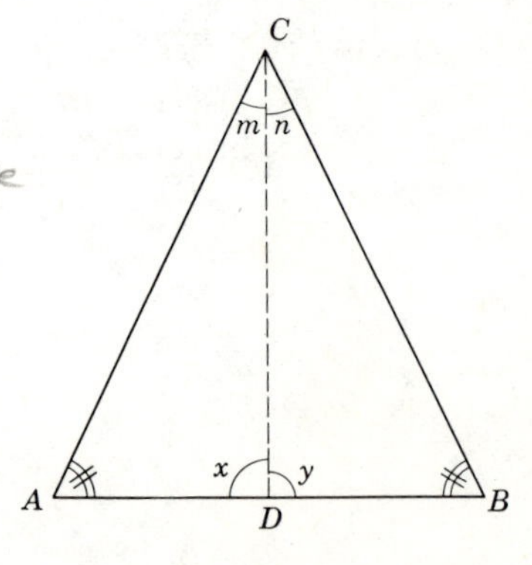

Theorem 5.18.

Given: $\triangle ABC$ with $\angle A \cong \angle B$.
Conclusion: $\overline{AC} \cong \overline{BC}$.

Proof:

STATEMENTS	REASONS
1. $\triangle ABC$ with $\angle A \cong \angle B$.	1. Given.
2. Draw $\overline{CD}$ bisecting $\angle C$.	2. An angle has one and only one ray which bisects it.
3. $\angle m \cong \angle n$.	3. A bisector divides an angle into two congruent angles.

4. $\angle x \cong \angle y$.	4. §5.28. Corollary.
5. $\overline{CD} \cong \overline{CD}$.	5. Reflexive property of congruence.
6. $\triangle ADC \cong \triangle BDC$.	6. A.S.A.
7. $\overline{AC} \cong \overline{BC}$.	7. Corresponding parts of congruent triangles are congruent.

5.31. Corollary: An equiangular triangle is equilateral.

Theorem 5.19

5.32. If two right triangles have a hypotenuse and an acute angle of one congruent respectively to the hypotenuse and an acute angle of the other, the triangles are congruent.

Given: $\triangle ABC$ and $\triangle DEF$ with $\angle B$ and $\angle E$ right ∠s; $\overline{AC} \cong \overline{DF}$; $\angle A \cong \angle D$.

Conclusion: $\triangle ABC \cong \triangle DEF$.

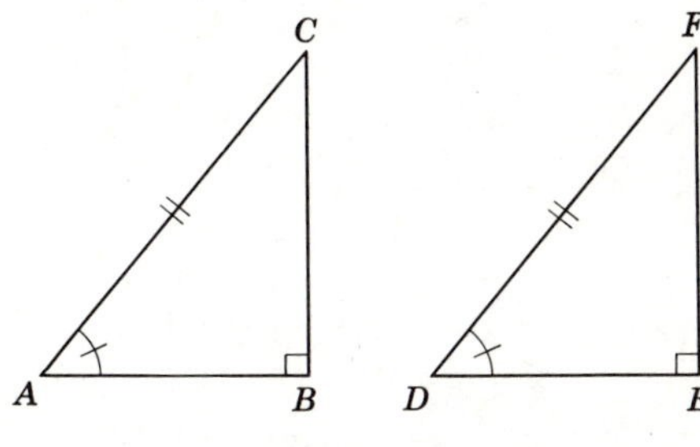

Theorem 5.19.

Proof:

STATEMENTS	REASONS
1. $\angle B$ and $\angle E$ are right ∠s.	1. Given.
2. $\angle B \cong \angle E$.	2. Right angles are congruent.
3. $\angle A \cong \angle D$.	3. Given.
4. $\angle C \cong \angle F$.	4. §5.28. Corollary.
5. $\overline{AC} \cong \overline{DF}$.	5. Given.
6. $\triangle ABC \cong \triangle DEF$.	6. A.S.A.

Theorem 5.20

5.33. If two right triangles have the hypotenuse and a leg of one congruent to the hypotenuse and a leg of the other, the triangles are congruent.

Given: Right $\triangle ABC$; right $\triangle DEF$; $\angle C$ and $\angle F$ are right $\angle$s; $\overline{BC} \cong \overline{EF}$, $\overline{AB} \cong \overline{DE}$.

Conclusion: $\triangle ABC \cong \triangle DEF$.

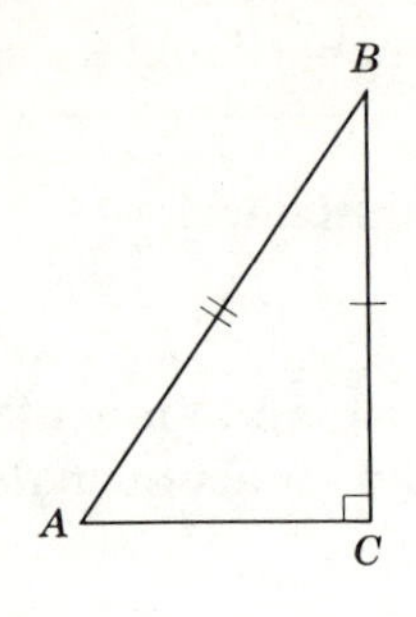

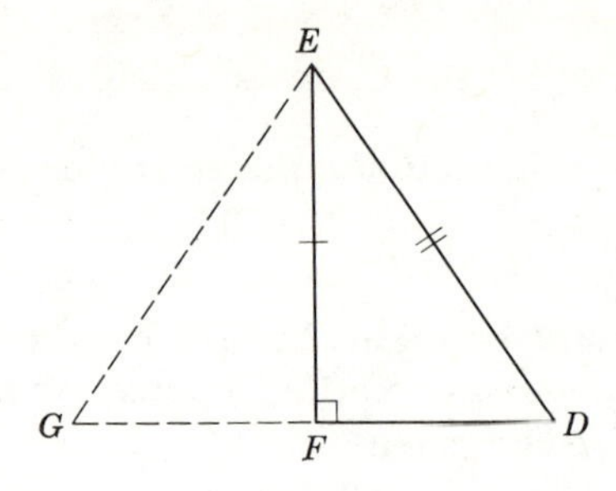

Theorem 5.20.

Proof:

STATEMENTS	REASONS
1. On the ray opposite $\overrightarrow{FD}$ construct $\overrightarrow{FG}$ such that $\overline{FG} \cong \overline{CA}$.	1. Point plotting postulate.
2. Draw $\overline{EG}$.	2. Postulate 2.
3. $\angle DFE$ is a right $\angle$.	3. Given.
4. $\overline{EF} \perp \overline{GD}$.	4. Definition of perpendicular.
5. $\angle GFE$ is a right $\angle$.	5. Definition of $\perp$ (reversible).
6. $\angle C$ is a right $\angle$.	6. Given.
7. $\angle C \cong \angle GFE$.	7. Right angles are congruent.
8. $\overline{BC} \cong \overline{EF}$.	8. Given.
9. $\triangle ABC \cong \triangle GEF$.	9. S.A.S.
10. $\overline{AB} \cong \overline{GE}$.	10. Corresponding sides of $\cong$ $\triangle$s are $\cong$.
11. $\overline{AB} \cong \overline{DE}$.	11. Given.
12. $\overline{GE} \cong \overline{DE}$.	12. Theorem 3.5.
13. $\angle GFD$ is a straight $\angle$.	13. Definition of straight angle.
14. $\angle G \cong \angle D$.	14. Base angles of an isosceles $\triangle$ are $\cong$.
15. $\triangle GEF \cong \triangle DEF$.	15. Theorem 5.19.
16. $\overline{FG} \cong \overline{FD}$.	16. Reason 10.
17. $\overline{CA} \cong \overline{FD}$.	17. Theorem 3.5 (from Statements 1 and 16).
18. $\triangle ABC \cong \triangle DEF$.	18. S.A.S. or S.S.S.

Theorem 5.21

5.34. If the measure of one acute angle of a right triangle equals 30, the length of the side opposite this angle is one-half the length of the hypotenuse. The proof of this theorem is left as an exercise for the student. (*Hint:* Extend $\overline{AB}$ to D, making $m\overline{BD} = m\overline{AB}$. Draw $\overline{CD}$. Prove $\overline{CB}$ bisects $\overline{AD}$ of equilateral $\triangle ADC$.)

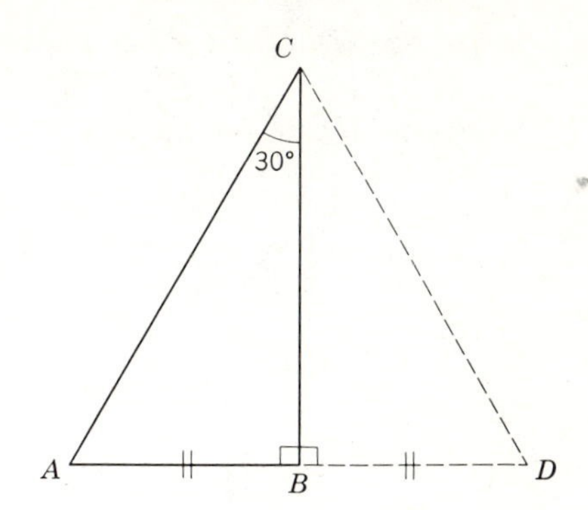

Theorem 5.21.

Exercises

1. *Given:* $\overline{CD} \perp \overline{AB}$;
   $\overline{BC} \perp \overline{AC}$;
   Prove: $\angle A \cong \angle BCD$.

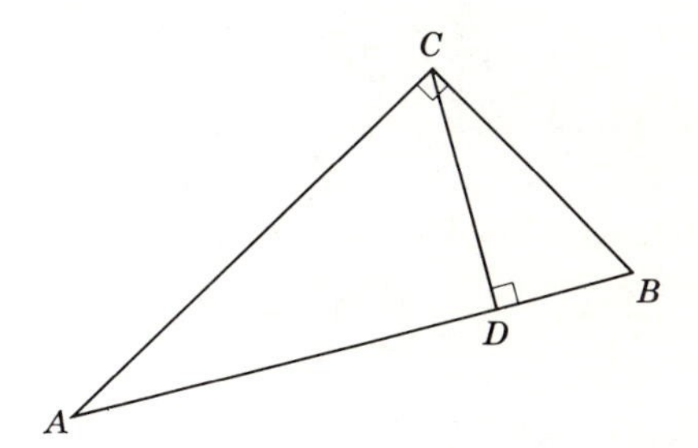

Ex. 1.

2. *Given:* $\overline{TM} \perp \overline{LM}$;
   $\overline{TK} \perp \overline{LK}$;
   $\overline{TM} \cong \overline{TK}$.
   Prove: $\overline{TL}$ bisects $\angle KLM$.

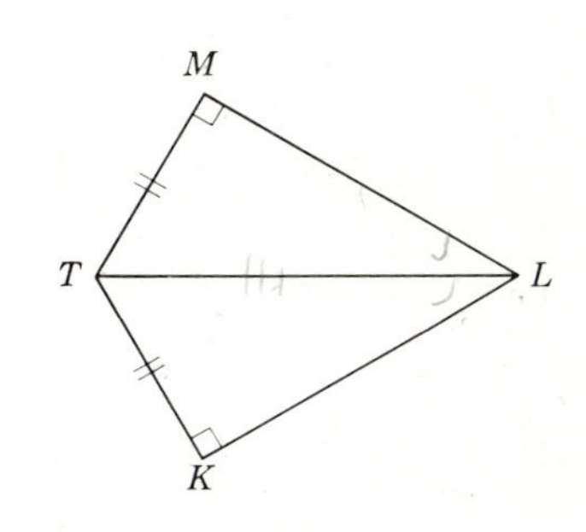

Ex. 2.

3. *Given:* $\overline{AD} \perp \overline{DC}$;
   $\overline{BC} \perp \overline{DC}$;
   M is the midpoint of $\overline{DC}$;
   $\overline{AM} \cong \overline{BM}$.
   Prove: $\overline{AD} \cong \overline{BC}$.

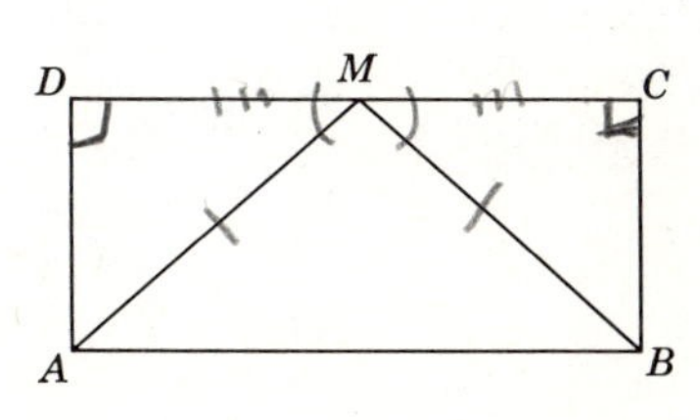

Ex. 3.

4. *Given:* $\overline{SQ}$ bisects $\overline{RL}$ at T;
$\overline{RS} \perp \overline{SQ}$; $\overline{LQ} \perp \overline{SQ}$.
Prove: $\overline{RL}$ bisects $\overline{SQ}$ at T.

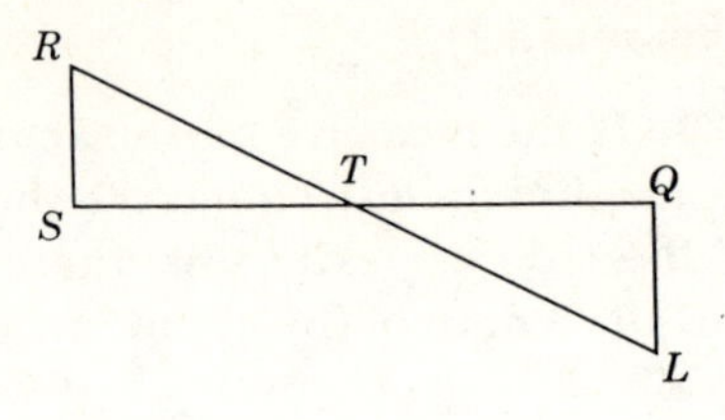

Ex. 4.

5. *Given:* $\overline{AE} \perp \overline{BC}$;
$\overline{CD} \perp \overline{AB}$;
$\overline{AE} \cong \overline{CD}$.
Prove: $\overline{BA} \cong \overline{BC}$.

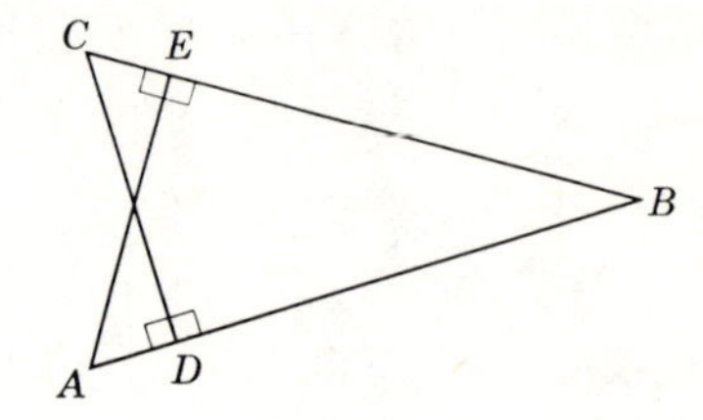

Ex. 5.

6. *Given:* L, M, R, T are collinear;
$\overline{RS} \perp \overline{LS}$; $\overline{LM} \cong \overline{TR}$;
$\overline{NM} \perp \overline{TN}$; $\angle L \cong \angle T$.
Prove: $\overline{RS} \cong \overline{MN}$.

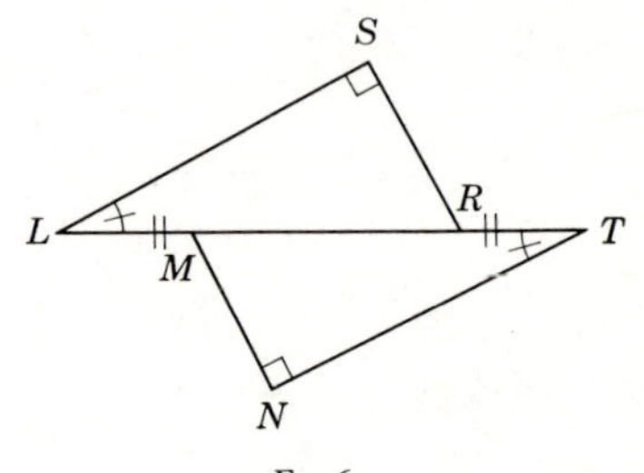

Ex. 6.

7. *Given:* $\overline{RT} \cong \overline{ST}$;
$\overline{RS} \perp \overline{TQ}$.
Prove: $\overline{RQ} \cong \overline{SQ}$.

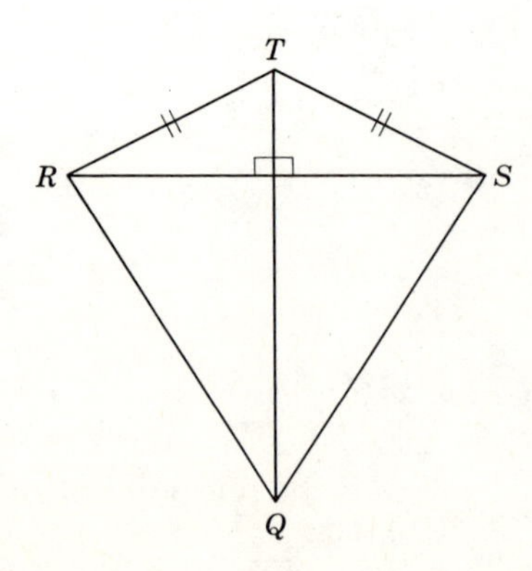

Ex. 7.

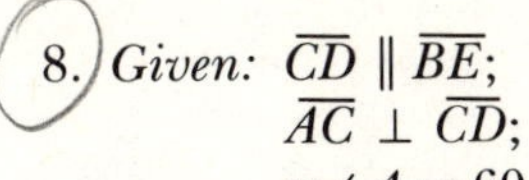

8. *Given:* $\overline{CD} \parallel \overline{BE}$;
$\overline{AC} \perp \overline{CD}$;
$m\angle A = 60$.
Prove: $m(\overline{AB}) = \frac{1}{2}m(\overline{AE})$.

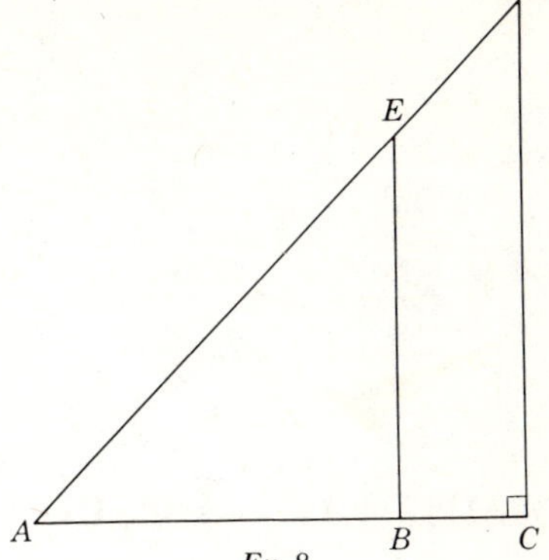

Ex. 8.

9. *Prove:* If the bisector of an exterior angle of a triangle is parallel to the opposite side, the triangle is isosceles.

10. *Prove:* The bisectors of the base angles of an isosceles triangle intersect at a point equidistant from the ends of the base.

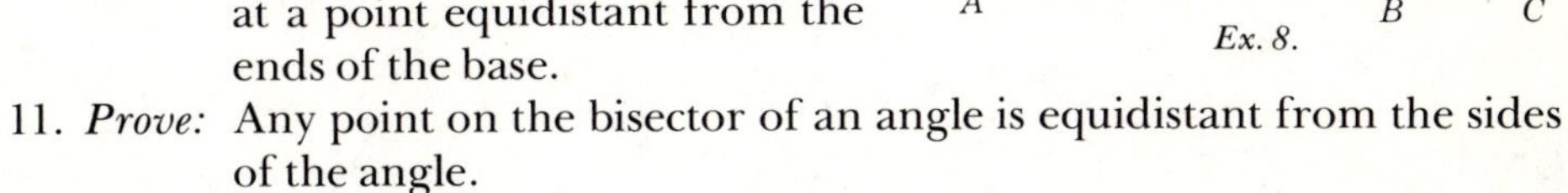

11. *Prove:* Any point on the bisector of an angle is equidistant from the sides of the angle.

Summary Tests

Test 1

COMPLETION STATEMENTS

1. The sum of the measures of the angles of any triangle is ______ .
2. Angles in the same half-plane of the transversal and between parallel lines are ______ .
3. Two lines parallel to the same line are ______ to each other.
4. The measure of an exterior angle of a triangle is ______ than the measure of either nonadjacent interior angle.
5. A proof in which all other possibilities are proved wrong is called ______ proof.
6. A line cutting two or more lines is called a(n) ______ .
7. If two isosceles triangles have a common base, the line joining their vertices is ______ to the base.
8. A line parallel to the base of an isosceles triangle cutting the other sides cuts off an ______ triangle.
9. A triangle is ______ if two of its altitudes are congruent.
10. The statement that through a point not on a given line there is one and only one line perpendicular to that given line asserts the ______ and ______ properties of that line.
11. The acute angles of a right triangle are ______ .
12. If two parallel lines are cut by a transversal, the interior angles on the same side of the transversal are ______ .
13. If the sum of the measures of any two angles of a triangle equals the measure of the third angle, the triangle is a(n) ______ triangle.
14. If from any point of the bisector of an angle a line is drawn parallel to one side of the angle, the triangle formed is a(n) ______ triangle.
15. Two planes are ______ if their intersection is a null set.

16. Two planes are perpendicular iff they form congruent adjacent _______ angles.
17. Two planes perpendicular to the same line are _______ .
18. Through a point outside a plane (how many?) lines can be drawn parallel to the plane.
19. No right triangle can have a(n) _______ angle.
20. Two lines perpendicular to the same plane are _______ to each other.
21. The two exterior angles at a vertex of a triangle are _______ angles and are therefore _______ angles.
22. The geometry which does not assume Playfair's postulate is sometimes called _______ geometry.

Test 2

TRUE-FALSE STATEMENTS

1. An isosceles triangle has three acute angles.
2. A line which bisects the exterior angle at the vertex of an isosceles triangle is parallel to the base.
3. The median of a triangle is perpendicular to the base.
4. If two lines are cut by a transversal, the alternate exterior angles are supplementary.
5. The perpendicular bisectors of two sides of a triangle are parallel to each other.
6. In an acute triangle the sum of the measures of any two angles must be greater than a right angle.
7. If any two angles of a triangle are congruent, the third angle is congruent.
8. If two parallel planes are cut by a third plane, the lines of intersection are skew lines.
9. To prove the existence of some thing, it is necessary only to prove that there is at least one of the things.
10. The acute angles of a right triangle are supplementary.
11. The expressions "exactly one" and "at most one" mean the same thing.
12. Two planes perpendicular to the same plane are parallel.
13. A plane which cuts one of two parallel planes cuts the other also.
14. Two lines perpendicular to the same line are parallel to each other.
15. Two lines parallel to the same line are parallel to each other.
16. Two lines parallel to the same plane are parallel to each other.
17. Two lines skew to the same line are skew to each other.
18. An exterior angle of a triangle has a measure greater than that of any interior angle of the triangle.
19. If two lines are cut by a transversal, there are exactly four pairs of alternate interior angles formed.

20. If l, m, and n are three lines such that $l \perp m$ and $m \perp n$, then $l \perp n$.
21. An exterior angle of a triangle is the supplement of at least one interior angle of the triangle.
22. In a right triangle with an acute angle whose measure is 30, the measure of the hypotenuse is one-half the measure of the side opposite the 30 angle.
23. When two parallel lines are cut by a transversal the two interior angles on the same side of the transversal are complementary.
24. If l, m, and n are lines, $l \parallel m$, $l \perp n$, then $n \perp m$.
25. If l, m, n, and p are lines, $l \parallel m$, $n \perp l$, $p \perp m$, and $n \neq p$, then $n \parallel p$.
26. Line l passes through P and is parallel to line m if and only if $P \in l$ and $l \cap m = \emptyset$.
27. If transversal t intersects line l at A and line m at B, then $t \cap (l \cup m) = \{A, B\}$, where $A \neq B$.

Test 3

PROBLEMS

1–8. Solve for $m\angle\alpha$:

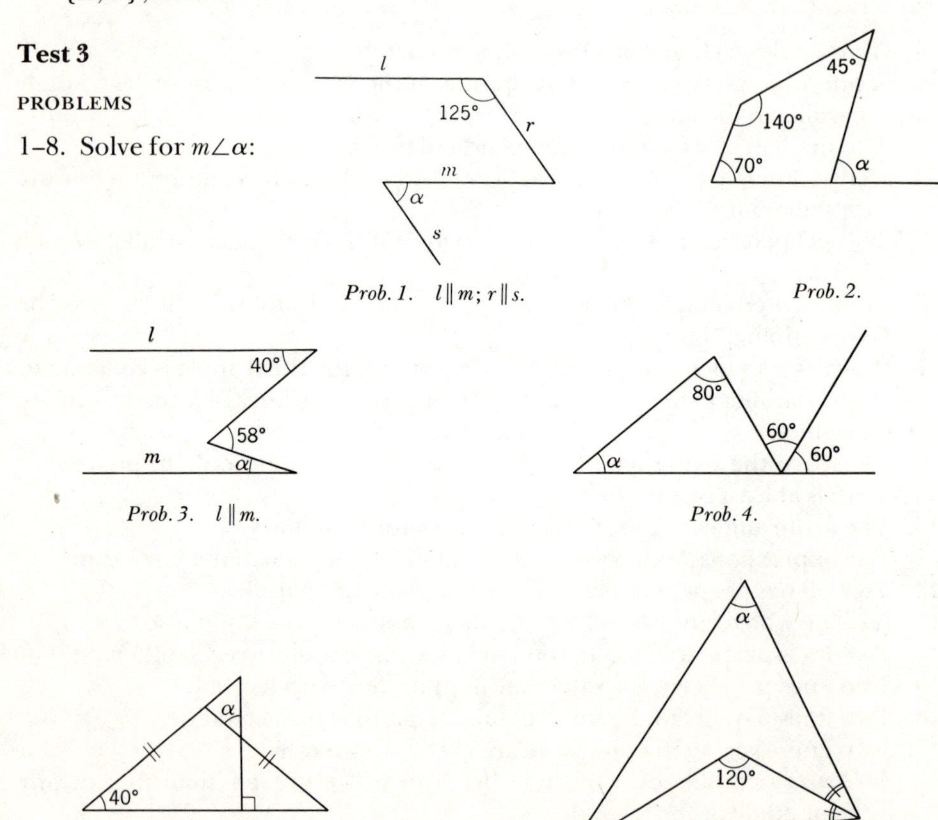

Prob. 1. $l \parallel m$; $r \parallel s$.

Prob. 2.

Prob. 3. $l \parallel m$.

Prob. 4.

Prob. 5.

Prob. 6.

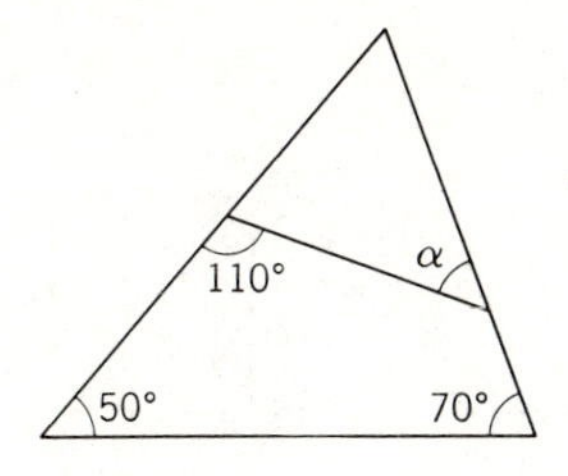

Prob. 7.

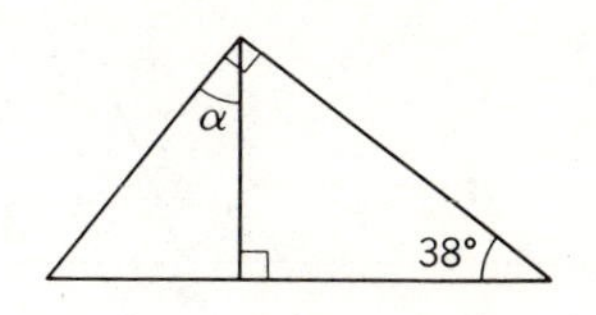

Prob. 8.

Test 4

EXERCISES

1. Supply the reasons for the statements in the following proof:

Given: $\overline{AC} \cong \overline{BC}$; $\overline{CD} \cong \overline{CE}$.
Prove: $\overline{DF} \perp \overline{AB}$.

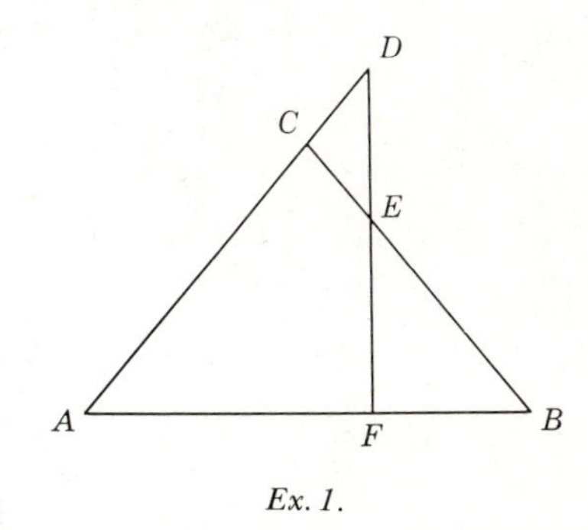

Ex. 1.

Proof:

STATEMENTS	REASONS
1. $\overline{AC} \cong \overline{BC}$; $\overline{CD} \cong \overline{CE}$.	1.
2. $m\angle A = m\angle B$; $m\angle CDE = m\angle CED$.	2.
3. $m\angle AFD = m\angle FEB + m\angle B$.	3.
4. $m\angle FEB = m\angle CED$.	4.
5. $m\angle FEB = m\angle CDE$.	5.
6. $\therefore m\angle AFD = m\angle CDE + m\angle A$.	6.
7. $m\angle AFD + m\angle CDE + m\angle A = 180$.	7.
8. $m\angle AFD + m\angle AFD = 180$.	8.
9. $m\angle AFD = 90$.	9.
10. $\therefore \overline{DF} \perp \overline{AB}$.	10.

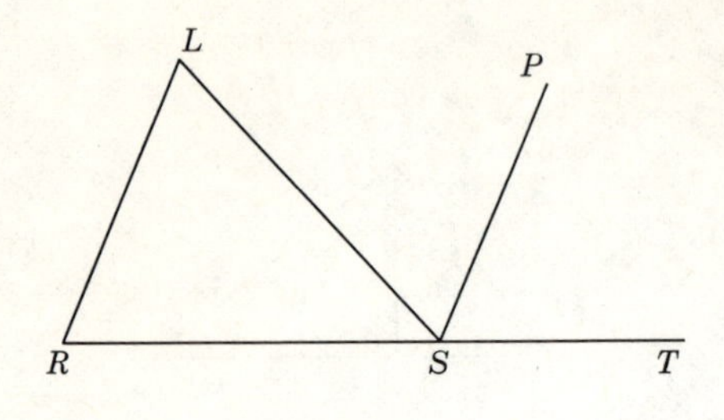

Ex. 2.

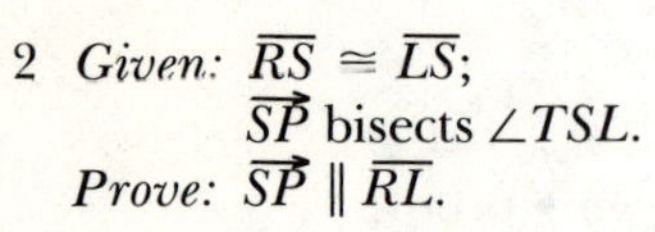

2 *Given:* $\overline{RS} \cong \overline{LS}$;
$\overrightarrow{SP}$ bisects $\angle TSL$.
Prove: $\overrightarrow{SP} \parallel \overline{RL}$.

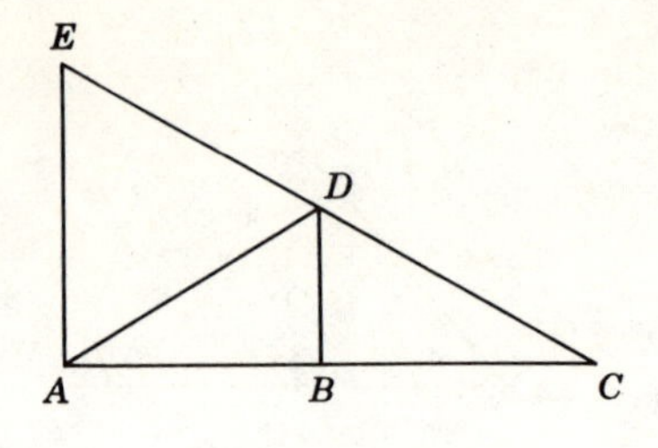

Ex. 3.

3. *Given:* $\overrightarrow{DB}$ bisects $\angle ADC$;
$\overline{BD} \parallel \overline{AE}$.
Prove: $\triangle ADE$ is isosceles.

6

Polygons – Parallelograms

6.1. Polygons. Many man-made and natural objects are in the shape of polygons. We see polygons in our buildings, the windows, the tile on our floors and walls, the flag, and the ordinary pencil. Many snowflakes under a microscope would be recognized as polygons. The cross section of the bee honeycomb is a polygon. Figure 6.1 illustrates various polygons.

Definitions: A *polygon* is a set of points which is the union of segments such that: (1) each endpoint is the endpoint of just two segments; (2) no two segments intersect except at an endpoint; and (3) no two segments with the same endpoint are collinear. The segments are called *sides* of the polygon. The endpoints are called *vertices* of the polygon. *Adjacent sides* of the polygon are those pairs of sides that share a vertex. Two vertices are called *adjacent vertices* if they are endpoints of the same side. Two angles of a polygon are *adjacent angles* if their vertices are adjacent.

A less rigorous definition for a polygon could be that it is a closed figure whose sides are segments. If each of the sides of a polygon is extended and the extensions intersect no other side, the polygon is a *convex polygon.* Figure 6.1a, b, c, d, e illustrate convex polygons. Figure 6.1f illustrates a polygon that is not convex.

In this text we will confine our study to convex polygons.

6.2. Kinds of polygons. A polygon can be named according to the number of its sides. The most fundamental subset of the set of polygons is the set of polygons having the least number of sides – the set of triangles. Every polygon of more than three sides can be subdivided, by properly drawing segments, into a set of distinct triangles.

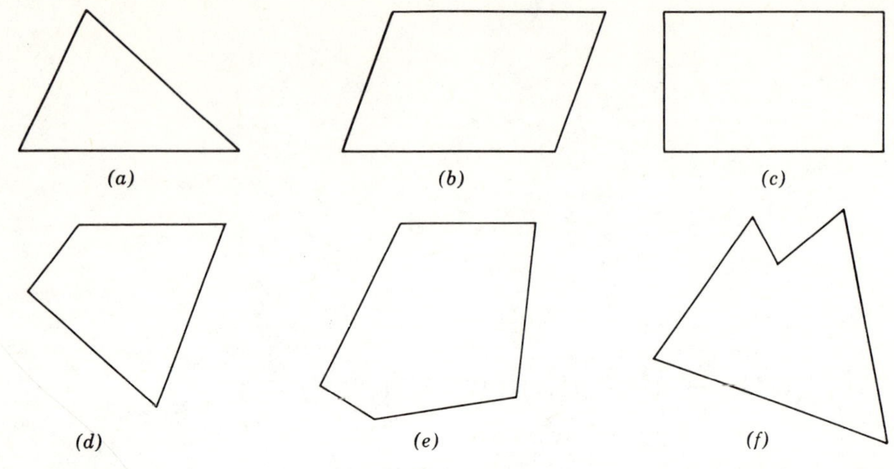

Fig. 6.1. Polygons.

Definitions: A polygon is a *quadrilateral* iff it has four sides; it is a *pentagon* iff it has five sides, a *hexagon* iff it has six sides, an *octagon* iff it has eight sides, a *decagon* iff it has ten sides, and an *n-gon* iff it has *n* sides.

Definitions: A polygon is *equilateral* iff all its sides are congruent. A polygon is *equiangular* iff all its angles are congruent. A polygon is a *regular polygon* iff it is both equilateral and equiangular.

Definitions: The sum of the measures of the sides of a polygon is called the *perimeter* of the polygon. The perimeter will always be a positive number. A *diagonal* of a polygon is a segment whose endpoints are nonadjacent vertices of the polygon. The side upon which the polygon appears to rest is called the *base* of the polygon. In Fig. 6.2, *ABCDE* is a polygon of five

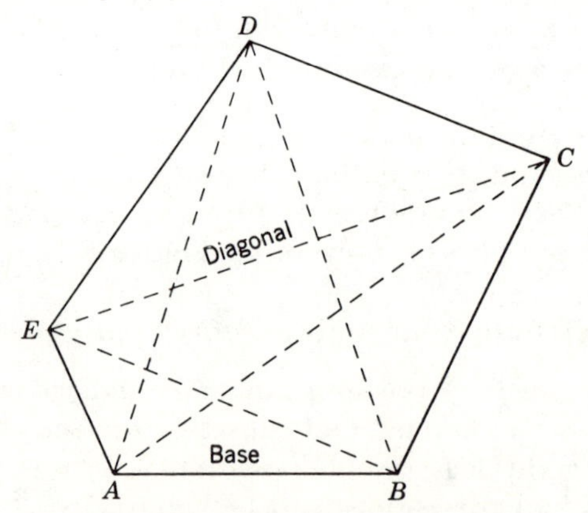

Fig. 6.2. Diagonals and base of a polygon.

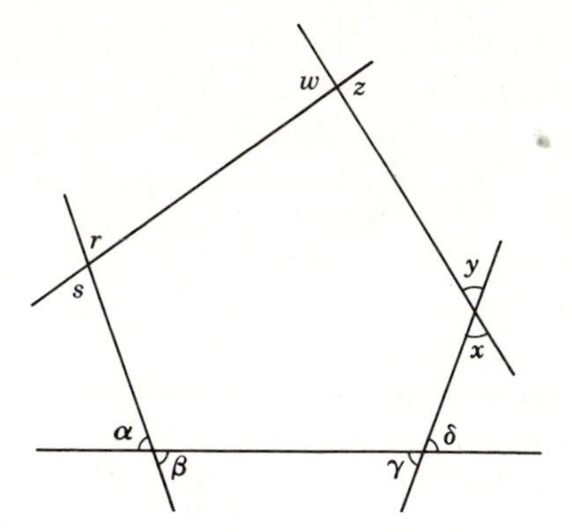

Fig. 6.3 Exterior angles of a polygon.

sides; A, B, C, D, E are vertices of the polygon; $\angle A$, $\angle B$, $\angle C$, $\angle D$, $\angle E$ are the angles of the polygon. There are two diagonals drawn from each vertex of the figure.

Definition: An *exterior angle* of a polygon is an angle that is adjacent to and supplementary to an angle of the polygon (Fig. 6.3).

6.3. Quadrilaterals. Unlike the triangle, the quadrilateral is not a rigid figure. The quadrilateral may assume many different shapes. Some quadrilaterals with special properties are referred to by particular names. We will define a few of them.

Definitions: A quadrilateral is a *trapezoid* (symbol ⏢) iff it has one and only one pair of parallel sides (Fig. 6.4). The parallel sides are the *bases* (upper and lower) of the trapezoid. The nonparallel sides are the *legs.* The *altitude* of a trapezoid is a segment, as $\overline{DE}$, which is perpendicular to one of the bases and whose endpoints are elements of the lines of which the bases are subsets. Often the word altitude is used to mean the distance between the bases. The *median* is the line segment connecting the midpoints of the nonparallel sides. An *isosceles trapezoid* is one the legs of which are congruent (Fig. 6.5). A pair of angles which share a base is called *base angles.*

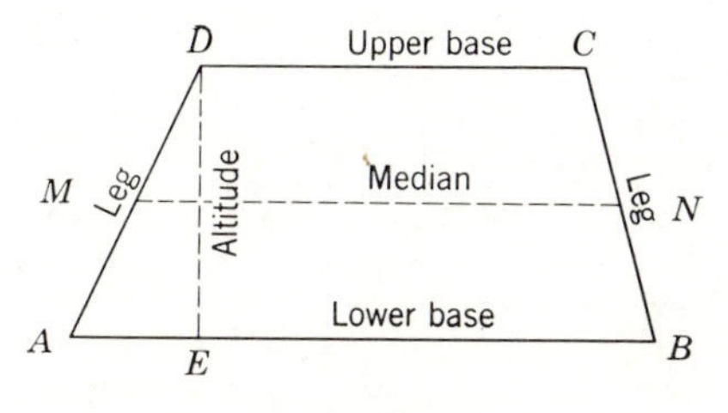

Fig. 6.4.

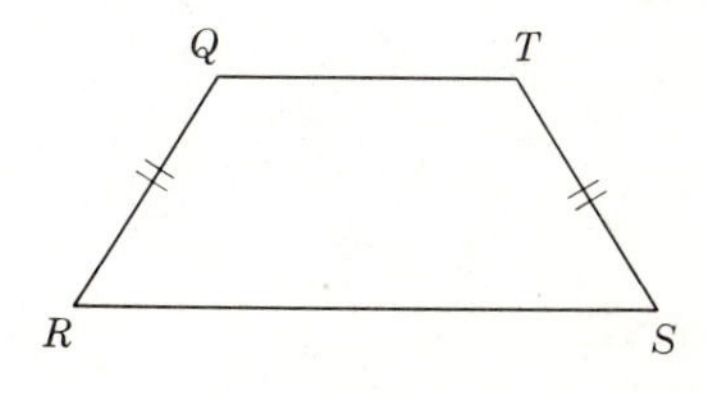

Fig. 6.5.

Definitions: A quadrilateral is a *parallelogram* (symbol ▱) iff the pairs of opposite sides are parallel. Any side of the parallelogram may be called the base, as $\overline{AB}$ of Fig. 6.6. The distance between two parallel lines is the perpendicular distance from any point on one of the lines to the other line. An *altitude* of a parallelogram is the segment perpendicular to a side of the parallelogram and whose endpoints are in that side and the opposite side or the line of which the opposite side is a subset. $\overline{DE}$ of Fig. 6.6 is an altitude of ▱ $ABCD$. Here, again, "altitude" is often referred to as the distance between the two parallel sides. A parallelogram has two altitudes.

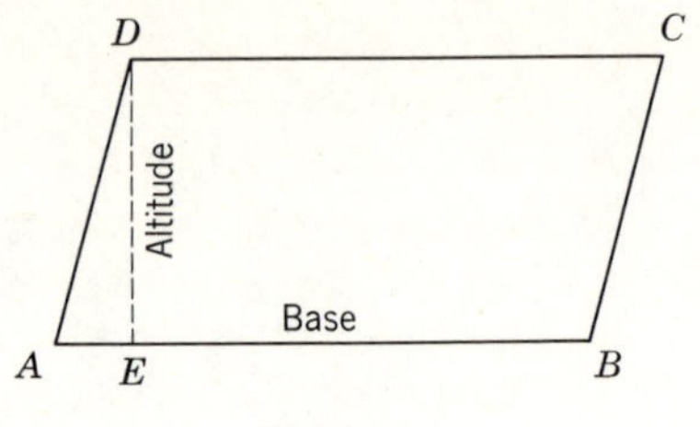

Fig. 6.6.

A rhombus is an equilateral parallelogram (Fig. 6.7).

A rectangle (symbol ▭) is a parallelogram that has a right angle (Fig. 6.7).

A rectangle is a square iff it has four congruent sides. Thus it is an equilateral rectangle (Fig. 6.7).

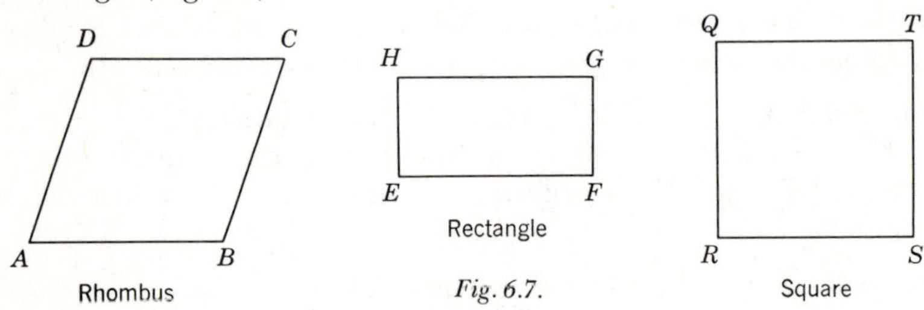

Fig. 6.7.

Theorem 6.1

6.4. All angles of a rectangle are right angles.

Given: $ABCD$ is a rectangle with $\angle A$ a right angle.
Prove: $\angle B$, $\angle C$, and $\angle D$ are right angles.
(*The proof of this theorem is left to the student. Hint: Use Theorem 5.15.*)

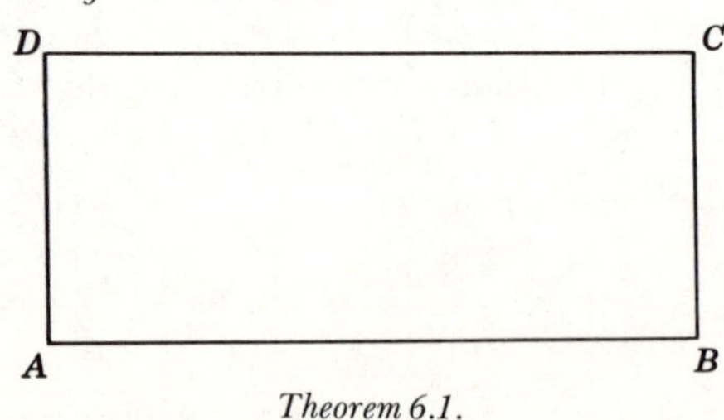

Theorem 6.1.

Exercises (A)

Indicate which of the following statements are *always true* and which are *not always true*.

1. The sides of polygons are segments.
2. The opposite sides of a trapezoid are parallel.
3. Every quadrilateral has two diagonals.
4. Some trapezoids are equiangular.
5. All rectangles are equiangular.
6. The set of parallelograms are subsets of rectangles.
7. An octagon has eight angles.
8. An octagon has five diagonals.
9. The set of diagonals of a given triangle is a null set.
10. The diagonals of a polygon need not be coplanar.
11. Every polygon has at least three angles.
12. If a polygon does not have five sides it is not a pentagon.
13. A rhombus is a regular polygon.
14. Each exterior angle of a polygon is supplementary to its adjacent angle of the polygon.
15. Only five exterior angles can be formed from a given pentagon.
16. A square is a rectangle.
17. A square is a rhombus.
18. A square is a parallelogram.
19. A rectangle is a square.
20. A rectangle is a rhombus.
21. A rectangle is a parallelogram.
22. A quadrilateral is a polygon.
23. A quadrilateral is a trapezoid.
24. A quadrilateral is a rectangle.
25. A polygon is a quadrilateral.

Exercises (B)

1. Draw a convex quadrilateral and a diagonal from one vertex. Determine the sum of the measures of the four angles of the quadrilateral.
2. Draw a convex pentagon and as many diagonals as possible from one of its vertices. (a) How many triangles are formed? (b) What will be the sum of the measures of the angles of the pentagon?
3. Repeat problem 2 for a hexagon.
4. Repeat problem 2 for an octagon.
5. Using problems 2–5 as a guide, what would be the sum of the measures of the angles of a polygon of 102 sides?
6. What is the measure of each angle of a regular pentagon?
7. What is the measure of each angle of a regular hexagon?
8. What is the measure of each exterior angle of a regular octagon?
9. What is the measure of each angle of a regular decagon?

10. Using the set of polygons as the Universal set, draw a Venn diagram relating polygons, rhombuses, quadrilaterals, and parallelograms.
11. Using the set of quadrilaterals as the Universal set, draw a Venn diagram relating quadrilaterals, squares, parallelograms, rhombuses, and trapezoids.
12. Using the set of parallelograms as the Universal set, draw a Venn diagram relating parallelograms, squares, rectangles, and rhombuses.

Theorem 6.2

6.5. The opposite sides and the opposite angles of a parallelogram are congruent.

Given: $\square ABCD$.
Conclusion: $\overline{AB} \cong \overline{DC}$; $\overline{AD} \cong \overline{BC}$;
$\angle A \cong \angle C$; $\angle B \cong \angle D$.

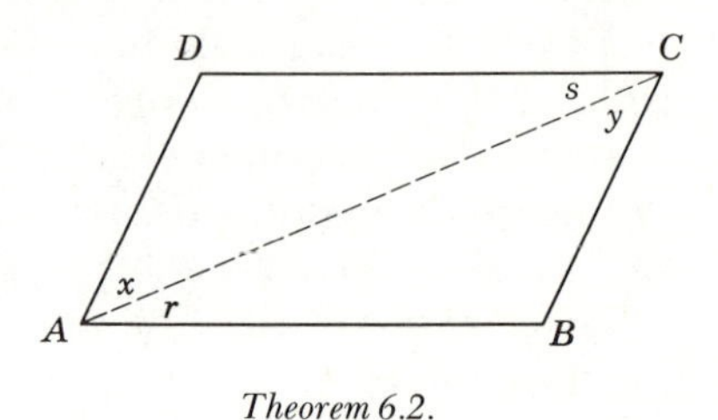

Theorem 6.2.

Proof:

STATEMENTS	REASONS
1. *ABCD* is a $\square$.	1. Given.
2. Draw the diagonal *AC*.	2. Postulate 2.
3. $\overline{AB} \parallel \overline{DC}$; $\overline{AD} \parallel \overline{BC}$.	3. Definition of $\square$.
4. $\angle x \cong \angle y$; $\angle r \cong \angle s$.	4. Theorem 5.13.
5. $\overline{AC} \cong \overline{AC}$.	5. Reflexive property of congruent segments.
6. $\triangle ABC \cong \triangle CDA$.	6. A.S.A.
7. $\overline{AB} \cong \overline{DC}$; $\overline{AD} \cong \overline{BC}$.	7. Corresponding parts of $\cong$ ⨻ are congruent.
8. $\angle B \cong \angle D$.	8. Corresponding parts of $\cong$ ⨻ are congruent.
9. $\angle A \cong \angle C$.	9. Angle addition theorem.

6.6. Corollary: Either diagonal divides a parallelogram into two congruent triangles.

6.7. Corollary: Any two adjacent angles of a parallelogram are supplementary.

6.8. Corollary: Segments of a pair of parallel lines cut off by a second pair of parallel lines are congruent.

6.9. Corollary: Two parallel lines are everywhere equidistant.

6.10. Corollary: The diagonals of a rectangle are congruent.

Theorem 6.3

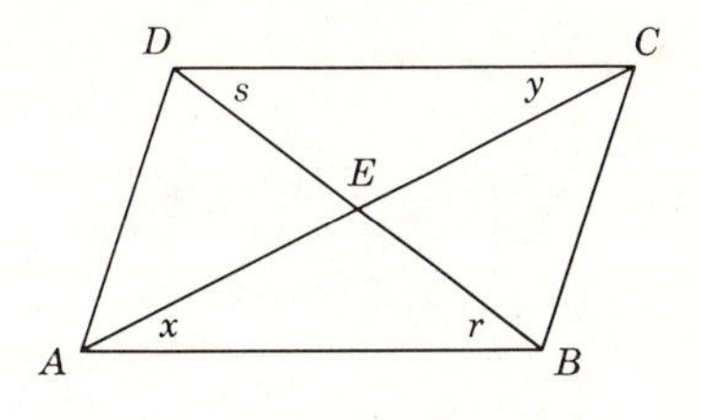

Theorem 6.3.

6.11. The diagonals of a parallelogram bisect each other.

Given: ▭$ABCD$ with diagonals intersecting at E.

Conclusion: $\overline{AC}$ and $\overline{BD}$ bisect each other.

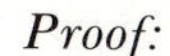

Proof:

STATEMENTS	REASONS
1. $ABCD$ is a ▭.	1. Given.
2. $\overline{AB} \parallel \overline{DC}$.	2. Definition of a ▭.
3. $\angle z \cong \angle y$, $\angle r \cong \angle s$.	3. Theorem 5.13.
4. $\overline{AB} \cong \overline{DC}$.	4. Theorem 6.2.
5. $\triangle ABE \cong \triangle CDE$.	5. A.S.A.
6. $\overline{AE} \cong \overline{EC}$, and $\overline{BE} \cong \overline{DE}$.	6. Corresponding sides of $\cong$ ⟁ are $\cong$.
7. $\overline{AC}$ and $\overline{BD}$ bisect each other.	7. Definition of bisector.

6.12. Corollary: The diagonals of a rhombus are perpendicular to each other.

Exercises (A)

Copy the chart below. Then put check marks (x) whenever the polygon has the indicated relationship.

Relationships	All sides are $\cong$	Oppossite sides are $\cong$	Oppossite sides are $\parallel$	Diagonals bisect each other	Diagonals bisect the ∠s of polygon	Opposite ∠s are $\cong$	Diagonals are $\cong$	Diagonals are $\perp$
Parallelogram								
Rectangle								
Rhombus								
Square								
Trapezoid								
Isosceles trapezoid								

Exercises (B)

1. *Given:* $ABCD$ is a ▱;
   $\overline{DR} \perp \overline{AC}$; $\overline{BT} \perp \overline{AC}$.
   Prove: $\overline{DR} \cong \overline{BT}$.

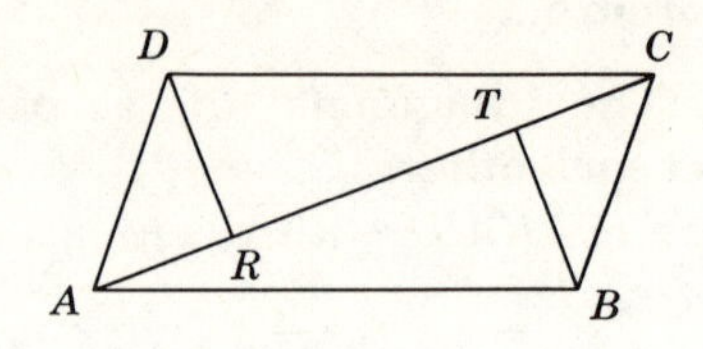

Ex. 1.

2. *Given:* $QRST$ is a ▱;
   $\overline{RM} \cong \overline{NT}$.
   Prove: $\overline{QM} \cong \overline{SN}$.

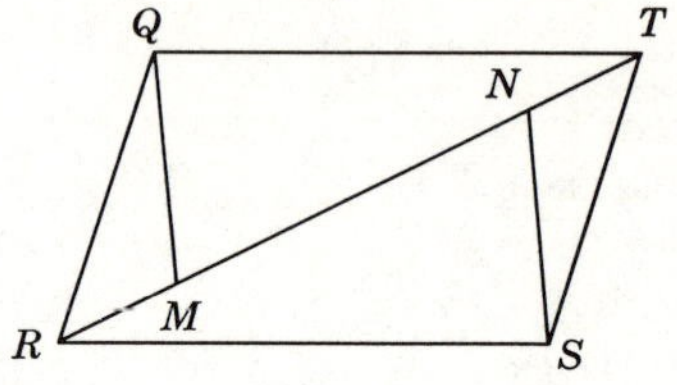

Ex. 2.

3. *Given:* $ABCD$ is a ▱;
   $\overline{DE} \perp \overline{AB}$; $\overline{CF} \perp \overline{AB}$ produced.
   Prove: $\overline{DE} \cong \overline{CF}$.

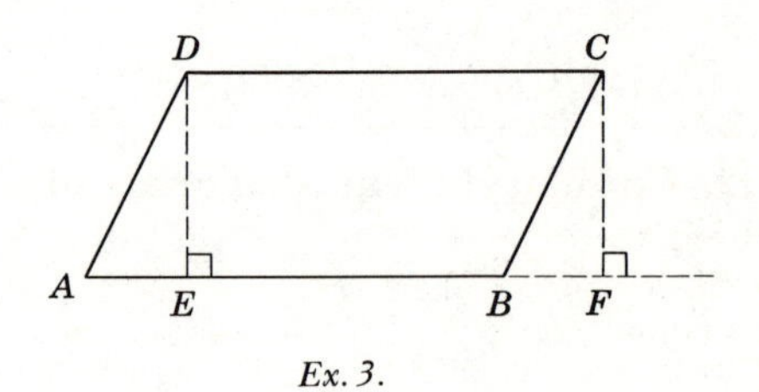

Ex. 3.

4. *Given:* $ABCD$ is an isosceles trapezoid with $\overline{AD} \cong \overline{BC}$.
   Prove: $\angle A \cong \angle B$.
   (*Hint:* Draw $\overline{CE} \parallel \overline{DA}$.)

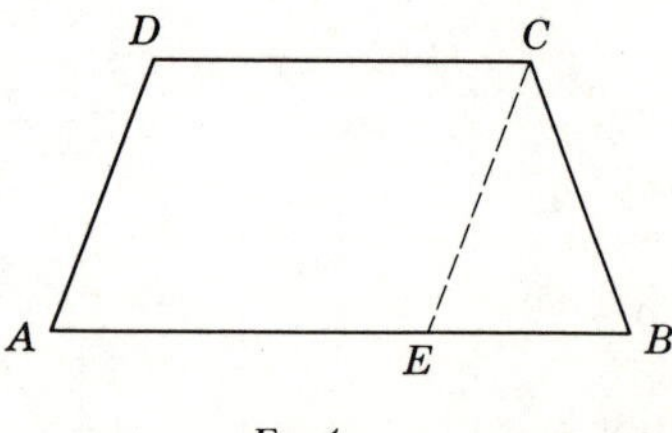

Ex. 4.

5. *Given:* $\overline{AB} \cong \overline{CD}$;
   $\overline{AD} \cong \overline{BC}$.
   Prove: $ABCD$ is a ▱.

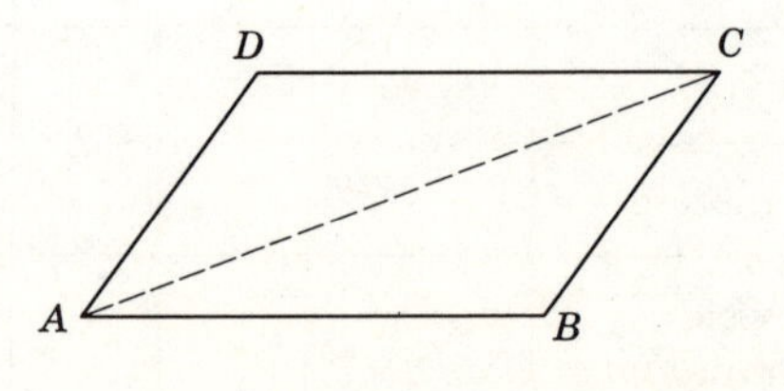

Ex. 5.

6. *Given:* $\overline{RS} \cong \overline{QT}$;
$\overline{RS} \parallel \overline{QT}$.
Prove: $QRST$ is a ▱.

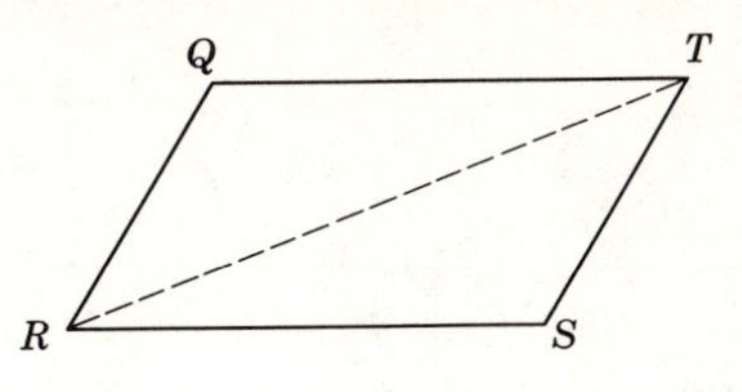

Ex. 6.

7. *Given:* $ABCD$ is a ▱;
$\overline{DE}$ bisects $\angle D$;
$\overline{BF}$ bisects $\angle B$.
Prove: $\overline{DE} \parallel \overline{BF}$.

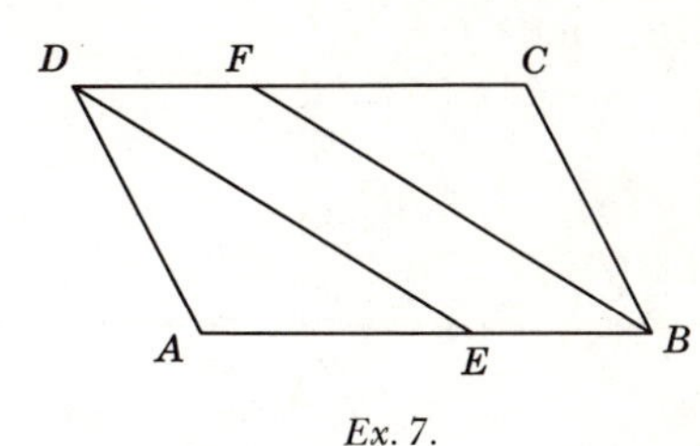

Ex. 7.

8. Prove that the diagonal QS of rhombus $QRST$ bisects $\angle Q$ and $\angle S$.

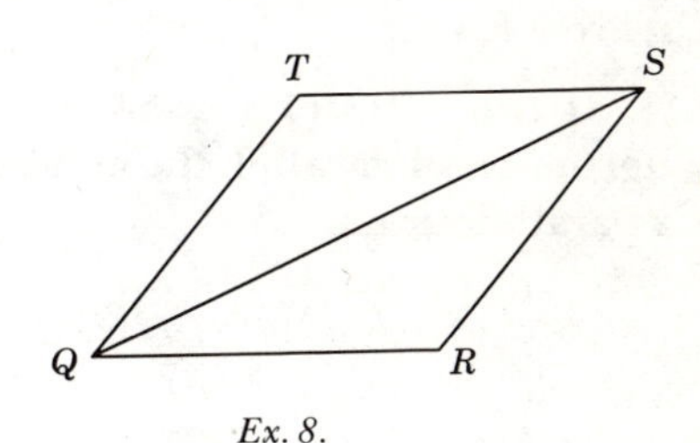

Ex. 8.

9. Prove that if the base angles of a trapezoid are congruent, the trapezoid is isosceles.
10. Prove that if the diagonals of a parallelogram are perpendicular to each other, the parallelogram is a rhombus.
11. Prove that if the diagonals of a parallelogram are congruent, it is a rectangle.
12. Prove that the bisectors of two consecutive angles of a parallelogram are perpendicular to each other.

Theorem 6.4

6.13. If the opposite sides of a quadrilateral are congruent, the quadrilateral is a parallelogram.

Given: Quadrilateral $ABCD$ with $\overline{AB} \cong \overline{CD}$; $\overline{AD} \cong \overline{BC}$.
Prove: $ABCD$ is a ▱.

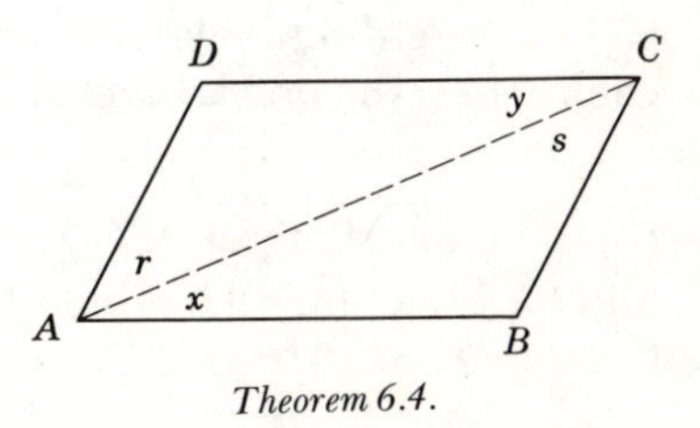

Theorem 6.4.

Proof:

STATEMENTS	REASONS
1. $\overline{AB} \cong \overline{CD}$; $\overline{AD} \cong \overline{BC}$.	1. Given.
2. Draw diagonal AC.	2. Postulate 2.
3. $\overline{AC} \cong \overline{AC}$.	3. Reflexive property of congruence.
4. $\triangle ABC \cong \triangle CDA$.	4. S.S.S.
5. $\angle x \cong \angle y$; $\angle r \cong \angle s$.	5. Corresponding parts of $\cong$ ⚠ are $\cong$.
6. $\overline{AB} \parallel \overline{CD}$; $\overline{AD} \parallel \overline{BC}$.	6. Theorem 5.11.
7. $\therefore ABCD$ is a ▱.	7. Definition of ▱.

Theorem 6.5

6.14. If two sides of a quadrilateral are congruent and parallel, the quadrilateral is a parallelogram.

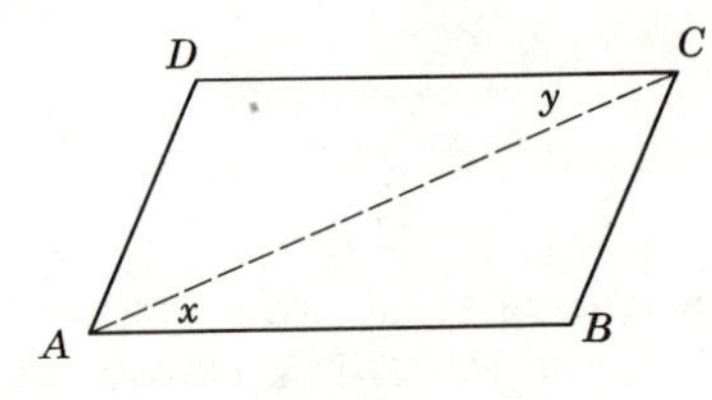

Theorem 6.5.

Given: Quadrilateral $ABCD$ with $\overline{AB} \cong \overline{CD}$; $\overline{AB} \parallel \overline{CD}$.
Conclusion: $ABCD$ is a ▱.

Proof:

STATEMENTS	REASONS
The proof is left to the student.	

Theorem 6.6

6.15. If the diagonals of a quadrilateral bisect each other, the quadrilateral is a parallelogram.

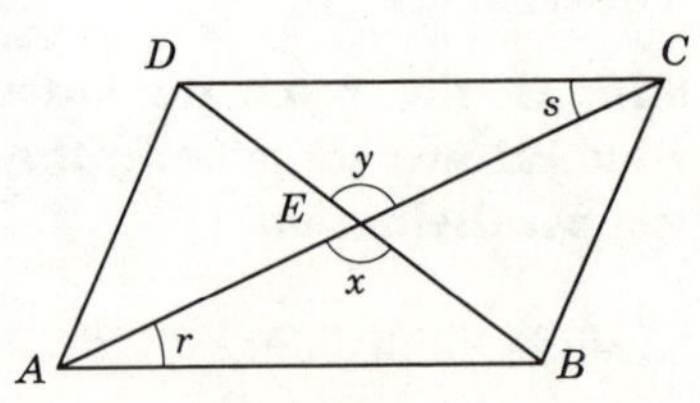

Theorem 6.6.

Given: Quadrilateral $ABCD$ with $\overline{AC}$ and $\overline{BD}$ bisecting each other at E.
Conclusion: $ABCD$ is a ▱.

Proof:

STATEMENTS	REASONS
1. $\overline{AC}$ and $\overline{BD}$ bisect each other at E.	1. Given.
2. $\overline{AE} \cong \overline{CE}$; $\overline{BE} \cong \overline{DE}$.	2. Definition of bisector.
3. $\angle x \cong \angle y$.	3. Vertical angles are congruent.
4. $\triangle ABE \cong \triangle CDE$.	4. S.A.S.
5. $\overline{AB} \cong \overline{CD}$.	5. Corresponding parts of $\cong$ ⚠ are congruent.
6. $\angle r \cong \angle s$.	6. Same as 4.
7. $\overline{AB} \parallel \overline{CD}$.	7. Theorem 5.11.
8. $ABCD$ is a ▱.	8. Theorem 6.5.

Theorem 6.7

6.16. If three or more parallel lines cut off congruent segments on one transversal, they cut off congruent segments on every transversal.

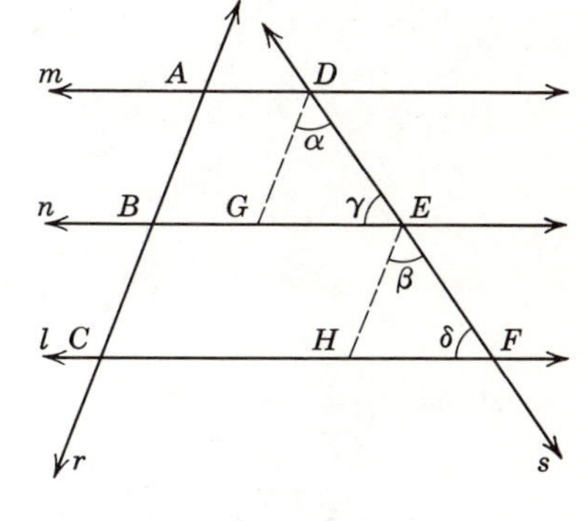

Theorem 6.7.

Given: Parallel lines l, m, and n cut by transversals r and s; $\overline{AB} \cong \overline{BC}$.
Conclusion: $\overline{DE} \cong \overline{EF}$.
Proof:

STATEMENTS	REASONS
1. Through D and E draw $\overline{DG} \parallel r$ and $\overline{EH} \parallel r$.	1. Postulate 18; Theorem 5.7.
2. $\overline{DG} \parallel \overline{EH}$.	2. Theorem 5.8.
3. $\overleftrightarrow{AD} \parallel \overleftrightarrow{BE} \parallel \overleftrightarrow{CF}$.	3. Given.
4. $\therefore ADGB$ and $BEHC$ are ▱s.	4. Definition of ▱.
5. $\overline{AB} \cong \overline{DG}$ and $\overline{BC} \cong \overline{EH}$.	5. Theorem 6.2.
6. $\overline{AB} \cong \overline{BC}$.	6. Given.
7. $\overline{DG} \cong \overline{EH}$.	7. Theorem 3.5 and transitive property of congruence.
8. $\angle\alpha \cong \angle\beta$ and $\angle\gamma \cong \angle\delta$.	8. Theorem 5.14.
9. $\angle DGE \cong \angle EHF$.	9. § 5.28.
10. $\triangle DGE \cong \triangle EHF$.	10. A.S.A.
11. $\overline{DE} \cong \overline{EF}$.	11. § 4.28.

Exercises

1. *Given:* $ABCD$ is a ▱;
   M is midpoint of $\overline{AD}$;
   N is midpoint of $\overline{BC}$.
   Prove: $MBND$ is a ▱.

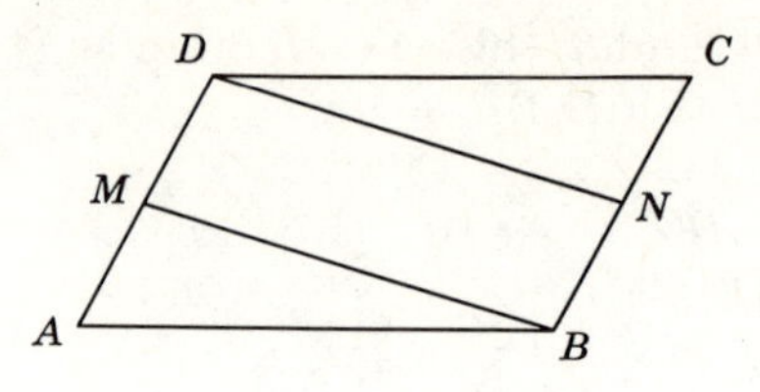

Ex. 1.

2. *Given:* $\overline{QR} \parallel \overline{ST}$;
   $\angle x \cong \angle y$.
   Prove: $QRST$ is a ▱.

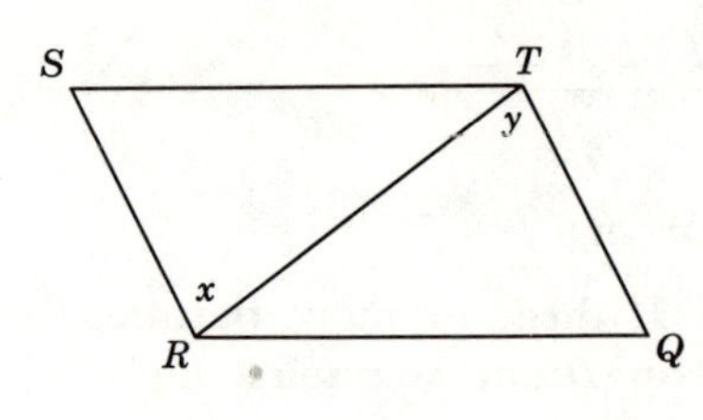

Ex. 2.

3. *Given:* $ABCD$ is a ▱;
   $\overline{AM} \cong \overline{CN}$.
   Prove: $MBND$ is a ▱.

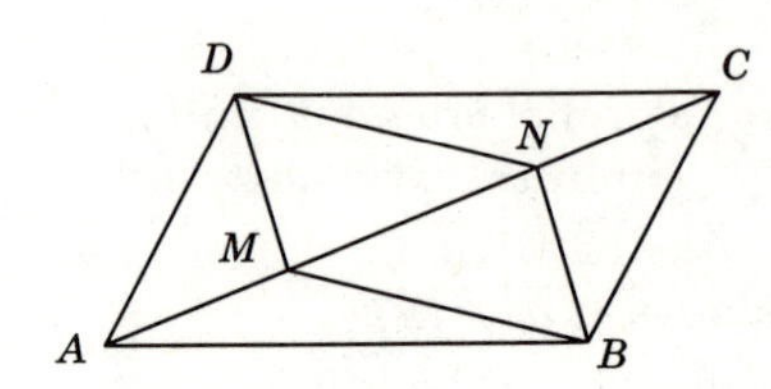

Ex. 3.

4. *Given:* $QRST$ is a ▱;
   $\overline{QL}$ bisects $\angle TQR$;
   $\overline{SM}$ bisects $\angle RST$.
   Prove: $QLSM$ is a ▱.

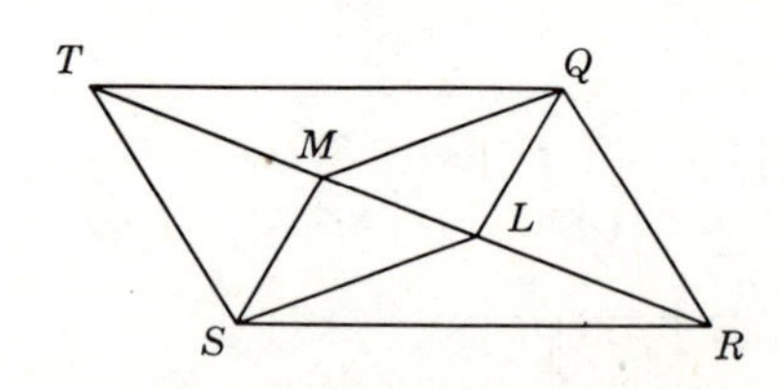

Ex. 4.

5. *Given:* ▱ $ABCD$ with diagonals intersecting at E.
   Prove: E bisects $\overline{FG}$.

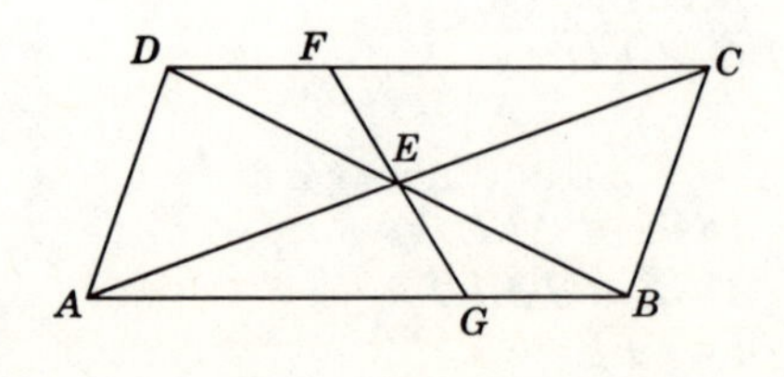

Ex. 5.

6. *Given:* $LMNP$ is a ▱.
   $\overline{PR} \perp \overline{LN}$;
   $\overline{MS} \perp \overline{LN}$.
   Prove: $RMSP$ is a ▱.

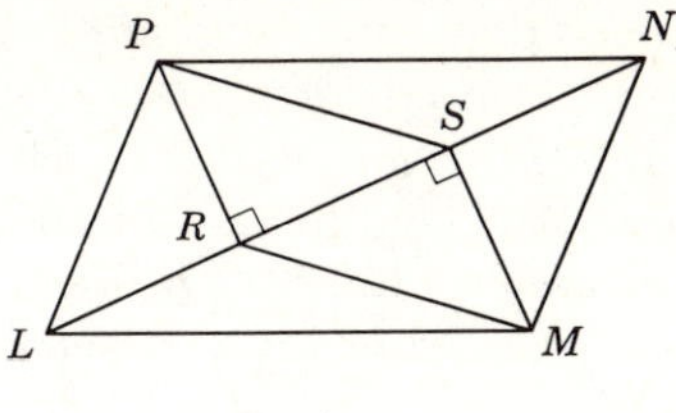

Ex. 6.

7. *Given:* $\triangle ABC$ with D midpoint of $\overline{AC}$; E midpoint of $\overline{BC}$; $\overline{DE} \cong \overline{EF}$.
   Prove: $ABFD$ is a ▱.
8. In Ex. 7, prove $m\overline{DE} = \frac{1}{2}m\overline{AB}$.

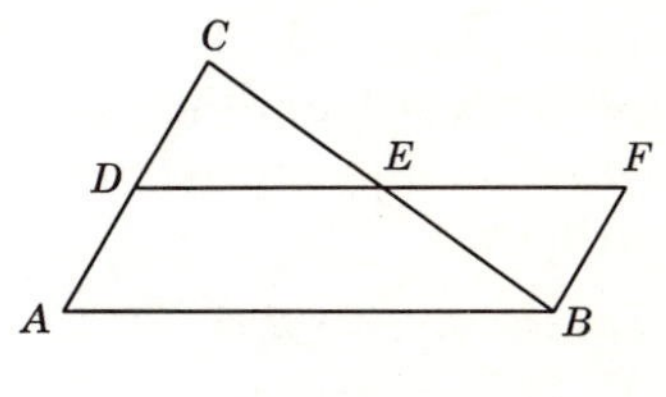

Exs. 7, 8.

9. Prove that two parallelograms are congruent if two sides and the included angle of one are congruent respectively to two sides and the included angle of the other.

10. *Given:* ▱ $QRST$ with $\overline{AQ} \cong \overline{SC}$;
    $\overline{RB} \cong DT$.
    Prove: $ABCD$ is a ▱.

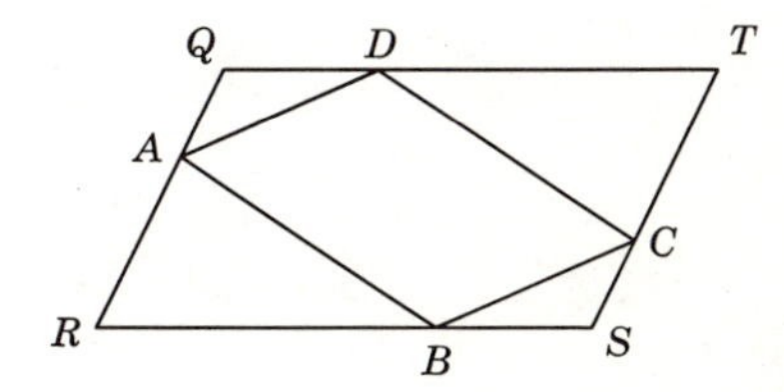

Ex. 10.

11. *Given:* Trapezoid $ABCD$ with $\overline{AB} \parallel \overline{DC}$;
    $\overline{AD} \cong \overline{DC}$.
    Prove: $\overline{AC}$ bisects $\angle A$.

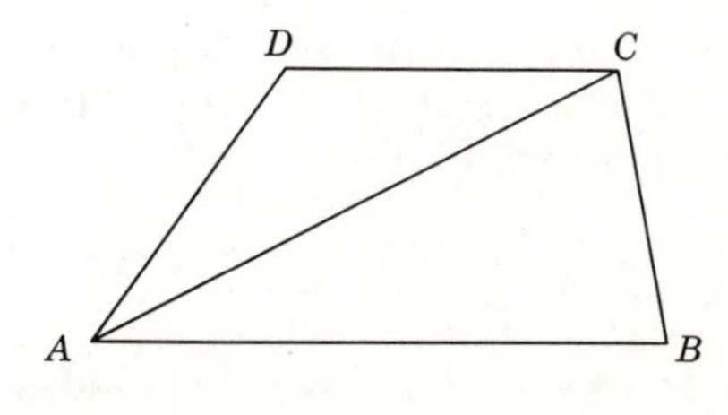

Ex. 11.

12. Prove that line segments drawn from A and B of $\triangle ABC$ to the opposite sides cannot bisect each other. (*Hint:* Use indirect method by assuming $\overline{AS}$ and $\overline{RB}$ bisect each other; then $ABSR$ is a $\square$, etc.)

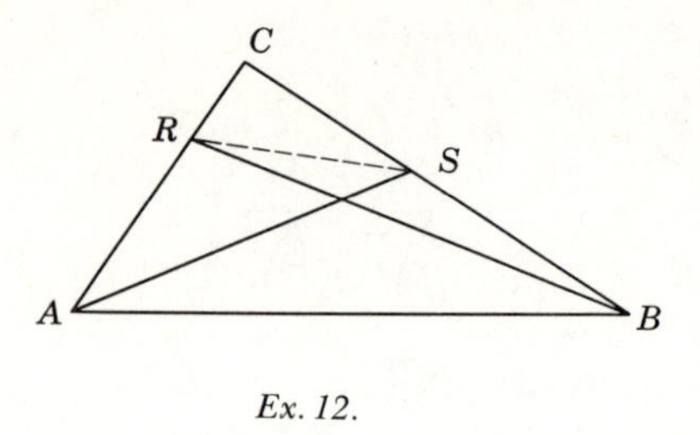

Ex. 12.

13. Prove that a quadrilateral is a rhombus if the diagonals bisect each other and are perpendicular to each other.
14. Prove that if from the point where the bisector of an angle of a triangle meets the opposite side parallels to the other sides are drawn a rhombus is formed.

6.17. Direction of rays. Two rays have the *same direction* if and only if either they are parallel and are contained in the same closed half-plane determined by the line through their endpoints *or* if one ray is a subset of the other (Fig. 6.8).

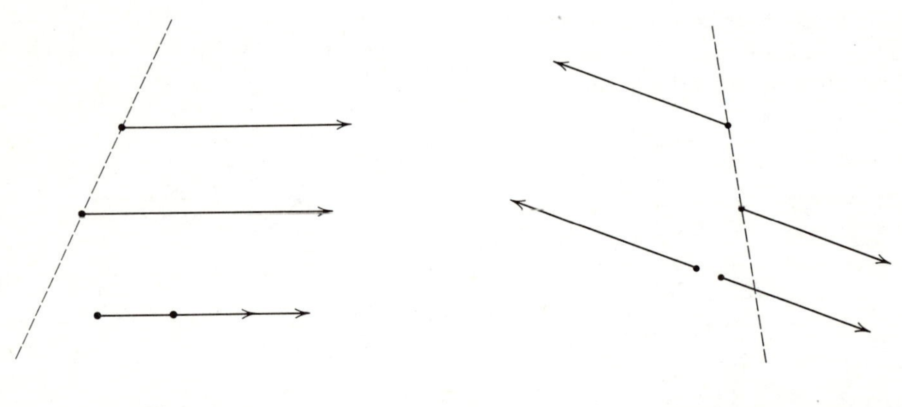

Fig. 6.8. *Fig. 6.9.*

Two rays have *opposite directions* if and only if either they are parallel and are contained in opposite closed half-planes determined by the line through their endpoints *or* are collinear and the intersection of the rays is a point, segment, or a null set (Fig. 6.9).

Theorem 6.8

6.18. If two angles have their sides so matched that corresponding sides have the same directions, the angles are congruent.

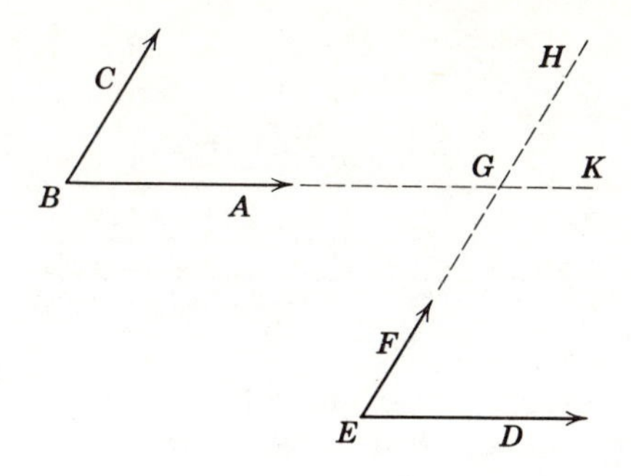

Theorem 6.8.

Given: $\angle ABC$ and $\angle DEF$ with
$\overrightarrow{BC} \parallel \overrightarrow{EF}$ and with same direction;
$\overrightarrow{BA} \parallel \overrightarrow{ED}$ and with same direction.
Conclusion: $\angle ABC \cong \angle DEF$.
Proof:

STATEMENTS	REASONS
1. Extend $\overrightarrow{BA}$ and $\overrightarrow{EF}$. Label their intersection G.	1. A ray has infinite length in one direction.
2. $\overrightarrow{BC} \parallel \overrightarrow{EF}$.	2. Given.
3. $\angle ABC \cong \angle KGH$.	3. Theorem 5.14.
4. $\overrightarrow{BA} \parallel \overrightarrow{ED}$.	4. Given.
5. $\angle DEF \cong \angle KGH$.	5. Theorem 5.14.
6. $\angle ABC \cong \angle DEF$.	6. Theorem 3.4.

Theorem 6.9

6.19. If two angles have their sides so matched that two corresponding sides have the same direction and the other two corresponding sides are oppositely directed, the angles are supplementary.

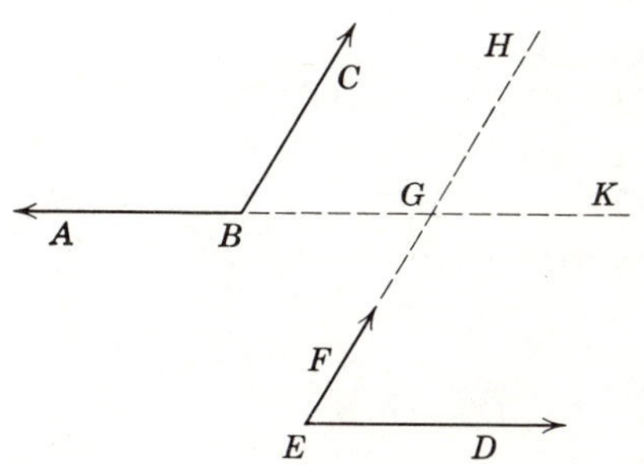

Theorem 6.9.

Given: $\angle ABC$ and $\angle DEF$ with
$\overrightarrow{BC} \parallel \overrightarrow{EF}$ and with the same direction;
$\overrightarrow{BA} \parallel \overrightarrow{ED}$ and with opposite directions.
Conclusion: $\angle ABC$ and $\angle DEF$ are supplementary.

Proof:

STATEMENTS	REASONS
The proof is left as an exercise for the student.	

Theorem 6.10

6.20. The segment joining the midpoints of two sides of a triangle is parallel to the third side, and its measure is one half the measure of the third side.

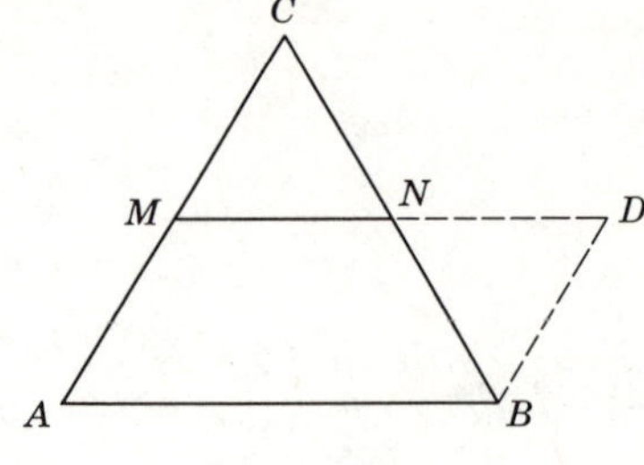

Theorem 6.10.

Given: $\triangle ABC$ with the M the midpoint of $\overline{AC}$ and N the midpoint of $\overline{BC}$.
Conclusion: $\overline{MN} \parallel \overline{AB}$, $m\overline{MN} = \frac{1}{2}m\overline{AB}$.
Proof:

STATEMENTS	REASONS
1. On the ray opposite $\overrightarrow{NM}$ construct $\overline{ND}$ such that $\overline{ND} \cong \overline{MN}$.	1. Point plotting postulate.
2. Draw $\overline{BD}$.	2. Postulate 2.
3. N is the midpoint of $\overline{BC}$.	3. Given.
4. $\overline{NB} \cong \overline{NC}$.	4. Definition of midpoint.
5. $\angle DNB \cong \angle MNC$.	5. Vertical angles are $\cong$.
6. $\triangle DNB \cong \triangle MNC$.	6. S.A.S.
7. $\overline{BD} \cong \overline{CM}$.	7. Corresponding sides of $\cong$ ⚠ are $\cong$.
8. M is the midpoint of $\overline{AC}$.	8. Given.
9. $\overline{CM} \cong \overline{AM}$.	9. Reason 4.
10. $\overline{BD} \cong \overline{AM}$.	10. Theorem 4.3.
11. $\angle DBN \cong \angle MCN$.	11. Reason 7.
12. $\overline{BD} \parallel \overline{AC}$.	12. Theorem 5.11.
13. $ABDM$ is a ▱.	13. Theorem 6.5.
14. $\overline{MN} \parallel \overline{AB}$.	14. Definition of a ▱.
15. $\overline{MD} \cong \overline{AB}$.	15. The opposite sides of a ▱ are $\cong$.
16. $m\overline{MD} = m\overline{AB}$.	16. Definition of congruent segments.
17. $m\overline{MD} = m\overline{MN} + m\overline{ND}$.	17. Postulate 13; symmetric property.

18. $m\overline{AB} = m\overline{MN} + m\overline{MN}$.	18. Theorem 3.5 and substitution property.
19. $m\overline{MN} + m\overline{MN} = m\overline{AB}$.	19. Symmetric property of equality.
20. $m\overline{MN} = \frac{1}{2}m\overline{AB}$.	20. Division property of equality.

Theorem 6.11

6.21. A line that bisects one side of a triangle and is parallel to a second side bisects the third side.

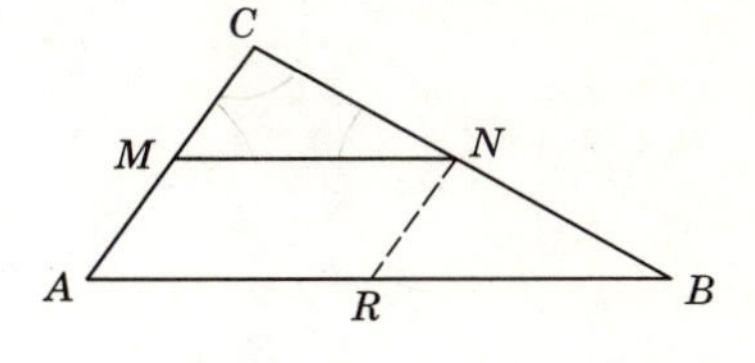

Theorem 6.11.

Given: $\overleftrightarrow{MN}$ bisects $\overline{AC}$,
$\overleftrightarrow{MN} \parallel \overline{AB}$.
Conclusion: $\overleftrightarrow{MN}$ bisects $\overline{BC}$.
Proof:

STATEMENTS	REASONS

The proof is left as an exercise for the student.

Theorem 6.12

6.22. The midpoint of the hypotenuse of a right triangle is equidistant from its vertices.

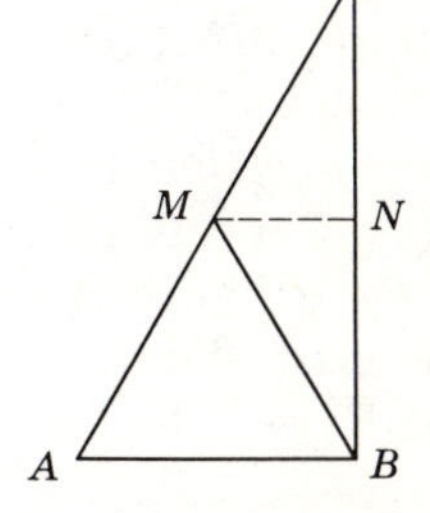

Theorem 6.12.

Given: Right $\triangle ABC$ with $\angle ABC$ a right $\angle$;
M is the midpoint of $\overline{AC}$.
Conclusion: $\overline{AM} \cong \overline{BM} \cong \overline{CM}$.
Proof:

STATEMENTS	REASONS

The proof is left as an exercise for the student. (*Hint:* Draw $\overrightarrow{MN} \parallel \overline{AB}$. Then prove $\triangle BMN \cong \triangle CMN$.)

Exercises

1. *Given:* $\triangle ABC$ with R, S, T midpoints of $\overline{AC}, \overline{BC}$, and $\overline{AB}$ respectively;

$m\overline{AC} = 6$ inches;
$m\overline{BC} = 8$ inches;
$m\overline{AB} = 12$ inches.
Find: the value of $m\overline{RS} + m\overline{ST} + m\overline{RT}$.

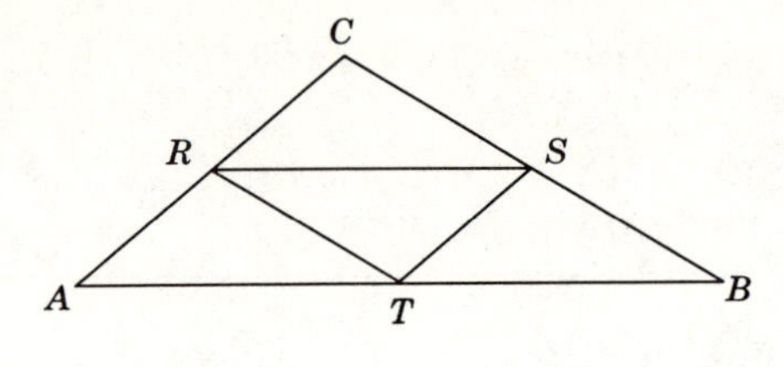

Ex. 1.

2. *Given:* $\angle C$ is a right $\angle$; $m\angle A = 60$; M is the midpoint of $\overline{AB}$, $m\overline{AC} = 8$ inches.
   Find: the value of $m\overline{AB}$.

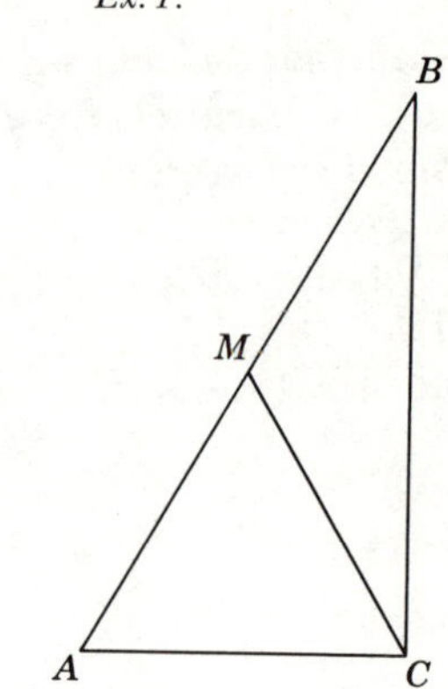

Ex. 2.

3. *Given:* $\overline{AB} \parallel \overline{DC}$; M is the midpoint of $\overline{AD}$; N is the midpoint of $\overline{BC}$.
   Prove: $\overline{MN} \parallel \overline{AB}$; $\overline{MN} \parallel \overline{DC}$; $m\overline{MN} = \frac{1}{2}(m\overline{AB} + m\overline{CD})$.
   (*Hint:* Draw $\overrightarrow{DN}$ until it meets $\overrightarrow{AB}$ at, say, P.)
4. In the figure for Ex. 3, find the length of $\overline{AB}$ if $m\overline{DC} = 8$ feet, $m\overline{MN} = 11$ feet.

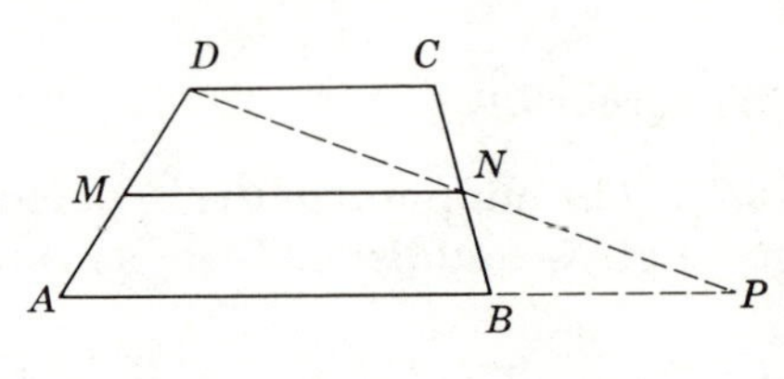

Exs. 3, 4.

5. *Given:* Quadrilateral $ABCD$ with Q, R, S, P, the midpoints of $\overline{AB}$, $\overline{BC}$, $\overline{CD}$, and $\overline{DA}$ respectively.
   Prove: $PQRS$ is a ▱.
   (*Hint:* You will need to draw the diagonals of $ABCD$.)

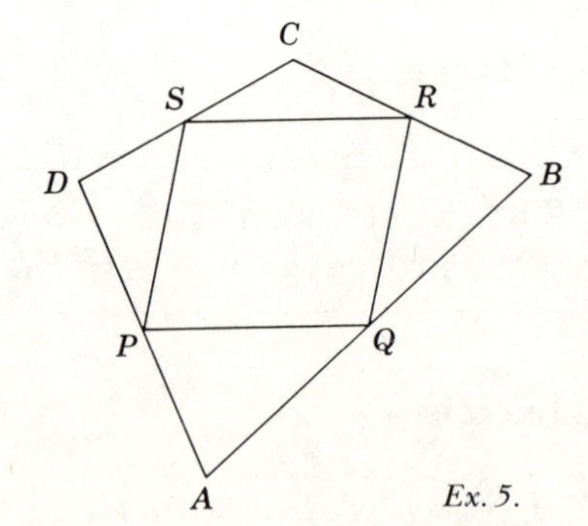

Ex. 5.

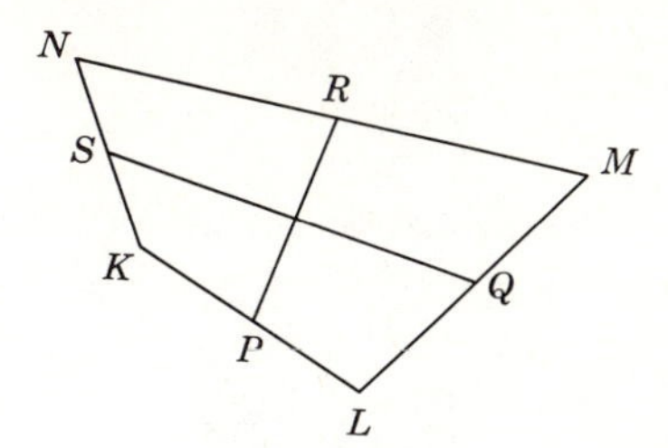

6. *Given:* Quadrilateral $KLMN$ with P, Q, R, S the midpoints of $\overline{KL}$, $\overline{LM}$, $\overline{MN}$, and $\overline{NK}$.
Prove: $\overline{PR}$ and $\overline{QS}$ bisect each other.

Ex. 6.

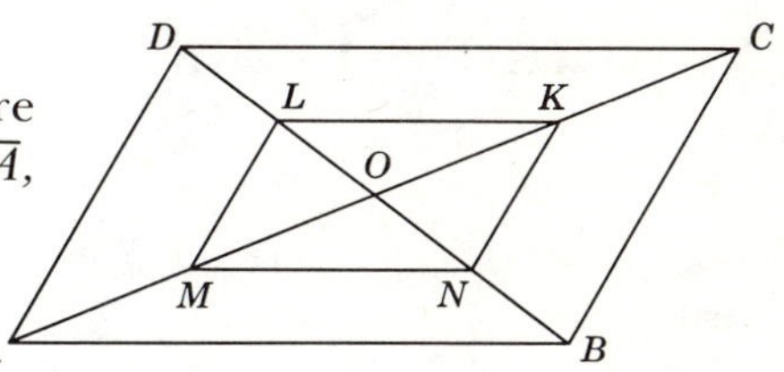

7. *Given:* $ABCD$ is a ▱; K, L, M, N are midpoints of $\overline{OC}, \overline{OD}, \overline{OA}, \overline{OB}$.
Prove: $KLMN$ is a ▱.

Ex. 7.

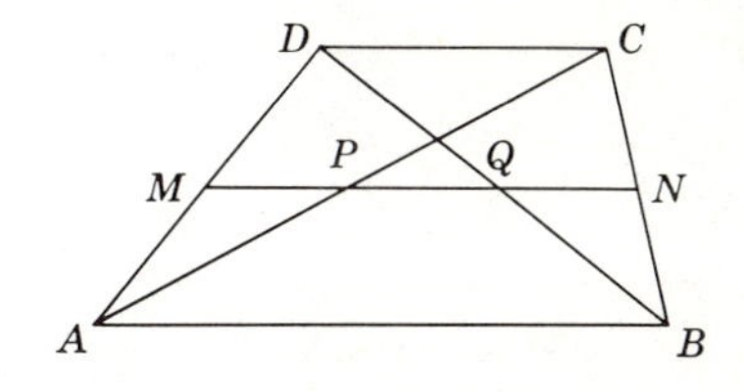

8. *Given:* $\overline{MN}$ is the median of trapezoid $ABCD$; $\overline{AC}$ and $\overline{BD}$ are diagonals.
Prove: $\overline{MN}$ bisects the diagonals.

Ex. 8.

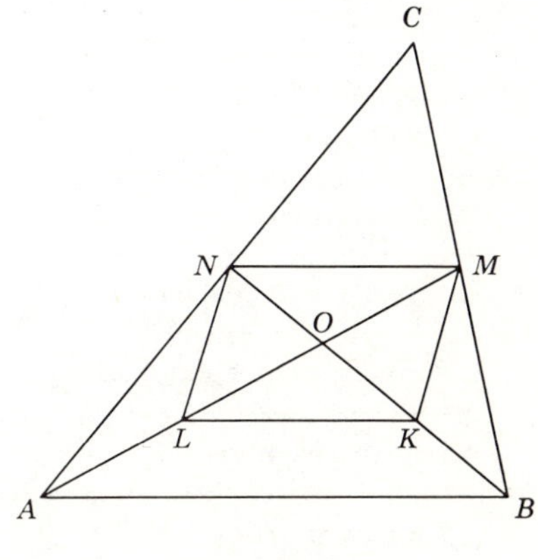

9. *Given:* $\overline{BN}$ and $\overline{AM}$ are medians of $\triangle ABC$; L is the midpoint of $\overline{OA}$; K is the midpoint of $\overline{OB}$.
Prove: $KMNL$ is a ▱.

10. Prove that the line joining the midpoints of two opposite sides of a parallelogram bisects the diagonal of the parallelogram.
11. Prove that the lines joining the midpoints of the sides of a rectangle form a rhombus.

Ex. 9.

Summary Tests

Test 1

Copy the following chart and then place a check mark in the space provided if the figure has the given property.

	Parallelogram	Rectangle	Square	Rhombus
Both pairs of opposite sides are parallel.				
Both pairs of opposite sides are congruent.				
Both pairs of opposite angles are congruent.				
Diagonals are of equal length.				
Diagonals bisect each other.				
Diagonals are perpendicular.				
All sides are congruent.				
All angles are congruent.				

Test 2

TRUE-FALSE STATEMENTS

1. The diagonals of a parallelogram bisect each other.
2. A quadrilateral that has two and only two parallel sides is a rhombus.
3. The bases of a trapezoid are parallel to each other.
4. An equilateral parallelogram is a square.
5. The diagonals of a parallelogram are congruent.
6. An equiangular rhombus is a square.
7. If a polygon is a parallelogram, it has four sides.
8. The diagonals of a quadrilateral bisect each other.
9. A parallelogram is a rectangle.
10. The diagonals of a rhombus are perpendicular to each other.
11. The measure of the line segment joining the midpoints of two sides of a triangle is equal to the measure of the third side.
12. If two angles have their corresponding sides respectively parallel to each other, they are either congruent or supplementary.
13. If the diagonals of a parallelogram are congruent, the parallelogram is a rectangle.
14. If the diagonals of a parallelogram are perpendicular, the parallelogram is a square.
15. A parallelogram is defined as a quadrilateral the opposite sides of which are congruent.
16. If the diagonals of a quadrilateral are perpendicular to each other, the quadrilateral is a parallelogram.
17. The nonparallel sides of an isosceles trapezoid make congruent angles with either base.
18. The line segments joining the midpoints of opposite sides of a quadrilateral bisect each other.
19. If two sides of a quadrilateral are congruent, it is a parallelogram.
20. The median of a trapezoid bisects each diagonal.
21. The diagonals of a parallelogram divide it into four congruent triangles.
22. The lines through the vertices of a parallelogram parallel to the diagonals form another parallelogram.
23. If the diagonals of a rectangle are perpendicular, the parallelogram is a square.
24. The segments joining the consecutive midpoints of the sides of a rectangle form a rhombus.
25. The segments joining the consecutive midpoints of a trapezoid form a parallelogram.
26. A quadrilateral is a parallelogram if its diagonals are perpendicular to each other.

27. A trapezoid is equilateral if it has two congruent sides.
28. The sum of the measures of the angles of a quadrilateral is 360.
29. The bisectors of the opposite angles of a rectangle are parallel.
30. The bisectors of the adjacent angles of a parallelogram are perpendicular.

Test 3

PROBLEMS

1. $\overline{AC}$ is the diagonal of rhombus $ABCD$. If $m\angle B = 120$, find $m\angle BAC$.
2. In $\square ABCD$, $m\overline{AB} = 10$ inches, $m\angle B = 30$, and $\overline{AH} \perp \overline{BC}$. Find $m\overline{AH}$.
3. In $\square ABCD$, $m\angle A = 2m\angle B$. Find $m\angle A$.
4. In $\square ABCD$, diagonal $AC \perp \overline{BC}$ and $\overline{AC} \cong \overline{BC}$. Find $m\angle D$.
5. In $\triangle ABC$, $\overline{AD} \cong \overline{DB}$; $m\angle C - 90$; $m\angle B = 30$; $m\overline{AC} = 14$ inches. Find $m\overline{BD}$.
6. In $\triangle ABC$, $\overline{AD} \cong \overline{DB}$; $m\angle C = 90$; $m\angle A = 60$; $\mathrm{m}\overline{CD} = 12$ inches. Find $m\overline{AC}$.
7. In $\triangle ABC$, $\overline{AD} \cong \overline{DB}$; $m\angle C = 90$; $m\angle A = 60$; $m\overline{AB} = 26$ inches. Find $m\overline{CD}$.

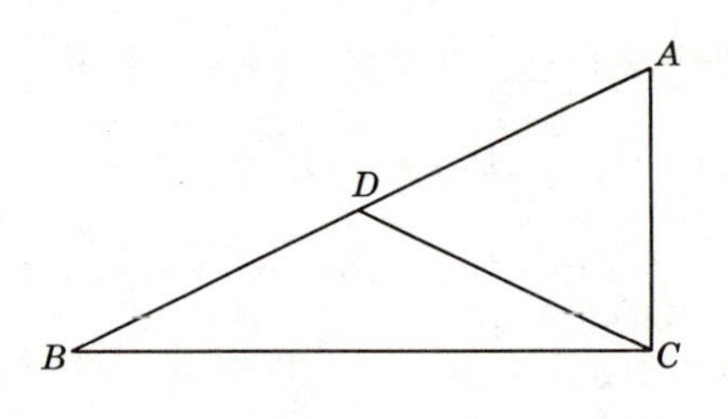

Probs. 5–7.

8–13. Median EF of trapezoid $ABCD$ intersects $\overleftrightarrow{AC}$ in R and $\overleftrightarrow{BD}$ in S.

8. Find $m\overline{EF}$ if $m\overline{DC} = 10$ and $m\overline{AB} = 14$.
9. Find $m\overline{ER}$ if $m\overline{DC} = 10$ and $m\overline{AB} = 14$.
10. Find $m\overline{DC}$ if $m\overline{EF} = 24$ and $m\overline{AB} = 30$.
11. Find $m\overline{RS}$ if $m\overline{EF} = 24$ and $m\overline{DC} = 20$.
12. Find $m\overline{AB}$ if $m\overline{ER} = 5$ and $m\overline{RF} = 7$.
13. Find $m\overline{SF}$ if $m\overline{ER} = 5$ and $m\overline{RF} = 7$.

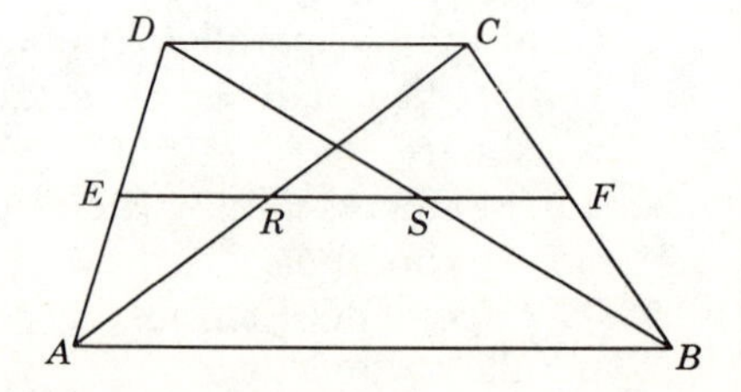

Probs. 8–13.

14–18. *ABCDE* is a regular pentagon.

14. Find $m\angle CBF$.
15. Find $m\angle AED$.
16. Find $m\angle ACE$.
17. Find $m\angle DAC$.
18. Find $m\angle DGC$.

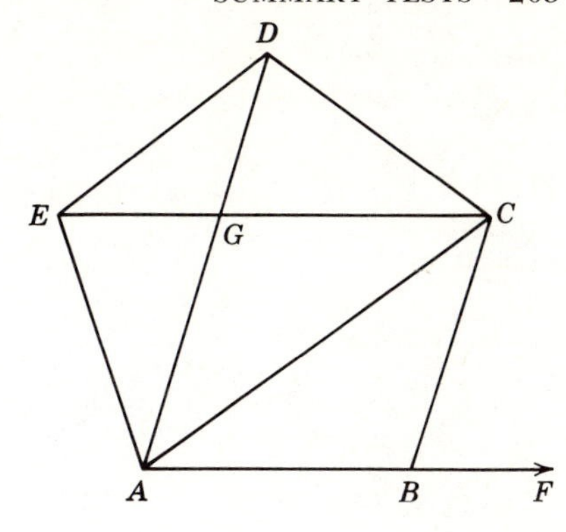

Probs. 14–18.

Test 4

EXERCISES

1. *Given:* *ABCD* is a ▱;
   $m\overline{DF} = m\overline{BE}$.
   Prove: $\overline{AE} \parallel \overline{CF}$.

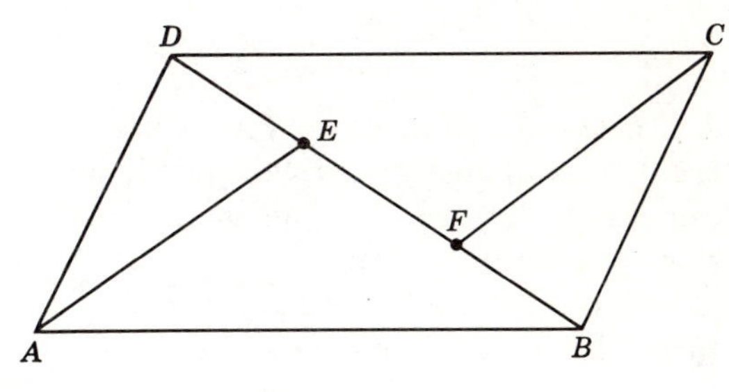

Ex. 1.

2. *Given:* $\overline{AM}$ is a median of $\triangle ABC$; $\angle DCB \cong \angle MBE$.
   Prove: *DBEC* is a ▱.

3. *Given:* Right triangle *ABC* with $\angle ACB$ a right angle; *M, R, S* are midpoints of $\overline{AB}, \overline{BC}$, and $\overline{AC}$.
   Prove: $\overline{MC} \cong \overline{RS}$.

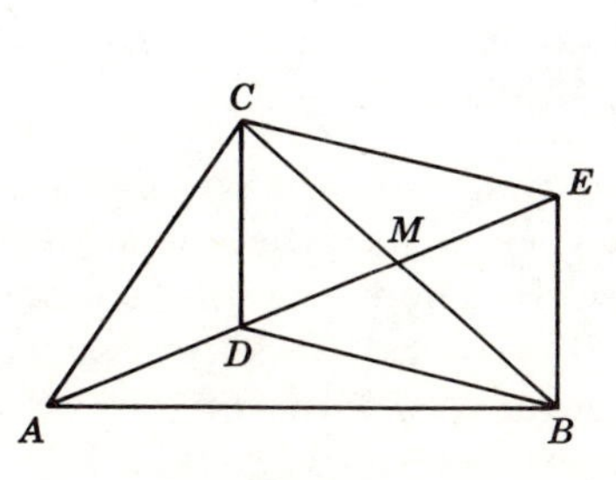

Ex. 2.

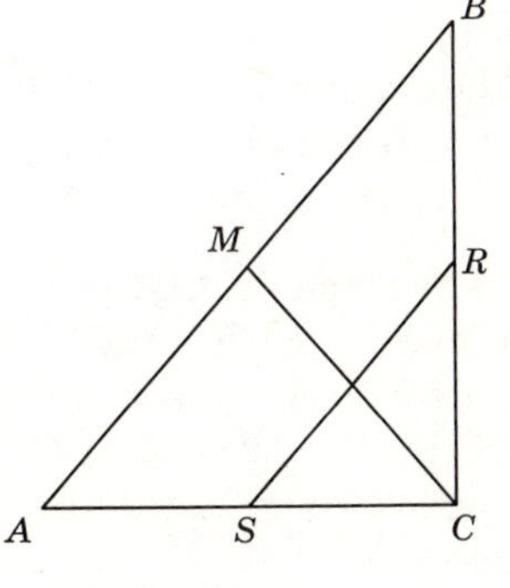

Ex. 3.

|7|

Circles

7.1. Uses of the circle. The history of civilization's continually improving conditions of living and working is intimately related to the use of the properties of the circle. One of the most important applications of the circle that man has invented is the wheel. Without the wheel most of the world's work would cease. Industry would be completely crippled without the circle in the form of wheels, gears, and axles. Transportation would revert back to conditions of prehistoric times. Without the wheel there would be no bicycles, no automobiles, no trains, no streetcars, and no airplanes. Farm machines, factory and mine equipment, without the wheel, would exist only in the form of useless metal, plastics, and wood.

Industry applies the properties of the circle when it uses ball bearings to reduce friction and builds spherical tanks for strength. Every year millions of feet of circular pipe and wire are manufactured. Countless manufactured articles of furniture, dishes, and tools are circular in form. Most of the cans of food on the grocery shelves have circular cross sections. Tanks with circular cross sections have many uses (see Fig. 7.1).

Circular shapes are found in such ornamental designs as rose windows, architectural columns, traffic circles, and landscape designs.

7.2. The circle in history. The invention and use of the circular wheel date back to very early times. No one knows when the wheel was invented or who invented it. Some authorities believe that the wheel was invented somewhere in Asia about 10,000 years ago. The oldest wheel in existence was discovered in Mesopotamia in 1927 when archeologists uncovered a four-wheeled chariot known to have existed about 5,500 years ago (Fig. 7.2).

The circle had an aesthetic appeal to the Greeks. To them it was the most

Fig. 7.1. The world's largest sewage treatment system at Chicago has been termed by the American Society of Civil Engineers "the seventh wonder of America." More than 1,100,000,000 gallons of waste are treated daily in the system. In this view can be seen the preliminary settling tanks, the aeration tanks, and the final settling tanks. Each of the final settling tanks is 126 feet in diameter. (Chicago Aerial Industries, Inc.)

Fig. 7.2. View of one side of a complete chariot found in a brick tomb 20 feet below plain level at Kish. Circa 4000 B.C. Skeleton of one of the oxen appears in original position beside the pole. (Chicago Natural History Museum.)

perfect of all plane figures. Thales, Pythagoras, Euclid, and Archimedes each contributed a great deal to the geometry of the circle.

Thales probably is best known for the deductive character of his geometric propositions. One of the most remarkable of his geometrical achievements was proving that any angle inscribed in a semicircle must be a right angle (§ 7.16).

Pythagoras was the founder of the Pythagorean school, a brotherhood of people with common philosophical and political beliefs. They were bound by oath not to reveal the teachings or secrets of the school. Pythagoras was primarily a philosopher. Members of his school boasted that they sought knowledge, not wealth. They were probably the first to arrange the various propositions on geometric figures in a logical order. No attempt at first was made to apply this knowledge to practical mechanics. Much of the early work on geometric constructions involving circles was attributed to this group.

Euclid published systematic, rigorous proofs of the leading propositions of the geometry known at his time. His treatise, entitled "Elements," was, to a large extent, a compilation of works of previous philosophers and mathematicians. However, the form in which the propositions was presented, consisting of statement, construction, proof and conclusion, was the work of Euclid. Much of Euclid's work was done when he served as a teacher in Alexandria. It is probable that his "Elements" was written to be used as a text in schools of that time. The Greeks at once adopted the work as their standard textbook in their studies on pure mathematics. Throughout the succeeding more than 2000 years, Euclid's works have served as the basis for most other textbooks in this field.

Archimedes, like his contemporaries, held that it was undesirable for a philosopher to apply the results of mathematical science to any practical use. However, he did introduce a number of new inventions.

Most readers are familiar with the story of his detection of the fraudulent goldsmith who diluted the gold in the king's crown. The Archimedean screw was used to take water out of the hold of a ship or to drain lands inundated by the floodwaters of the Nile. Burning glasses and mirrors to destroy enemy ships and large catapults to keep Romans besieging Syracuse at bay are devices attributed to the remarkable mechanical ingenuity of this man.

Science students today are referred to Archimede's principles dealing with the mechanics of solids and fluids. His work in relating the radius and the circumference of a circle and in finding the area of the circle has stood the test of time.

It is told that Archimedes was killed by an enemy soldier while studying geometric designs he had drawn in the sand.

7.3. Basic definitions. To develop proofs for various theorems on circles, we must have a foundation of definitions and postulates. Many of the terms to be defined the student will recognize from his previous studies in mathematics.

A *circle* is the set of points lying in one plane each of which is equidistant from a given point of the plane. The given point is called the *center* of the circle (Fig. 7.3). Circles are often drawn with a compass (Fig. 7.4). The symbol for circle is $\odot$. In Fig. 7.3, O is the center of $\odot ABC$, or simply $\odot O$.

A line segment one of whose endpoints is the center of the circle and the other one a point on the circle is a *radius* (plural, radii) of the circle. $\overline{OA}$, $\overline{OB}$, and $\overline{OC}$ are radii of $\odot O$. Thus, we can say radii of the same circle are congruent.

A *chord* of a circle is a segment whose endpoints are points of the circle. A *diameter* is a chord containing the center of the circle. $\overline{ED}$ is a chord of the circle in Fig. 7.3.

It will be noted that we defined "radius" and "diameter" as a segment; that is, as a set of points. Common usage, however, often lets the words denote their measures. Thus we speak of a circle with radius of, say, 7 inches. Or we speak of a diameter equaling twice the radius. No confusion should arise because the context of the statement should clearly indicate whether a set of points or a number is being referred to.

Circles are *congruent* iff they have congruent radii. *Concentric* circles are coplanar circles having the same center and noncongruent radii (see Fig. 7.5).

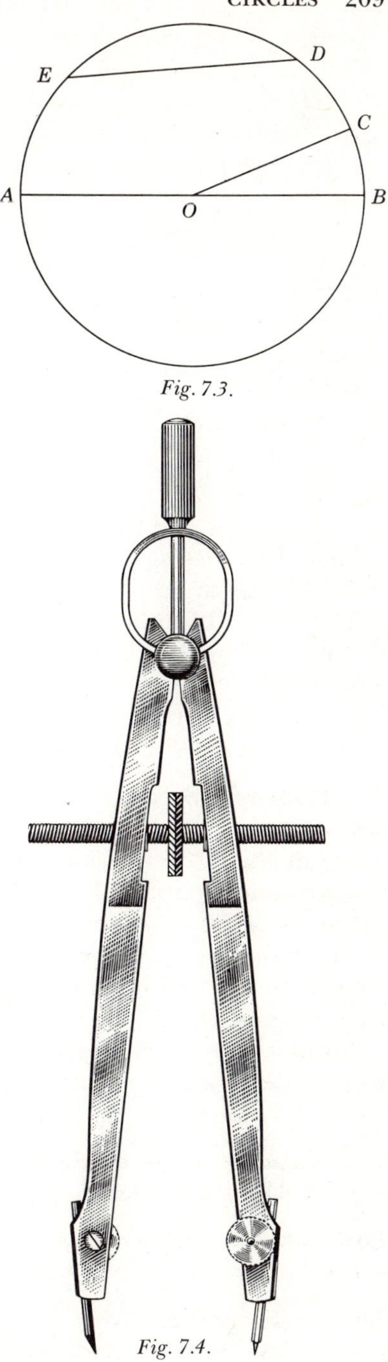

Fig. 7.3.

Fig. 7.4.

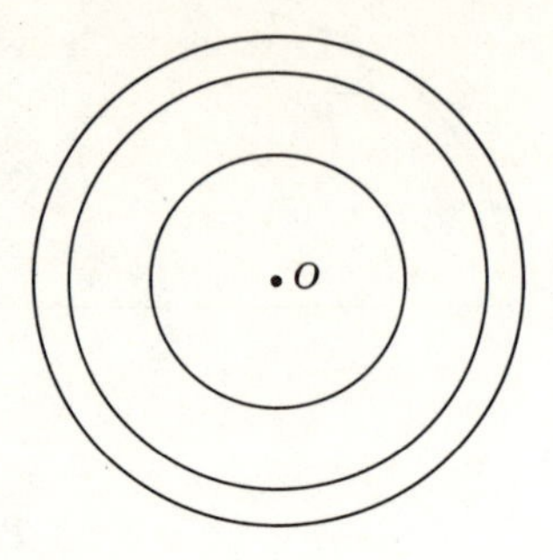

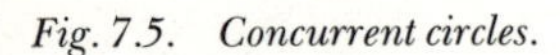
Fig. 7.5. Concurrent circles.

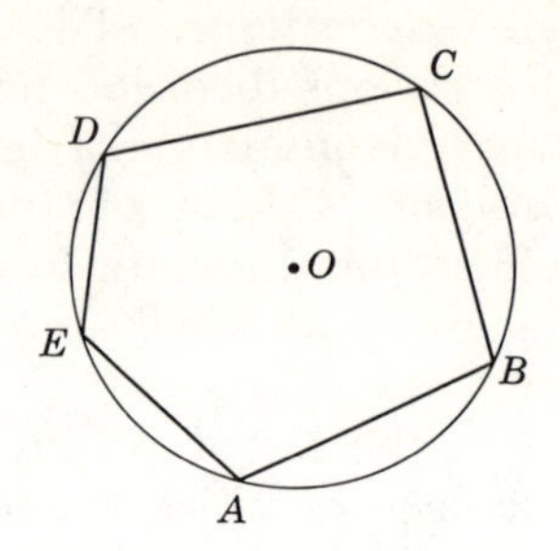

Fig. 7.6. Inscribed polygon.

A circle is said to be *circumscribed* about a polygon when it contains all the vertices of the polygon. In Fig. 7.6, the circle is circumscribed about polygon $ABCDE$.

A polygon is said to be inscribed in a circle if each of its vertices lies on the circle. Thus its sides will be chords of the circle. Polygon $ABCDE$ is inscribed in the circle.

The *interior* of a circle is the union of its center and the set of all points in the plane of the circle whose distances from the center are less than the radius. The *exterior* of a circle is the set of points in the plane of the circle such that their distances from the center are greater than the radius. Frequently the words "inside" and "outside" are used for "interior" and "exterior."

7.4. Tangent. Secant. A line is *tangent* to a circle if it lies in the plane of the circle and intersects it in only one point. This point is called the *point of tangency*, and we say that the line and the circle *are tangent* at this point. In Fig. 7.7, $\overleftrightarrow{PT}$ is tangent to $\odot O$.

A line or ray containing a chord of a circle is a *secant* of the circle. In Fig. 7.7, $\overleftrightarrow{AB}$ and $\overrightarrow{BC}$ are secants.

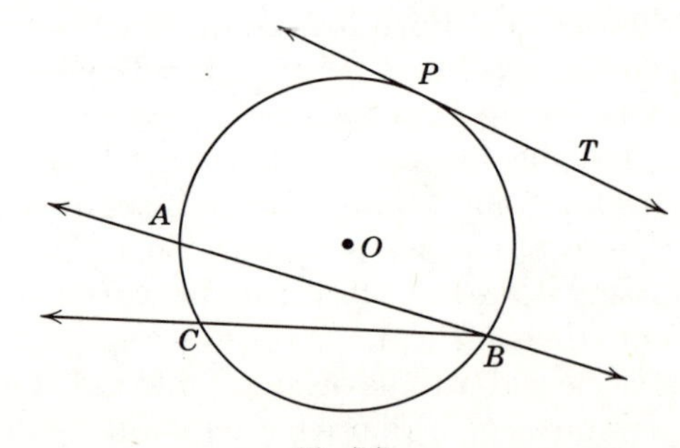

Fig. 7.7.

7.5. Postulate on the circle. The following fundamental assumption may be stated relating to circles.

Postulate 19. *In a plane one, and only one, circle can be drawn with a given point as center and a given line segment as radius.*

7.6. Angles and arcs. It will be noted that we defined a circle as a set of points. We avoided the use of the word "curve" in our definition. The word "curve" is defined in a more precise manner in more advanced courses. We will not attempt the definition here.

A *central angle* of a circle is an angle whose vertex is the center of the circle. In Fig. 7.8, $\angle AOB$, $\angle BOC$, and $\angle COD$ are central angles. The central angle AOB is said to *intercept* or to *cut off arc AB*, and the arc AB is said to *subtend* or to *have central angle AOB*.

If A and B are points on a circle O, not the endpoints of a diameter, the union of A, B, and the set of points of the circle which are interior to $\angle AOB$ is a *minor arc* of the circle. The union of A, B, and the set of points of the circle O which are exterior to $\angle AOB$ is a *major arc* of the circle. If A and B are endpoints of a diameter of circle O, the union of A, B, and the set of points of the circle in one of the two half-planes of $\overleftrightarrow{AB}$ is called a *semicircle* or *semicircular arc.* In each of the foregoing cases, A and B are called *endpoints* of the arc.

If A and B are points of a circle O, the circle is divided into a minor and a major arc or into two semicircles. Arc AB is usually abbreviated $\widehat{AB}$. Usually it is clear from the context of a statement whether $\widehat{AB}$ refers to the minor or major arc. However, to make clear exactly which arc is referred to, another point of the arc can be selected. Thus, in Fig. 7.9, $\widehat{AEB}$ refers to the minor arc.

7.7. Sphere. Many terms and definitions given for a circle are associated with like terms and definitions for a sphere except that we omit the restriction "in a given plane."

Definition: A *sphere* is the set of all points in space for which each is equidistant from a fixed point, called the *center* (see Fig. 7.10).

A *radius* of a sphere is a line segment joining the center and any point of the sphere. Spheres are *congruent* iff they have congruent radii. Spheres are *concentric* iff they have the same center and noncongruent radii.

A point is in the *interior* or *exterior* of a sphere according to whether its distance from the center of the sphere is less than or more than the measure of the radius.

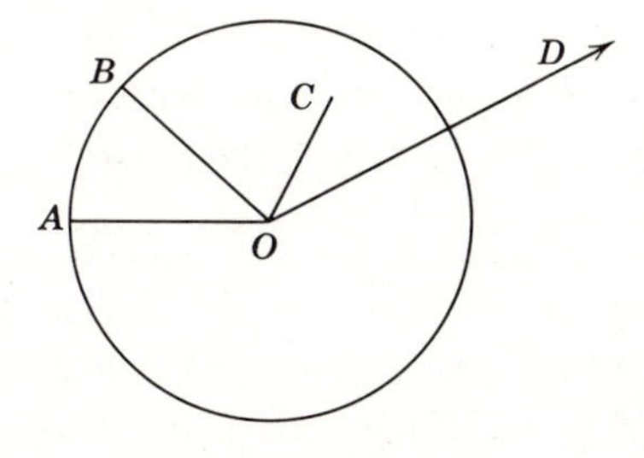

Fig. 7.8.

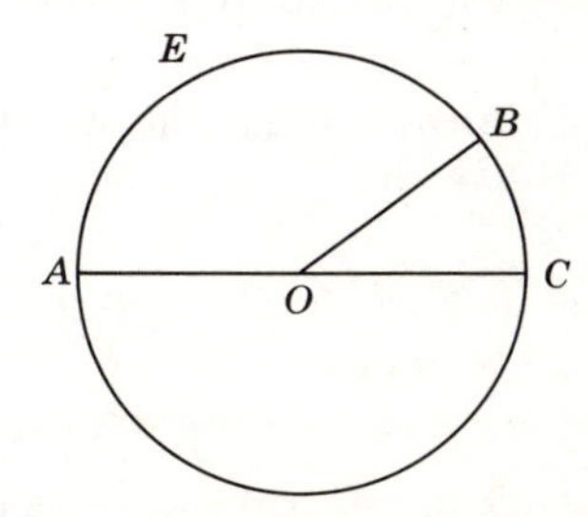

Fig. 7.9.

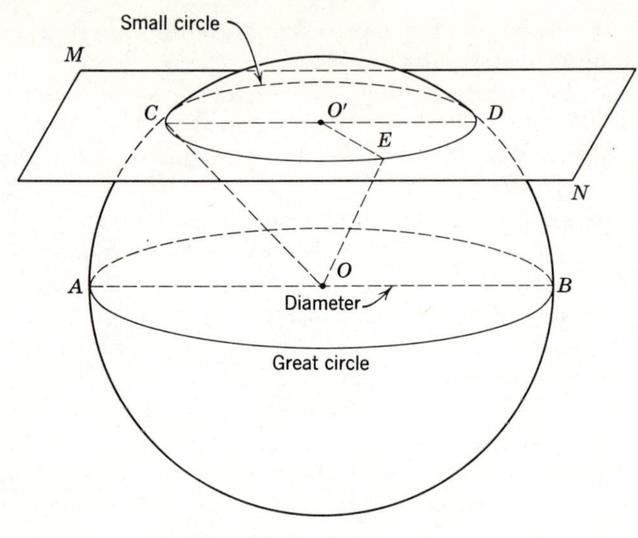

Fig. 7.10.

A *diameter* of a sphere is a line segment passing through the center and having its endpoints on the sphere. The diameter of a sphere is twice as long as the radius.

If a plane intersects a sphere, the intersection is a circle. If the plane contains a diameter of the sphere, the circle is called a *great circle*; otherwise the circle is a *small circle*. All great circles of a sphere are congruent. Every great circle bisects the sphere into surfaces called *hemispheres*. A hemisphere includes its bounding circle.

A plane is *tangent* to a sphere if it intersects the sphere in exactly one point.

The spherical tank (see Fig. 7.11) is the strongest tank per given volume that can be manufactured from a given substance.

7.8. Properties of the sphere. The following facts about spheres can be proved (we will not attempt them in this text).

1. *Through three points of a sphere, one and only one small circle can be drawn.*
2. *Through the ends of a diameter of a sphere, any number of great circles can be drawn.*
3. *Through two points which are not ends of a diameter of a sphere, exactly one great circle can be drawn.*
4. *A plane perpendicular to a radius at its point on the sphere is tangent to the sphere.*

7.9. Measuring central angles and arcs. A quantity is *measured* by finding how many times it contains another quantity of the same kind, called the *unit*

Fig. 7.11. Special tanks for propane and butane. The gases are liquified under high pressure and then stored in these tanks. (Standard Oil Company of California)

of measurement. In Chapter 1 we learned that the degree is the unit for measuring angles.

The unit of arc measure is the arc which is intercepted by a central angle of one degree. It is likewise called a degree. Thus, the number of angle degrees about a point and the number of arc degrees in a circle each total 360. It should be clear that, although the angle degree is not the same as the arc degree, the numerical measure of angles is closely related to the numerical measure of arcs. This relationship is expressed in the following definition.

Definition: If $\widehat{ACB}$ is a minor arc, then $m\widehat{ACB}$ is equal to the measure of the corresponding central angle. If $\widehat{ACB}$ is a semicircle, then $m\widehat{ACB} = 180$. If $\widehat{ADB}$ is a major arc and $\widehat{ACB}$ is the corresponding minor arc, then $m\widehat{ADB} = 360 - m\widehat{ACB}$.

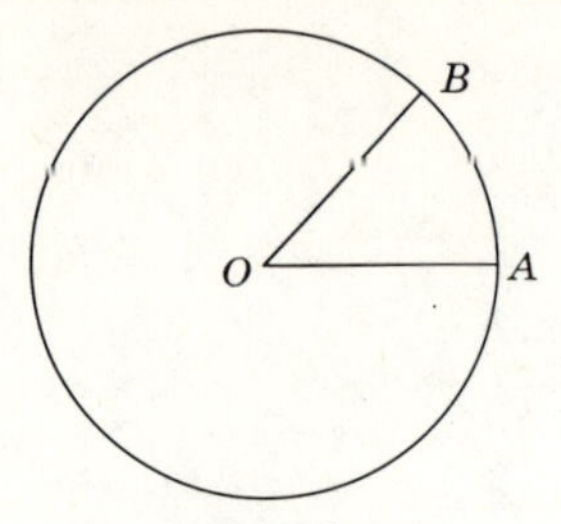

Fig. 7.12.

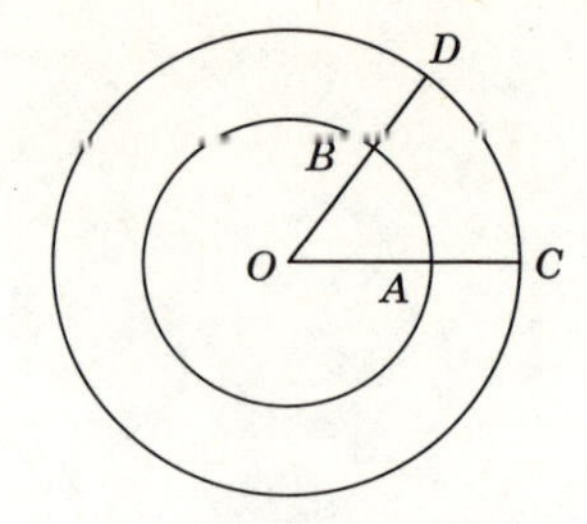

Fig. 7.13.

Thus, in Fig. 7.12, if $m\angle O = 40$, then $m\widehat{AB} = 40$. This relationship is written $m\angle O = m\widehat{AB}$. For brevity we sometimes say that the "central angle O is measured by its arc AB." The equation should not be read "Angle O equals arc AB" because an angle cannot equal an arc. In like manner, the student should not confuse the number of degrees in an arc with the length of the arc. The arc degree is not a unit of length. Two arcs of the same, or congruent circles, are *congruent* iff they have the same measure.

7.10. Comparison of arcs. The relationship between the arc length and arc degrees can be illustrated by referring to Fig. 7.13. In this figure we have two unequal circles with the same center. Since both $\widehat{AB}$ and $\widehat{CD}$ are cut off by the same central angle ($\angle COD$), they must contain the same number of arc degrees. Thus $m\widehat{AB} = m\widehat{CD}$ or $\widehat{AB} \cong \widehat{CD}$. However, the arc lengths are unequal. Here we have two arcs of different circles with the same number of arc degrees but of unequal lengths.

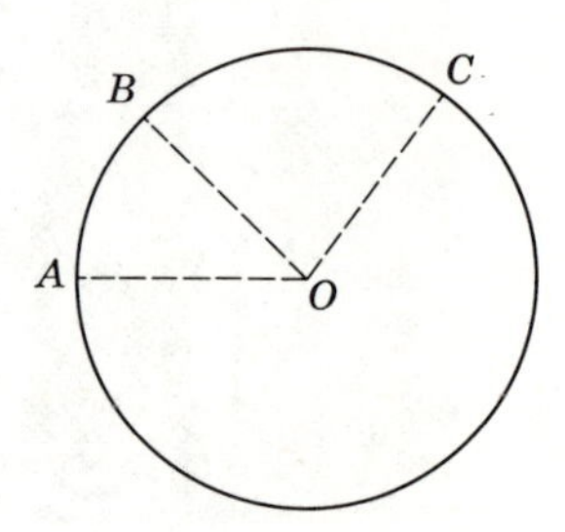

Fig. 7.14.

The following postulate is analogous to the segment addition postulate (Postulate 13).

Postulate 20. (arc-addition postulate). *If the intersection of $\widehat{AB}$ and $\widehat{BC}$ of a circle is the single point B, then $m\widehat{AB} + m\widehat{BC} = m\widehat{AC}$.* (Fig. 7.14).

Theorem 7.1

7.11. If two central angles of the same or congruent circles are congruent, then their intercepted arcs are congruent.
Given: Circle O with $\angle AOB \cong \angle COD$.
Conclusion: $\widehat{AB} \cong \widehat{CD}$.

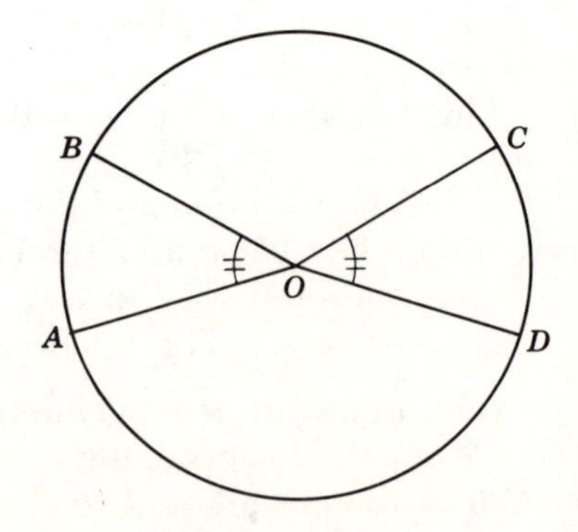

Theorem 7.1.

Proof:

STATEMENTS	REASONS
1. $\angle AOB \cong \angle COD$.	1. Given.
2. $m\angle AOB = m\angle COD$.	2. Definition of $\cong$ $\angle$s.
3. $m\angle AOB = m\widehat{AB}$; $m\angle COD = m\widehat{CD}$.	3. Definition of measure of arc.
4. $m\widehat{AB} = m\widehat{CD}$.	4. Substitution property.
5. $\widehat{AB} \cong \widehat{CD}$.	5. Definition of $\cong$ arcs.

Theorem 7.2

7.12. If two arcs of a circle or congruent circles are congruent, then the central angles intercepted by these arcs are congruent.
(The proof of this theorem is left as an exercise for the student.)

Exercises (A)

Indicate which of the following statements are *always true* (mark T) and which are *not always true* (mark F).

1. All central angles of the same circle are congruent.
2. If the radii of two circles are not congruent, the circles are not congruent.
3. Two circles each with radii of 10 inches have congruent diameters.
4. A point is outside of a circle if its distance from the center of the circle equals the measure of the diameter of the circle.
5. The measure of a major arc is more than the measure of a minor arc.
6. The vertex of a central angle is on the circle.
7. Every circle has exactly two semicircles.
8. Every arc of a circle subtends a central angle.
9. All semicircles are congruent.
10. If two arcs of the same circle are congruent, their central angles must be congruent.
11. A chord is a diameter.
12. Some radii of a circle are chords of that circle.
13. Every diameter is a chord.
14. In a given circle it is possible for a chord to be congruent to a radius.
15. The intersection of a circle and one of its chords is a null set.
16. The intersection of a plane and a sphere can be a point.
17. The intersection of two diameters of a given circle is four points.
18. No chord of a circle can equal a diameter.
19. A sphere is a set of points.
20. Every sphere has only one great circle.

Exercises (B)

Find the number of degrees asked for in each of the following. The letter O will indicate centers of circles.

21. *Given:* $m\widehat{BC} = 70$; $\overline{AC}$ is a diameter.
 Find: $m\angle AOB$.

22. *Given:* $m\angle OAB = 36$.
 Find: $m\widehat{AB}$.

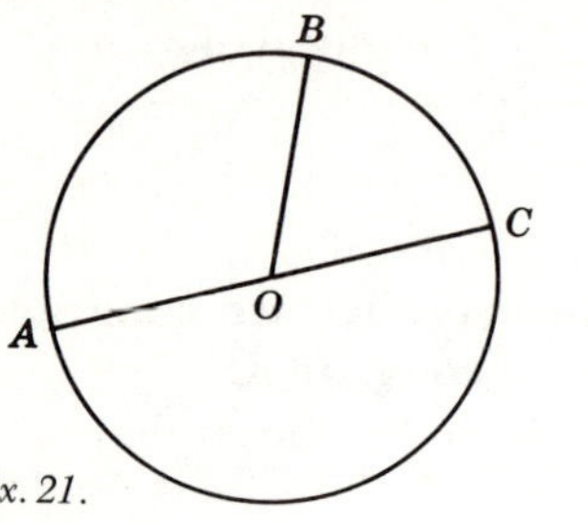
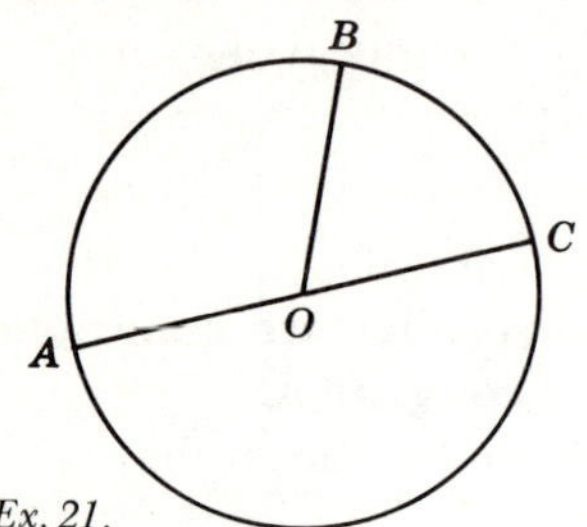

Ex. 21.

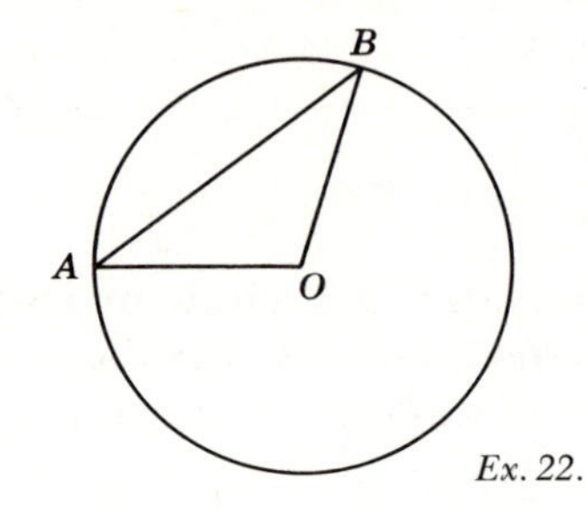

Ex. 22.

23. *Given:* $m\angle OAB = 30$; $\overline{AC}$ is a diameter.
 Find: $m\widehat{BC}$.

24. *Given:* $m\widehat{AB} = 70$; $\overline{AC}$ is a diameter.
 Find: $m\angle OBC$.

25. *Given:* $m\angle AOB = 60$; $\overline{AC}$ is a diameter.
 Find: (a) $m\angle ABC$; (b) $m\widehat{BC}$.

26. *Given:* $m\widehat{AD} = 140$; $\overline{BD}$ and $\overline{AC}$ are diameters.
 Find: $m\angle OBC$.

Ex. 23.

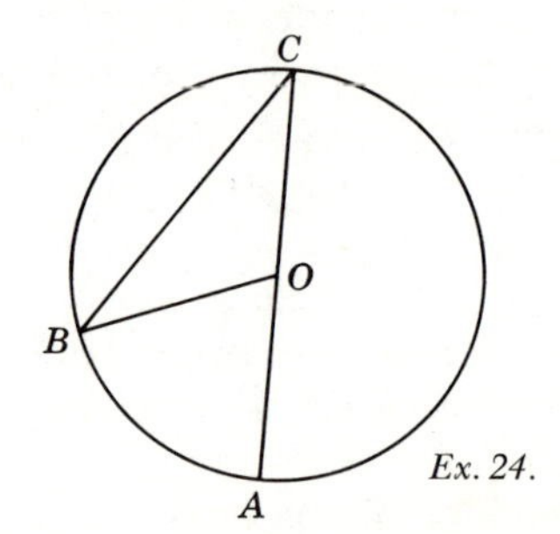

Ex. 24.

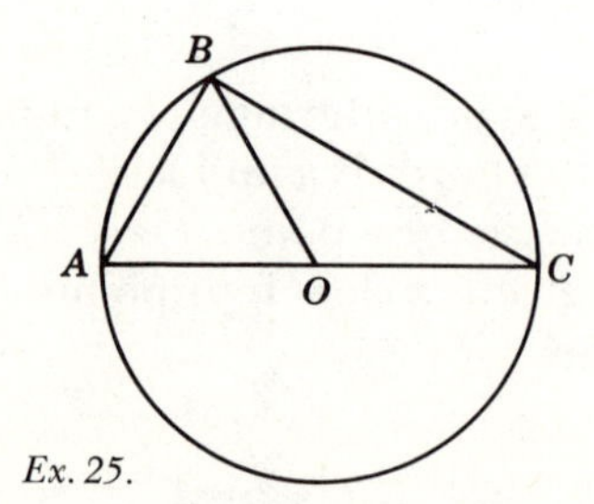

Ex. 25.

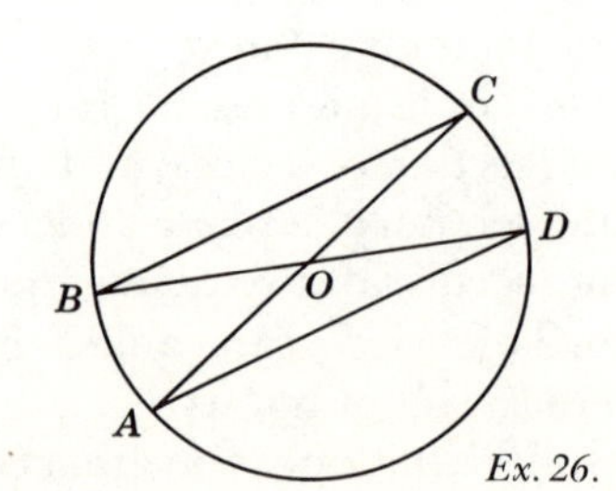

Ex. 26.

7.13. Inscribed angle. An *angle is inscribed in an arc* of a circle if the endpoints of the arc are points on the sides of the angle and if the vertex of the angle is a point, but not an endpoint, of the arc. In Fig. 7.15*a*, $\angle ABC$ is inscribed in the minor arc ABC. In Fig. 7.15*b*, $\angle DEF$ is inscribed in major $\widehat{DEF}$. The angles are called *inscribed angles.*

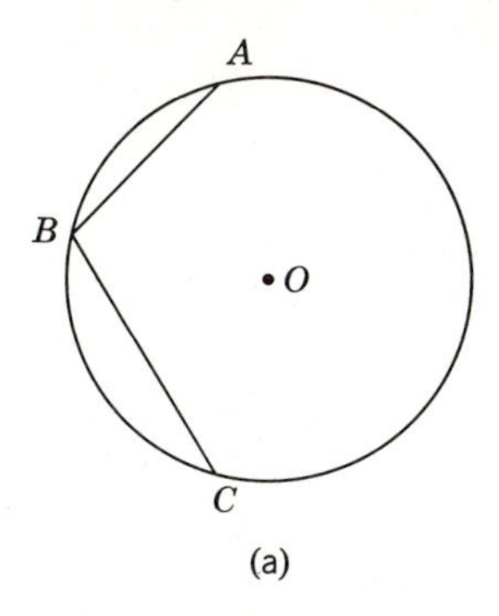

(a)

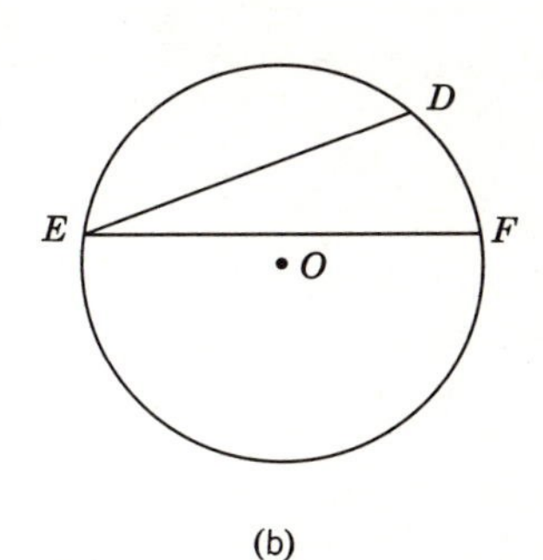

(b)

Fig. 7.15.

7.14. Intercepted arcs. An angle *intercepts* an arc if the endpoints of the arc lie on the sides of the angle, and if each side of the angle contains at least one endpoint of the arc, and if, except for the endpoints, the arc lies in the interior of the angle. Thus, in Fig. 7.15, $\angle ABC$ intercepts $\widehat{AC}$ and $\angle DEF$ intercepts $\widehat{DF}$. In Fig. 7.16, $\angle APB$ intercepts $\widehat{AE}$ and $\widehat{AB}$, and $\angle EPD$ intercepts $\widehat{DE}$ and $\widehat{BC}$.

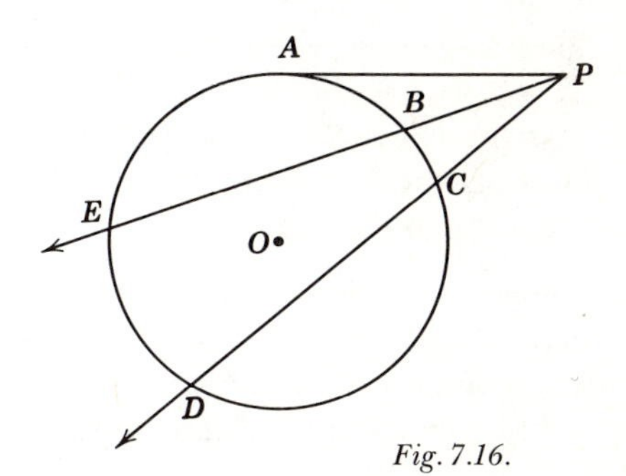

Fig. 7.16.

Theorem 7.3

7.15. The measure of an inscribed angle is equal to half the measure of its intercepted arc.

Given: $\angle BAC$ inscribed in $\odot O$.

Conclusion: $m\angle BAC = \frac{1}{2}m\widehat{BC}$.

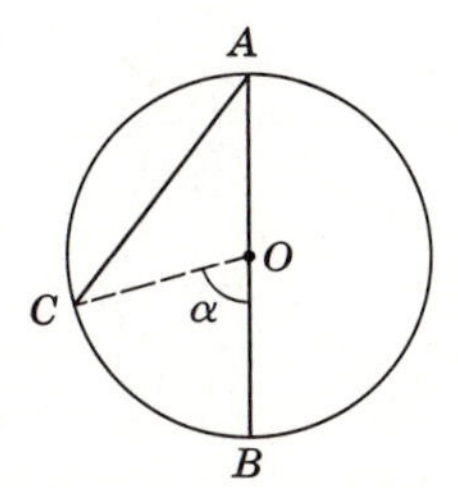

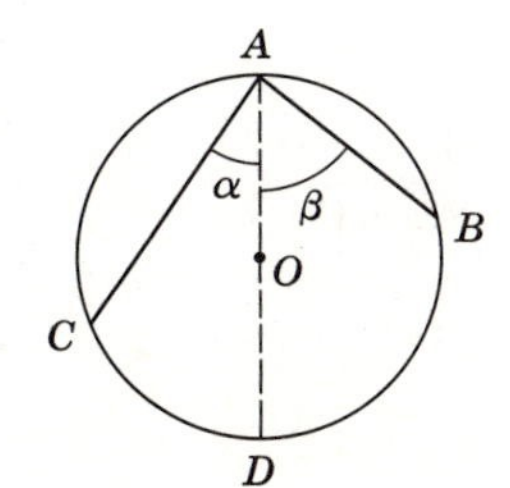

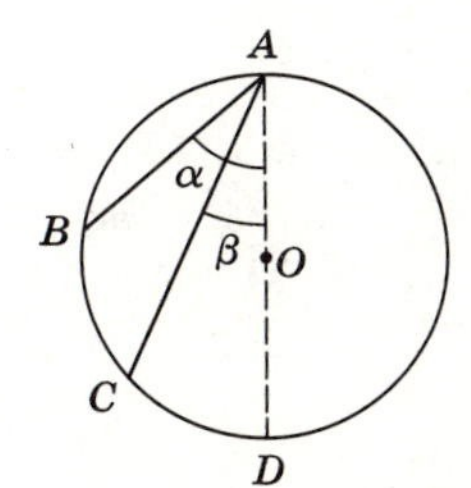

Theorem 7.3.

CASE I: *When one side of the angle is a diameter.*

Proof:

STATEMENTS	REASONS
1. $\angle BAC$ is inscribed in $\odot O$.	1. Given.
2. Draw $\overline{OC}$.	2. Postulate 2.
3. $\overline{OA} \cong \overline{OC}$.	3. Definition of a $\odot$.
4. $\angle A \cong \angle C$.	4. Theorem 4.16.
5. $m\angle BOC = m\angle A + m\angle C$.	5. Theorem 5.16.
6. $m\angle BOC = m\angle A + m\angle A$.	6. Symmetric and substitution properties of equality. (With Statements 4 and 5).
7. $m\angle BOC = m\widehat{BC}$.	7. Definition of measure of arc.
8. $m\angle A + m\angle A = m\widehat{BC}$.	8. Theorem 3.5.
9. $m\angle A = \frac{1}{2}m\widehat{BC}$.	9. Division property of equality.

CASE II: *When the center of the circle lies within the interior of the angle.*
Plan: Draw diameter AD and apply CASE I to $\angle\alpha$ and $\angle\beta$. Use the additive property of equality.

CASE III: *When the center of the circle lies in the exterior of the angle.*
Plan: Draw diameter AD. Apply CASE I to $\angle\alpha$ and $\angle\beta$. Use the subtractive property of equality.

The proofs of the following corollaries are left to the student.

7.16. Corollary: An angle inscribed in a semicircle is a right angle.

7.17. Corollary: Angles inscribed in the same arc are congruent.

7.18. Corollary: Parallel lines cut off congruent arcs on a circle.

Plan of proof: $\angle ABC \cong \angle BCD$.
$m\angle ABC = \frac{1}{2}m\widehat{AC}$. $m\angle BCD = \frac{1}{2}m\widehat{BD}$.
Then $\widehat{AC} \cong \widehat{BD}$.

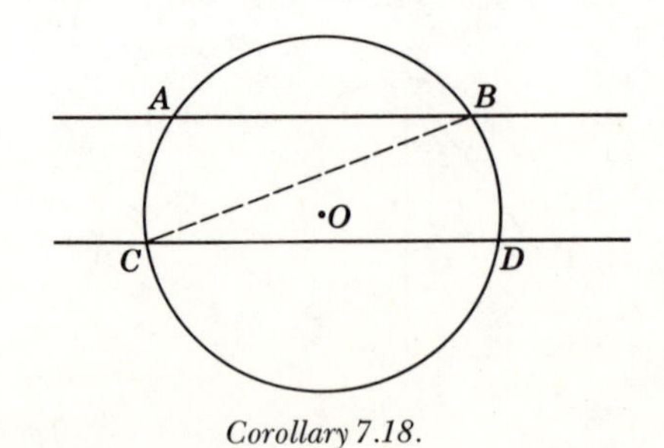

Corollary 7.18.

Exercises

In each of the following problems, O is the center of a circle.

1. *Given:* O center of $\odot$;
   $m\widehat{AB} = 100$;
   $m\widehat{AD} = 140$;
   $m\widehat{DC} = 66$.

   Find: the measure of each of the four central $\angle$s.

Ex. 1.

2. *Given:* Inscribed quadrilateral $WRST$ with diagonals RT and WS.
   Which angles have the same measures as
   (*a*) $\angle WRT$? (*b*) $\angle WTR$?
   (*c*) $\angle RTS$?

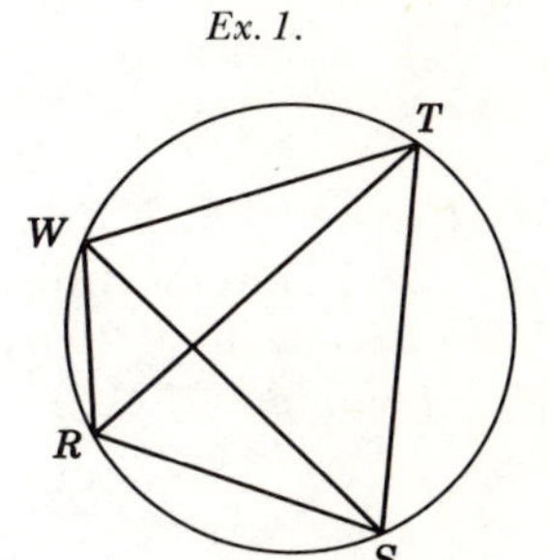

Ex. 2.

3. *Given:* Chord $LM \parallel$ chord UV;
   $m\angle VLM = 25$.

   Find: (*a*) $m\widehat{VM}$; (*b*) $m\widehat{UL}$.

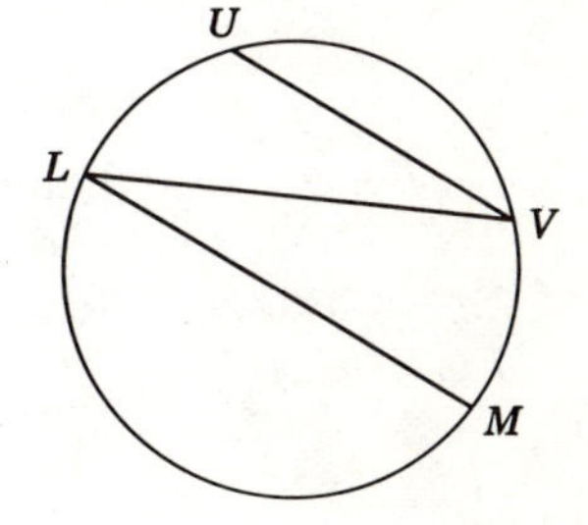

Ex. 3.

4. *Given:* $m\angle A = 68$.
   Find: $m\angle C$.

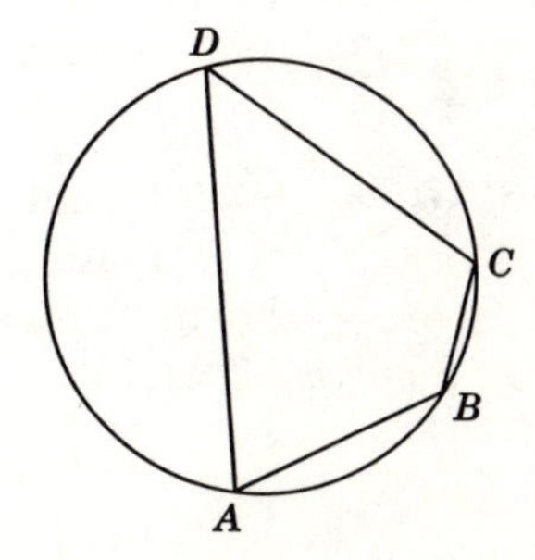

Ex. 4.

5. *Given:* $m\angle SOR = 80$.
   Find: $m\angle T$.

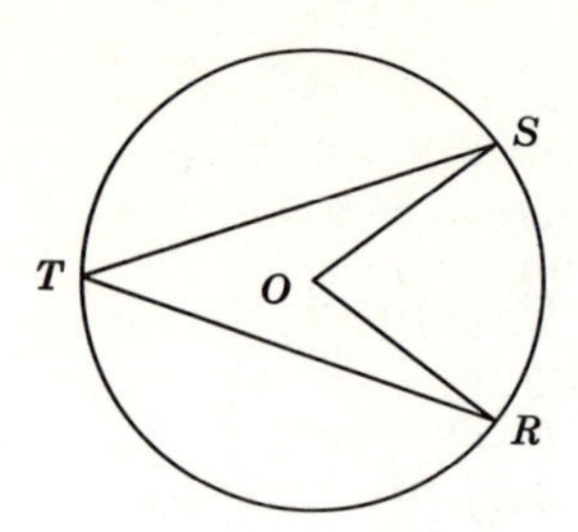

Ex. 5.

6. *Given:* Chords AB and CD intersecting at E; $m\widehat{AC} = 40$; $m\widehat{BD} = 70$.
   Find: $m\angle AEC$.

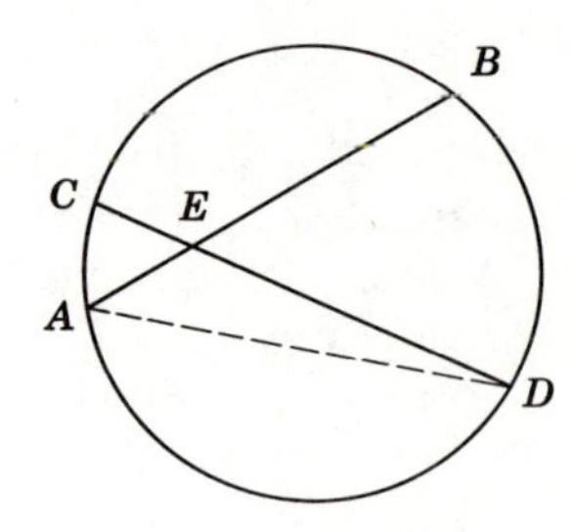

Ex. 6.

7. *Given:* $\overleftrightarrow{AB} \perp$ to diameter TD; $\overline{TC}$ is a chord; $m\widehat{TC} = 100$.
   Find: $m\angle BTC$.

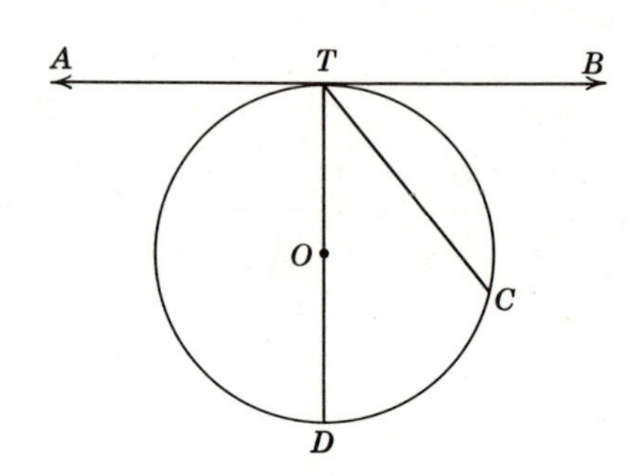

Ex. 7.

8. *Given:* $\overleftrightarrow{AB} \perp$ to diameter RS; $\overline{OT}$ is a radius; $\overline{TS}$ is a chord; $m\angle ROT = 110$.
   Find: $m\angle TSA$.

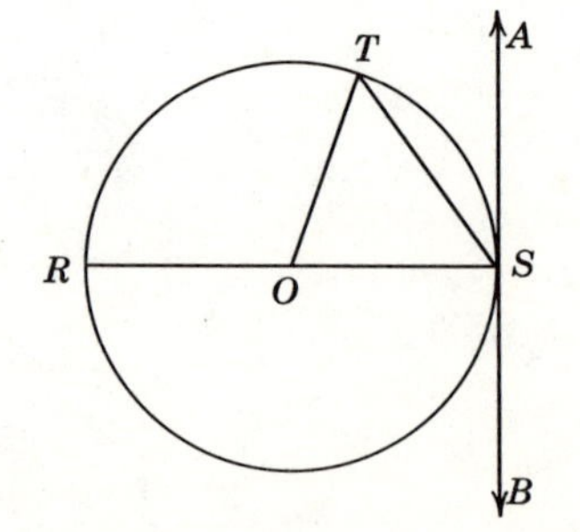

Ex. 8.

9. *Given:* $m\angle AEC = 80$;
$m\widehat{AC} = 100$.
Find: $m\widehat{BD}$.

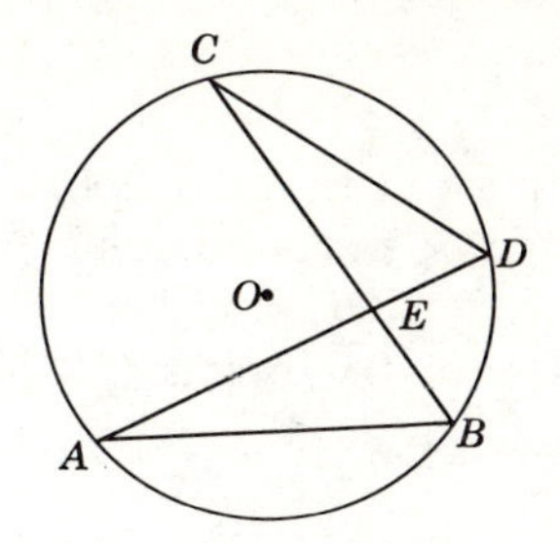

Ex. 9.

10. *Given:* $\overline{WS}$ bisects $\angle RST$;
$m\widehat{RS} = 120$;
$m\widehat{WR} = 62$.
Find: (*a*) $m\angle TKS$; (*b*) $m\widehat{TS}$.

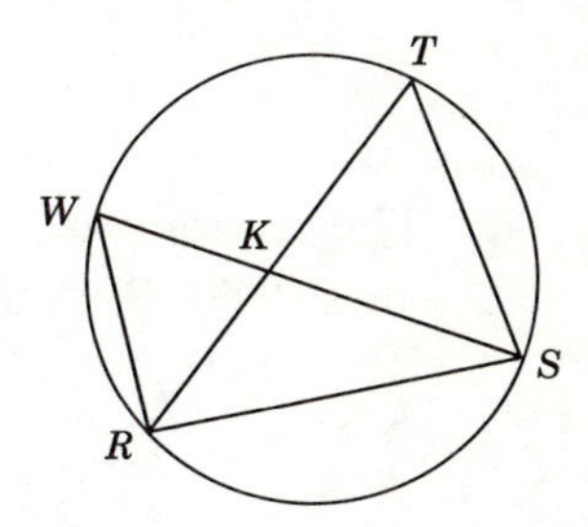

Ex. 10.

11. *Given:* Chord $AD \perp$ diameter ST;
$m\widehat{RS} = 50$.
Find: (*a*) $m\angle RST$; (*b*) $m\angle ABR$.

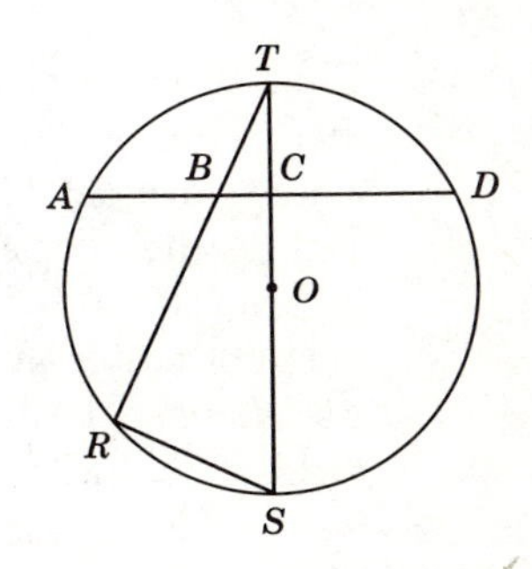

Ex. 11.

12. *Given:* $\odot O$ with diameter AB; chord BD.
$m\widehat{BD} = m\widehat{DE}$.
Prove: $\overline{DA}$ bisects $\angle BAC$.

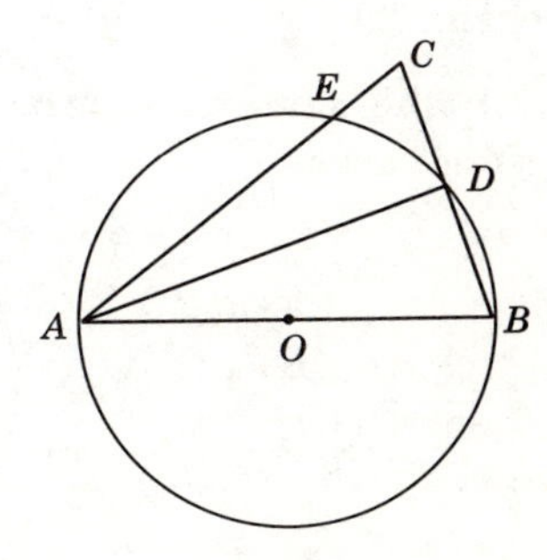

Ex. 12.

13. *Given:* Inscribed $\triangle ABC$; radius OE is the perpendicular bisector of chord AB.
Prove: $\overline{CE}$ bisects $\angle ACB$.

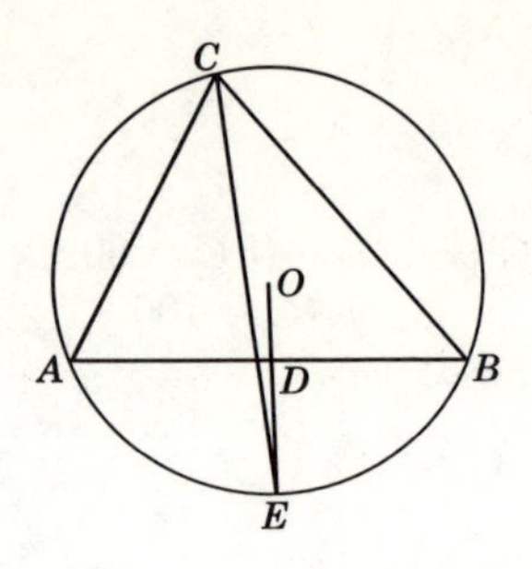

Ex. 13.

14. *Prove:* The opposite angles of an inscribed quadrilateral are supplementary.
Hint: $m\angle A = \frac{1}{2}?$; $m\angle C = \frac{1}{2}?$; $m\angle A + m\angle C = \frac{1}{2}(? + ?)$.

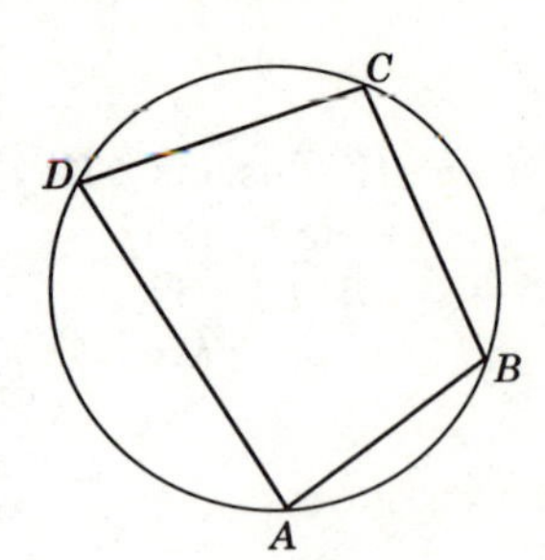

Ex. 14.

15. *Prove:* That, if two circles the centers of which are at O and Q intersect at A and B, $\overleftrightarrow{OQ}$ is perpendicular to $\overline{AB}$. (*Hint:* Draw radii to A and B.)

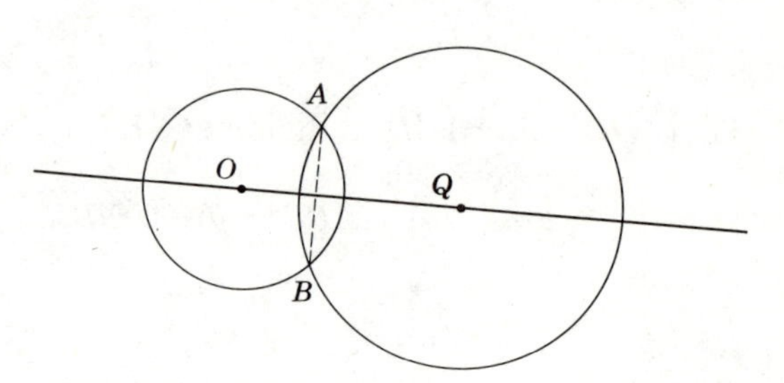

Ex. 15.

Theorem 7.4

7.19. In the same circle, or in congruent circles, congruent chords have congruent arcs.

Given: $\odot O \cong \odot Q$ *and* chord $AB \cong$ chord CD.
Conclusion: $\widehat{AB} \cong \widehat{CD}$.

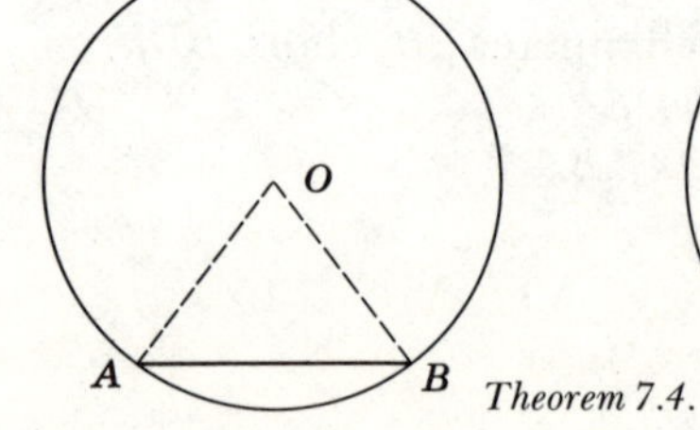

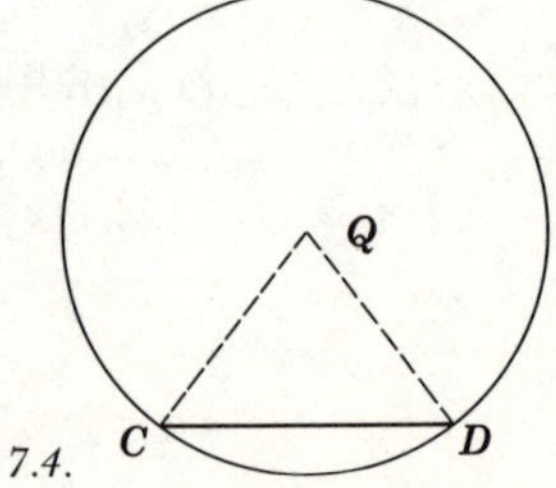

Theorem 7.4.

Proof:

STATEMENTS	REASONS
1. $\odot O \cong \odot Q$.	1. Given.
2. Draw radii OA, OB, QC, QD.	2. Postulate 2.
3. $\overline{OA} \cong \overline{QC}$; $\overline{OB} \cong \overline{QD}$.	3. Definition of $\cong$ Ⓢ.
4. $\overline{AB} \cong \overline{CD}$.	4. Given.
5. $\triangle AOB \cong \triangle CQD$.	5. S.S.S.
6. $\angle O \cong \angle Q$.	6. Corresponding parts of $\cong$ ⚠ are congruent.
7. $m\widehat{AB} = m\widehat{CD}$.	7. Definition of measure of minor arcs; substitution property.
8. $\widehat{AB} \cong \widehat{CD}$.	8. Definition of congruence of arcs.

Theorem 7.5

7.20. In the same circle, or in congruent circles, congruent arcs have congruent chords.

Given: $\odot O \cong \odot Q$ and $\widehat{AB} \cong \widehat{CD}$.
Conclusion: Chord $AB \cong$ chord CD.

Proof: (The proof is left for the student.)

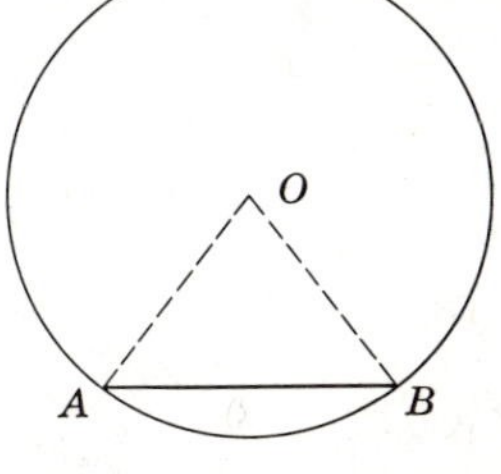

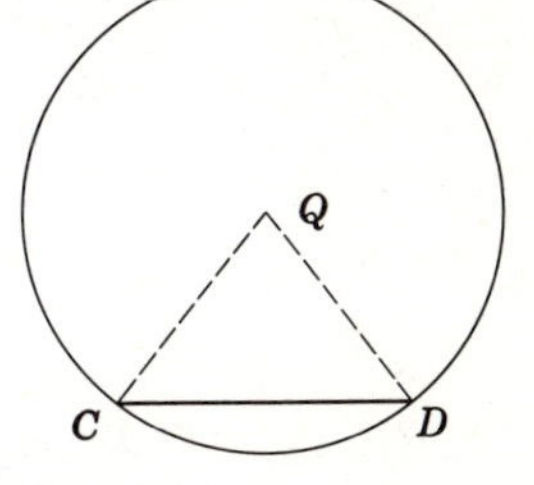

Theorem 7.5.

Theorem 7.6

7.21. In the same circle, or in congruent circles chords are congruent iff they have congruent central angles.

(The proof is left to the student.)

7.22. Congruent arcs, chords, and angles. We now have two methods of proving that two arcs are congruent. To prove that arcs are congruent, prove that they are arcs of the same or congruent circles and that they have congruent central angles. Or prove that they are arcs of the same or congruent circles and that their chords are congruent. To prove that chords are congruent, prove that their central angles (of the same or congruent circles) are congruent. To prove that central angles are congruent, prove that their arcs or chords are congruent. The following exercises will utilize these methods.

Exercises

In each of the following problems, O is the center of a circle.

1. In the figure for Ex. 1, $\overline{AB} \cong \overline{CD}$. State what other parts of the figure are congruent.

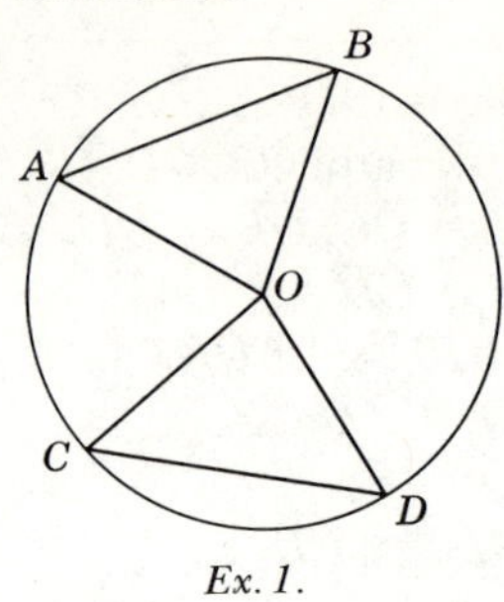

Ex. 1.

2. *Given:* $\overline{AB} \cong \overline{CD}$.
   Prove: $\widehat{AC} \cong \widehat{BD}$.

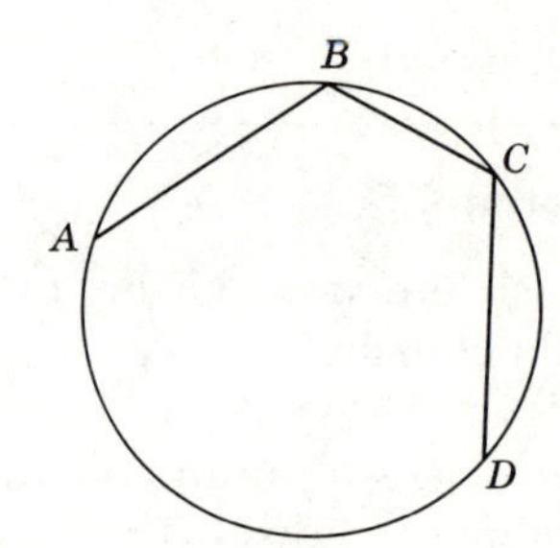

Ex. 2.

3. *Given:* $\angle ROS \cong \angle KOT$.
   Prove: $\widehat{RT} \cong \widehat{KS}$.

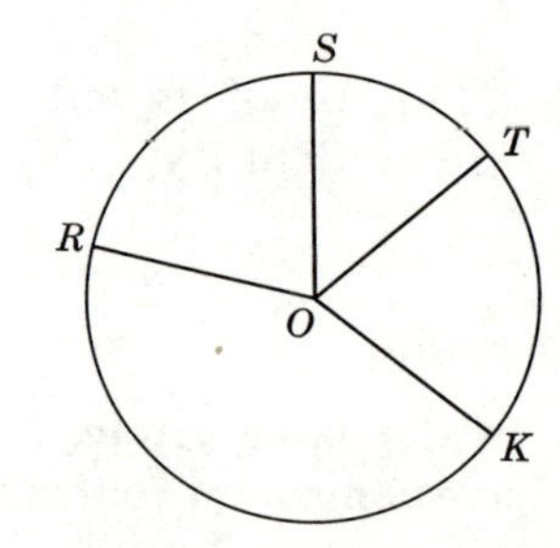

Ex. 3.

4. *Given:* $\widehat{RS} \cong \widehat{KT}$.
   Prove: $\overline{RT} \cong \overline{KS}$.

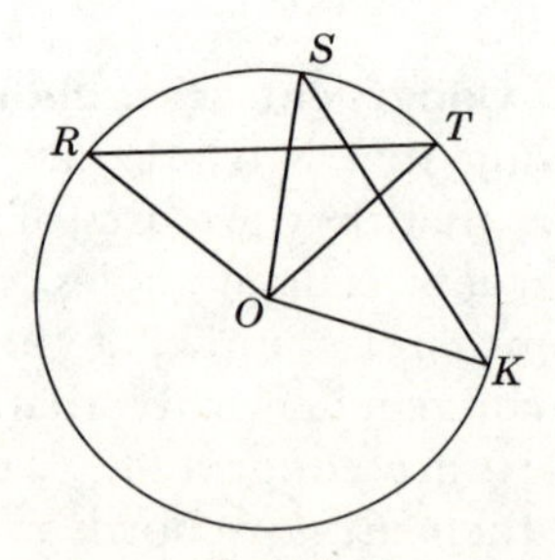

Ex. 4.

5. *Given:* $\widehat{AC} \cong \widehat{BD}$.
Prove: $\angle AOB \cong \angle COD$.

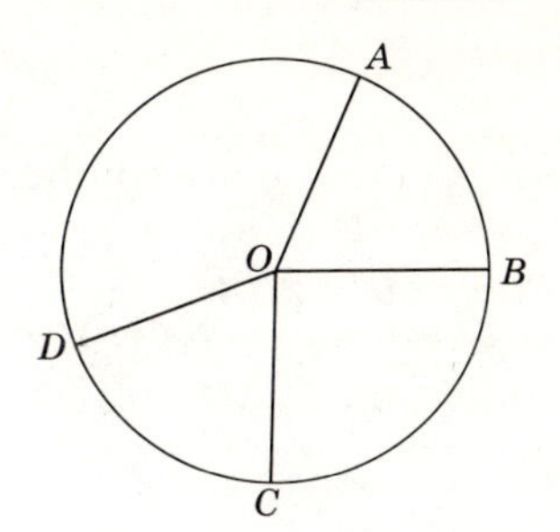

Ex. 5.

6. *Given:* $\overline{AC}$ is a diameter; $\widehat{BC} \cong \widehat{DC}$.
Prove: $\overline{AB} \cong \overline{AD}$.
7. *Given:* $\overline{AC}$ is a diameter; $\overline{AB} \cong \overline{AD}$.
Prove: $\angle BOC \cong \angle DOC$.

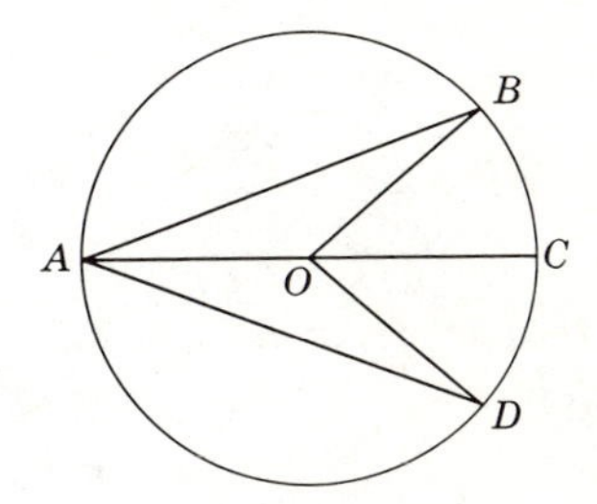

Ex. 6, 7.

8. *Given:* $\widehat{AB} \cong \widehat{BC}$.
Prove: $\triangle ADO \cong \triangle CDO$.

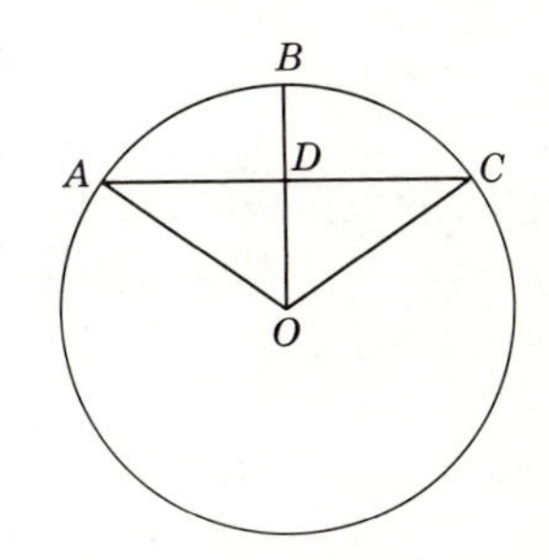

Ex. 8.

9. *Given:* $\odot B \cong \odot S$;
$\angle A \cong \angle R$.
Prove: $\overline{AC} \cong \overline{RT}$.

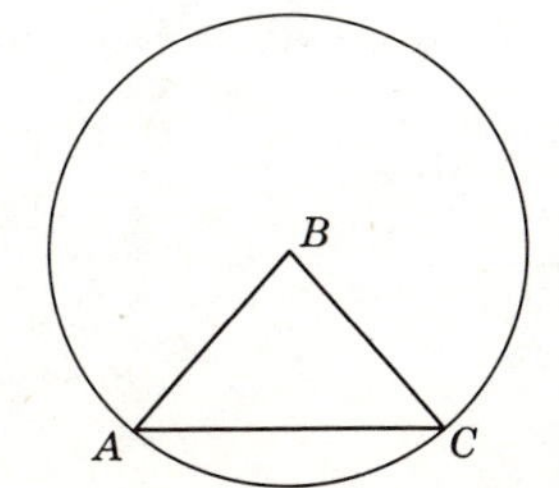

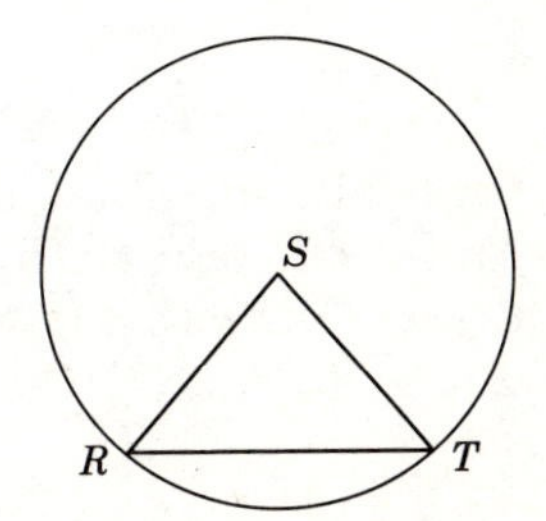

Ex. 9.

10. *Given:* A, O, B are collinear; $\overline{AD} \parallel \overline{OC}$.
Prove: $\widehat{DC} \cong \widehat{BC}$.

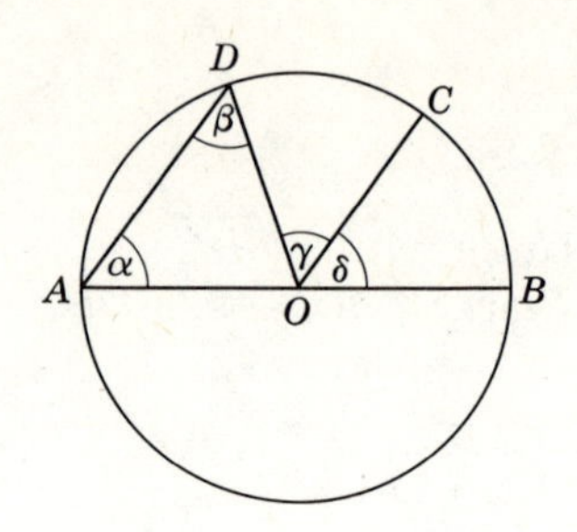

Ex. 10.

11. *Given:* $\overline{CD} \perp \overline{OB}$; $\overline{CE} \perp \overline{OA}$; $\overline{CD} \cong \overline{CE}$.
Prove: $\widehat{AC} \cong \widehat{BC}$.

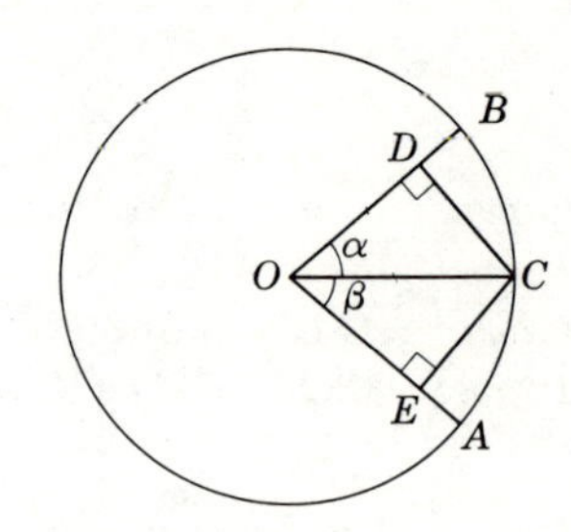

Ex. 11.

12. *Given:* Chord $CD \parallel$ diameter AB.
Prove: $\angle AOC \cong \angle BOD$.

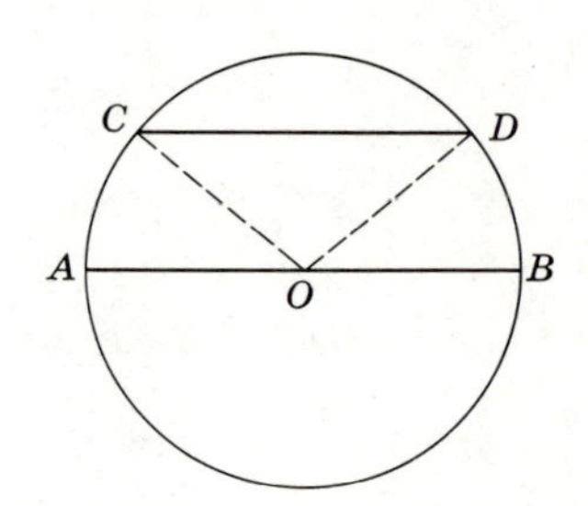

Ex. 12.

13. *Given:* Radius $OC \perp$ chord AB.
Prove: $\overline{OC}$ bisects $\overline{AB}$.

14. *Given:* $\odot O$ with $\overline{AM} \cong \overline{MB}$.
Prove: Radius $OMC \perp \overline{AB}$.

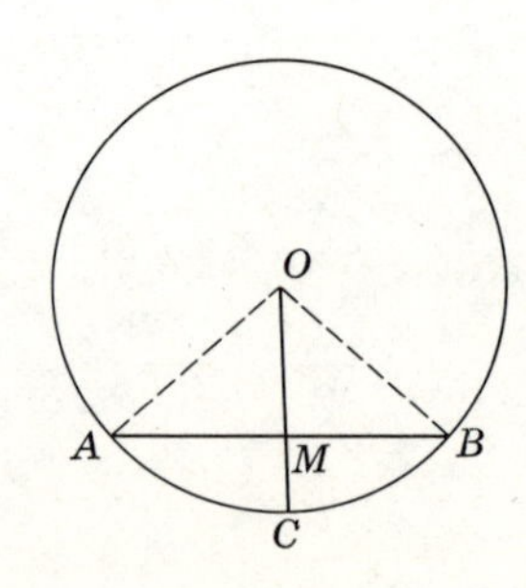

Exs. 13, 14.

Theorem 7.7

7.23. A line through the center of a circle and perpendicular to a chord bisects the chord and its arc.

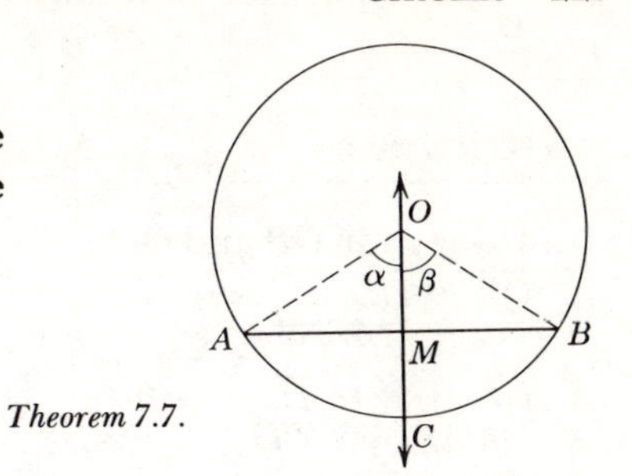

Theorem 7.7.

Given: $\odot O$ with $\overleftrightarrow{OC} \perp \overline{AB}$ at M.
Conclusion: $\overleftrightarrow{OC}$ bisects $\overline{AB}$;
$\overleftrightarrow{OC}$ bisects $\widehat{AB}$.

Proof:

STATEMENTS	REASONS
1. Draw radii OA and OB.	1. Postulate 2.
2. $\overline{OA} \cong \overline{OB}$.	2. Definition of a $\odot$.
3. $\overline{OC} \perp \overline{AB}$.	3. Given.
4. $\angle AMO$ and $\angle BMO$ are right ∠s.	4. $\perp$ lines form right ∠s.
5. $\overline{OM} \cong \overline{OM}$.	5. Reflexive property.
6. $\triangle AMO \cong \triangle BMO$.	6. Theorem 5.20.
7. $\overline{AM} \cong \overline{BM}$.	7. Corresponding parts of $\cong$ △s are congruent.
8. $\angle\alpha \cong \angle\beta$.	8. Same as 7.
9. $m\widehat{AC} = m\widehat{BC}$.	9. Definition of measures of minor arcs and substitution.
10. $\widehat{AC} \cong \widehat{BC}$.	10. Definition of congruence of arcs.
11. $\overleftrightarrow{OC}$ bisects both $\overline{AB}$ and $\widehat{AB}$.	11. Definition of bisector.

Theorem 7.8

7.24. If a line through the center of a circle bisects a chord that is not a diameter, it is perpendicular to the chord.

The proof of this theorem is left to the student.

7.25. Corollary: The perpendicular bisector of a chord of a circle passes through the center of the circle.

Theorem 7.9

7.26. In a circle, or in congruent circles, congruent chords are equidistant* from the center.

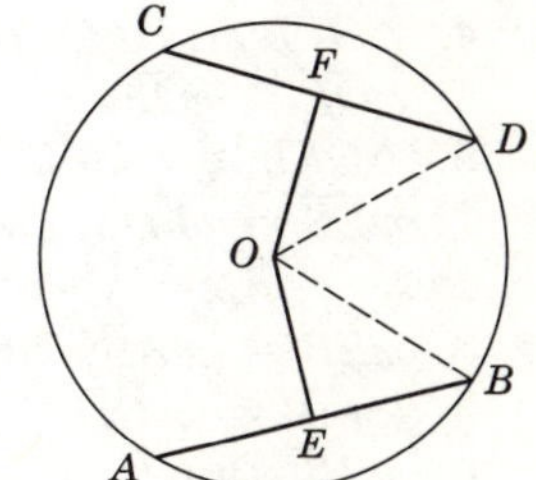

Theorem 7.9.

Given: $\odot O$ with chord $AB \cong$ chord CD;
$\overline{OE} \perp \overline{AB}$; $\overline{OF} \perp \overline{CD}$.
Conclusion: $m\overline{OE} = m\overline{OF}$.

**Recall that the distance from a point to a line is measured on the perpendicular from the point to the line.*

Proof:

STATEMENTS	REASONS
1. Draw radii OB and OD.	1. Postulate 2.
2. $\overline{OB} \cong \overline{OD}$.	2. Definition of a circle.
3. $\overline{OE} \perp \overline{AB}$; $\overline{OF} \perp \overline{CD}$.	3. Given.
4. $\overline{OE}$ bisects $\overline{AB}$.	4. $\perp$ lines form right $\angle$s.
5. OF bisects $\overline{CD}$.	5. Theorem 7.7.
6. $\overline{AB} \cong \overline{CD}$.	6. Theorem 7.7.
7. $m\overline{AB} = m\overline{CD}$.	7. Given.
8. $\overline{EB} \cong \overline{FD}$.	8. Definition of congruent segments.
9. $\angle OFD$ and $\angle OEB$ are right $\angle$s.	9. Segment bisector theorem.
10. Right $\triangle OEB \cong$ right $\triangle OFD$.	10. Theorem 5.20.
11. $m\overline{OE} = m\overline{OF}$.	11. Corresponding parts of $\cong$ $\triangle$s have the same measure.

Theorem 7.10

7.27. In a circle, or in congruent circles, chords equidistant from the center are congruent.

Given: $\odot O$ with $\overline{OE} \perp \overline{AB}$; $\overline{OF} \perp \overline{CD}$; $m\overline{OE} = m\overline{OF}$.

Conclusion: $\overline{AB} \cong \overline{CD}$.

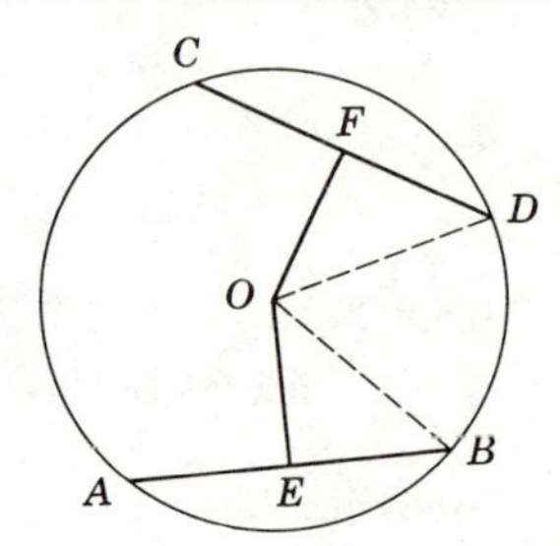

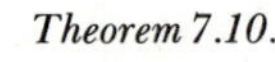
Theorem 7.10.

Proof:

STATEMENTS	REASONS
1. Draw radii OB and OD.	1. Postulate 2.
2. $m\overline{OB} = m\overline{OD}$.	2. Definition of a circle.
3. $\overline{OE} \perp \overline{AB}$; $\overline{OF} \perp \overline{CD}$.	3. Given.
4. $\angle OEB$ and $\angle OFD$ are right $\angle$s.	4. $\perp$ lines form right $\angle$s.
5. $m\overline{OE} = m\overline{OF}$.	5. Given.
6. $\triangle OEB \cong \triangle OFD$.	6. Theorem 5.20.
7. $m\overline{EB} = m\overline{FD}$.	7. Corresponding parts of $\cong$ $\triangle$s have the same measure.
8. $m\overline{EB} = \frac{1}{2}m\overline{AB}$; $m\overline{FD} = \frac{1}{2}m\overline{CD}$.	8. Theorem 7.7.
9. $\frac{1}{2}m\overline{AB} = \frac{1}{2}m\overline{CD}$.	9. Substitution property.
10. $m\overline{AB} = m\overline{CD}$.	10. Multiplication property.
11. $\overline{AB} \cong \overline{CD}$.	11. Definition of congruence of segments.

Exercises

In each of the following problems, O is the center of a circle.

1. *Given:* $\overline{OE} \perp \overline{AB}$;
   $\overline{OF} \perp \overline{CD}$; $\overline{AB} \cong \overline{CD}$.
   Prove: $\angle OEF \cong \angle OFE$.
2. *Given:* $\overline{OE} \perp \overline{AB}$;
   $\overline{OF} \perp \overline{CD}$; $\angle OEF \cong \angle OFE$.
   Prove: $\overline{AB} \cong \overline{CD}$.

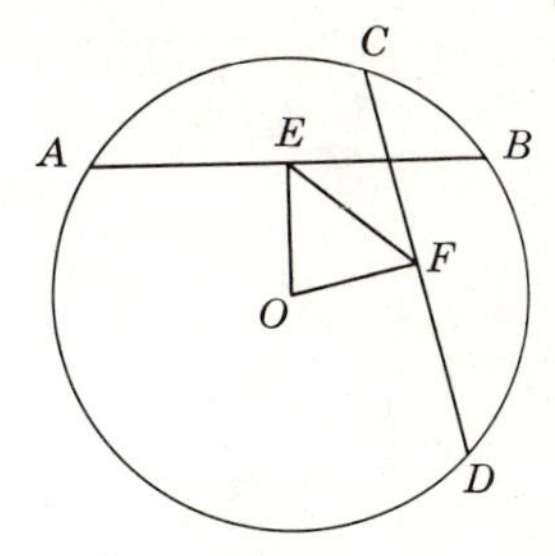

Exs. 1, 2.

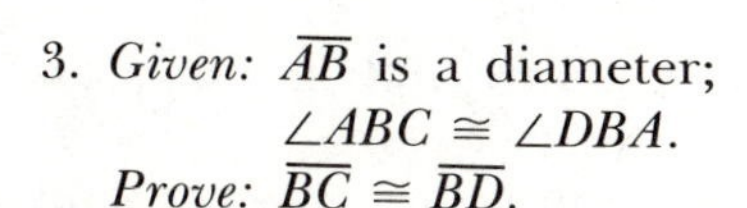

3. *Given:* $\overline{AB}$ is a diameter;
   $\angle ABC \cong \angle DBA$.
   Prove: $\overline{BC} \cong \overline{BD}$.

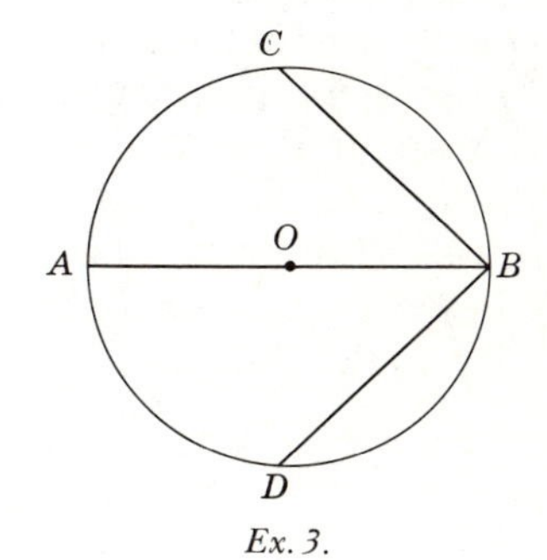

Ex. 3.

4. *Given:* Chord $AB \cong$ chord ED
   extending to meet at C.
   Prove: $\overline{EC} \cong \overline{AC}$.

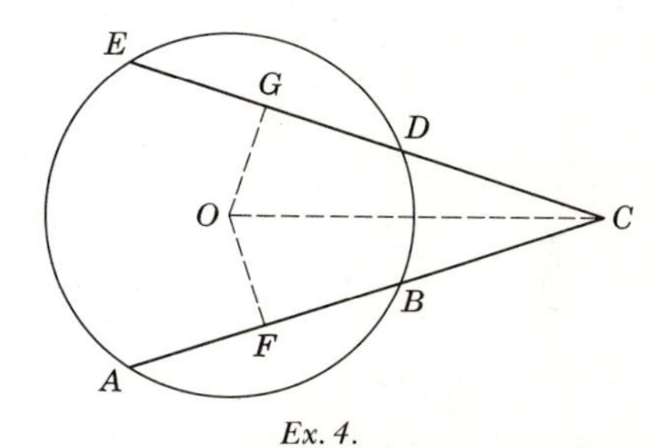

Ex. 4.

5. *Given:* $\angle OGB \cong \angle OGD$.
   Prove: $\overline{AB} \cong \overline{CD}$.
6. *Given:* Chord $AB \cong$ chord
   CD.
   Prove: $\angle OGB \cong \angle OGD$.

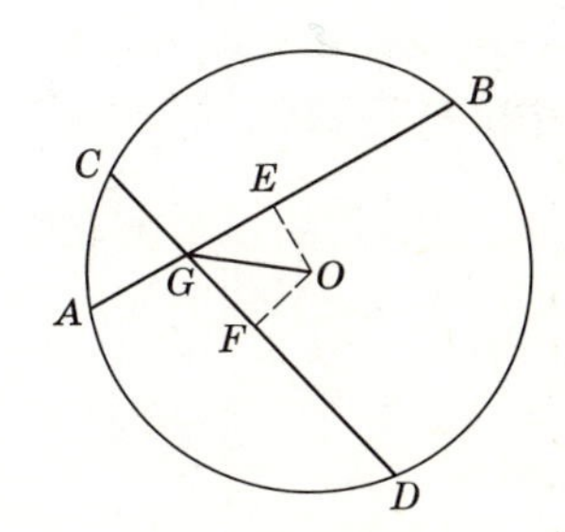

Exs. 5, 6.

7. *Given:* $\overline{AB}$ a diameter;
   $\overline{AC} \parallel \overline{BD}$.
   Prove: $\overline{AC} \cong \overline{BD}$.

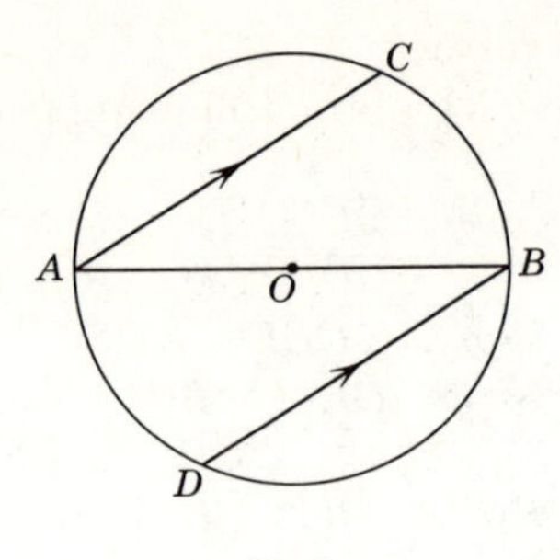

Ex. 7.

8. *Prove:* If a parallelogram is inscribed in a circle, the opposite sides are equally distant from the center.
9. *Prove:* If perpendiculars from the center of two chords of a circle are congruent, the minor arcs of the chords are congruent.
10. *Prove:* The line joining the midpoints of a chord and its arc passes through the center of a circle.
11. *Prove:* If a line joins the midpoint of a chord and its arc, it is perpendicular to the chord.
12. *Prove:* If chord CD is parallel to diameter AB, then $\overline{AC} \cong \overline{BD}$.

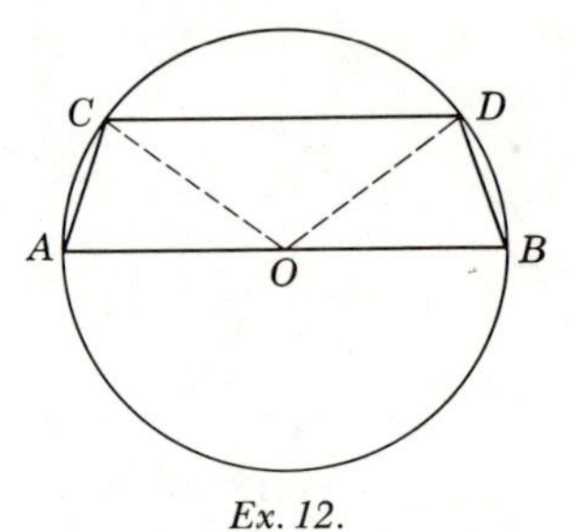

Ex. 12.

Theorem 7.11

7.28. If a line is tangent to a circle, it is perpendicular to the radius drawn to the point of tangency.

Given: $\overleftrightarrow{AB}$ tangent to $\odot O$ at C,
$\overline{OC}$ is a radius.
Conclusion: $\overleftrightarrow{AB} \perp \overline{OC}$.

Proof:

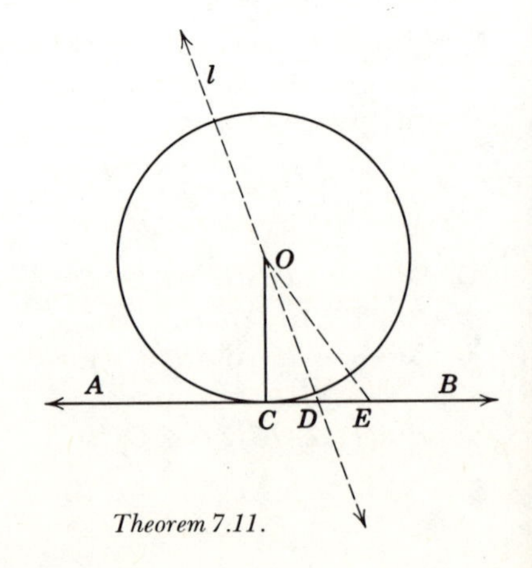

Theorem 7.11.

STATEMENTS	REASONS
1. $\overleftrightarrow{AB}$ is tangent of $\odot O$ at C.	1. Given.
2. $\overline{OC}$ is a radius.	2. Given.
3. Either $\overleftrightarrow{AB} \perp \overline{OC}$ or $\overleftrightarrow{AB}$ is not $\perp \overline{OC}$.	3. Law of excluded middle.
4. Assume $\overleftrightarrow{AB}$ is not $\perp$ to $\overline{OC}$.	4. Temporary assumption.
5. Let l be a line passing through O and $\perp$ to $\overleftrightarrow{AB}$. Let D be the intersection of l and $\overleftrightarrow{AB}$.	5. Theorem 5.4.
6. Let E be a point on $\overleftrightarrow{AB}$ on the opposite side of D from C, and such that $\overline{DE} \cong \overline{CD}$.	6. Postulate 11.
7. $\angle CDO$ and $\angle EDO$ are right ∠s.	7. $\perp$ lines form right ∠s.
8. $\angle CDO \cong \angle EDO$.	8. Theorem 3.7.
9. $\overline{OD} \cong \overline{OD}$.	9. Reflexive property.
10. $\triangle CDO \cong \triangle EDO$.	10. S.A.S.
11. $\overline{OC} \cong \overline{OE}$.	11. § 4.28.
12. E is on $\odot O$.	12. Definition of a circle.
13. $\overleftrightarrow{AB}$ intersects the $\odot$ twice.	13. Statements 1 and 12.
14. This is impossible.	14. Definition of tangent.
15. $\therefore \overleftrightarrow{AB} \perp \overline{OC}$.	15. Rule for denying the alternative.

7.29. Corollary: If a line, lying in the plane of a circle, is perpendicular to a tangent at the point of tangency, it passes through the center of the circle.

Theorem 7.12

7.30. If a line, lying in the plane of a circle, is perpendicular to a radius at its point on the circle, it is tangent to the circle.

Given: $\odot O$ with $\overleftrightarrow{AB} \perp \overline{OC}$ at C and lying in the plane of $\odot O$.

Conclusion: $\overleftrightarrow{AB}$ is tangent to $\odot O$ at C.

Proof:

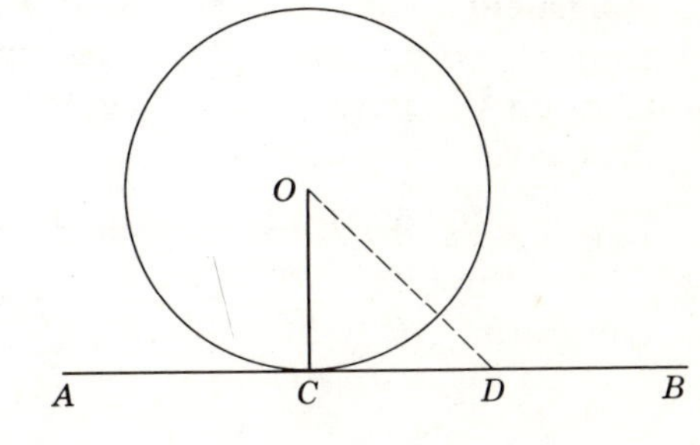

Theorem 7.12.

STATEMENTS	REASONS
1. $\overleftrightarrow{AB} \perp \overline{OC}$ of $\odot O$ and lies in the plane of $\odot O$.	1. Given.
2. Either C is the only point $\overleftrightarrow{AB}$ has in common with $\odot O$ or C is not the only point $\overleftrightarrow{AB}$ has in common with $\odot O$.	2. Law of the excluded middle.
3. Assume C is not the only point common to $\overleftrightarrow{AB}$ and $\odot O$. Let D be another such point.	3. Temporary assumption.
4. Draw $\overline{OD}$.	4. Postulate 2.
5. $\overline{OD} \cong \overline{OC}$.	5. All radii of the same circle are $\cong$.
6. $\angle ODC \cong \angle OCD$.	6. Theorem 4.16.
7. $\angle OCD$ is a right $\angle$.	7. Theorem 3.13.
8. $\angle ODC$ is a right $\angle$.	8. Substitution property.
9. $\overline{OD} \perp \overleftrightarrow{AB}$.	9. Definition of perpendicular lines.
10. $\overline{OD} \parallel \overline{OC}$.	10. Theorem 5.5.
11. Statement 10 contradicts Playfair's postulate.	11. Statement 10; Postulate 18.
12. The assumption is false and C is the only point in common with $\odot O$ and $\overleftrightarrow{AB}$.	12. Rule for denying the alternative.
13. $\overleftrightarrow{AB}$ is tangent to $\odot O$ at C.	13. Definition of tangent.

Theorem 7.13

7.31. Tangent segments from an external point to a circle are congruent and make congruent angles with the line passing through the point and the center of the circle.

Given: $\overrightarrow{PA}$ and $\overrightarrow{PB}$ are tangents from P to $\odot O$.

Conclusion: $\overline{AP} \cong \overline{BP}$ and $\angle APO \cong \angle BPO$.

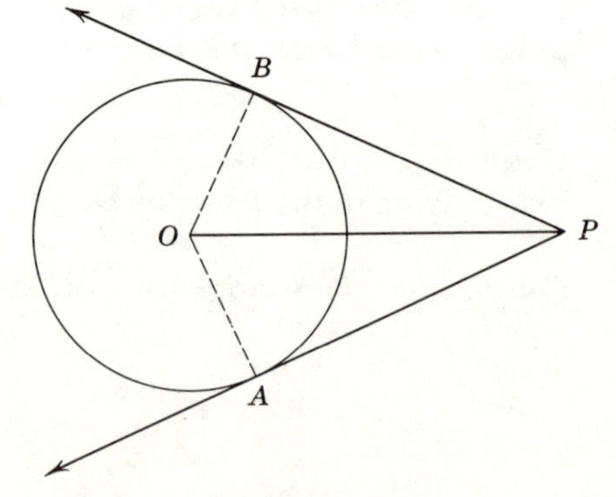

Theorem 7.13.

Proof:

STATEMENTS	REASONS
1. $\overrightarrow{PA}$ is tangent to $\odot O$; $\overrightarrow{PB}$ is tangent to $\odot O$.	1. Given.
2. Draw radii OA and OB.	2. Postulate 2.
3. $\overline{OA} \cong \overline{OB}$.	3. Radii of the same circle are $\cong$.
4. $\overline{AP} \perp \overline{OA}$; $\overline{BP} \perp \overline{OB}$.	4. Theorem 7.11.
5. $\angle OAP$ and $\angle OBP$ are right $\angle$s.	5. §1.20.
6. $\overline{OP} \cong \overline{OP}$.	6. Reflexive property of congruence.
7. $\triangle OAP \cong \triangle OBP$.	7. Theorem 5.20.
8. $\overline{AP} \cong \overline{BP}$.	8. §4.28.
9. $\angle APO \cong \angle BPO$.	9. §4.28.

Exercises

1. *Given:* $\overrightarrow{PA}$ and $\overrightarrow{PB}$ tangent to $\odot O$.
   Prove: $\overline{OP}$ bisects chord AB.

2. *Given:* $\overrightarrow{PA}$ and $\overrightarrow{PB}$ are tangent to $\odot O$; chord AB.
   Prove: $\angle PAM \cong \angle PBM$.

3. *Given:* $\overrightarrow{PA}$ and $\overrightarrow{PB}$ are tangent to $\odot O$.
   Prove: $\overline{OP} \perp$ chord AB.

4. *Given:* $\overleftrightarrow{AB}$ is tangent to $\odot O$ and $\odot Q$.
   Prove: $\angle AOC \cong \angle BQC$.

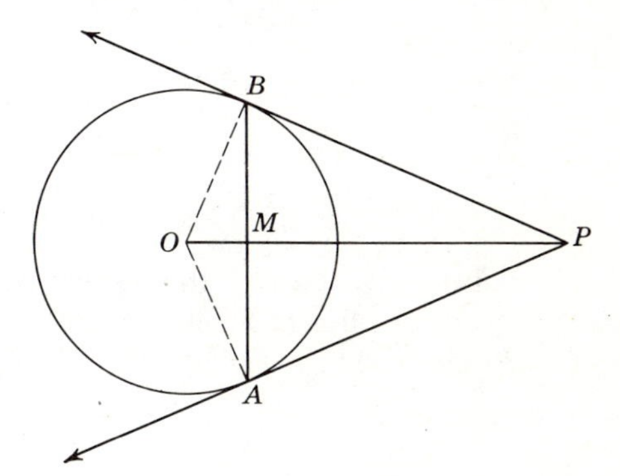

Exs. 1–3.

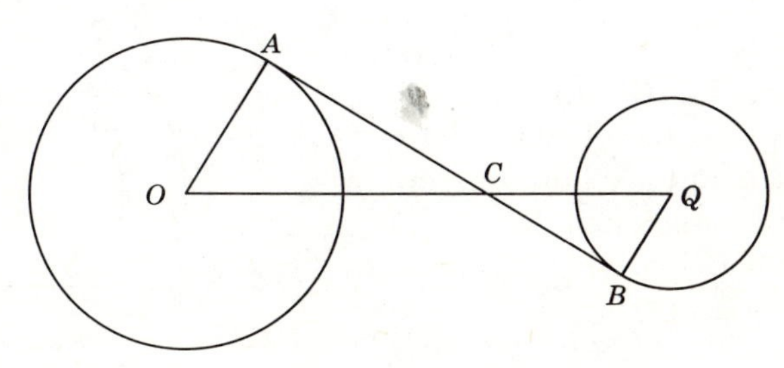

Ex. 4.

5. *Given:* $\overleftrightarrow{AB}$ and $\overleftrightarrow{CD}$ common tangents to ⓈO and Q.
Prove: $\overline{AB} \cong \overline{CD}$.

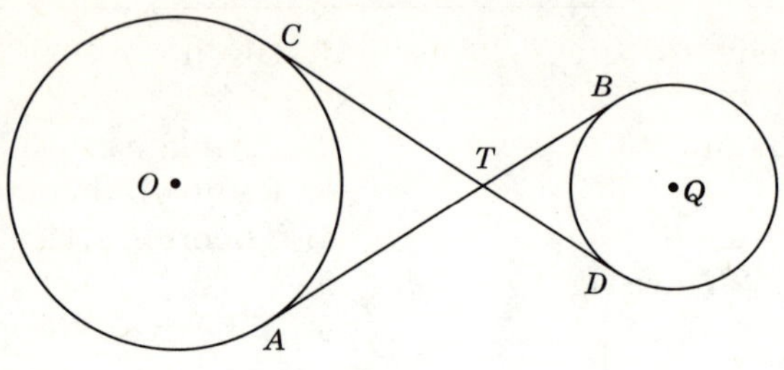

Ex. 5.

6. *Given:* ⊙O with diameter AB bisecting chords CD and EF at G and H respectively.
Prove: $\overline{CD} \parallel \overline{EF}$.

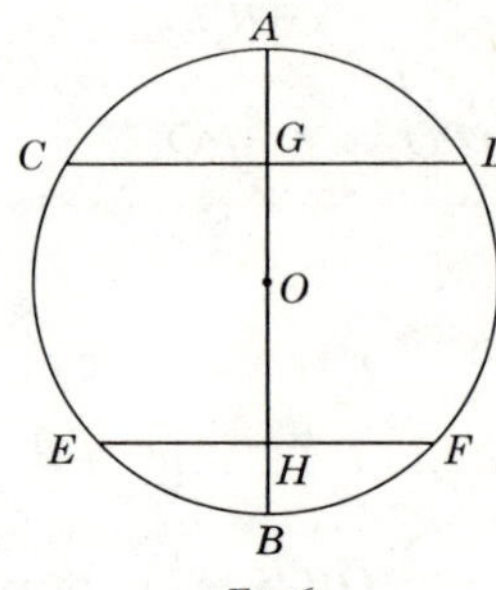

Ex. 6.

7. *Given:* ⊙O with diameter AB; radius $OD \parallel$ chord AC.
Prove: $\overline{OD}$ bisects $\overparen{CB}$.

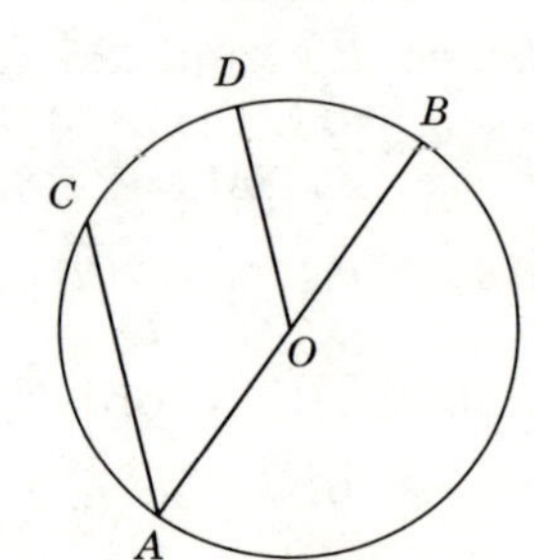

Ex. 7.

8. *Given:* Two concentric Ⓢ with center at O; $\overline{AB}$ and $\overline{CD}$ are chords of the larger circle and are both tangent to the smaller circle.
Prove: $\overline{AB} \cong \overline{CD}$.

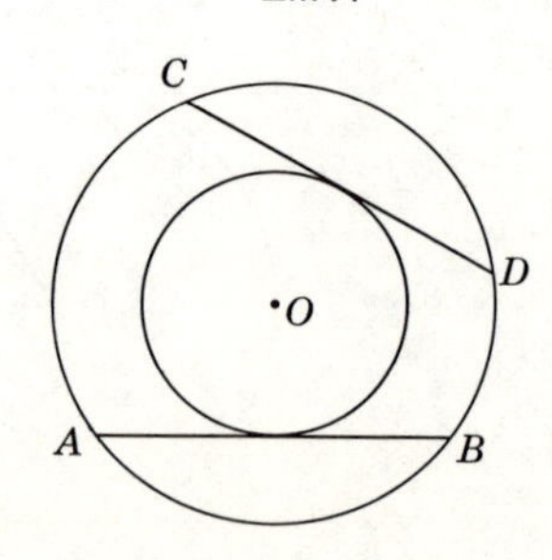

Ex. 8.

9. *Given:* $\overleftrightarrow{PA}$ and $\overleftrightarrow{PB}$ tangent to $\odot O$.
Prove: $m\angle APB = 2m\angle OAB$.

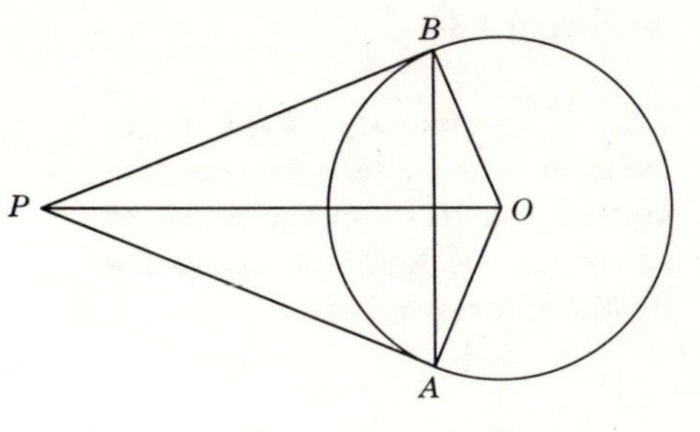

Ex. 9.

10. *Prove:* The sum of the measures of one pair of opposite sides of a circumscribed quadrilateral is equal to the sum of the measures of the other pair.

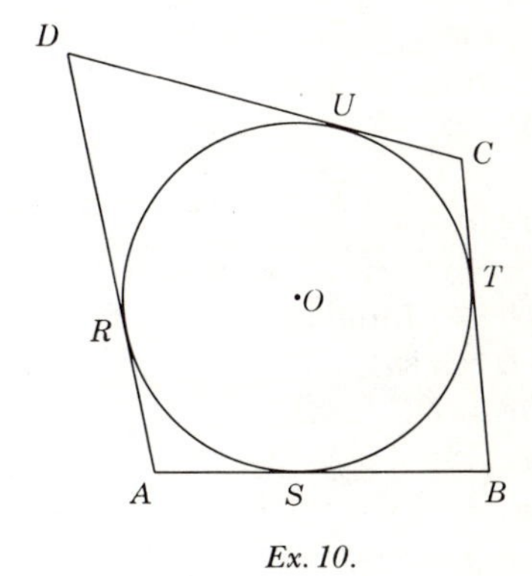

Ex. 10.

11. *Prove:* A point on a circle which is equidistant from two radii bisects the arc cut off by the radii.
12. *Prove:* The sum of the measures of the legs of a right $\triangle$ is equal to the measure of the hypotenuse of the triangle plus the measure of the diameter of the inscribed circle.

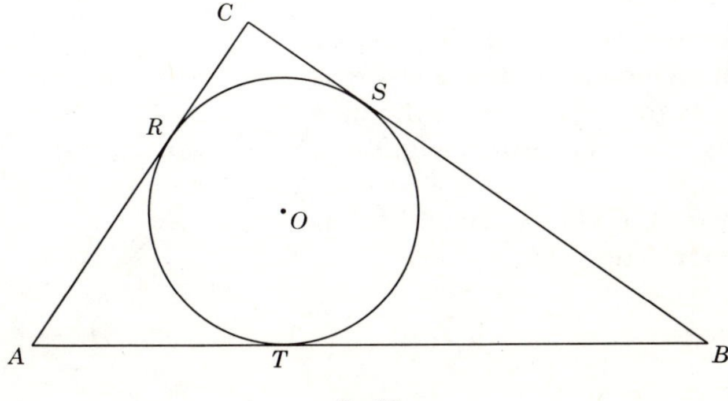

Ex. 12.

Theorem 7.14

7.32. The measure of the angle formed by a tangent and a secant drawn from the point of tangency is half the measure of its intercepted arc.

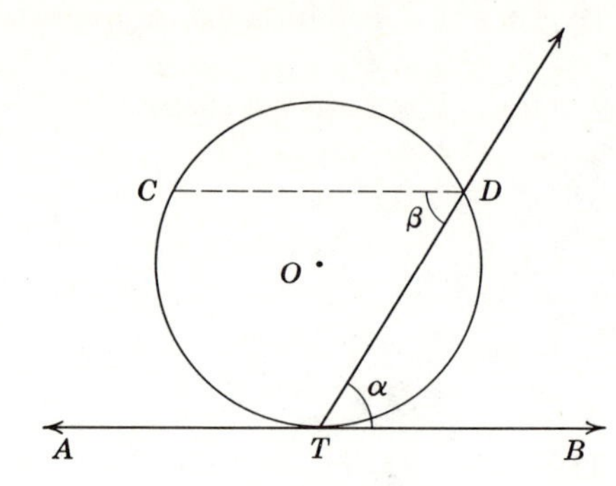

Theorem 7.14.

Given: $\overleftrightarrow{AB}$ is tangent to $\odot O$ at T; TD is a ray.

Conclusion: $m\angle\alpha = \frac{1}{2}m\widehat{DT}$.

Proof:

STATEMENTS	REASONS
1. $\overleftrightarrow{AB}$ is tangent to $\odot O$ at T; TD is a ray.	1. Given.
2. Draw $\overline{DC} \parallel \overleftrightarrow{AB}$.	2. Postulate 18; Theorem 5.7.
3. $m\angle\alpha = m\angle\beta$.	3. Theorem 5.13.
4. $m\widehat{CT} = m\widehat{DT}$.	4. § 7.18.
5. $m\angle\beta = \frac{1}{2}m\widehat{CT}$.	5. Theorem 7.3.
6. $m\angle\alpha = \frac{1}{2}m\widehat{CT}$.	6. Transitive property (with Statements 3 and 5).
7. or $m\angle\alpha = \frac{1}{2}m\widehat{DT}$.	7. Substitution property (with Statements 4 and 6).

Theorem 7.15

7.33. The measure of an angle formed by two chords intersecting within a circle is half the sum of the measures of the arcs intercepted by it and its vertical angle.

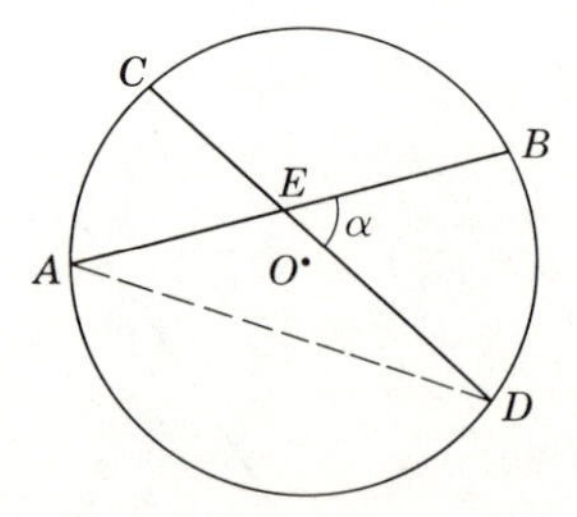

Theorem 7.15.

Given: $\angle\alpha$ formed by chords AB and CD of $\odot O$ intersecting at E.

Conclusion: $m\angle\alpha = \frac{1}{2}[m\widehat{AC} + m\widehat{BD}]$.

Proof:

STATEMENTS	REASONS
1. $\angle\alpha$ is formed by chords AB and CD of $\odot O$ intersecting at E.	1. Given
2. Draw $\overline{AD}$ forming $\triangle ADE$.	2. Postulate 2; definition of a $\triangle$.
3. $m\angle\alpha = m\angle A + m\angle D$.	3. Theorem 5.16.
4. $m\angle A = \frac{1}{2}m\widehat{BD}$.	4. Theorem 7.3.
5. $m\angle D = \frac{1}{2}m\widehat{AC}$.	5. Theorem 7.3.
6. $m\angle A + m\angle D = \frac{1}{2}[m\widehat{AC} + m\widehat{BD}]$.	6. Additive property.
7. $m\angle\alpha = \frac{1}{2}[m\widehat{AC} + m\widehat{BD}]$.	7. Substitution property.

Theorem 7.16

7.34. The measure of the angle formed by two secants intersecting outside a circle is half the difference of the measures of the intercepted arcs.

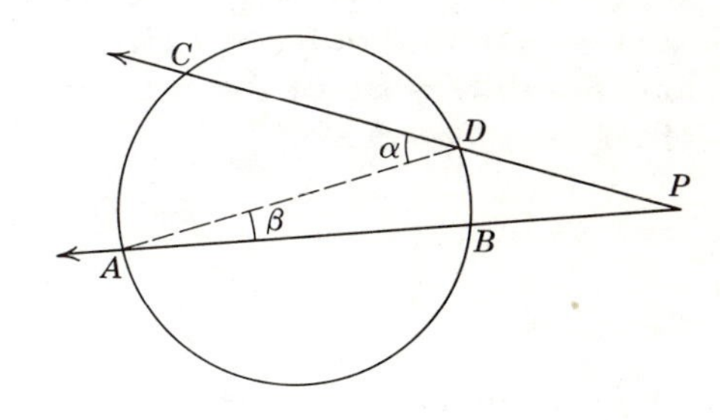

Theorem 7.16.

Given: Secants ABP and CDP intersecting at P outside $\odot O$.

Conclusion: $m\angle P = \frac{1}{2}[m\widehat{AC} - m\widehat{BD}]$.

Proof:

STATEMENTS	REASONS
1. Secants ABP and CDP intersect at P outside $\odot O$.	1. Given.
2. Draw $\overline{AD}$.	2. Postulate 2.
3. $m\angle P + m\angle\beta = m\angle\alpha$.	3. Theorem 5.16; symmetric property.
4. $m\angle P = m\angle\alpha - m\angle\beta$.	4. Subtraction property.
5. $m\angle\alpha = \frac{1}{2}m\widehat{AC}$; $m\angle\beta = \frac{1}{2}m\widehat{BD}$.	5. Theorem 7.3.
6. $m\angle\alpha - m\angle\beta = \frac{1}{2}[m\widehat{AC} - m\widehat{BD}]$.	6. Subtraction property.
7. $m\angle P = \frac{1}{2}[m\widehat{AC} - m\widehat{BD}]$.	7. Substitution property.

7.35. Corollary: The measure of the angle formed by a secant and a tangent intersecting outside a circle is half the difference of the measures of the intercepted arcs.

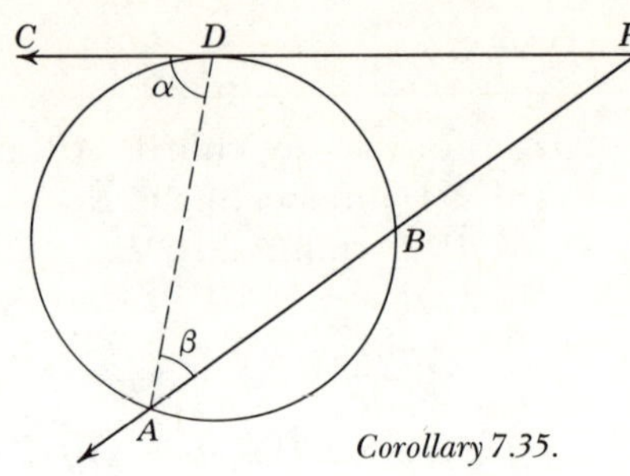

Corollary 7.35.

7.36. Corollary: The measure of the angle formed by two tangents drawn from an external point to a circle is half the difference of the measures of the intercepted arcs.

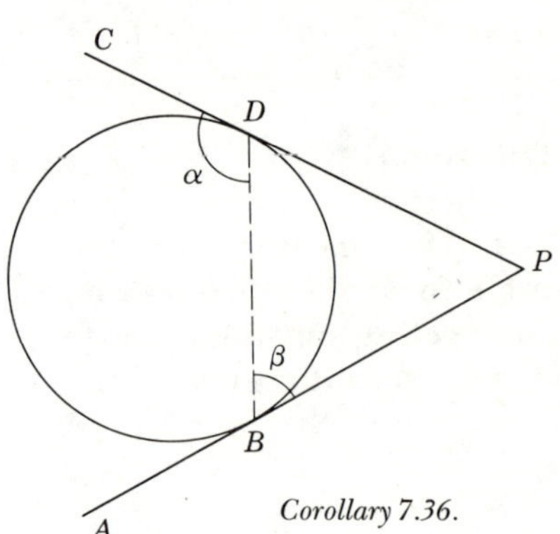

Corollary 7.36.

Exercises

Find the number of degrees measure in $\angle\alpha$, $\angle\beta$, and in arc s. O is the center of a circle.

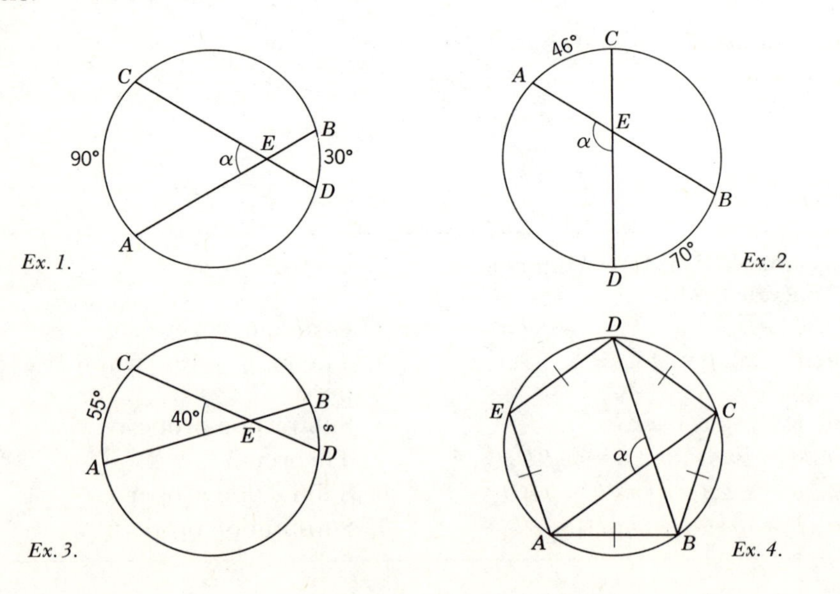

Ex. 1. *Ex. 2.* *Ex. 3.* *Ex. 4.*

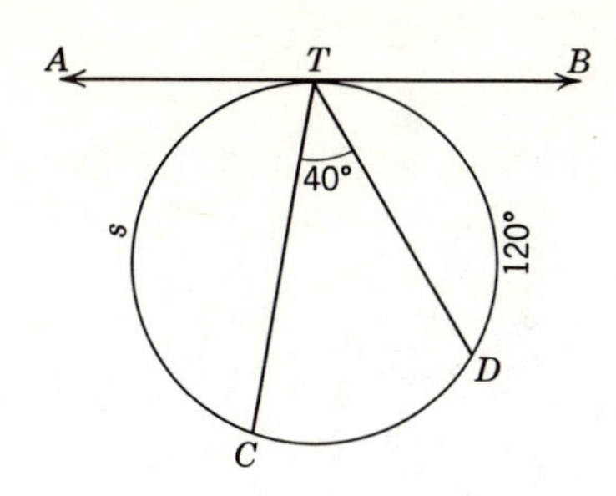

Ex. 5.

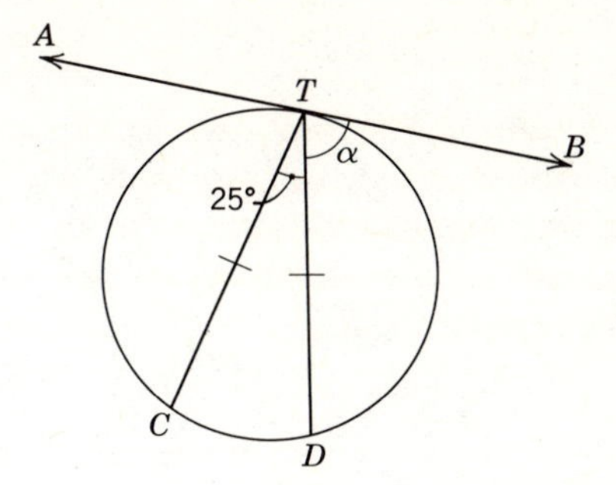

Ex. 6.

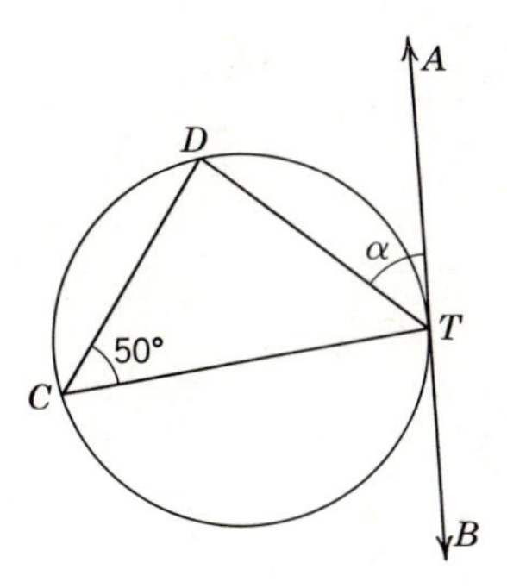

Ex. 7.

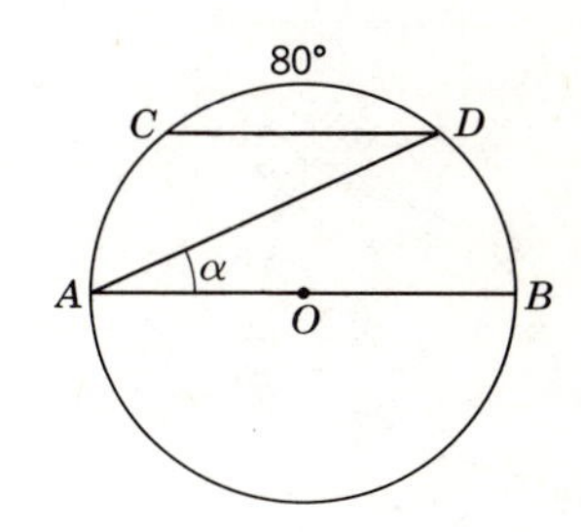

Ex. 8. $\overline{CD} \parallel \overline{AB}$.

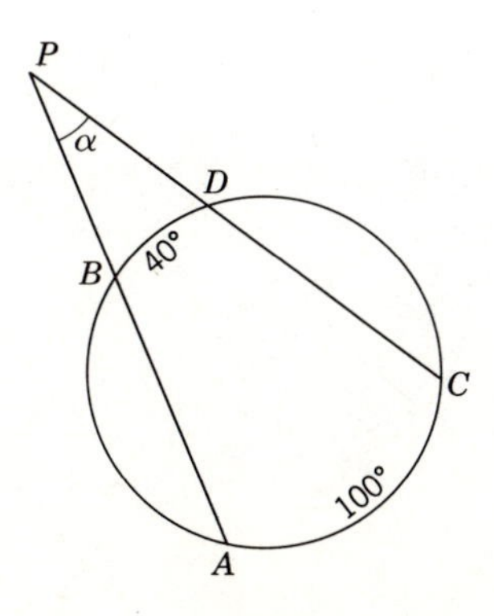

Ex. 9.

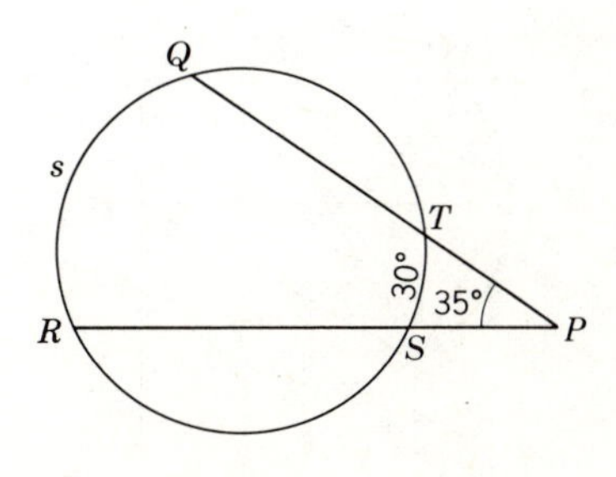

Ex. 10.

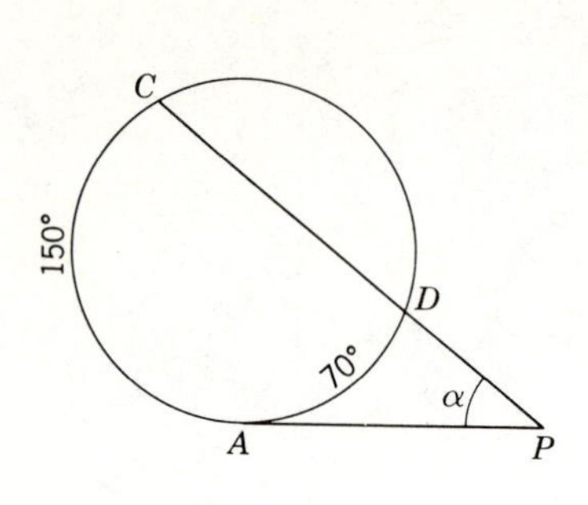

Ex. 11.

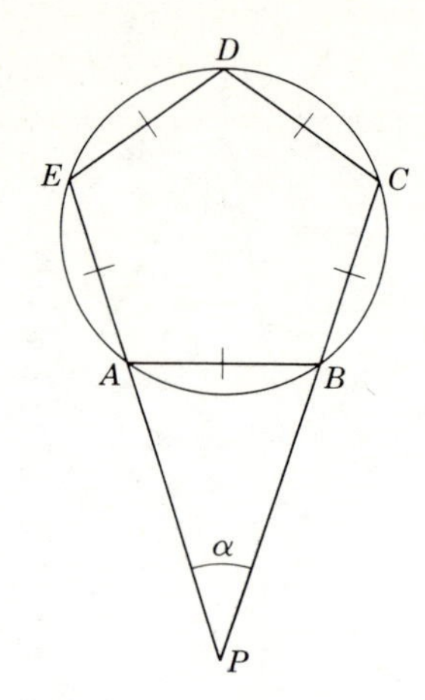

Ex. 12.

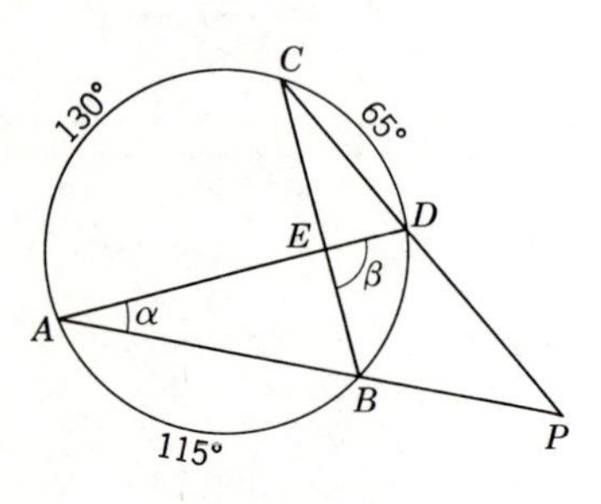

Ex. 13.

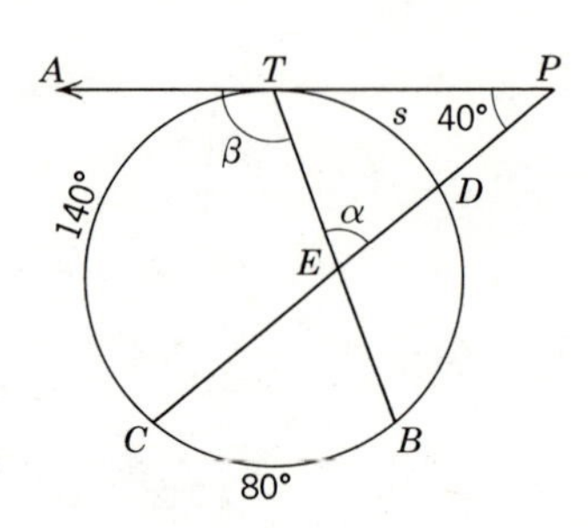

Ex. 14.

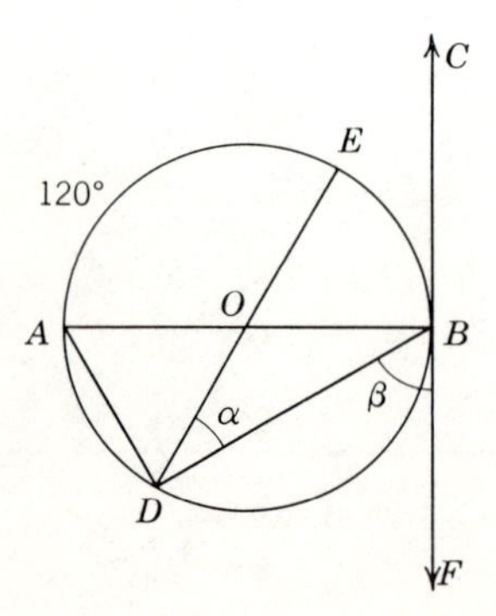

Ex. 15.

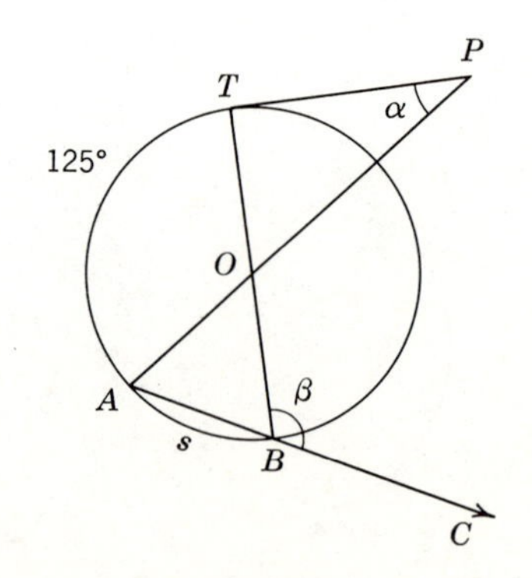

Ex. 16.

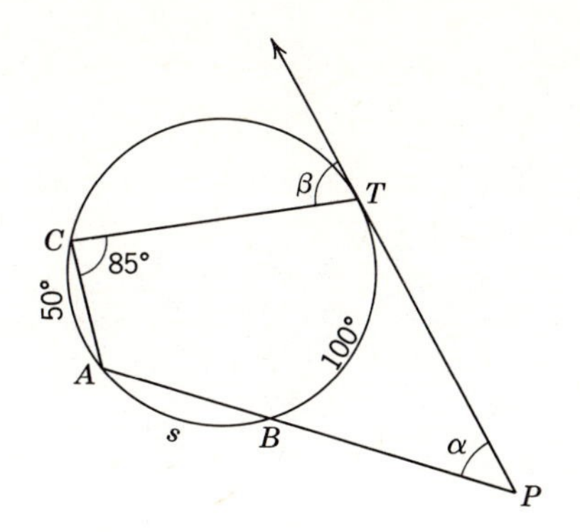

Ex. 17.

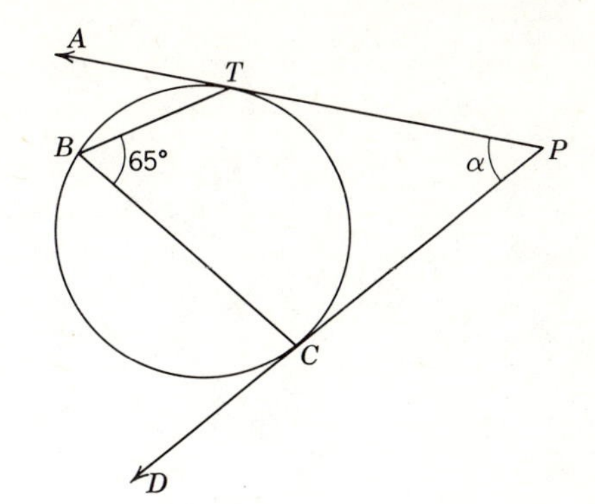

Ex. 18.

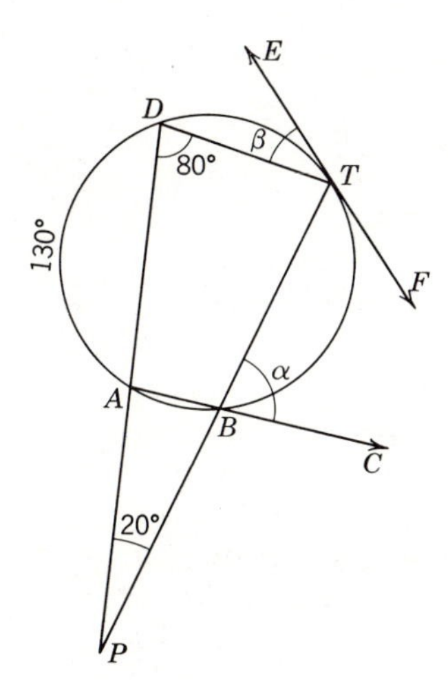

Ex. 19.

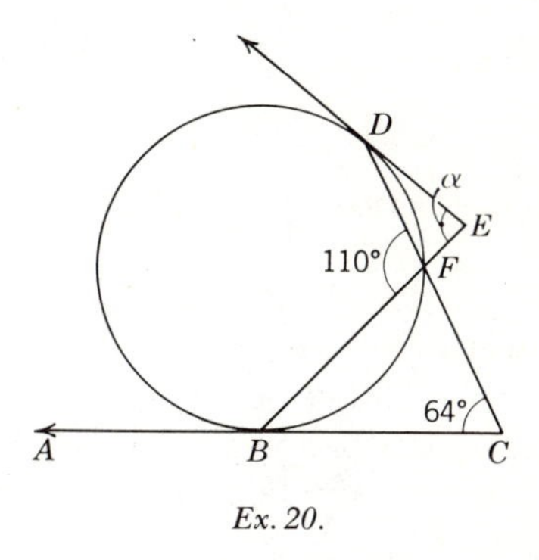

Ex. 20.

Summary Tests

Test 1

COMPLETION STATEMENTS

1. In $\odot O$ the diameter AOB and tangent AT are ______.
2. A central angle of a circle is formed by two ______.
3. An inscribed angle of a circle is formed by two ______.
4. An angle inscribed in a semicircle is a(n) ______ angle.
5. The greatest number of obtuse angles an inscribed triangle can have is ______.
6. Tangent segments drawn to a circle from an outside point are ______.
7. The largest chord of a circle is the ______ of the circle.
8. An angle is inscribed in an arc. If the intercepted arc is increased by 10, the inscribed angle is increased by ______.
9. The opposite angles of an inscribed quadrilateral are ______.
10. A line through the center of a circle and perpendicular to a chord ______ the chord and its arc.
11. If a line is ______ to a radius at its point on the circle, it is tangent to the circle.
12. If two circles intersect, the line joining their centers is the ______ of their common chord.
13. In a circle, or in congruent circles, chords equidistant from the center of a circle are ______.
14. An angle formed by two tangents drawn from an external point to a circle is equal in degrees to one-half the ______ of its intercepted arcs.

Test 2

TRUE-FALSE STATEMENTS

1. If a parallelogram is inscribed in a circle, it must be a rectangle.
2. Doubling the minor arc of a circle will double the chord of the arc.
3. On a sphere, exactly two circles can be drawn through two points which are not ends of a diameter.
4. An equilateral polygon inscribed in a circle must be equiangular.
5. A radius of a circle is a chord of the circle.
6. If an inscribed angle and a central angle subtend the same arc, the measure of the inscribed angle is twice the measure of the central angle.
7. A straight line can intersect a circle in three points.
8. A rectangle circumscribed about a circle must be a square.
9. The angle formed by two chords intersecting in a circle equal in degrees to half the difference of the measures of the intercepted arcs.
10. A trapezoid inscribed in a circle must be isosceles.
11. All the points of an inscribed polygon are on the circle.
12. Angles inscribed in the same arc are supplementary.
13. A line perpendicular to a radius is tangent to the circle.
14. The angle formed by a tangent and a chord of a circle is equal in degrees to one-half the measure of the intercepted arc.
15. The line joining the midpoint of an arc and the midpoint of its chord is perpendicular to the chord.
16. The angle bisectors of a triangle meet in a point that is equidistant from the three sides of the triangle.
17. Two arcs are congruent if they have equal lengths.
18. If two congruent chords intersect within a circle, the measurements of the segments of one chord respectively equal the measurements of the segments of the other.
19. The line segment joining two points on a circle is a secant.
20. An angle inscribed in an arc less than a semicircle must be acute.
21. The angle formed by a secant and a tangent intersecting outside a circle is measured by half the sum of the measures of the intercepted arcs.
22. If two chords of a circle are perpendicular to a third chord at its endpoints, they are congruent.
23. An acute angle will intercept an arc whose measure is less than 90.
24. A chord of a circle is a diameter.
25. The intersection of a line and a circle may be an empty set.
26. Spheres are congruent iff they have congruent diameters.
27. If a plane and a sphere have more than one point in common, these points lie on a circle.

Test 3

PROBLEMS

Find the number of degrees in $\angle\alpha$, $\angle\beta$, and s in each of the following:

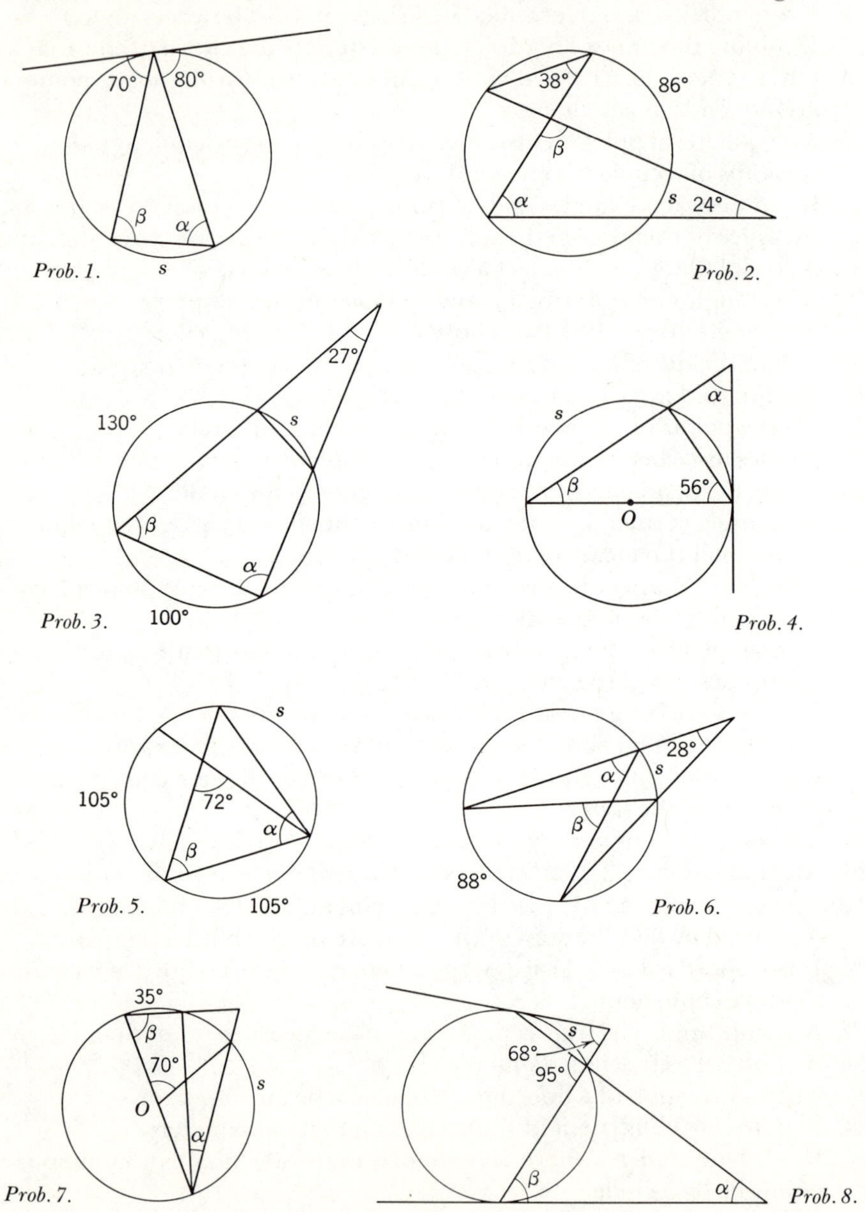

Prob. 1. Prob. 2. Prob. 3. Prob. 4. Prob. 5. Prob. 6. Prob. 7. Prob. 8.

8

Proportion – Similar Polygons

8.1. Ratio. The communication of ideas today is often based upon comparing numbers and quantities. When you describe a person as being 6 feet tall, you are comparing his height to that of a smaller unit, called the foot. When a person describes a commodity as being expensive, he is referring to the cost of this commodity as compared to that of other similar or different commodities. If you say that the dimensions of your living room are 18 by 24 feet, a person can judge the general shape of the room by comparing the dimensions. When the taxpayer is told that his city government is spending 42 per cent of each tax dollar for education purposes, he knows that 42 cents out of every 100 cents are used for this purpose.

The chemist and the physicist continually compare measured quantities in the laboratory. The housewife is comparing when measuring quantities of ingredients for baking. The architect with his scale drawings and the machine draftsman with his working drawings are comparing lengths of lines in the drawings with the actual corresponding lengths in the finished product.

Definition: The *ratio* of one quantity to another like quantity is the quotient of the first divided by the second.

It is important for the student to understand that a ratio is a quotient of measures of *like* quantities. The ratio of the measure of a line segment to that of an angle has no meaning; they are not quantities of the same kind. We can find the ratio of the measure of one line segment to the measure of a second line segment or the ratio of the measure of one angle to the measure of a second angle. However, no matter what unit of length is used for measuring two segments, the ratio of their measures is the same number as long as the same unit is used for each. In like manner, the ratio of the measures of two angles does not depend upon the unit of measure, so long as the same unit is used for both angles. The measurements must be expressed in the same units.

A ratio is a fraction and all the rules governing a fraction apply to ratios. We write a ratio either with a fraction bar, a solidus, division sign, or with the symbol : (which is read "is to"). Thus the ratio of 3 to 4 is $\frac{3}{4}$, 3/4, $3 \div 4$, or 3:4. The 3 and 4 are called *terms* of the ratio.

The ratio of 2 yards to 5 feet is 6/5. The ratio of three right angles to two straight angles is found by expressing both angles in terms of a common unit (such as a right angle). The ratio then becomes 3/4.

A ratio is always an *abstract* number; i.e., it has no units. It is a number considered apart from the measured units from which it came. Thus in Fig. 8.1, the ratio of the width to the length is 15 to 24 or 5:8. Note this does *not*

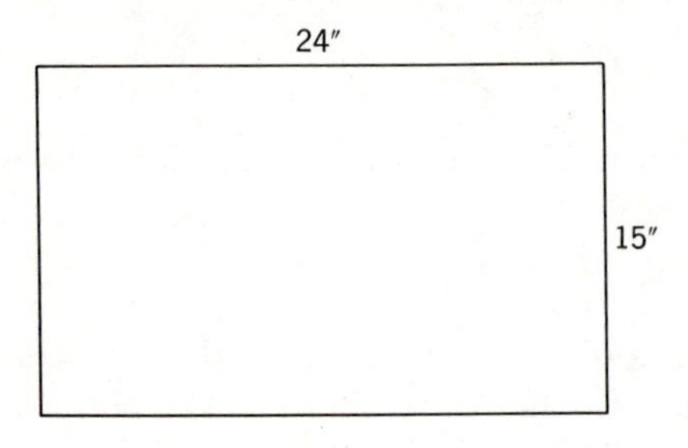

Fig. 8.1.

mean $\frac{5}{8}$ of an inch. If the dimensions of the figure were 15 by 24 feet or 15 by 24 miles, the ratio of width to length would still be 5:8.

Unless there is an important reason to the contrary, a ratio should be expressed in its simplest form. In the previous example, note that the final ratio was expressed as 5:8, not 15:24.

Exercises

1. Express in lowest terms the following ratios:
   (*a*) 8 to 12.
   (*b*) 15 to 9.
   (*c*) $\frac{42}{70}$.
   (*d*) $2x$ to $3x$.
   (*e*) $\frac{4}{15}$ to $\frac{1}{3}$.

2. What is the ratio of:
   (*a*) 1 right $\angle$ to 1 straight $\angle$?

(*b*) 3 inches to 2 feet?
(*c*) 3 hours to 15 minutes?
(*d*) 4 degrees to 20 minutes?

3. Mary is 5 years and 4 months old. Her mother is 28 years and 9 months old. What is the ratio of Mary's age to her mother's?
4. What is the ratio of the lengths of two lines which are 7 feet 8 inches and 4 feet 4 inches long?
5. What two complementary angles have the ratio 4 : 1?
6. What two supplementary angles have the ratio 1 : 3?
7. Gear A has 36 teeth. Gear B has 12 teeth. What is the ratio of the circumference length of gear A to that of B?

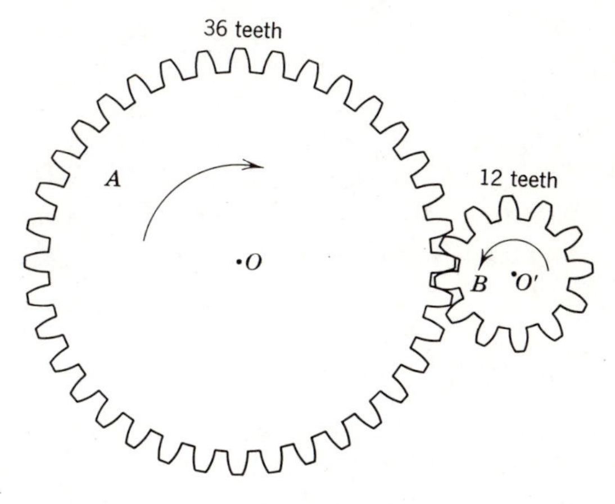

Exs. 7, 8.

8. If gear A turns 400 times a minute, how many times a minute will gear B turn?
9. In a school there are 2200 pupils and 105 teachers. What is the pupil-teacher ratio?
10. Express 37 per cent as a ratio.
11. The specific gravity of a substance is defined as the ratio of the weight of a given volume of that substance to the weight of an equal volume of water. If one gallon of alcohol weighs 6.8 pounds and one gallon of water weighs 8.3 pounds, what is the specific gravity of the alcohol?
12. One mile = 5280 feet. A kilometer = 3280 feet. What is the ratio of a kilometer to a mile?
13. What is the ratio of the length of the circumference of a circle to the length of its diameter?
14. The measures of the acute angles of a right triangle are in the ratio of 7 to 8. How large are the measures of the angles?

15. Draw $\triangle ABC$, as in the figure for Ex. 15, with $m\overline{AB} = 8$ centimeters, $m\overline{CB} = 6$ centimeters, $m\angle C = 1$ right $\angle$ and $\overline{CD} \perp \overline{AB}$. Measure $\overline{AB}$, $\overline{CD}$, $\overline{AD}$, and $\overline{BD}$ accurately to $\frac{1}{10}$ centimeter. Express the ratio of $m\overline{AB}$ / $m\overline{AC}$ and $m\overline{BC}$ / $m\overline{CD}$ to the nearest tenth.

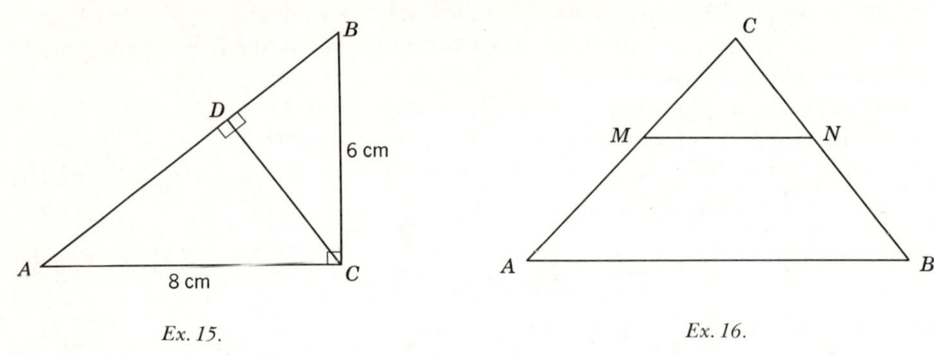

Ex. 15. *Ex. 16.*

16. Draw $\triangle ABC$ with $m\angle A = 50$, $m\overline{AC} = 8$ centimeters, $m\overline{AB} = 10$ centimeters. Then draw $\overline{MN} \parallel \overline{AB}$. Measure $\overline{AM}$, $\overline{MC}$, $\overline{BN}$, and $\overline{NC}$ to the nearest tenth of a centimeter. Express the ratios of $m\overline{AM}$ /$m\overline{MC}$ and $m\overline{BN}$ /$m\overline{NC}$ to the nearest tenth.

17. Draw $\odot O$ with radius $= 5$ centimeters. Draw chords $AD = 6$ centimeters, $CB = 8$ centimeters any place on the circle. Draw $\overline{AB}$ and $\overline{CD}$. Measure $\overline{AE}, \overline{EB}, \overline{CD}$, and $\overline{DE}$ accurately to $\frac{1}{10}$ centimeter. Express the ratios of $m\overline{DE}$ /$m\overline{BE}$ and $m\overline{AE}$ /$m\overline{CE}$ to the nearest tenth.

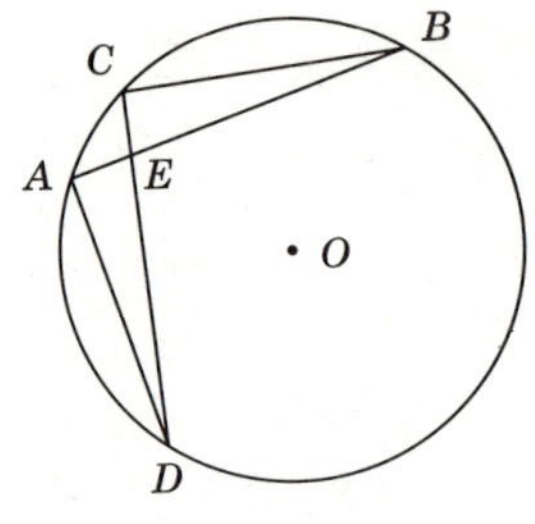

Ex. 17.

8.2. Proportion. A *proportion* is an expression of equality of two ratios. For example, since 6/8 and 9/12 have the same value, the ratios can be equated as a proportion, $6/8 = 9/12$ or $6:8 = 9:12$. Thus, if ratios $a:b$ and $c:d$ are equal, the expression $a:b = c:d$ is a proportion. This is read, "a is to b as c is to d" or "a and b are proportional to c and d". In the proportion, a is referred to as the first term, b as the second term, c as the third term, and d as the fourth term.

The first and fourth terms of a proportion are often called the *extremes*, and the second and third terms are often called the *means*. It should be noted that four terms are necessary to form a proportional. Therefore care should be taken not to use such a meaningless expression as "a is proportional to b."

The *fourth proportional* to three quantities is the fourth term of the proportion, the first three terms of which are taken in order. Thus in the proportion $a:b = c:d$, d is the fourth proportional to a, b, and c.

When the second and third terms of a proportion are equal, either is said to be the *mean proportional* between the first and fourth terms of the proportion. Thus, if $x:y = y:z$, y is the mean proportional between x and z.

If three or more ratios are equal, they are said to form a series of equal ratios. Thus $a/x = b/y = c/z$ is a series of equal ratios and may also be written in the form $a:b:c = x:y:z$.

8.3. Theorems about proportions. Since a proportion is an equation, all the axioms which deal with equalities can be applied to a proportion. Algebraic manipulation of proportions which change the form occur so frequently wherever proportions are used that it will be useful to list them as follows:

Theorem 8.1. In a proportion, the product of the extremes is equal to the product of the means.

Given: $\frac{a}{b} = \frac{c}{d}$.

Conclusion: $ad = bc$.

Proof:

STATEMENTS	REASONS
1. $\frac{a}{b} = \frac{c}{d}$.	1. Given.
2. $bd = bd$.	2. Reflexive property.
3. $\frac{a}{b} \times bd = \frac{c}{d} \times bd$, or $ad = bc$.	3. Multiplication property.

Theorem 8.2*. In a proportion, the second and third terms may be interchanged to obtain another valid proportion.

This property is called proportion by alternation. It can easily be proved by using Theorem 8.1. Thus, if $a:b = c:d$, then $a:c = b:d$. For example, since $2:3 = 8:12$, then $2:8 = 3:12$.

*The first and fourth terms in a given proportion also may be interchanged.

Theorem 8.3. In a proportion, the ratios may be inverted.

This transformation can be proved by using the division property of equality. Thus if $a/b = c/d$, then $b/a = d/c$. For example, since $\frac{2}{3} = \frac{8}{12}$, then $\frac{3}{2} = \frac{12}{8}$.

Theorem 8.4. If the product of two quantities is equal to the product of two other quantities, either pair of quantities can be used as the means and the other pair as the extremes of a proportion.

Given: $ab = cd$.

Conclusion: $\frac{a}{c} = \frac{d}{b}$.

Proof:

STATEMENTS	REASONS
1. $ab = cd$.	1. Given.
2. $bc = bc$.	2. Reflexive property.
3. $\frac{ab}{bc} = \frac{cd}{bc}$, or $\frac{a}{c} = \frac{d}{b}$.	3. Division property.

Theorem 8.5. If the numerators of a proportion are equal but not equal to zero, the denominators are equal. Thus, $a/x = b/y \wedge a = b \rightarrow x = y$.

Theorem 8.6. If three terms of one proportion are equal to the corresponding three terms of another proportion, the remaining terms are equal.

Given: $\frac{a}{b} = \frac{c}{x}$ and $\frac{a}{b} = \frac{c}{y}$.

Conclusion: $x = y$.

Theorem 8.7. In a series of equal ratios the sum of the numerators is to the sum of the denominators as the numerator of any one of the ratios is to the denominator of that ratio.

Given: $\frac{a}{b} = \frac{c}{d} = \frac{e}{f} = \ldots.$

Conclusion: $\dfrac{a+c+e+\ldots}{b+d+f+\ldots}=\dfrac{a}{b}.$

Proof:

STATEMENTS	REASONS
1. Let $\dfrac{a}{b}=k.$	1. Definition of k.
2. $\therefore \dfrac{c}{d}=k, \dfrac{e}{f}=k.$	2. Transitive property.
3. $a=kb, c=kd, e=kf, \ldots$	3. Multiplication property.
4. $a+c+e+\ldots=kb+kd+kf+\ldots=k(b+d+f\ldots).$	4. Addition property.
5. $\dfrac{a+c+e+\ldots}{b+d+f+\ldots}=k.$	5. Division property.
6. $\dfrac{a+c+e+\ldots}{b+d+f+\ldots}=\dfrac{a}{b}.$	6. Substitution property.

Theorem 8.8. If four quantities are in proportion, the terms are in proportion by addition or subtraction; that is, the sum (or difference) of the first and second terms is to the second term as the sum (or difference) of the third and fourth terms is to the fourth term.

Given: $\dfrac{a}{b}=\dfrac{c}{d}.$

Conclusion: $\dfrac{a+b}{b}=\dfrac{c+d}{d}$ and $\dfrac{a-b}{b}=\dfrac{c-d}{d}.$

Proof: (The proof is left to the student.)

Suggestion: $\dfrac{a}{b}+1=\dfrac{c}{d}+1, \dfrac{a+b}{b}=\dfrac{c+d}{d}.$

Exercises

1. Find the value of x which satisfies the following proportions:
   (*a*) $2:x=5:8.$ (*d*) $3:5=x:8.$

(*b*) $2x:3=4:5$.

(*c*) $8:3=4:x$.

(*e*) $1\frac{3}{4}$ feet: 3 inches = 2 yards: x inches.

(*f*) 30 inches : x feet = 3 yards: 2 feet

2. Change the following proportions to another whose first and second terms are respectively x and y:

(*a*) $\frac{y}{x}=\frac{3}{4}$.

(*b*) $\frac{x}{5}=\frac{y}{7}$.

(*c*) $\frac{2}{x}=\frac{3}{y}$.

(*d*) $\frac{4}{y}=\frac{8}{x}$.

(*e*) $x:3=y:7$.

(*f*) $y:2=x:6$.

3. Find the ratio of x to y in each of the following:

(*a*) $2x=5y$. (*c*) $\frac{3}{4}x=1\frac{1}{4}y$. (*e*) $ax=by$.

(*b*) $9x=4y$. (*d*) $x=\frac{1}{3}y$. (*f*) $ry=sx$.

4. Find the fourth proportional to 3, 5, and 8.
5. Find the mean proportional between:

(*a*) 9 and 16.

(*b*) 14 and 9.

6. Form five different proportions from each of the following equal products:

(*a*) $3\times 12=4\times 9$.

(*b*) $xy=rs$.

7. In $\triangle ABC$ it is given that $m\overline{AD} : m\overline{DC} = m\overline{BE} : m\overline{EC}$. *Prove:*

(*a*) $m\overline{CD} : m\overline{DA} = m\overline{CE} ; m\overline{EB}$.

(*b*) $m\overline{AD} : m\overline{BE} = m\overline{DC} : m\overline{EC}$.

(*c*) $m\overline{AC} : m\overline{AD} = m\overline{BC} : m\overline{BE}$.

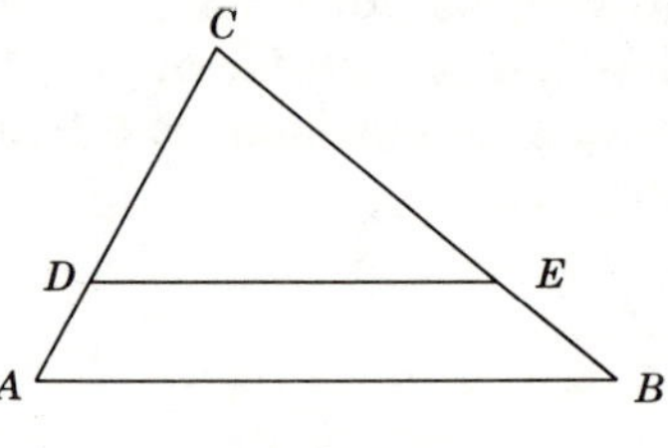

Ex. 7.

8. In a draftsman's scale drawing $\frac{1}{4}$ inch represents 1 foot. What length will be represented by a line $2\frac{3}{4}$ inches long on the drawing?
9. If a car can travel 120 miles on 8 gallons of gasoline, how many gallons will be required for a journey of 450 miles?
10. If a machine can manufacture 300 objects in 40 minutes, how many objects can the machine manufacture in 8 hours?
11. The model of a church is to be built to a scale of $\frac{1}{8}$ inch to 1 foot. How high will the model of the church spire need to be if the actual spire is 90 feet high?

12. At noon a 6-foot man casts a shadow 10 feet long. How high must a tree be at noon to cast a shadow of 240 feet?

IMPORTANT. It has been the conscious effort of this text thus far to emphasize the distinction between a geometric figure and its measure. Often, however, the symbolism involving the measures of line segments can be so involved as to be distracting. This would be true in this and in subsequent chapters if we were not to introduce another way of indicating the measure of a segment. Each time the letters AB has been used thus far, it has had an added symbolism above it. Thus we have written $\overleftrightarrow{AB}$, $\overrightarrow{AB}$, $\overline{AB}$, $\overset{\circ\;\;\circ}{AB}$, and $\overset{\circ\;\;}{\overline{AB}}$ to represent a line, ray, segment, interval, and an interval closed at one end.

Hereafter, when AB appears in this text with no mark above it, it will represent the measure of the line segment AB. Thus,

$$AB = m\overline{AB}$$

The student will note that, hereafter, the following are equivalent statements: $m\overline{AB} = m\overline{CD}$; $\overline{AB} \cong \overline{CD}$; $AB = CD$.

Theorem 8.9

8.3. If a line parallel to one side of a triangle cuts a second side into segments which have a ratio with integer terms, the line will cut the third side into segments which have the same ratio.

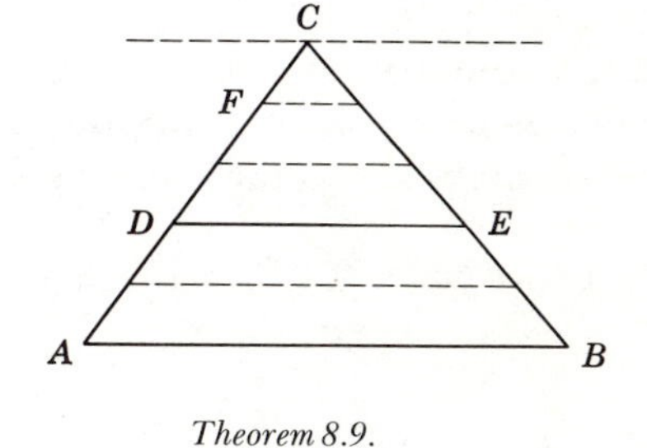

Theorem 8.9.

Given: $\triangle ABC$ with $\overleftrightarrow{DE} \parallel \overleftrightarrow{AB}$.

Conclusion: $\dfrac{CD}{DA} = \dfrac{CE}{EB}$.

Proof:

STATEMENTS	REASONS
1. Let CF be a unit of measure, common to $\overline{CD}$ and $\overline{DA}$, contained m times in $\overline{CD}$ and n times in $\overline{DA}$. $\therefore \dfrac{CD}{DA} = \dfrac{m}{n}$.	1. Definition of ratio.

2. At points of division on $\overline{AC}$, draw lines $\parallel$ to $\overline{AB}$.	2. Postulate 18.
3. These lines divide $\overline{CE}$ into m congruent parts and $\overline{EB}$ into n congruent parts.	3. Theorem 6.7.
4. $\frac{CE}{EB} = \frac{m}{n}$.	4. Substitution property.
5. $\frac{CD}{DA} = \frac{CE}{EB}$.	5. Theorem 3.4.

Note: Statement 1 assumes there is a common unit which will be contained integral times in $\overline{CD}$ and $\overline{DA}$. When this is true, the segments are said to be *commensurable* with each other. The proof of the incommensurable case is difficult, since it requires a knowledge about *limits*.

Mathematics courses in calculus can prove the more general theorem:

Theorem 8.10. A line parallel to one side of a triangle and intersecting the other two sides divides these sides into proportional segments.

8.5. Corollary: If a line is parallel to one side of a triangle and intersects the other two sides, it divides these sides so that either side is to one of its segments as the other is to its corresponding segment.

Suggested proof: Use Theorem 8.9 and Theorem 8.8.

8.6. Corollary: Parallel lines cut off proportional segments on two transversals.

Suggested proof: Draw $\overleftrightarrow{DH} \parallel \overleftrightarrow{AC}$;

$$\frac{DG}{GH} = \frac{DE}{EF}.$$

$$\text{Then } \frac{AB}{BC} = \frac{DE}{EF}.$$

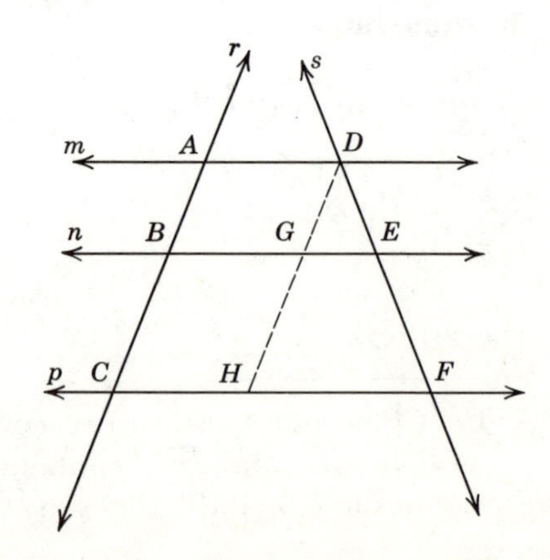

Corollary 8.6.

Theorem 8.11

8.7. If a line divides two sides of a triangle proportionally, it is parallel to the third side.

Given: $\triangle ABC$ with $\overline{DE}$ intersecting $\overline{AC}$ and $\overline{BC}$ so that $\dfrac{CD}{DA} = \dfrac{CE}{EB}$.

Conclusion: $\overline{DE} \parallel \overline{AB}$.

Proof:

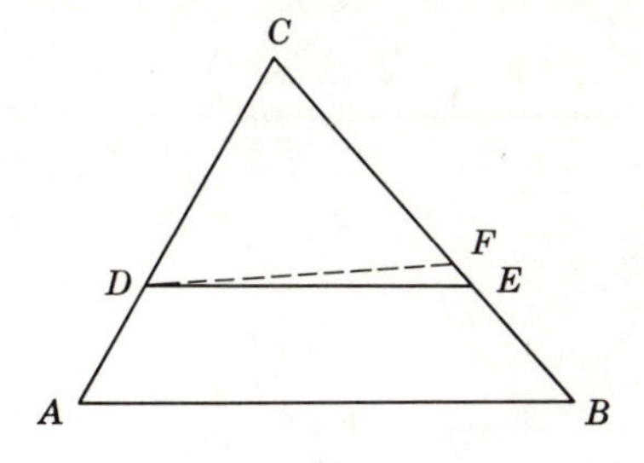

Theorem 8.11.

STATEMENTS	REASONS
1. Either $\overline{DE} \parallel \overline{AB}$ or $\overline{DE} \parallel \overline{AB}$.	1. Law of the excluded middle.
2. Assume $\overline{DE} \nparallel \overline{AB}$.	2. Temporary assumption.
3. Then let $\overline{DF} \parallel \overline{AB}$ and intersecting $\overline{BC}$ at F.	3. Postulate 18; theorem 5.7.
4. $\dfrac{CA}{DA} = \dfrac{CB}{FB}$.	4. §8.5. Corollary.
5. $\dfrac{CD}{DA} = \dfrac{CE}{EB}$.	5. Given.
6. $\dfrac{CA}{DA} = \dfrac{CB}{EB}$.	6. Theorem 8.8.
7. $FB = EB$.	7. Theorem 8.6.
8. F falls on E.	8. Definition of $\cong$ segments.
9. $\overline{DF}$ coincides with $\overline{DE}$.	9. Postulate 2.
10. This is impossible.	10. Statements 2 and 3.
11. $\therefore \overline{DE} \parallel \overline{AB}$.	11. Substitution property.

8.8. Corollary: If a line divides two sides of a triangle so that either side is to one of its segments as the other side is to its corresponding segment, the line is parallel to the third side.

Exercises

1–8. In the following exercises it is given that $\overline{DE} \parallel \overline{AB}$. In each exercise the lengths of three segments are given. Find the value of x in each exercise.

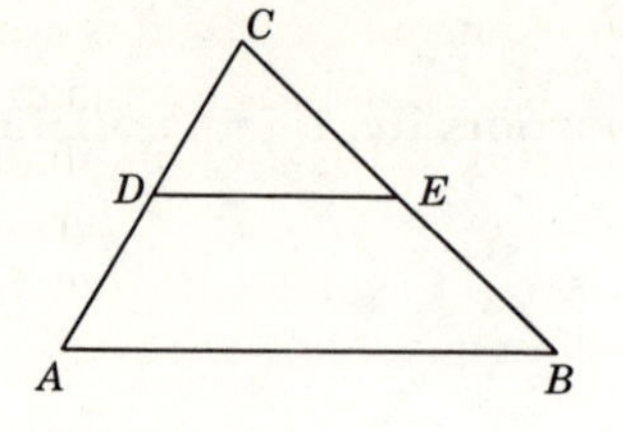

	1	2	3	4	5	6	7	8
CD	*x*	6	8	12	8	9	*x*	10
DA	12	*x*	16	15				15
CE	9	6	*x*	18	10		20	12
EB	15	8	20	*x*	15	18	30	
AC					*x*	15	40	
BC						*x*		*x*

Exs. 1–8.

9. If $MT = 12$, $RM = 8$, $TN = 15$, and $TS = 25$, is $\overline{MN} \parallel \overline{RS}$? Why?
10. If $RM = 5\frac{1}{3}$, $MT = 24$, $TN = 36$, and $TS = 44$, is $\overline{MN} \parallel \overline{RS}$? Why?

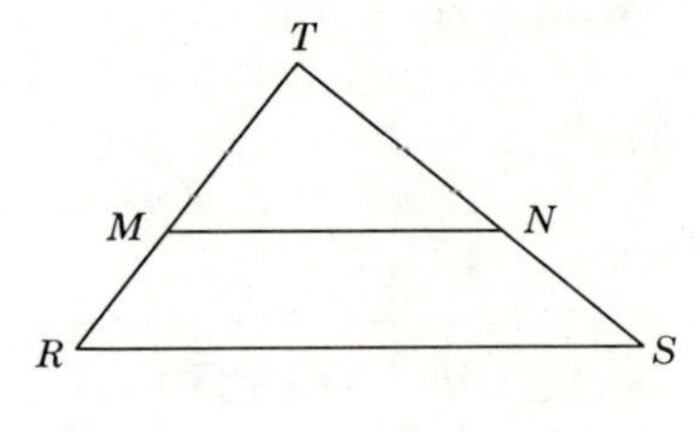

Exs. 9, 10.

11–16. *Given:* $m \parallel n \parallel p$. Find the value of x in each of the following.

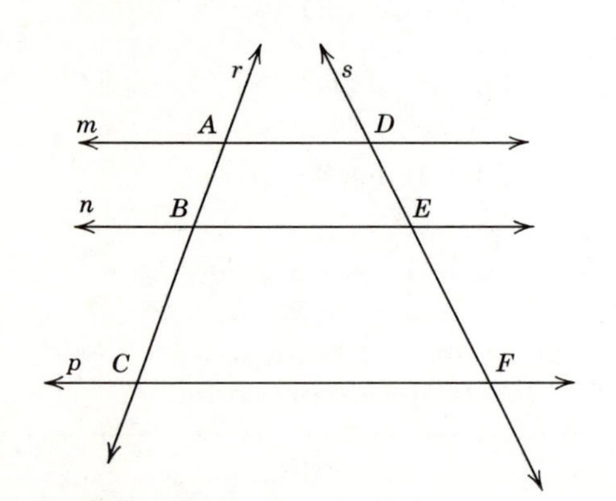

	11	12	13	14	15	16
AB	*x*	8		9	12	
BC	15	*x*	9	15	18	6
AC			*x*			10
DE	24	10	12	*x*		
EF	24			18	*x*	8
DF		25	30		40	*x*

Exs. 11–16.

8.9. Similar polygons. In Chapter 4 we studied a relationship between figures, called congruency. Congruent figures are alike in every respect; they have the same shape and the same size. Now we will consider figures that have the same shape, but may differ in size. Such figures are called *similar figures.*

A photograph of a person or of a structure shows an image which is considerably smaller than the object photographed, but the shape of the image is just like that of the object. And when a photograph is enlarged, this shape is maintained (see Fig. 8.2); that is, all parts of the photograph are enlarged by the same factor. In mathematical terms, we say the images in the two photographs are similar.

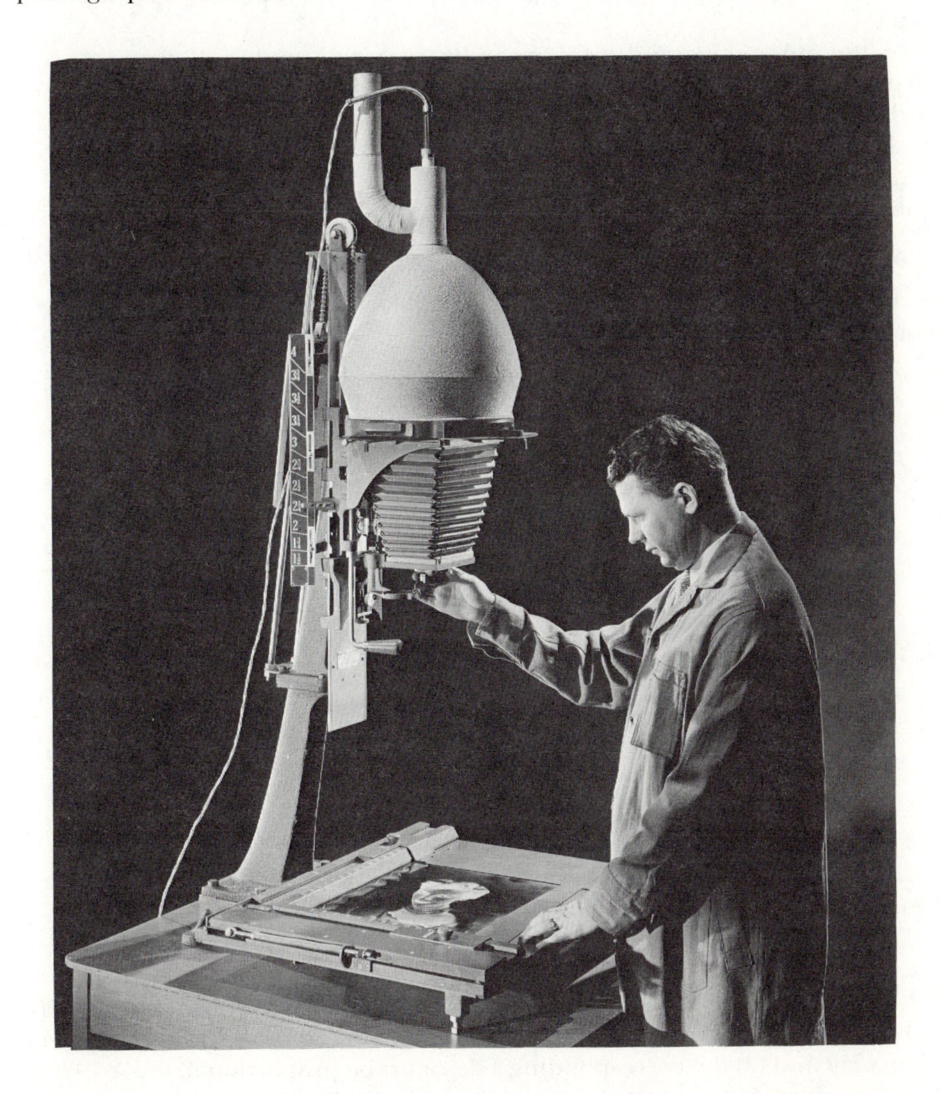

Fig. 8.2. The professional photographer is using an autofocus enlarger.

Design engineers and architects are continually dealing with similar figures. A newly designed structure is first drawn to scale on paper. The design is much smaller than the structure itself, but all parts have the shape of the finished product. Blueprints of these drawings are made. The blueprint can be read by the manufacturer. By using a ruler and a scale, he can determine the true dimensions of any part of the structure represented in the blueprint.

In the automotive and airplane industry, small models of new cars and airplanes are generally first constructed. These models will match in shape and detail the final product. The surveyor continually uses the properties of similarity of triangles in his work.

Definition: Two polygons are *similar* if there is a matching of their vertices for which the corresponding angles are congruent and the corresponding sides are proportional.

The symbol for "similar to" or "is similar to" is $\sim$. Thus in Fig. 8.3, polygon $ABCDE \sim$ polygon $PQRST$ if:

1. $\angle A \cong \angle P, \angle B \cong \angle Q, \angle C \cong \angle R, \angle D \cong \angle S, \angle E \cong \angle T.$
2. $\dfrac{AB}{PQ} = \dfrac{BC}{QR} = \dfrac{CD}{RS} = \dfrac{DE}{ST} = \dfrac{EA}{TP}.$

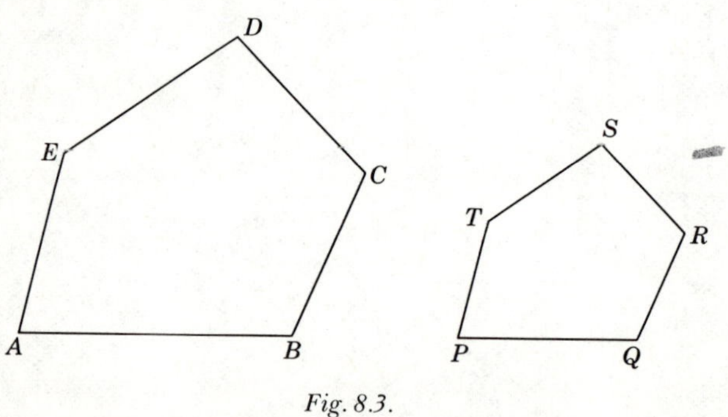

Fig. 8.3.

Conversely, if two polygons are similar, their corresponding angles are equal, and their corresponding sides are proportional.

The ratio of any two corresponding sides of two similar polygons is called the *ratio of similitude*.

It is important to note that the definition of similar polygons has two parts. In order for two polygons to be similar, (1) the corresponding angles must be congruent and (2) the corresponding sides must be proportional.

In general, when one of these conditions is fulfilled, it does not necessarily follow that the second condition is also fulfilled. Consider Fig. 8.4. Square

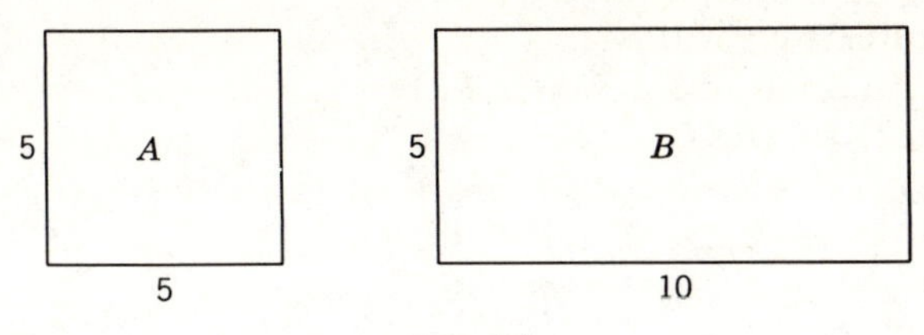

Fig. 8.4.

A and rectangle *B* have the angles of one congruent to the corresponding angles of the other, but obviously are not similar. In Fig. 8.5, the ratio of similitude of the two polygons is 2:1, but the corresponding angles are not congruent. They are not similar.

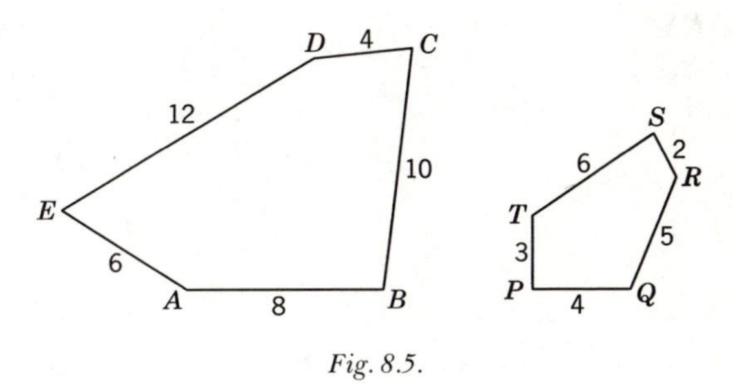

Fig. 8.5.

We will prove, however, that in the case of triangles the angles of one triangle cannot be congruent to the angles of a second triangle without the corresponding sides being in proportion. Conversely, we will prove that two triangles cannot have their corresponding sides proportional without the corresponding angles being congruent.

Theorem 8.12

8.10. If two triangles have the three angles of one congruent respectively to the three angles of the other, the triangles are similar. (A.A.A. Similarity Theorem.)

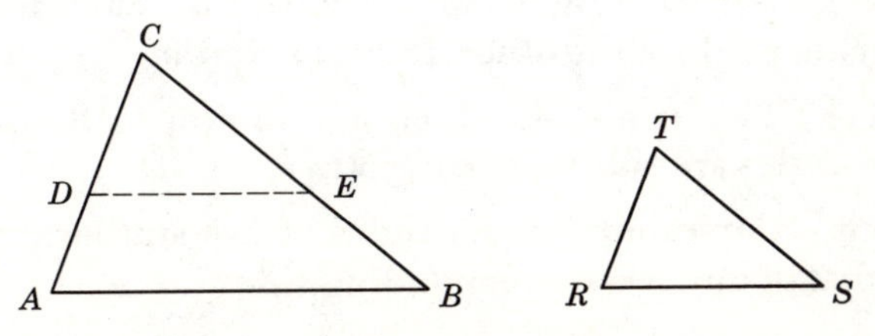

Theorem 8.12.

Given: $\triangle ABC$ and $\triangle RST$ with
$\angle A \cong \angle R, \angle B \cong \angle S, \angle C \cong \angle T.$
Conclusion: $\triangle ABC \sim \triangle RST.$

Proof:

STATEMENTS	REASONS
1. $\angle A \cong \angle R, \angle B \cong \angle S, \angle C \cong \angle T.$	1. Given.
2. Let D and E be points of $\overrightarrow{CA}$ and $\overrightarrow{CB}$ such that $\overline{DC} \cong \overline{RT}$ and $\overline{EC} \cong \overline{ST}$.	2. Postulate 11.
3. Draw $\overline{DE}$.	3. Postulate 2.
4. $\triangle CDE \cong \triangle RST.$	4. S.A.S.
5. $\angle CDE \cong \angle R.$	5. § 4.28.
6. $\angle CDE \cong \angle A.$	6. Theorem 3.4.
7. $\overline{DE} \parallel \overline{AB}.$	7. Theorem 5.12.
8. $\frac{AC}{DC} = \frac{BC}{EC}.$	8. § 8.5.
9. $\frac{AC}{RT} = \frac{BC}{ST}.$	9. E-8 property.
10. In like manner, by taking points F and G on $\overline{AB}$ and $\overline{BC}$ such that $\overline{BF} \cong \overline{SR}$ and $\overline{BG} \cong \overline{ST}$, we can prove $\frac{AB}{RS} = \frac{BC}{ST}.$	10. Reasons 3 through 8.
11. Then $\frac{AC}{RT} = \frac{BC}{ST} = \frac{AB}{RS}$, and	11. E-8 property; Theorem 3.4.
12. $\triangle ABC \sim \triangle RST.$	12. Definition of similar polygons.

8.11. Corollary: If two triangles have two angles of one congruent to two angles of the other, the triangles are similar. (A.A. Similarity Corollary.)

8.12. Corollary: If two right triangles have an acute angle of one congruent to an acute angle of the other, they are similar.

8.13. Corollary: Two triangles which are similar to the same triangle or two similar triangles are similar to each other.

8.14. Corollary: Corresponding altitudes of two similar triangles have the same ratio as that of any two corresponding sides.

(*Suggested proof:* $\triangle ADC \sim \triangle A'D'C'$.)

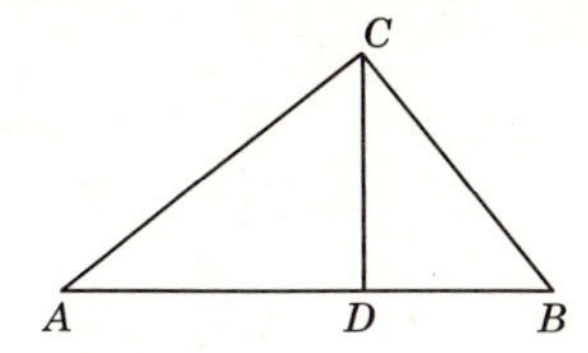

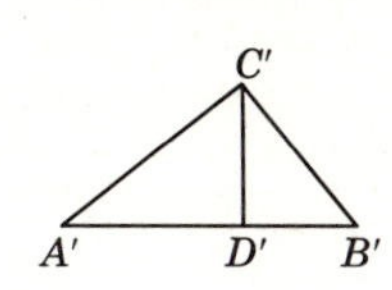

Corollary 8.14.

8.15. Method of proving line segments proportional. Thus far we have learned four ways to prove line segments proportional. The student at this time should review these methods under Theorem 8.10, its two corollaries, and § 8.9. The last method is very common and will be used extensively in this and succeeding chapters. It might be restated thus: "To prove that four segments are proportional, prove that they are corresponding sides of similar triangles."

The procedure then would be:

1. Find two triangles each of which has two of the four segments as sides.
2. Prove these two triangles are similar.
3. Form a proportion involving these four sides as pairs of corresponding sides of the two triangles.
4. If necessary, use theorems about proportion to transform the proportion to the desired form.

8.16. Illustrative Example 1:

Given: $\overline{AC} \perp \overline{AD}$ and $\overline{DE} \perp \overline{AD}$.

Prove: $AC:DE = AB:BD$.

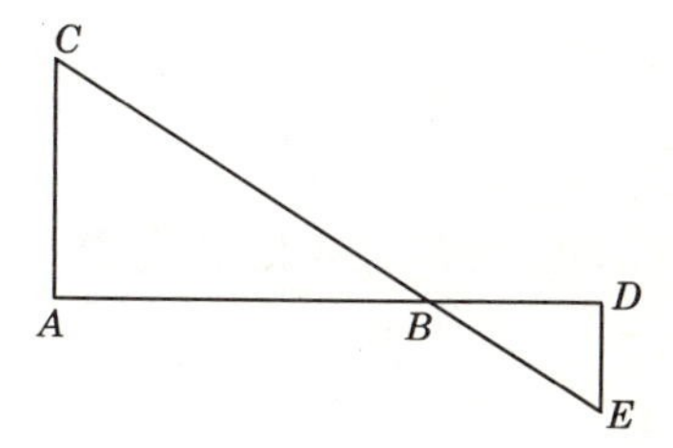

Illustrative Example 1.

Proof:

STATEMENTS	REASONS
1. $\overline{AC} \perp \overline{AD}$; $\overline{DE} \perp \overline{AD}$.	1. Given.
2. $\angle CAB$ and $\angle EDB$ are right $\triangle$s.	2. $\perp$ lines form right $\triangle$s.
3. $\angle ABC \cong \angle DBE$.	3. Theorem 3.12.
4. $\triangle ABC \sim \triangle DBE$.	4. § 8.12.
5. $\therefore AC:DE = AB:BD$.	5. If two $\triangle$s are $\sim$, their corresponding sides are proportional.

8.17. Illustrative Example 2:

Given: $\triangle ABC$ with right $\angle ACB$;

$\overline{CD} \perp \overline{AB}$.

Prove: $CD:CB = AC:AB$.

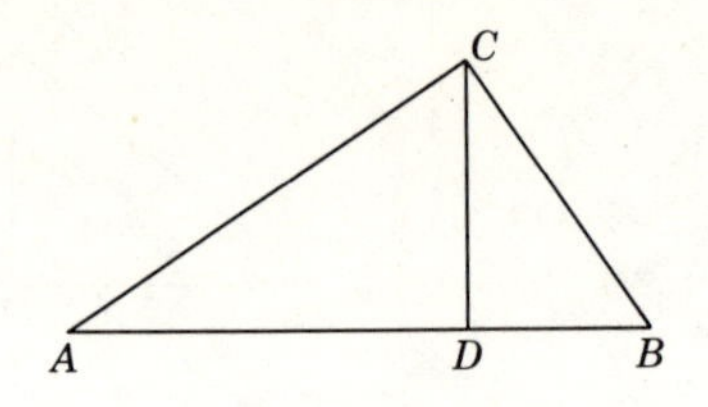

Illustrative Example 2.

Proof:

STATEMENTS	REASONS
1. $\angle ACB$ is a right $\angle$.	1. Given.
2. $\overline{CD} \perp \overline{AB}$.	2. Given.
3. $\angle ADC$ is a right $\angle$.	3. § 1.20.
4. In $\triangle$s ADC and ACB, $\angle A \cong \angle A$.	4. Reflexive property.
5. $\triangle ADC \sim \triangle ACB$.	5. § 8.12.
6. $\therefore CD:CB = AC:AB$.	6. § 8.9.

Exercises

1. *Given:* $\triangle ABC$ with $\overline{DE} \parallel \overline{AB}$.
   Prove: $\triangle DEC \sim \triangle ABC$.

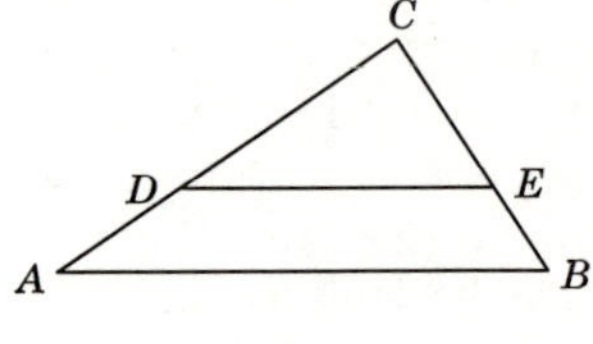

Ex. 1.

2. *Given:* $\triangle RST$ with $\overline{ST} \perp \overline{RS}$;
   $\overline{MN} \perp \overline{RT}$.
   Prove: $\triangle RMN \sim \triangle RST$.

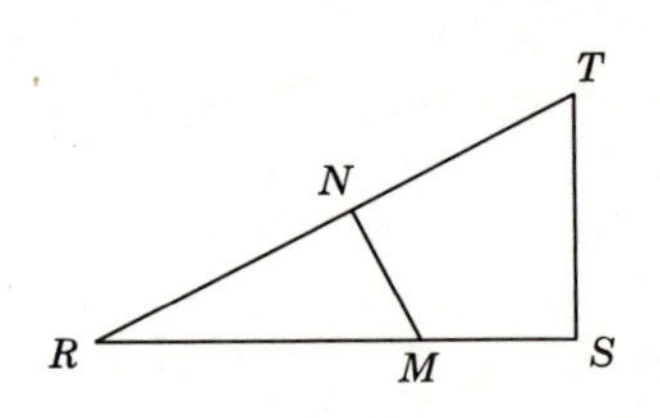

Ex. 2.

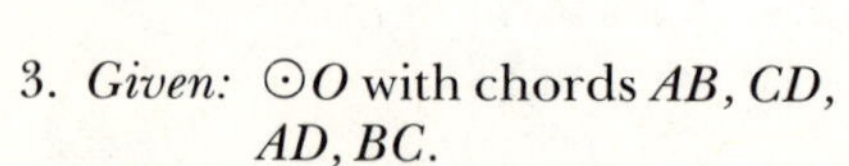

3. *Given:* $\odot O$ with chords AB, CD, AD, BC.
   Prove: $\triangle ABE \sim \triangle CDE$.

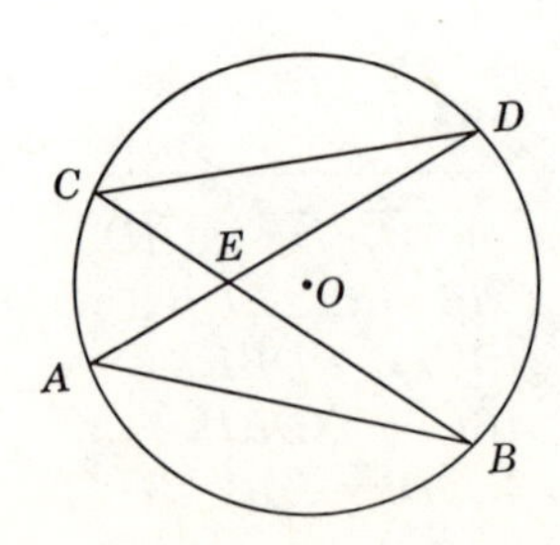

Ex. 3.

4. *Given:* $\odot O$ with $\overline{AC}$ a diameter; $\overline{DE} \perp \overline{AD}$.
Prove: $\triangle ABC \sim \triangle EDC$.

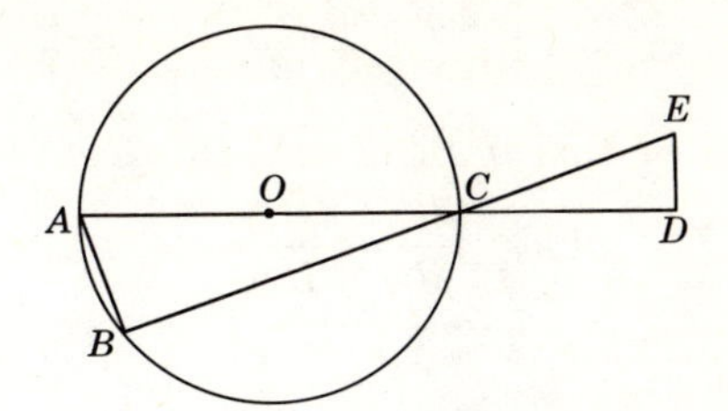

Ex. 4.

5. *Given:* $\odot O$, with diameter RS; chords RT and PS intersecting at Q.
Prove: $\triangle RPQ \sim \triangle STQ$.

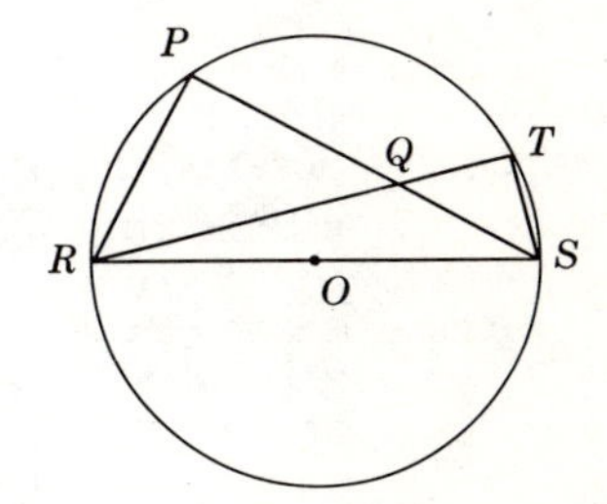

Ex. 5.

6. *Given:* $\triangle ABC$ with altitudes CD and BE.
Prove: $\triangle BDF \sim \triangle CEF$.

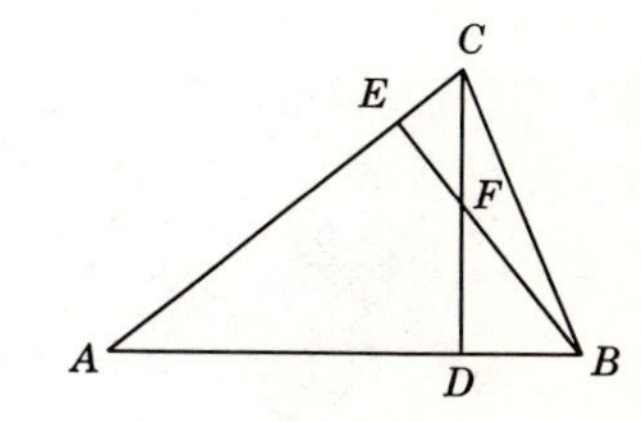

Ex. 6.

7. *Given:* $\odot O$ with diameter AC, tangent BC, secant AB.
Prove: $\triangle BDC \sim \triangle BCA$.

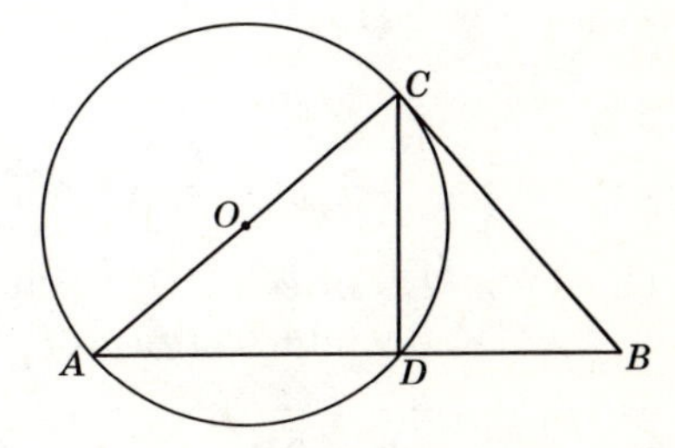

Ex. 7.

8. *Given:* $\overline{BC} \perp \overrightarrow{AX}$; $\overline{DE} \perp \overrightarrow{AX}$.

Prove: $\dfrac{BC}{AC} = \dfrac{DE}{AE}$.

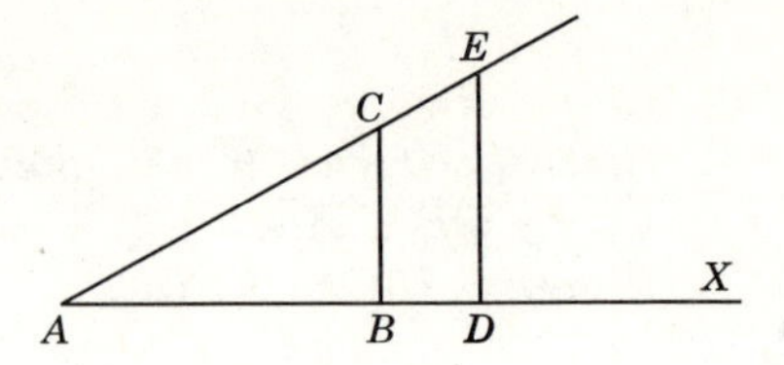

Ex. 8.

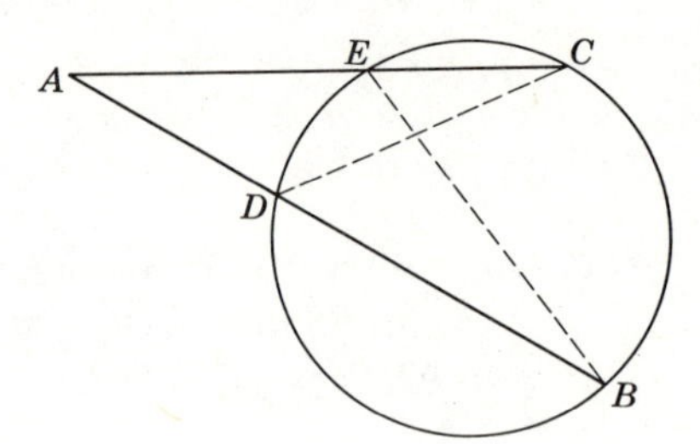

Ex. 9.

9. *Given:* Secants AB and AC intersecting at A.

Prove: $\dfrac{AC}{AB} = \dfrac{AD}{AE}$.

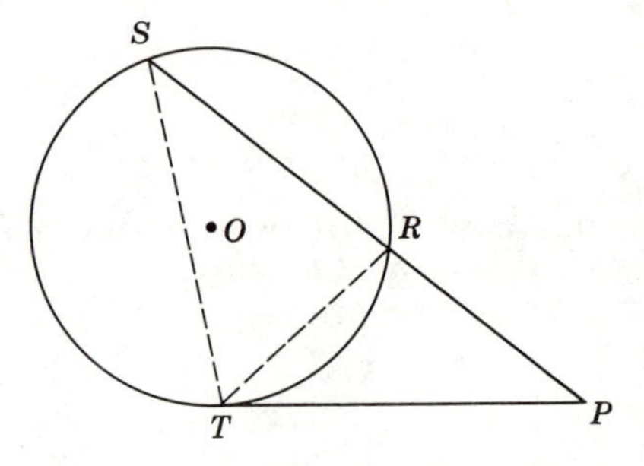

Ex. 10.

10. *Given:* $\odot O$ with tangent TP and secant SP intersecting at P.

Prove: $\dfrac{PS}{PT} = \dfrac{PT}{PR}$ and $(PT)^2 = PS \times PR$.

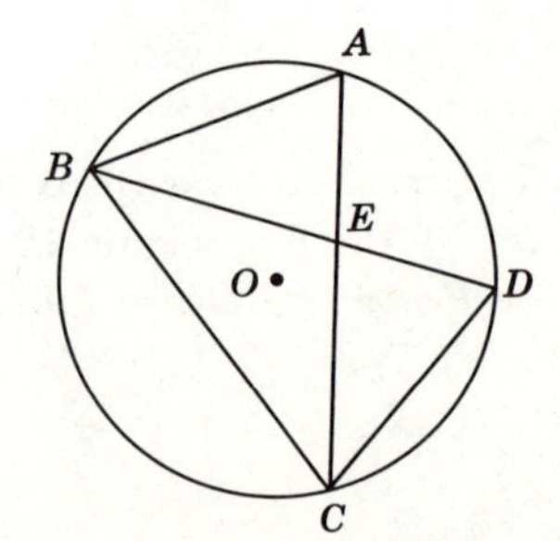

Ex. 11.

11. *Given:* $\overline{BD}$ bisects $\angle ABC$; chords AC and BD intersecting at E.

Prove: $\dfrac{AE}{AB} = \dfrac{CD}{BD}$.

12. *Given:* *Right* $\triangle ABC$ with $\angle ABC$ a right $\angle$; $\overline{BD} \perp \overline{AC}$.

Prove: $\dfrac{AD}{BD} = \dfrac{BD}{DC}$ and $(BD)^2 = AD \times DC$.

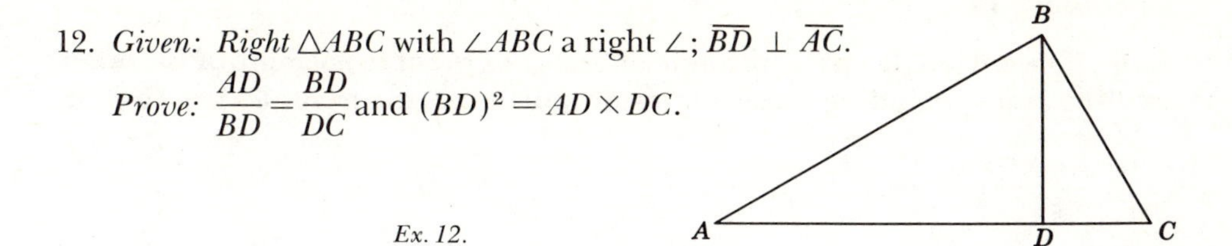

Ex. 12.

13. To find the height of a flag pole, a boy scout whose eyes are 5 feet 6 inches from the ground placed a 10-foot rod in the ground 50 feet from the flag pole. Then stepping back 8 feet 6 inches, he found that he could just sight the top of the flag pole in line with the top of the rod. How high is the flag pole?

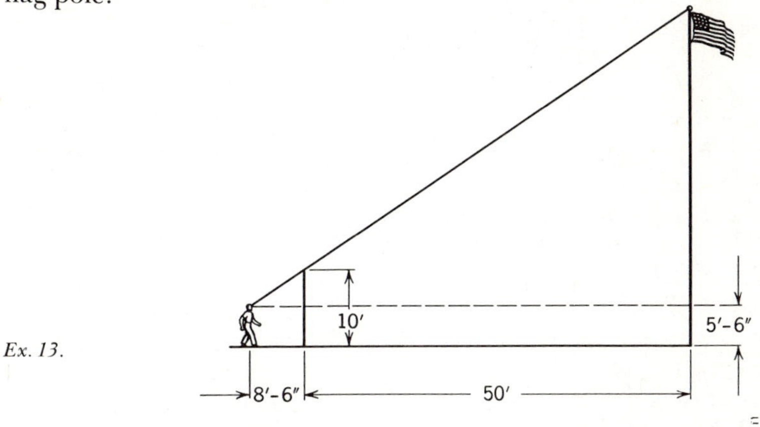

Ex. 13.

14. A boy notices that the shadow of a tree is 52 feet 3 inches long while his shadow is 6 feet 6 inches long. If the boy is 5 feet 9 inches tall, how tall is the tree? (*Note:* we assume the sun's rays are parallel.)

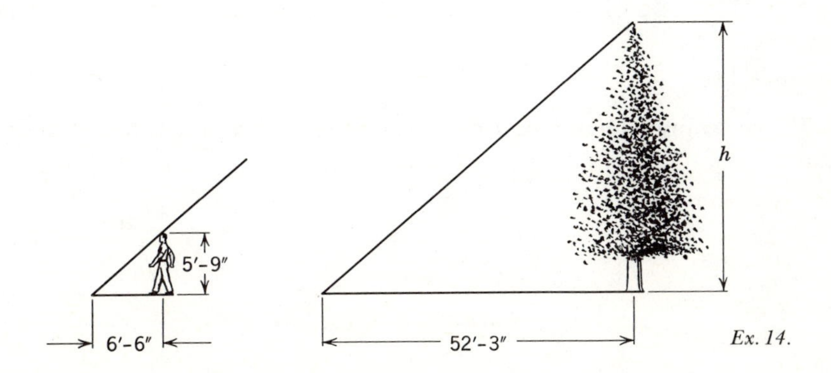

Ex. 14.

Theorem 8.13

8.18. If two triangles have an angle of one congruent to an angle of the other and the sides including these angles proportional, the triangles are similar.

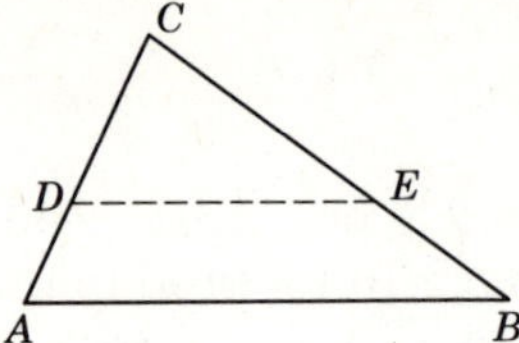
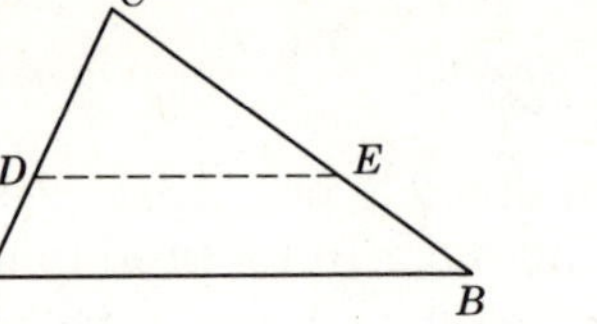

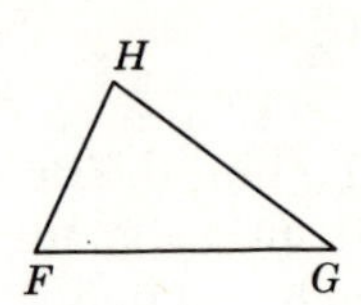

Theorem 8.13.

Given: $\triangle ABC$ and FGH with $\angle C \cong \angle H$; $AC:FH = BC:GH$.
Conclusion: $\triangle ABC \sim \triangle FGH$.
Proof:

STATEMENTS	REASONS
1. $\angle C \cong \angle H$.	1. Given.
2. Let D and E be points of $\overrightarrow{CA}$ and $\overrightarrow{CB}$ such that $\overline{DC} \cong \overline{FH}$ and $\overline{EC} \cong \overline{GH}$.	2. Postulate 11.
3. Draw $\overline{DE}$.	3. Postulate 2.
4. $\triangle CDE \cong \triangle FGH$.	4. S.A.S.
5. $\angle CDE \cong \angle F$.	5. § 4.28.
6. $AC:FH = BC:GH$.	6. Given.
7. $AC:DC = BC:EC$.	7. E-2 and E-8 properties.
8. $\overline{DE} \parallel \overline{AB}$.	8. § 8.8. Corollary.
9. $\angle A \cong \angle CDE$.	9. Theorem 5.14.
10. $\angle A \cong \angle F$.	10. E-3 property.
11. $\therefore \triangle ABC \sim \triangle FGH$.	11. § 8.11. Corollary.

Theorem 8.14

8.19. If two triangles have their corresponding sides proportional, they are similar.

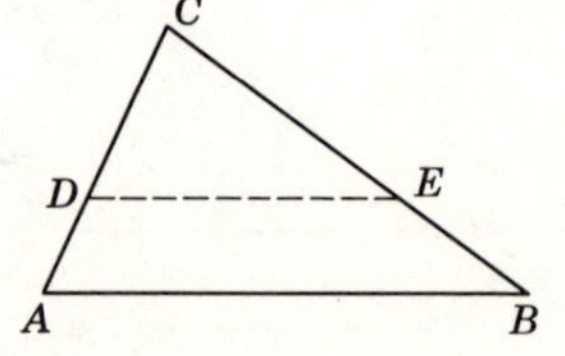

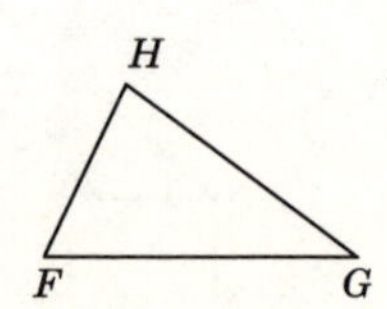

Theorem 8.14.

Given: ⊿s ABC and FGH with $\dfrac{AB}{FG}=\dfrac{BC}{GH}=\dfrac{AC}{FH}$.

Conclusion: $\triangle ABC \sim \triangle FGH$.

Proof:

STATEMENTS	REASONS
1. Let D and E be points on $\overrightarrow{CA}$ and $\overrightarrow{CB}$ such that $\overline{CD} \cong \overline{HF}$, $\overline{CE} \cong \overline{HG}$.	1. Postulate 11.
2. Draw $\overline{DE}$.	2. Postulate 2.
3. $\dfrac{AC}{FH}=\dfrac{BC}{GH}$.	3. Given.
4. $\dfrac{AC}{CD}=\dfrac{BC}{CE}$.	4. E-2 and E-8 properties.
5. $\angle C \cong \angle C$.	5. Reflexive property.
6. $\triangle ABC \sim \triangle DEC$.	6. Theorem 8.12.
7. $\dfrac{AC}{CD}=\dfrac{AB}{DE}$ or $\dfrac{AC}{FH}=\dfrac{AB}{DE}$.	7. § 8.9; E-2.
8. But $\dfrac{AC}{FH}=\dfrac{AB}{FG}$.	8. Given.
9. $FG = DE$.	9. Theorem 8.6.
10. $\triangle DEC \cong \triangle FGH$.	10. S.S.S.
11. $\therefore \triangle ABC \sim \triangle FGH$.	11. § 8.13. Corollary.

Exercises

In Exs. 1 through 10, prove two triangles similar and complete the proportions.

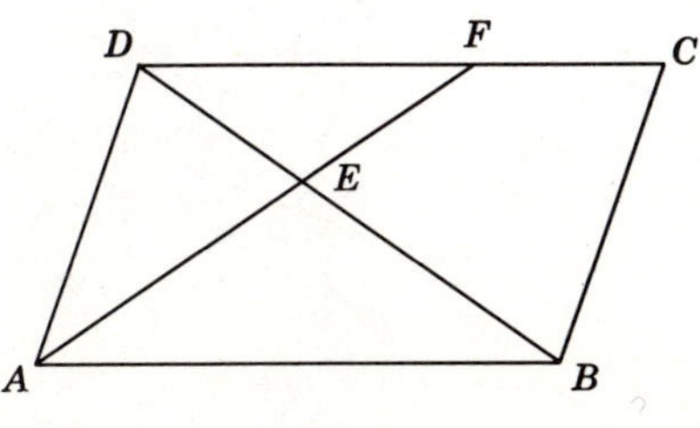

Ex. 1. $\overline{DC} \parallel \overline{AB}$; $\dfrac{DF}{?}=\dfrac{EF}{?}$.

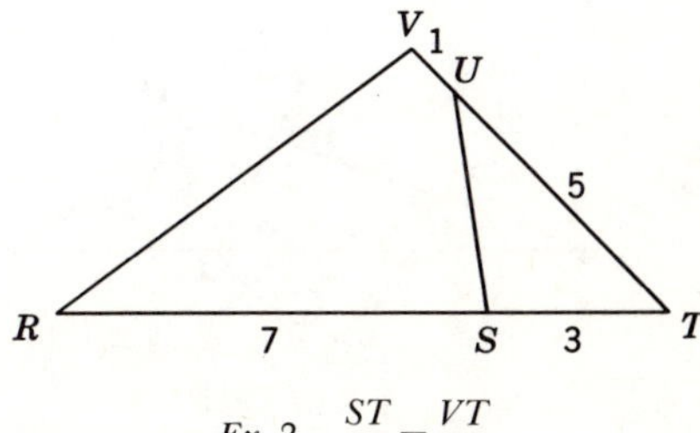

Ex. 2. $\dfrac{ST}{?}=\dfrac{VT}{?}$.

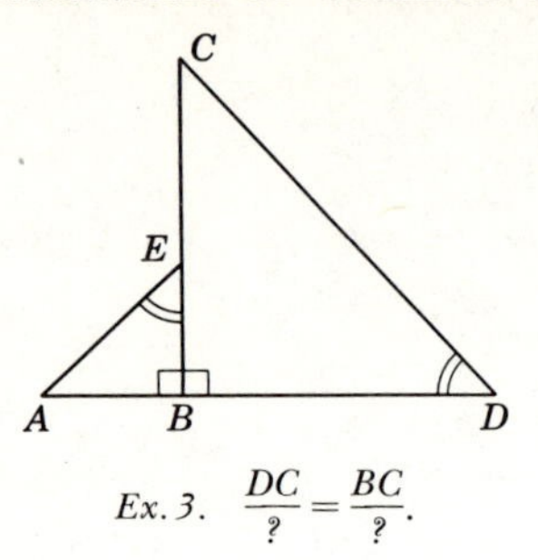

Ex. 3. $\frac{DC}{?} = \frac{BC}{?}$.

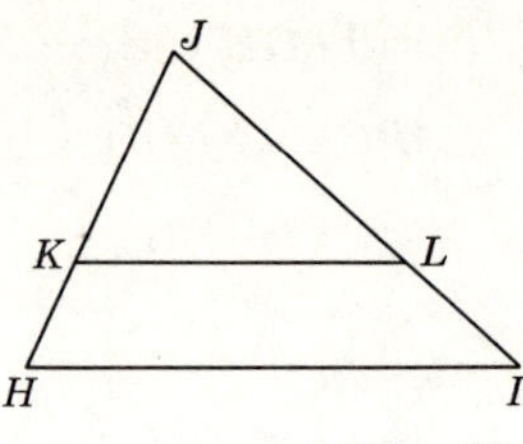

Ex. 4. $\overline{KL} \parallel \overline{HI}$; $\frac{KJ}{KL} = \frac{HJ}{?}$.

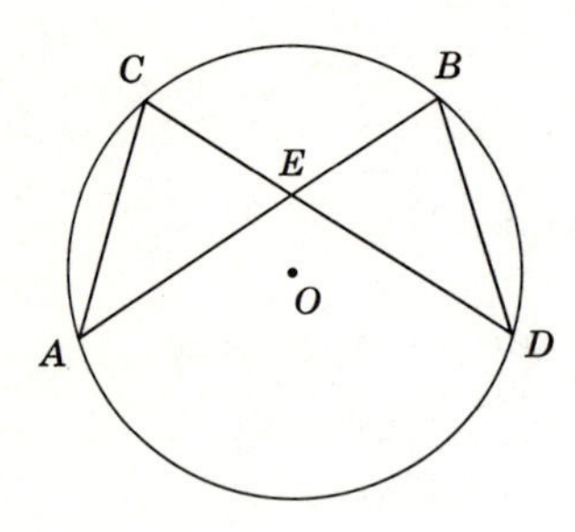

Ex. 5. $\frac{CE}{BE} = \frac{?}{BD}$.

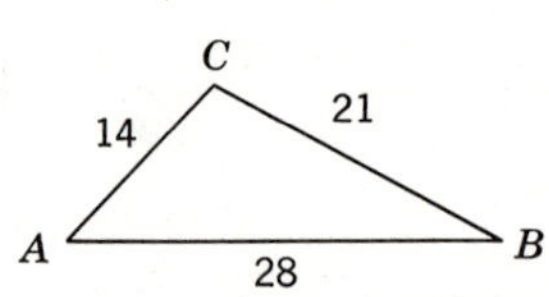

Ex. 6. $\frac{AC}{?} = \frac{BC}{?}$.

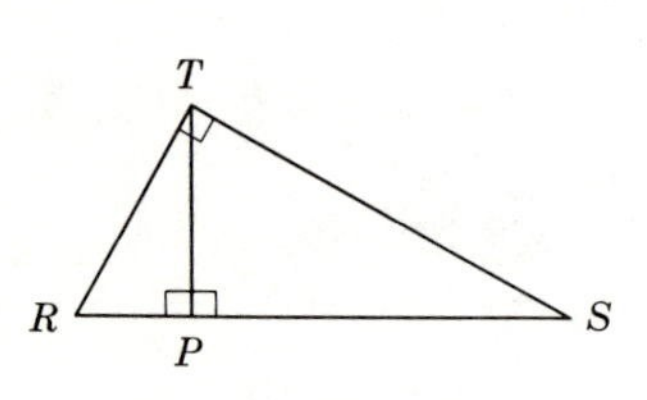

Ex. 7. $\frac{RP}{PT} = \frac{PT}{?}$.

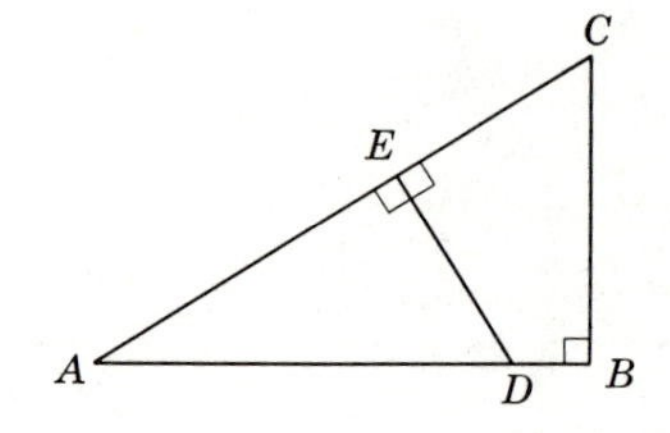

Ex. 8. $\frac{AE}{?} = \frac{ED}{?}$.

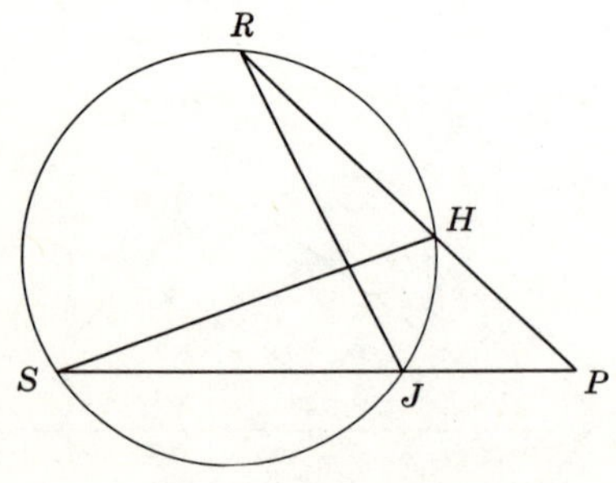

Ex. 9. $\frac{PJ}{HP} = \frac{?}{SH}$.

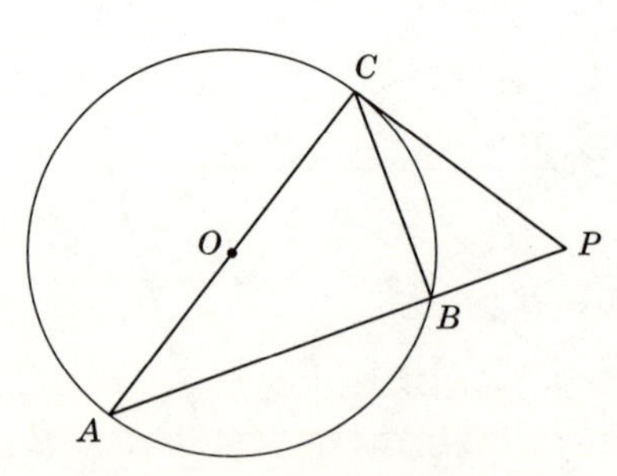

Ex. 10. $\frac{AP}{PC} = \frac{PC}{?}$.

In Exs. 11 through 16, find the length of line segment x.

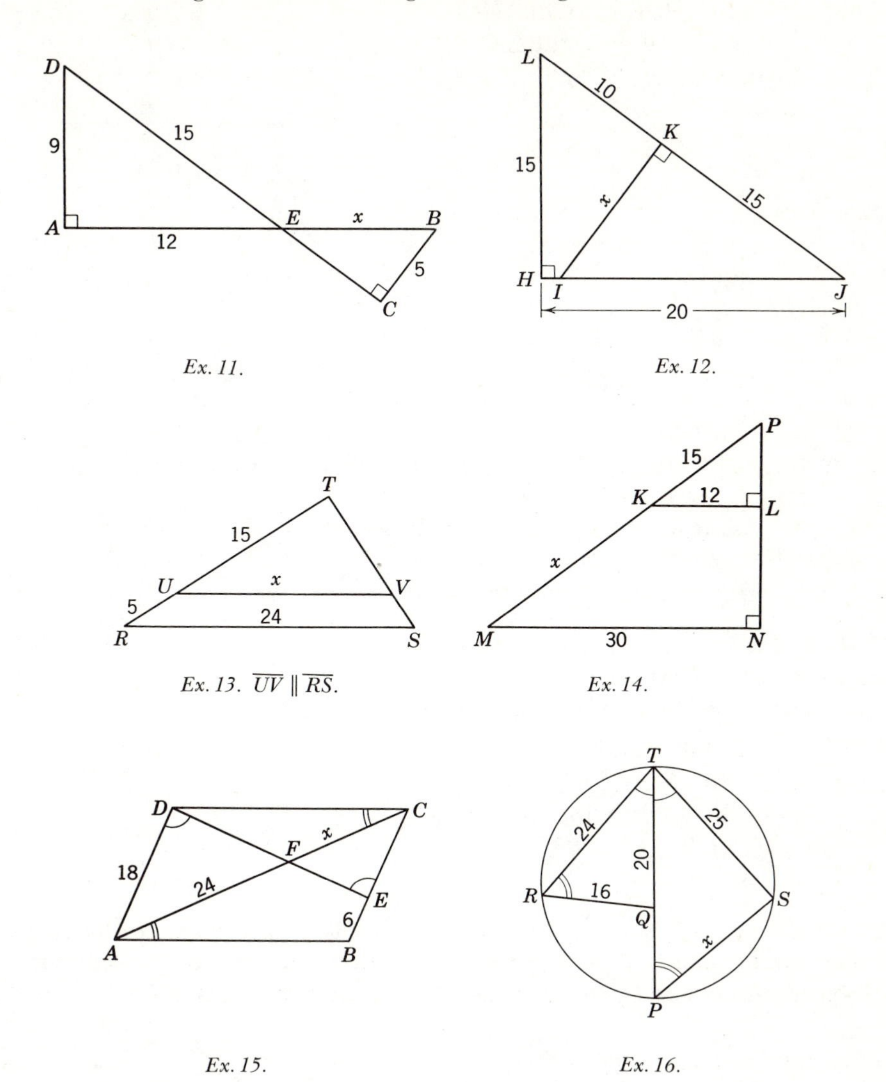

Ex. 11. *Ex. 12.*

Ex. 13. $\overline{UV} \parallel \overline{RS}$. *Ex. 14.*

Ex. 15. *Ex. 16.*

Theorem 8.15

8.20. The altitude on the hypotenuse of a right triangle forms two right triangles which are similar to the given triangle and similar to each other.

Given: $\triangle ABC$ with $\angle ACB$ a right $\angle$;
$\overline{CD} \perp \overline{AB}$.

Conclusion: Right $\triangle ADC \sim$ right $\triangle ACB$;
right $\triangle CDB \sim$ right $\triangle ACB$;
right $\triangle ADC \sim$ right $\triangle CDB$.

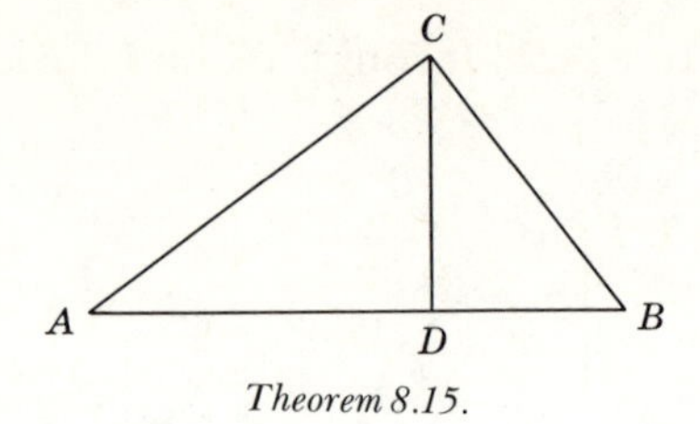

Theorem 8.15.

Proof:

STATEMENTS	REASONS
1. $\angle ACB$ is a right $\angle$.	1. Given.
2. $\overline{CD} \perp \overline{AB}$.	2. Given.
3. $\angle ADC$ and $\angle BDC$ are right $\angle$s.	3. § 1.20.
4. In right $\triangle$s ADC and ACB, $\angle A \cong \angle A$.	4. Reflexive property.
5. $\therefore \triangle ADC \sim \triangle ACB$.	5. § 8.12.
6. In right $\triangle$s BDC and ABC, $\angle B \cong \angle B$.	6. Reflexive property.
7. $\therefore \triangle CDB \sim \triangle ACB$.	7. Reason 5.
8. $\therefore \triangle ADC \sim \triangle CDB$.	8. § 8.13.

8.21. Corollary: The altitude on the hypotenuse of a right triangle is the mean proportional between the measures of the segments of the hypotenuse.

Suggestions: $\triangle ADC \sim \triangle CDB$ by Theorem 8.15. Then $AD:CD = CD:DB$.

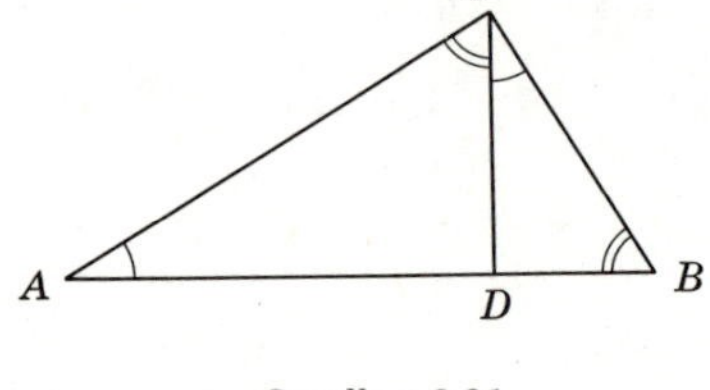

Corollary 8.21.

8.22. Corollary: Either leg of a right triangle is the mean proportional between the measure of the hypotenuse and the measure of the segment of the hypotenuse cut off by the altitude which is adjacent to that leg.

Suggestions: Drop a $\perp$ from C to $\overline{AB}$;
$\triangle BDC \sim \triangle BCA$ by Theorem 8.15.
Then $AB:BC = BC:BD$.

Theorem 8.16

8.23. The square of the measure of the hypotenuse of a right triangle is equal to the sum of the squares of the measures of the legs.

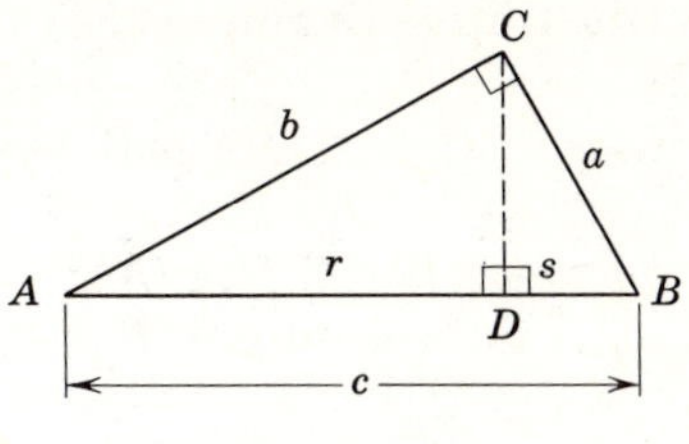

Given: $\triangle ABC$ with $\angle ACB$ a right $\angle$.
Conclusion: $c^2 = a^2 + b^2$.

Proof:

Theorem 8.16.

STATEMENTS	REASONS
1. Draw $\overline{CD} \perp \overline{AB}$.	1. Theorem 5.3; theorem 5.4.
2. $c:a = a:s$ and $c:b = b:r$.	2. § 8.22.
3. $a^2 = cs$, $b^2 = cr$.	3. Theorem 8.1.
4. $a^2 + b^2 = cs + cr$.	4. E-4.
5. $a^2 + b^2 = c(s + r)$.	5. Factoring (distributive law).
6. $s + r = c$.	6. § 1.13.
7. $a^2 + b^2 = c^2$.	7. E-8.
8. $c^2 = a^2 + b^2$.	8. E-2.

Theorem 8.16 is known as the Pythagorean theorem. Although the truths of the theorem were used for many years by the ancient Egyptians, the first formal proof of the theorem is attributed to the Pythagoreans, a mathematical society which was founded by the Greek philosopher Pythagoras. Since that time many other proofs of this famous theorem have been discovered.

8.24. Corollary: The square of the measure of the leg of a right triangle is equal to the square of the measure of the hypotenuse minus the square of the measure of the other leg.

Illustrative Example 1:

Given: Right $\triangle RST$ with $\overline{SQ} \perp$ to hypotenuse RT; $RQ = 3$ and $QS = 5$. Find QT.

Solution: From § 8.21, $RQ:QS = QS:QT$. Substituting, $3:5 = 5:QT$. Therefore, $3QT = 25$; $QT = \frac{25}{3}$.

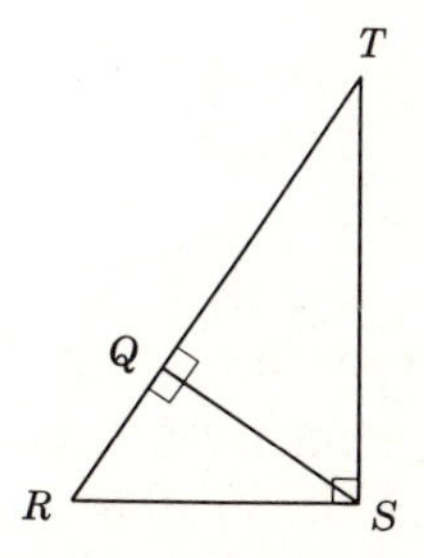

Illustrative Example 1.

Illustrative Example 2:

Given: Right $\triangle HJK$ with hypotenuse $HK = 17$, leg $HJ = 15$. Find JK.

Solution: By § 8.24, $(JK)^2 = (HK)^2 - (HJ)^2$.

Substituting, $(JK)^2 = (17)^2 - (15)^2$
$= 289 - 225$
$= 64$.
$\therefore JK = 8$.

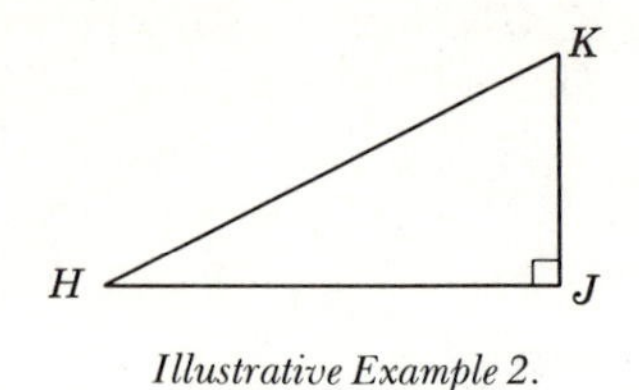

Illustrative Example 2.

Exercises

In $\triangle ABC$, $\angle ACB$ is a right $\angle$, and $\overline{CD} \perp \overline{AB}$.

1. Find CD if $AD = 9$ and $BD = 4$.
2. Find BC if $AB = 16$ and $BD = 4$.
3. Find BC if $AD = 12$, $AC = 15$, and $CD = 9$.
4. Find AC if $AD = 24$, $CD = 18$, and $BC = 22.5$.
5. Find AC if $BD = 9$, $BC = 15$, and $CD = 12$.
6. Find CD if $AC = 20$, and $BC = 15$.
7. Find BD if $AC = 21$, and $CD = 15$.
8. Find AC if $BD = 12$, $BC = 13$, and $CD = 5$.
9. Find BD if $AD = 2$ and $CD = 4$.
10. Find CD if $AD = 16$ and $BD = 4$.
11. Find BC if $AB = 20$ and $BD = 5$.
12. Find AC if $AB = 18$ and $AD = 8$.

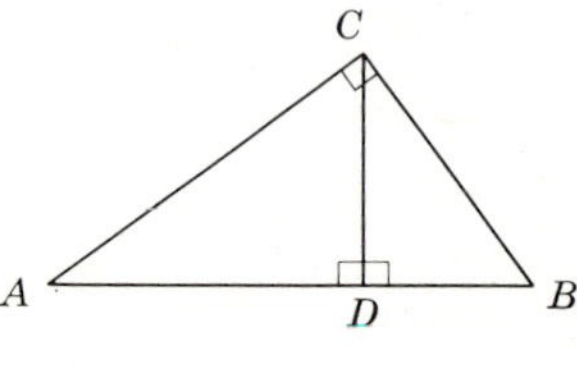

Exs. 1–12.

13. *Given:* $\overline{AB}$ is a diameter of $\odot O$, $\overline{CD} \perp \overline{AB}$.
If $AD = 3$ and $BD = 27$, find CD.
(*Hint:* Draw chords AC and BC.)

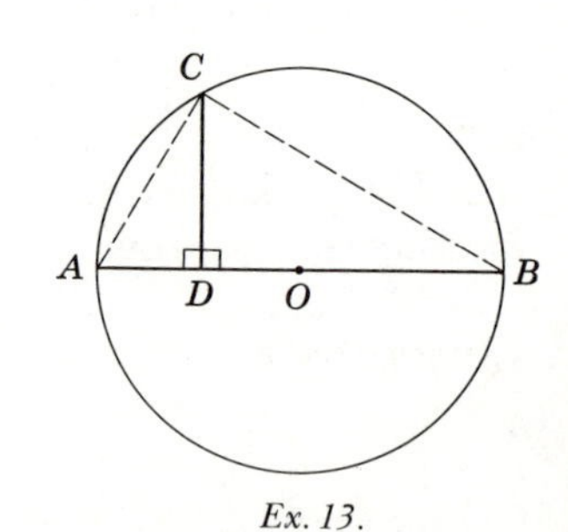

Ex. 13.

In $\triangle MNT$, $\angle MNT$ is a right $\angle$.

14. Find MT if $MN = 16$ and $NT = 12$.
15. Find NT if $MN = 24$ and $MT = 30$.
16. Find MN if $MT = 13$ and $NT = 5$.
17. Find NT if $MN = 15$ and $MT = 17$.

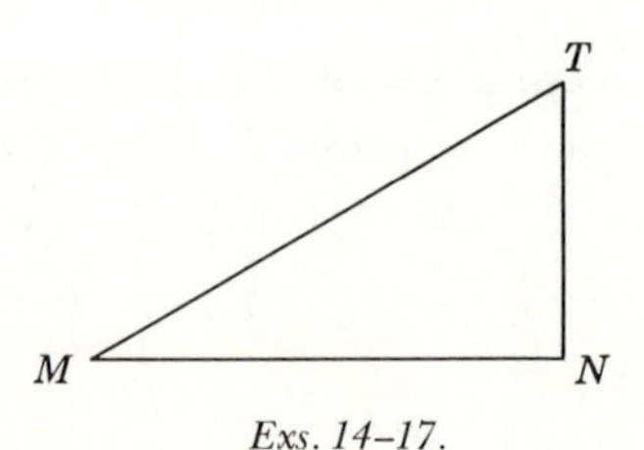

Exs. 14–17.

18–25. Find the length of segment x in each diagram. Draw a perpendicular if necessary.

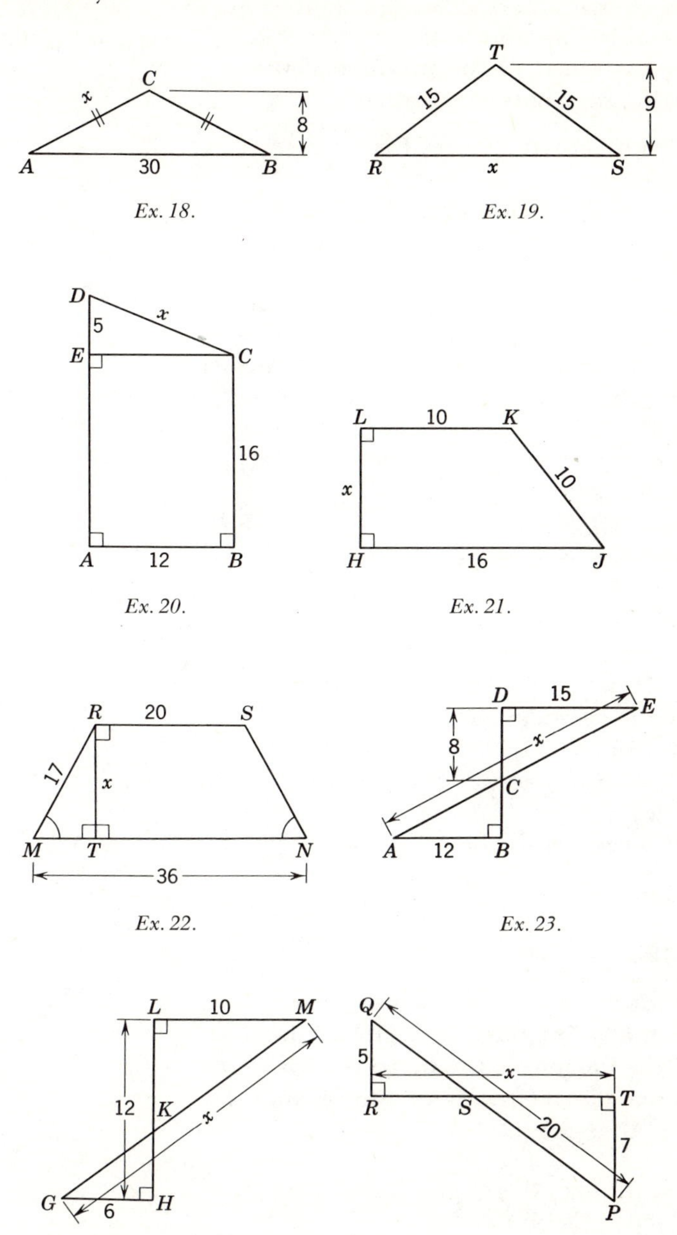

Ex. 18. *Ex. 19.*

Ex. 20. *Ex. 21.*

Ex. 22. *Ex. 23.*

Ex. 24. *Ex. 25.*

Theorem 8.17

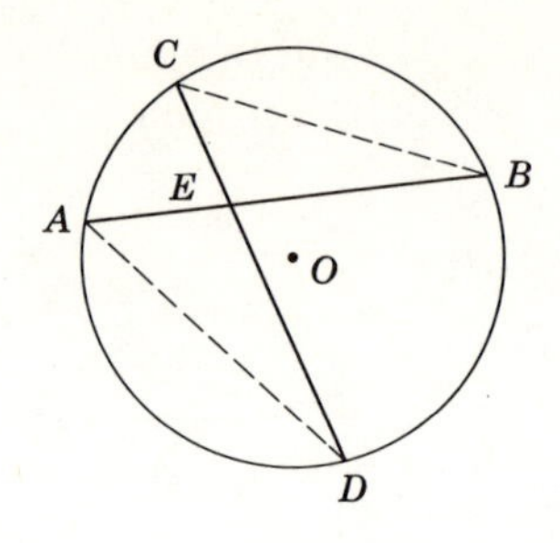

Theorem 8.17.

8.25. If two chords intersect within a circle, the product of the measures of the segments of one chord is equal to the product of the measures of the segments of the other.

Given: $\odot O$ with chords AB and CD intersecting at E.
Conclusion: $AE \times EB = CE \times ED$.

Proof:

STATEMENTS	REASONS
1. Draw chords CB and AD.	1. Postulate 2.
2. $\angle DAE \cong \angle BCE$.	2. § 7.17.
3. $\angle AED \cong \angle CEB$.	3. Theorem 3.12.
4. $\therefore \triangle AED \sim \triangle CEB$.	4. § 8.11.
5. $AE:CE = ED:EB$.	5. § 8.9.
6. $\therefore AE \times EB = CE \times ED$.	6. Theorem 8.1.

8.26. Segment of a secant. When a circle is cut by a secant, as $\overrightarrow{AP}$ in Fig. 8.6, we speak of the segment AP as a secant from P to $\odot O$. $\overline{PB}$ is the *external segment* of the secant and $\overline{BA}$ is the *internal segment* of the secant.

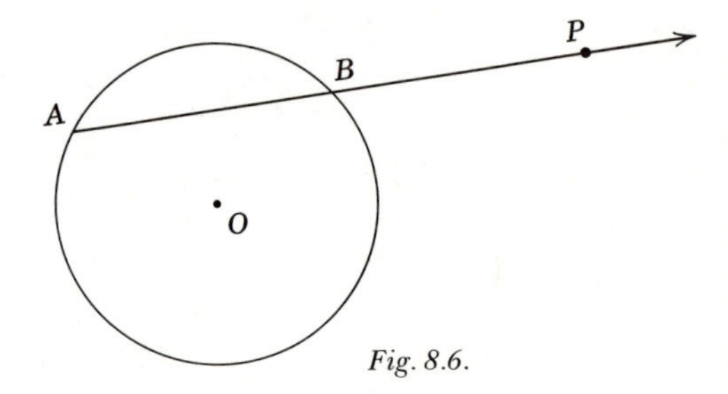

Fig. 8.6.

Theorem 8.18

8.27. If a tangent and a secant are drawn from the same point outside a circle, the measure of the tangent is the mean proportional between the measures of the secant and its external segment.

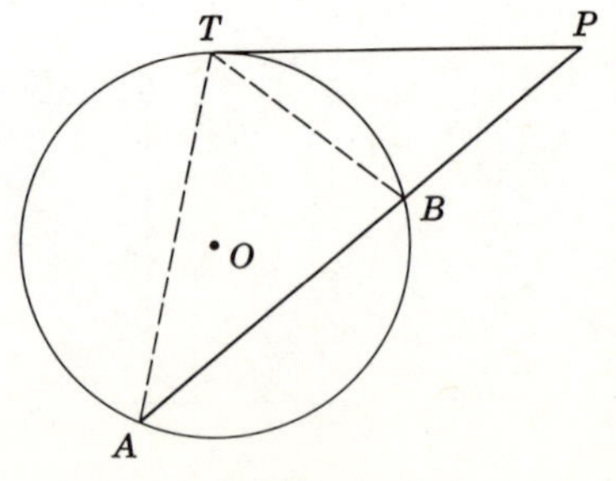

Theorem 8.18.

Given: $\odot O$ with tangent PT and secant PBA drawn from P.
Conclusion: $PB:PT = PT:PA$.

Proof:

STATEMENTS	REASONS
1. Draw chords TA and TB.	1. Postulate 2.
2. $m\angle TAP = \frac{1}{2}m\widehat{TB}$.	2. Theorem 7.3.
3. $m\angle BTP = \frac{1}{2}m\widehat{TB}$.	3. Theorem 7.14.
4. $\angle TAP \cong \angle BTP$.	4. Theorem 3.4.
5. $\angle P \cong \angle P$.	5. Reflexivity.
6. $\triangle TAP \sim \triangle BTP$.	6. § 8.11.
7. $PB:PT = PT:PA$.	7. § 8.9.

Theorem 8.19

8.28. If two secants are drawn from the same point outside a circle, the product of the measures of one secant and its external segment is equal to the product of the measures of the other secant and its external segment.

Given: $\odot O$ with secants PA and PC drawn from P.
Conclusion: $PA \times PB = PC \times PD$.

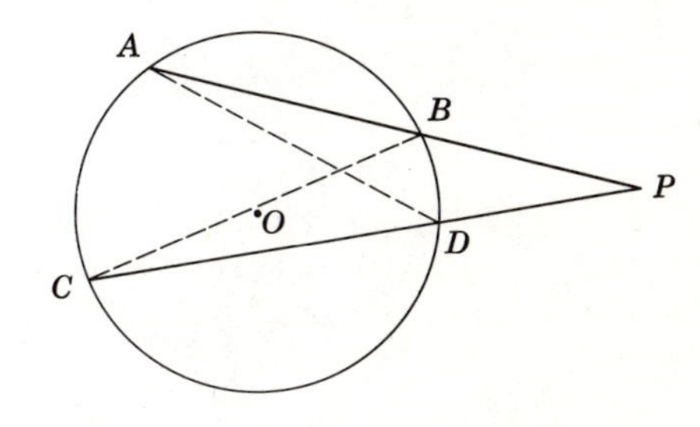

Theorem 8.19.

Proof:

STATEMENTS	REASONS
1. Draw chords AD and BC.	1. Postulate 2.
2. $m\angle DAP = \frac{1}{2}m\widehat{BD}$.	2. Theorem 7.3.
3. $m\angle BCP = \frac{1}{2}m\widehat{BD}$.	3. Theorem 7.3.
4. $\angle DAP \cong \angle BCP$.	4. Theorem 3.4.
5. $\angle P \cong \angle P$.	5. Reflexive property.
6. $\triangle DAP \sim \triangle BCP$.	6. § 8.11.
7. $PA:PC = PD:PB$.	7. § 8.9.
8. $PA \times PB = PC \times PD$.	8. Theorem 8.1.

Exercises

In the following exercises, O is the center of the circle.

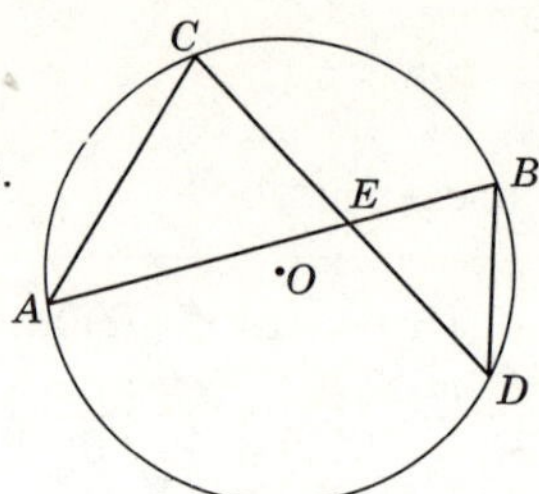

Exs. 1–5.

1. Find AE if $EB = 4$, $CE = 8$, $ED = 5$.
2. Find ED if $AE = 12$, $CE = 8$, $EB = 6$.
3. Find CE if $AB = 20$, $EB = 15$, $ED = 7$.
4. Find AC if $CE = 9$, $EB = 3$, $BD = 5$.
5. Find CD if $EB = 6$, $AB = 18$, $ED = 8$.

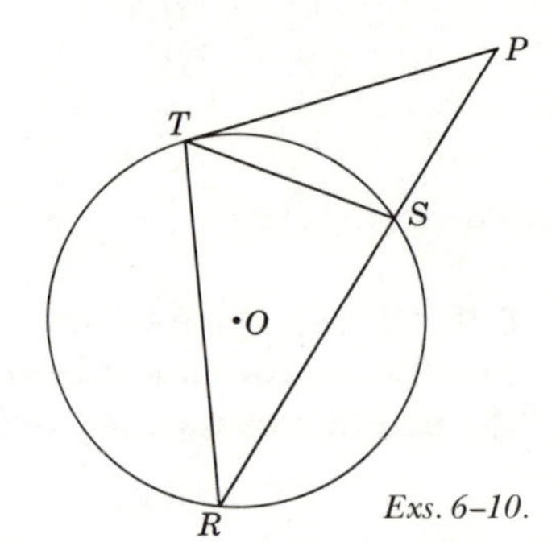

Exs. 6–10.

6. Find PT if $PS = 4$, $PR = 9$.
7. Find PR if $PS = 5$, $PT = 8$.
8. Find PT if $RS = 7$, $PR = 16$.
9. Find RT if $PT = 18$, $TS = 9$, $PS = 12$.
10. Find PT if $RS = 24$, $PS = 8$.

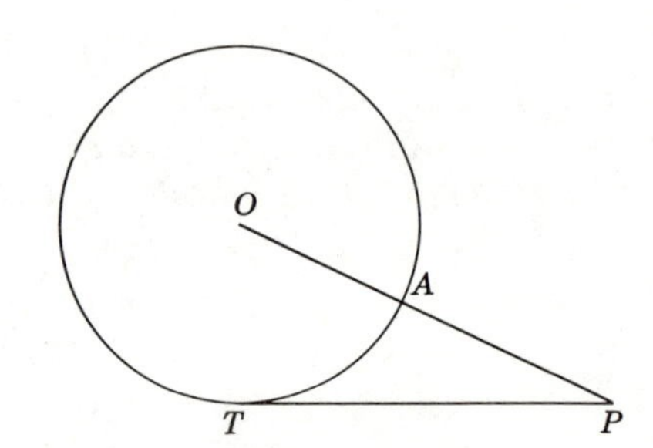

Exs. 11–13.

11. Find PT if $OA = 15$, $PA = 10$.
12. Find AP if $PT = 12$, $OA = 9$.
13. Find OA if $PT = 8$, $PA = 4$.

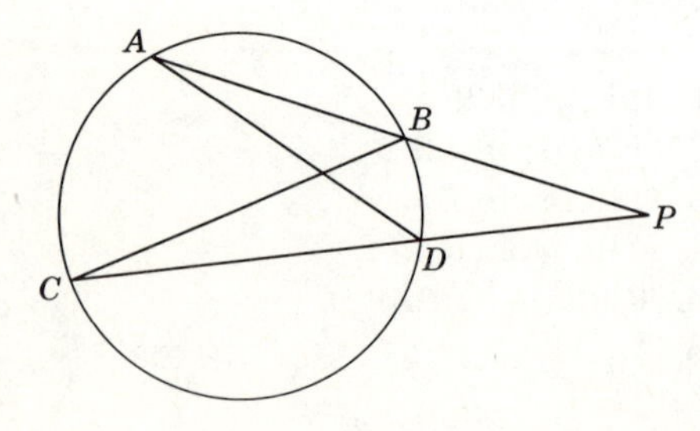

Exs. 14–18.

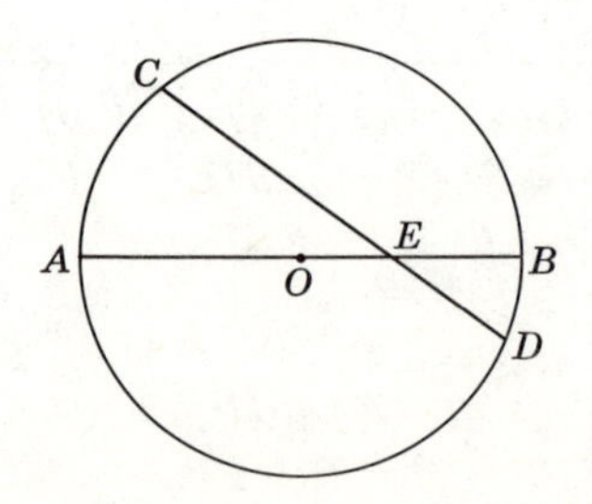

Ex. 19.

14. Find PA if $PC = 24$, $PB = 10$, $PD = 8$.
15. Find PB if $AP = 18$, $PC = 24$, $PD = 6$.
16. Find PC if $PD = 6$, $PB = 8$, $BA = 10$.
17. Find AD if $AP = 16$, $BC = 12$, $PC = 20$.
18. Find PD if $PB = 8$, $AD = 10$, $BC = 16$.
19. Find ED if $OA = 8$, $OE = 3$, $CE = 10$.
20. Find BD if $OA = 8$, $CD = 3$, $AD = 5$.
21. Find OA if $AD = 8$, $BD = 5$, $CD = 4$.

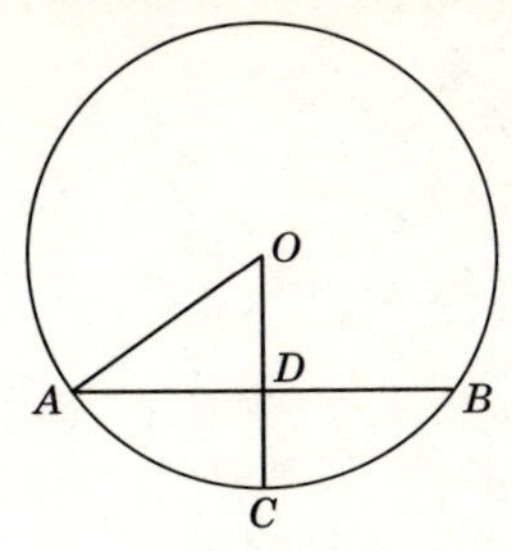

Exs. 20, 21.

Summary Tests

Test 1

COMPLETION STATEMENTS

1. Given right $\triangle MNP$ with $\angle N$ a right angle and $\overline{NT}$ the altitude on $\overline{MP}$. Then $\overline{NP}$ is the mean proportional between _______ and $\overline{MP}$.
2. A statement of equality of two ratios is termed a _______.
3. If two polygons have the same shape, they are _______.
4. If chords MN and RS of $\odot O$ intersect at P, then $RP:MP =$ _______.
5. If $\overleftrightarrow{ABC}$ and $\overleftrightarrow{EDC}$ are secants from external point C to $\odot O$, then $AC \times BC =$ _______ $\times$ _______.
6. If $\overleftrightarrow{PR}$ is a tangent and $\overleftrightarrow{PTS}$ is a secant of $\odot O$ drawn from external point P through points T and S of the $\odot$, then $PS \times PT =$ _______ $\times$ _______.
7. The perimeter of a rhombus having diagonals of 6 inches and 8 inches is _______ inches.
8. The square of a leg of a right triangle equals the square of the hypotenuse _______ the square of the other leg.
9. If $\frac{x}{y} = \frac{a}{b}$, then $\frac{x+y}{y} =$ _______.
10. If $xy = rs$, then $x:s =$ _______ : _______.
11. The mean proportional between 4 and 9 is _______.
12. The fourth proportional to 6, 8, 12 is _______.
13. If $7a = 3b$, then $a:b =$ _______.
14. $b:5 = a:10 \leftrightarrow a:b =$ _______.
15. $8:x = 5:y \leftrightarrow x:y =$ _______.
16. $ay = bx \leftrightarrow x:y =$ _______.

Test 2

TRUE-FALSE STATEMENTS

1. A proportion has four unequal terms.
2. If two triangles have their corresponding sides congruent, then their corresponding angles are congruent.
3. If two triangles have their corresponding angles congruent, then their corresponding sides are congruent.
4. The mean proportional between two quantities can be found by taking the square root of their product.
5. Two isosceles triangles are similar if an angle of one is congruent to a corresponding angle of the other.
6. The altitude on the hypotenuse of a right triangle is the mean proportional between segments of the hypotenuse cut off by the altitude.
7. Of two unequal chords of the same circle, the greater chord is the farther from the center.
8. If a line divides two sides of a triangle proportionately, it is parallel to the third side.
9. If two polygons have their corresponding sides proportional, they are similar.
10. Two isosceles right triangles are similar.
11. The square of the hypotenuse of a right triangle is equal to the sum of the legs.
12. If a line divides two sides of a triangle proportionately, it is equal to half the third side.
13. The diagonals of a trapezoid bisect each other.
14. If two triangles have two angles of one congruent respectively to two angles of the other, the triangles are similar.
15. Corresponding altitudes of similar triangles have the same ratio as that of any two corresponding sides.
16. Congruent polygons are similar.
17. If two chords intersect within a circle, the sum of the segments of one chord equals the sum of the segments of the other.
18. If a tangent and a secant are drawn from the same point outside a circle, the tangent is equal to one-half the difference of the secant and its external segment.
19. If two right triangles have an acute angle of one congruent to an acute angle of the other, the triangles are congruent.
20. Two triangles congruent to the same triangle are similar to each other.

Test 3

PROBLEMS

Find the value of x in each of the following:

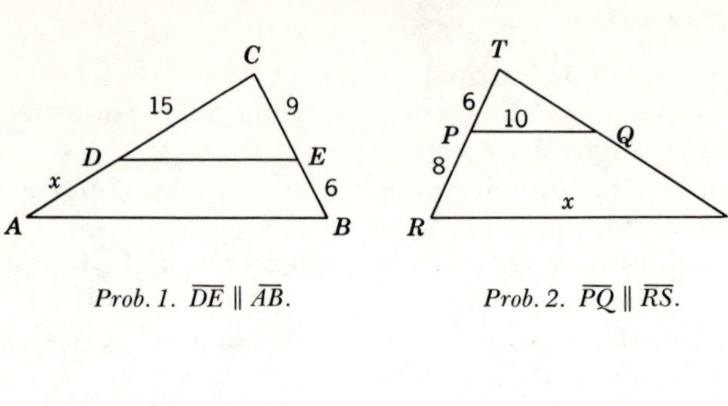

Prob. 1. $\overline{DE} \parallel \overline{AB}$. *Prob. 2.* $\overline{PQ} \parallel \overline{RS}$.

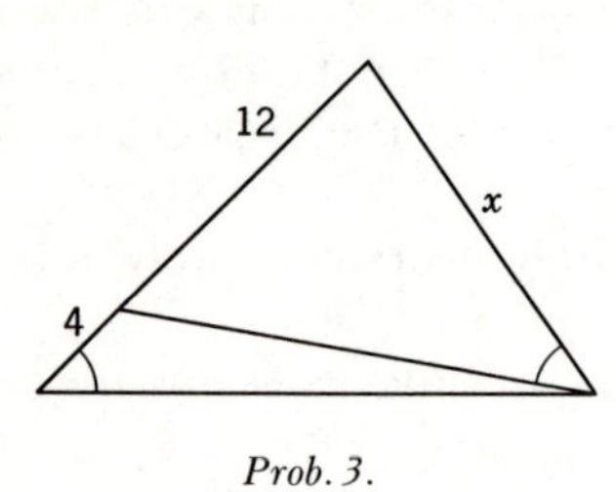

Prob. 3.

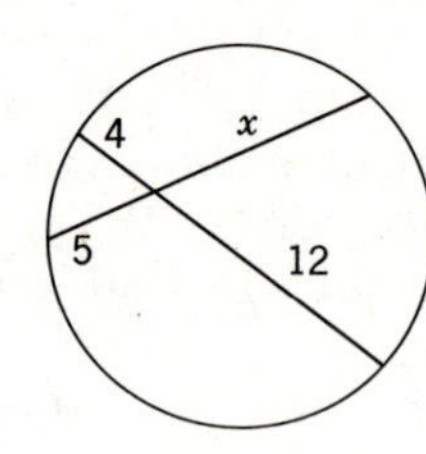

Prob. 4.

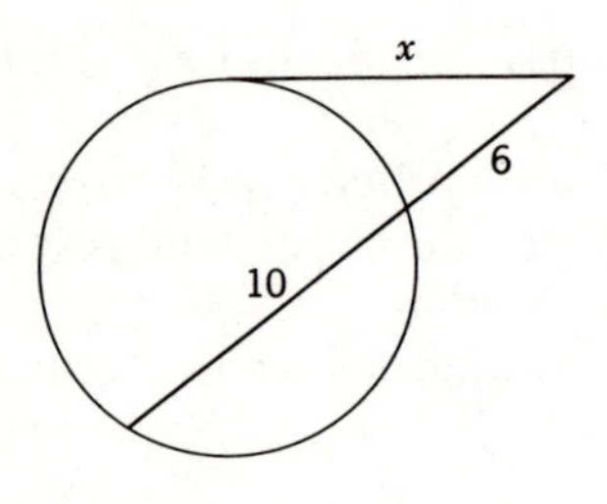

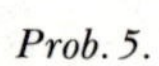

Prob. 5.

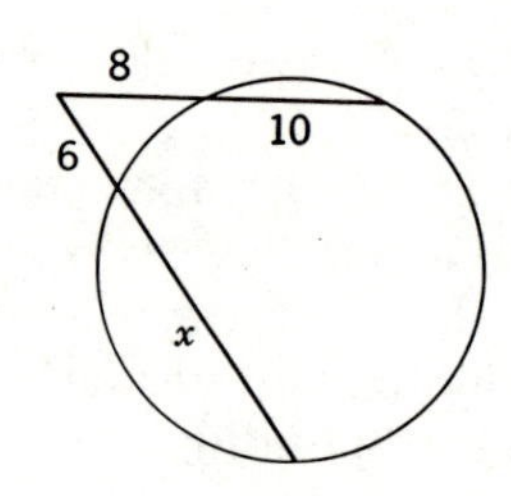

Prob. 6.

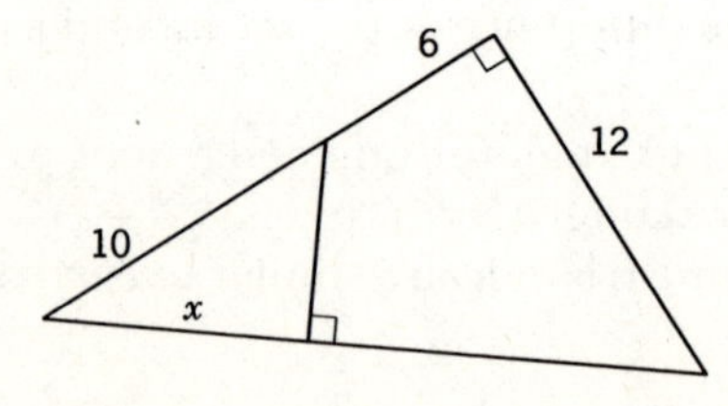

Prob. 7.

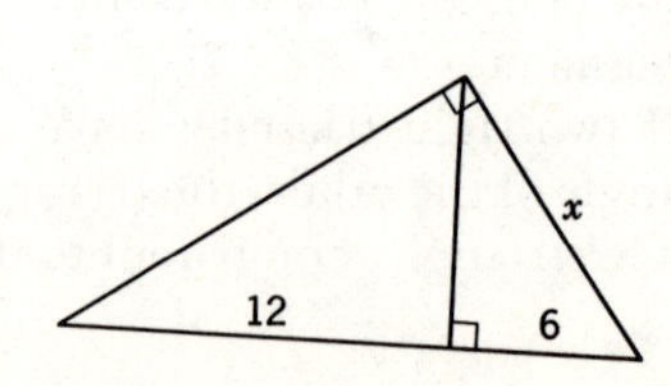

Prob. 8.

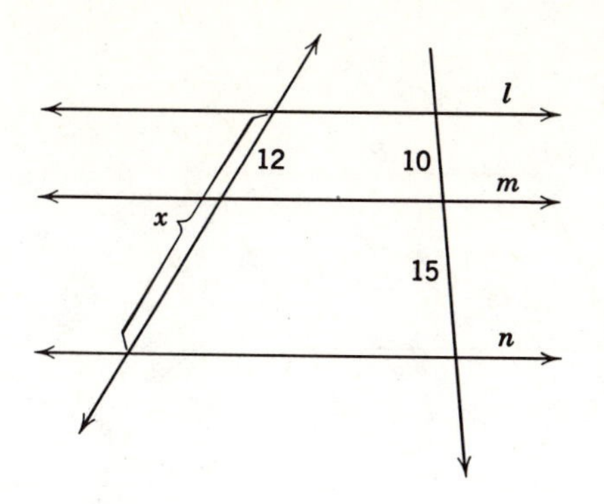

Prob. 9. $l \parallel m \parallel n$.

Prob. 10 figure:

Prob. 10. $ED = 20$; $BD = x$.

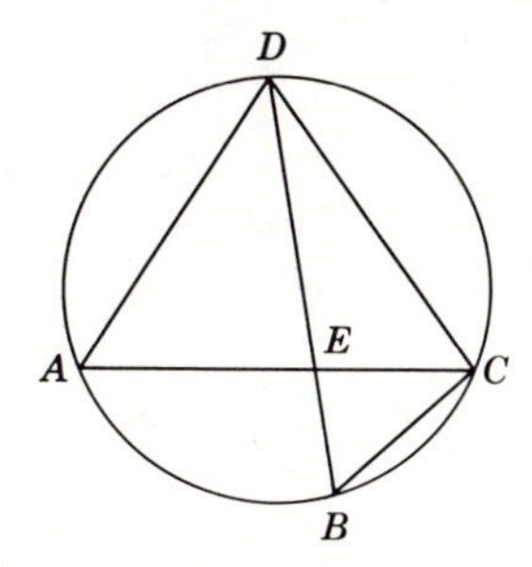

Prob. 11. $\overline{BD}$ bisects $\angle ADC$; $AD = 24$; $DC = 25$; $DE = 20$; $AE = 16$; $BC = x$.

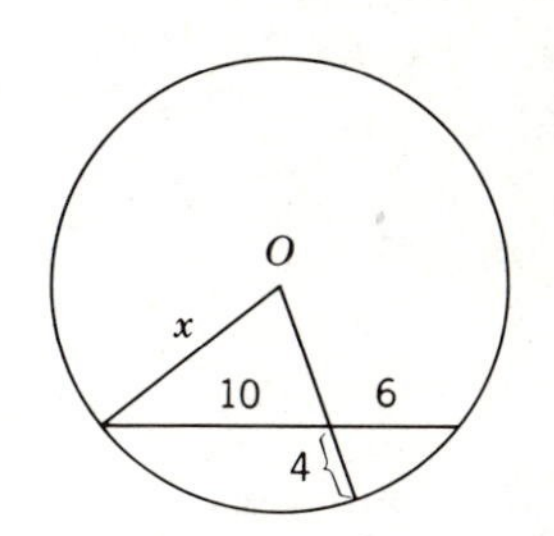

Prob. 12.

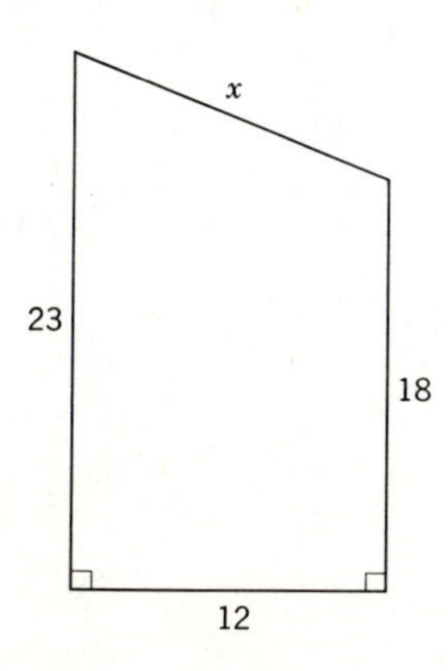

Prob. 13.

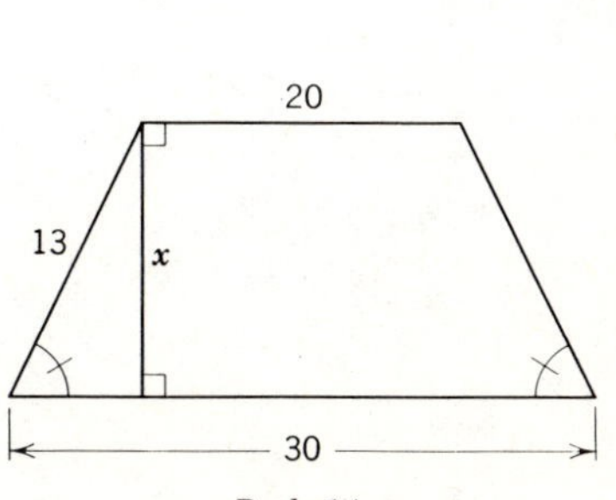

Prob. 14.

Test 4

EXERCISES

1. *Given:* ▱ $URST$ with diagonal RT.
   Prove: $PM:SM = MT:RM$.

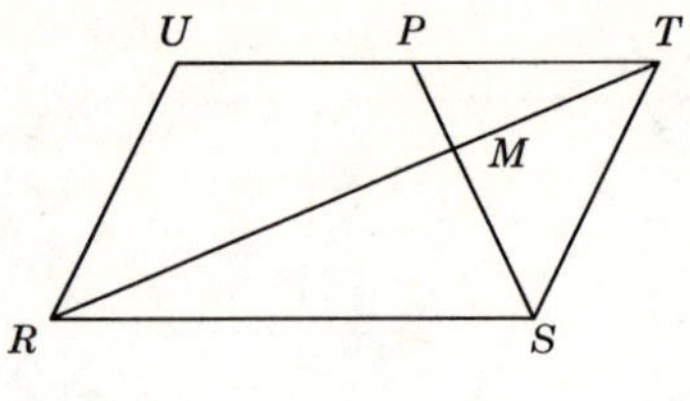

Ex. 1.

2. *Given:* $HK = LK$;
   $\overline{IK}$ bisects $\angle HKJ$.
   Prove: $LH:HJ = KI:IJ$.

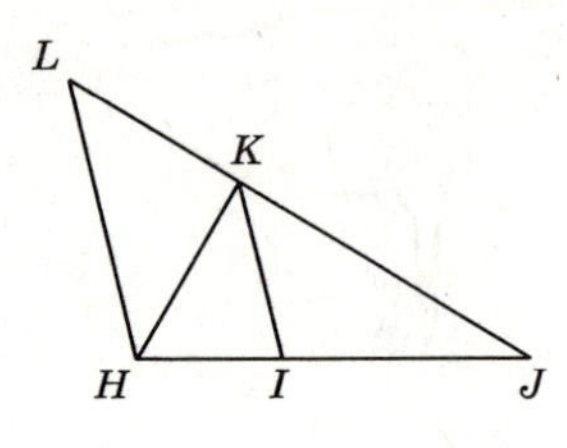

Ex. 2.

3. *Given:* $\overline{AOB}$ a diameter of $\odot O$;
   $\overline{DE} \perp \overline{AOB}$ extended.
   Prove: $\triangle ADE \sim \triangle ACB$.

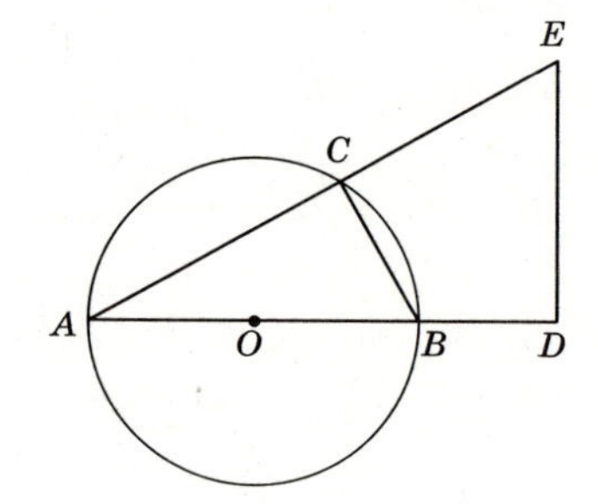

Ex. 3.

|9|

Inequalities

9.1. Inequalities are commonplace and important. In our study thus far we have found various ways of proving things equal. Often it is equally important to know when things are unequal. In this chapter we study relationships between unequal line segments, angles, and arcs.

9.2. Order relations. Since we will use the order relations given in Chapter 3 as operating principles in this chapter, the student is advised to review them at this time. The student will note the relation between Postulates 13 and 14 and the partition property (0–8). In Fig. 9.1, we use Postulate 13 to justify the relationship $m\overline{AB} + m\overline{BC} = m\overline{AC}$ (or $AB + BC = AC$). Postulate 14 can be cited to express the relation $m\angle ABC = m\angle ABD + m\angle DBC$.

By using the partition property, it immediately follows that

1. $m\overline{AC} > m\overline{AB}$ and $m\overline{AC} > m\overline{BC}$.
2. $m\angle ABC > m\angle ABD$ and $m\angle ABC > m\angle DBC$.

Often $\overline{AB}$ and $\overline{BC}$ are referred to as the *parts of* $\overline{AC}$, while $\angle ABD$ and $\angle DBC$ are the *parts of* $\angle ABC$. Thus the partition property could be stated as: "The whole is greater than its parts."

►IMPORTANT. Hereafter we will frequently follow the practice of referring to a given segment as being "equal to" or "greater than" another segment instead of stating that the "measure of one segment is equal to" or "the measure of one segment is greater than" the measure of a second segment. The student is reminded that we are now using $m\overline{AB}$ and AB interchangeably, and we are now accepting $m\overline{AB} = m\overline{CD}$, $\overline{AB} \cong \overline{CD}$, and $AB = CD$ as equivalent statements. Also "$AB > CD$" will be considered equivalent to $m\overline{AB} > m\overline{CD}$ and $AB + BC$ and $m\overline{AB} + m\overline{BC}$ will be two ways used to say the same thing.

This practice will be followed in order to shorten otherwise long and

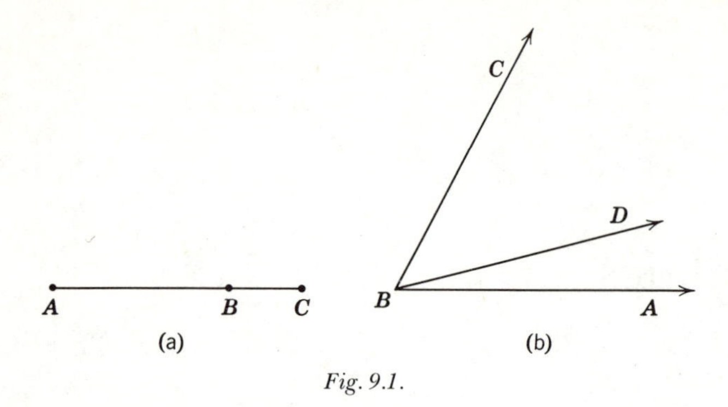

Fig. 9.1.

cumbersome statements. The student should keep in mind, however, that the *measures of geometric figures* are being compared in these instances.

9.3. Sense of inequalities. Two inequalities are of the *same sense* if the same symbol is used in the inequalities. Thus $a < b$ and $c < d$ are inequalities of the same sense. Two inequalities are of the *opposite sense* if the symbol of one inequality is the reverse of the symbol in the other. Thus $a < b$ and $c > d$ are of opposite sense.

A study of the basic properties of and theorems for inequalities will reveal processes which will transform an inequality to another inequality of the same sense. Some of them are:

(*a*) Adding equal real numbers to both sides of an inequality.
(*b*) Subtracting equal real numbers from both sides of an inequality.
(*c*) Multiplying both sides of an inequality by equal positive real numbers.
(*d*) Dividing both sides of an inequality by equal positive real numbers.
(*e*) Substituting a number for its equal in an inequality.

The following processes will transform an inequality to another inequality of opposite sense:

(*a*) Dividing the same (or equivalent) positive number by an inequality.
(*b*) Subtracting both sides of an inequality from the same real number.
(*c*) Multiplying both sides of an inequality by the same negative number.
(*d*) Dividing both sides of an inequality by the same negative number.

[*Note:* To divide by a number a is the same as to multiply by its multiplicative inverse $1/a$.]

Exercises (A)

Answer each question. If no answer is possible, indicate with "no answer possible."

1. Bill has more money than Tom. Each earned an additional 10 dollars. How do Bill's and Tom's total amounts compare?
2. Bill has more money than Tom and Frank has less than John. How do Bill's and John's compare?
3. Bill has the same amount of money as Alice. Alice spends more than Bill. How then do their remaining amounts compare?
4. John has more money than Tom. John loses half his money. How do their remaining amounts compare?
5. Bill has less money than Mary. Each decides to give half of his money to charity. How do the amounts they have left compare?
6. John has more money than Tom. Each doubles his amount. Who, then, has the more money?
7. Ann is older than Alice. Mary is younger than Alice. Compare Mary's and Ann's ages.
8. Mary and Alice together have as much money as Tom. Compare Tom's and Alice's amounts.
9. Ann and Bill are of different ages. Mary and Tom are also of different ages. Compare the ages of Ann and Tom.
10. John has twice as much money as Mary, and Mary has one-third as much as Alice. Compare the amounts of John and Alice.

Exercises (B)

Copy and complete the following exercises. If no conclusion is possible, write a question mark in place of the blank.

11. If $a > b$ and $c = d$, then $a + c$ ________ $b + d$.
12. If $r < s$ and $x > b$, then $r - b$ ________ $s - x$.
13. If $x = 2y$, $r = 2s$, and $y < s$, then x ________ r.
14. If $l > k$ and $k > m$, then l ________ m.
15. If $x + y = z$, then z ________ x.
16. If $AB + BC > AC$, then AB ________ $AC - BC$.

17. If $\angle x \cong \angle y$
and $m\angle r > m\angle s$, then
$m\angle ABC$ ______ $m\angle DEF$.

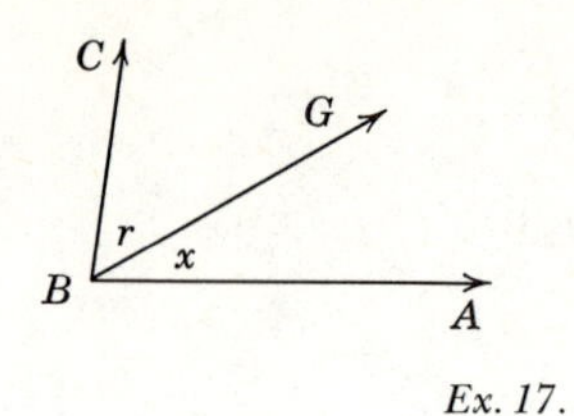

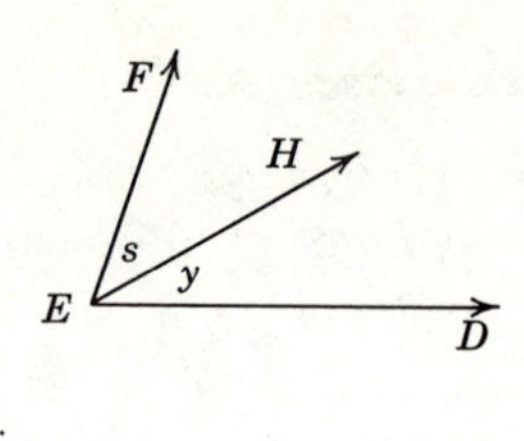

Ex. 17.

18. If $\overleftrightarrow{BC} \perp \overleftrightarrow{AB}$; $\overleftrightarrow{EF} \perp \overleftrightarrow{DE}$;
and $m\angle\beta < m\angle\delta$, then
$m\angle\alpha$ ______ $m\angle\gamma$.

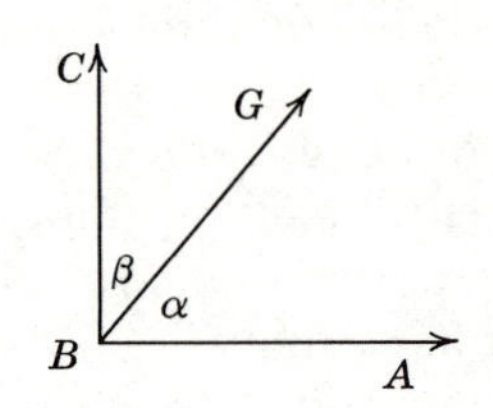

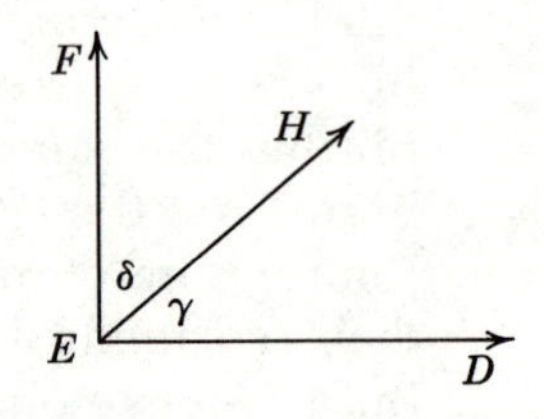

Ex. 18.

19. If $\angle A \cong \angle B$ and $CD < CE$,
then AD ______ BE.
20. If $AD > BE$ and $EC > CD$,
then AC ______ BC.

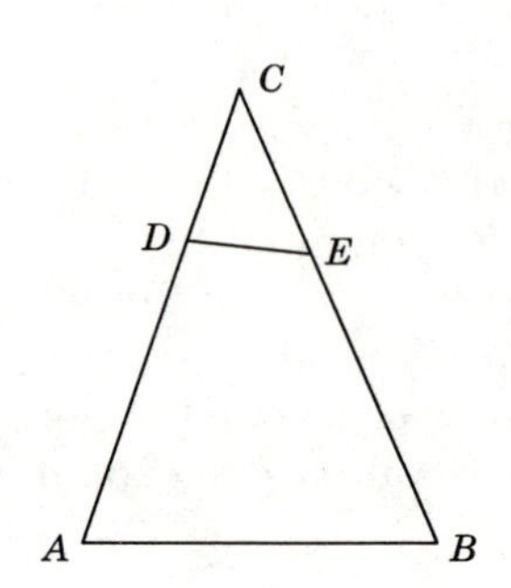

Exs. 19, 20.

21. If $m\angle CAB = m\angle ABC$,
then BD ______ AC.
22. If $AC = BC$,
then $m\angle ABC$ ______ $m\angle BAD$.

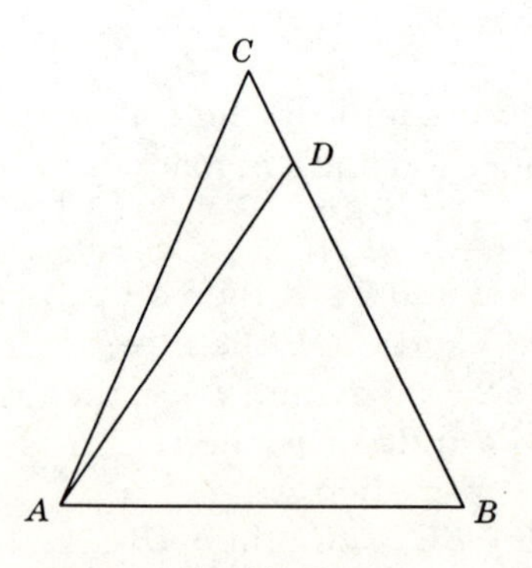

Exs. 21, 22.

23. If $\angle\alpha$ is an exterior $\angle$ of $\triangle ABC$, then $m\angle\alpha$ ______ $m\angle A$.

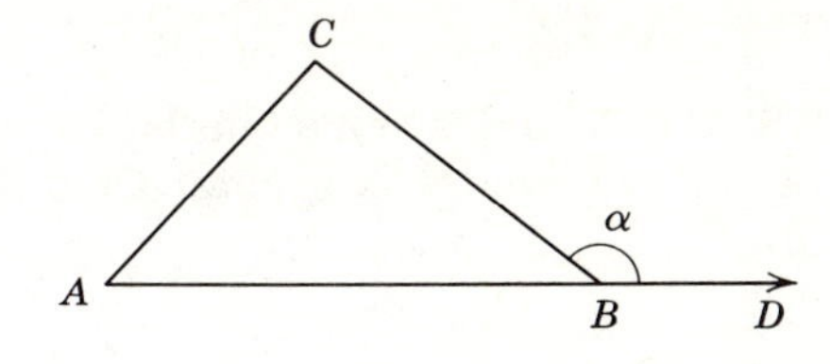

Ex. 23.

24. If $m\angle R > m\angle S$,
$\overline{RK}$ bisects $\angle SRT$,
$\overline{SK}$ bisects $\angle RST$,
then $m\angle\alpha$ ______ $m\angle\beta$.

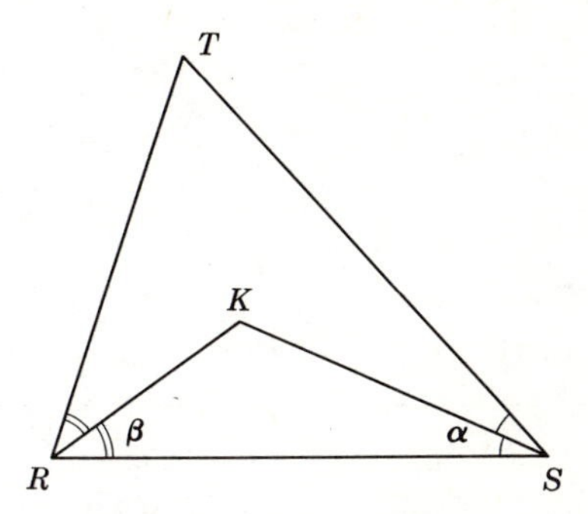

Ex. 24.

Theorem 9.1

9.4. If two sides of a triangle are not congruent, the angle opposite the longer of the two sides has a greater measure than does the angle opposite the shorter side.

Given: $\triangle ABC$ *with* $BC > AC$.
Conclusion: $m\angle BAC > m\angle B$.
Proof:

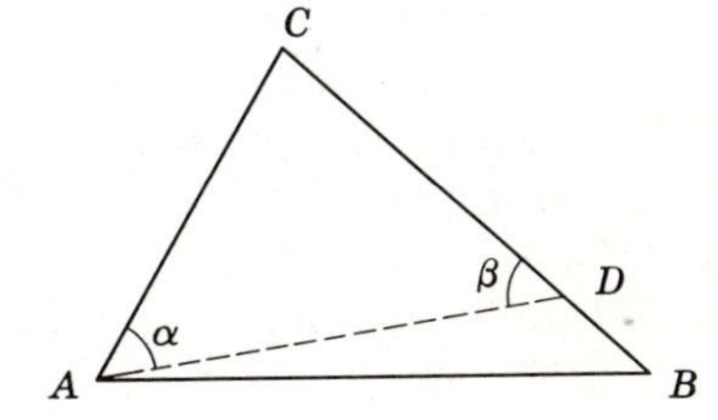

Theorem 9.1.

STATEMENTS	REASONS
1. $BC > AC$.	1. Given.
2. On $\overline{CB}$ let D be a point such that $CD = AC$.	2. Postulate 11.
3. Draw $\overline{AD}$.	3. Postulate 2.
4. $m\angle\alpha = m\angle\beta$.	4. Theorem 4.16.
5. $m\angle BAC = m\angle BAD + m\angle\alpha$.	5. Postulate 14.
6. $m\angle BAC > m\angle\alpha$.	6. O-8.
7. $m\angle BAC > m\angle\beta$.	7. O-7.
8. $m\angle\beta > m\angle B$.	8. Theorem 4.17.
9. $\therefore m\angle BAC > m\angle B$.	9. O-6.

Theorem 9.2

9.5. If two angles of a triangle are not congruent, the side opposite the larger of the two angles is greater than the side opposite the smaller of the two angles.

Given: $\triangle ABC$ with $m\angle A > m\angle B$.
Conclusion: $BC > AC$.

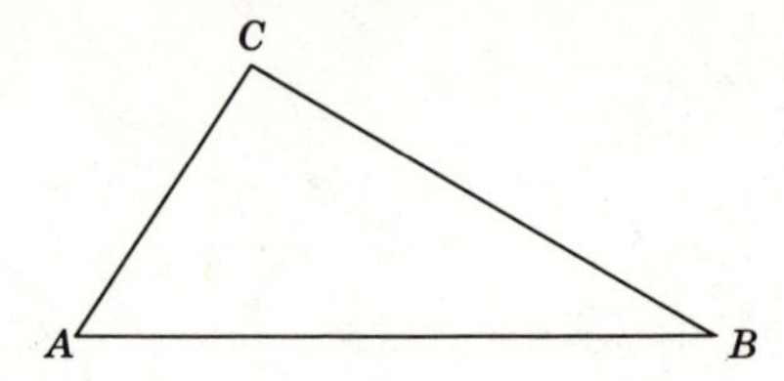

Theorem 9.2.

Proof:

STATEMENTS	REASONS
1. $m\angle A > m\angle B$.	1. Given.
2. In $\triangle ABC$, since BC and AC are real numbers, there are only the following possibilities: $BC = AC$, $BC < AC, BC > AC$.	2. Trichotomy property.
3. Assume $BC = AC$.	3. Temporary Assumption 1.
4. Then $m\angle A = m\angle B$.	4. Theorem 4.16.
5. Statement 4 is false.	5. Contradicts Statement 1 (the given).
6. Next assume $BC < AC$.	6. Temporary Assumption 2.
7. Then $m\angle A < m\angle B$.	7. Theorem 9.1.
8. Statement 7 is false.	8. Contradicts Statement 1.
9. The only possibility remaining is is $BC > AC$.	9. Rule for denying the alternatives.

9.6. Corollary: The shortest segment joining a point to a line is the perpendicular segment.

Note. Here we can prove what we stated in §1.20.

9.7. Corollary: The measure of the hypotenuse of a right triangle is greater than the measure of either leg.

Theorem 9.3

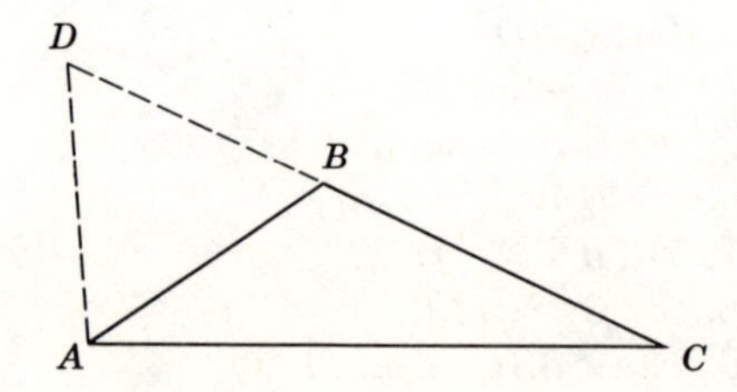

Theorem 9.3.

9.8. The sum of the measures of two sides of a triangle is greater than the measure of the third side.
Given: $\triangle ABC$.
Conclusion: $AB + BC > AC$.
Proof:

STATEMENTS	REASONS
1. Let D be the point on the ray opposite $\overrightarrow{BC}$ such that $DB = AB$.	1. Postulate 11.
2. Draw $\overline{AD}$.	2. Postulate 2.
3. $DC = DB + BC$.	3. Postulate 13.
4. $DC = AB + BC$.	4. E-8.
5. $m\angle DAC = m\angle DAB + m\angle BAC$.	5. Postulate 14.
6. $m\angle DAC > m\angle DAB$.	6. O-8.
7. $m\angle DAB = m\angle ADB$.	7. Theorem 4.16.
8. $m\angle DAC > m\angle ADB$.	8. O-7.
9. $DC > AC$.	9. Theorem 9.2.
10. $AB + BC > AC$.	10. Substitution property.

This theorem may be used to show that the shortest route between two points is the straight line route.

Theorem 9.4

9.9. If two triangles have two sides of one congruent respectively to two sides of the other and the measure of the included angle of the first greater than the measure of the included angle of the second triangle, the third side of the first is greater than the third side of the second.

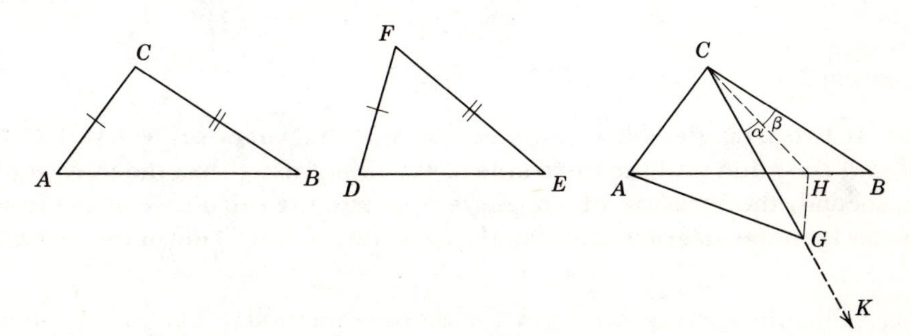

Theorem 9.4.

Given: $\triangle ABC$ and $\triangle DEF$ with $AC = DF$, $CB = FE$, and $m\angle C > m\angle F$.
Conclusion: $AB > DE$.
Proof:

STATEMENTS	REASONS
1. Draw $\overrightarrow{CK}$ (see third figure) with K on the same side of $\overleftrightarrow{BC}$ as A and such that $\angle ACK \cong \angle DFE$.	1. Postulate 12.
2. $\overrightarrow{CK}$ take a point G such that $CG = FE$.	2. Postulate 11.
3. Draw $\overline{AG}$.	3. Postulate 2.
4. $AC = DF$.	4. Given.
5. $\triangle ACG \cong \triangle DFE$.	5. S.A.S.
6. $AG = DE$.	6. §4.28.
7. Bisect $\angle BCK$ and let H be the point where the bisector intersects $\overleftrightarrow{AB}$.	7. O-1.
8. $\angle\alpha \cong \angle\beta$.	8. §1.19.
9. Draw $\overline{HG}$.	9. Postulate 2.
10. $CH = CH$.	10. Reflexive property.
11. $CB = FE$.	11. Given.
12. $CG = CB$.	12. Theorem 3.4.
13. $\triangle CHG \cong \triangle CHB$.	13. S.A.S.
14. $GH = BH$.	14. §4.28.
15. $GH + AH > AG$.	15. Theorem 9.3.
16. $BH + AH > AG$.	16. O-7.
17. $BH + AH = AB$.	17. Postulate 13.
18. $AB > AG$.	18. O-7.
19. $AB > DE$.	19. O-7.

Theorem 9.5

9.10. If two triangles have two sides of one congruent respectively to two sides of the other and the third side of the first greater than the third side of the second, the measure of the angle opposite the third side of the first is greater than the measure of the angle opposite the third side of the second.

(*Note:* This theorem is proved by the indirect method. The proof is left to the student.)

9.11. Illustrative Example 1:

Given: D a point in the interior of $\triangle ABC$; $AC = CD$.
Prove: $DB < AB$.
Proof:

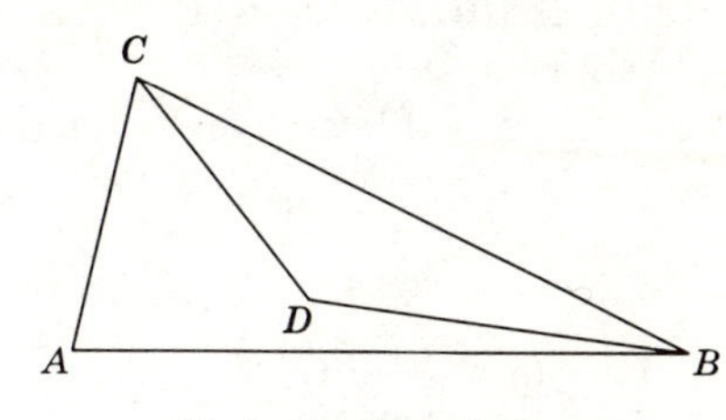

Illustrative Example 1.

STATEMENTS	REASONS
1. AC (of $\triangle ABC$) $= CD$ (of $\triangle DBC$).	1. Given.
2. BC (of $\triangle ABC$) $= BC$ (of $\triangle DBC$).	2. Reflexive property.
3. D is in the interior of $\angle ACB$.	3. Given.
4. $m\angle ACB = m\angle DCB + m\angle ACD$.	4. Postulate 14.
5. $m\angle DCB < m\angle ACB$.	5. O-8.
6. $\therefore DB < AB$.	6. Theorem 9.4.

9.12. Illustrative Example 2:

Given: $ST = RT$;
K any point on $\overline{RS}$.
Prove: $ST > KT$.

Proof:

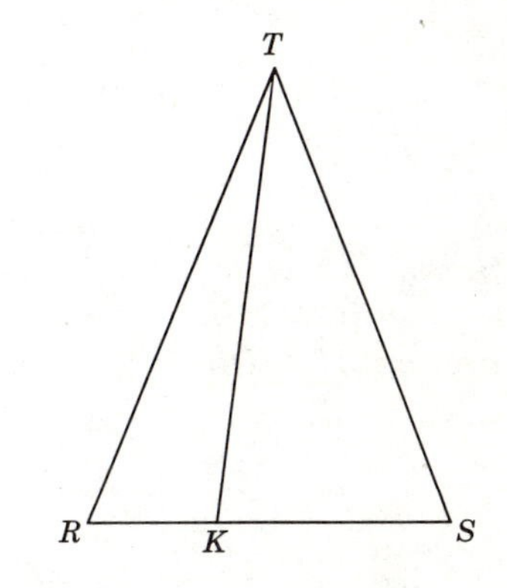

Illustrative Example 2.

STATEMENTS	REASONS
1. $ST = RT$.	1. Given.
2. $m\angle S = m\angle R$.	2. 4.16.
3. $m\angle RKT > m\angle S$.	3. 4.17.
4. $m\angle RKT > m\angle R$.	4. O-7.
5. $RT > KT$.	5. Theorem 9.2.
6. $ST > KT$.	6. O-7.

Exercises

1. In $\triangle ABC$, $m\angle A = 60$, $m\angle B = 70$. Which is (*a*) the longest side; (*b*) the shortest side?
2. Is it possible to construct triangles with sides the lengths of which are: (*a*) 6, 8, 10; (*b*) 1, 2, 3; (*c*) 6, 7, 8; (*d*) 7, 5, 1.
3. *Given:* $AC = BC$.
   Prove: $AC > DC$.

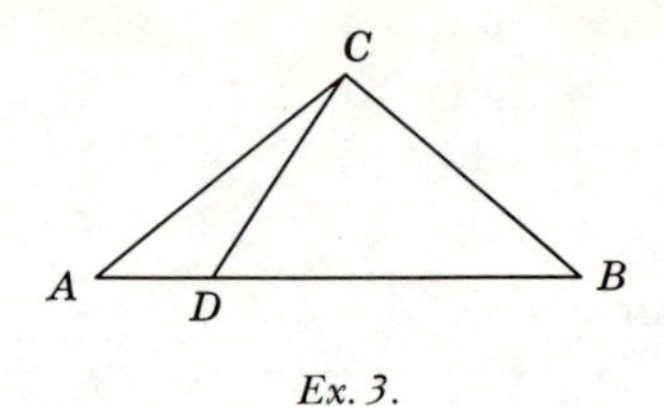

Ex. 3.

4. *Given:* $BC > AC$;
   $\overline{AD}$ bisects $\angle BAC$;
   $\overline{BD}$ bisects $\angle ABC$.
   Prove: $BD > AD$.

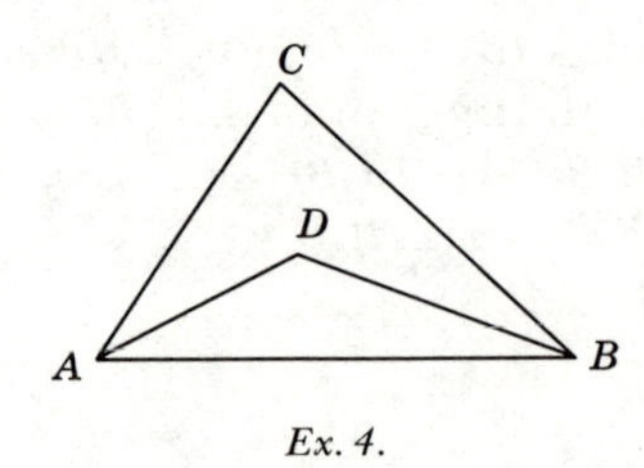

Ex. 4.

5. *Given:* $\overline{DA} \perp \overleftrightarrow{RS}$;
   $AC > AB$.
   Prove: $DC > DB$.

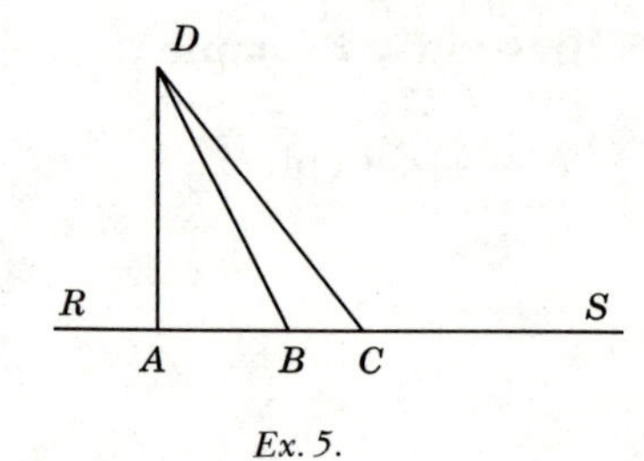

Ex. 5.

6. *Given:* $\triangle RST$ with $RT = ST$.
   Prove: $QS > QR$.

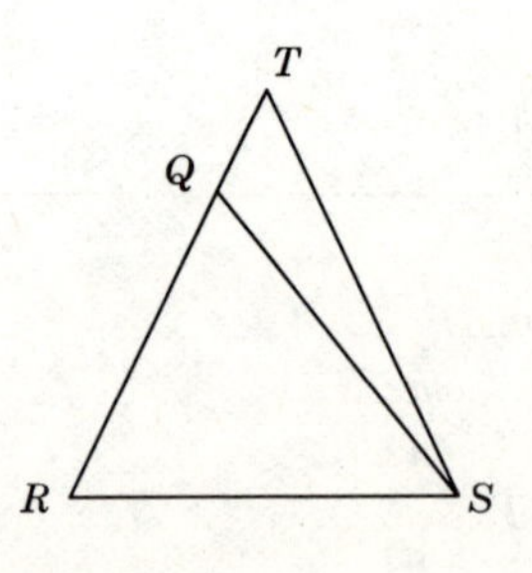

Ex. 6.

7. *Given:* $DC = BC$.
   Prove: $m\angle ADC > m\angle A$.
8. *Given:* $DC = BC$.
   Prove: $AD > BD$.
9. *Given:* $DC = BC$.
   Prove: $m\angle CDB > m\angle A$.
10. *Given:* $DC = BC$.
    Prove: $AC > DC$.

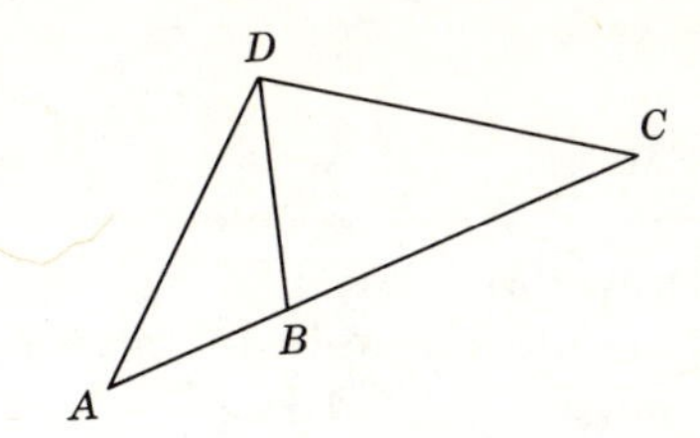

Exs. 7–10.

11. *Given:* $RP = RS$;
    $PT = ST$;
    $PT > RP$.
    Prove: $m\angle PRS > m\angle PTS$.

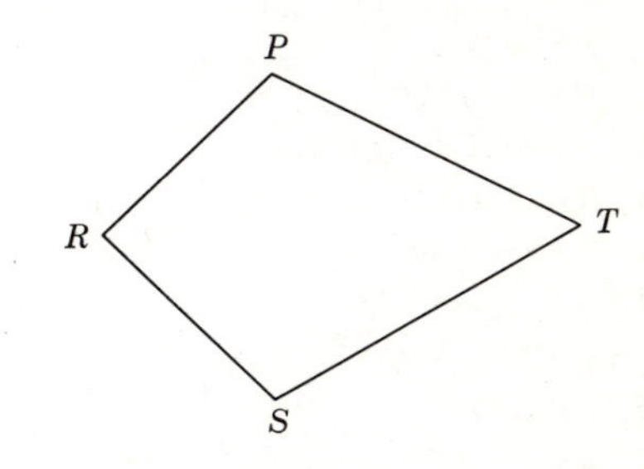

Ex. 11.

12. *Given:* $\overline{AM}$ is a median of $\triangle ABC$.
    Prove: $AM < \frac{1}{2}(AB + AC)$.
    (*Hint:* Extend $\overline{AM}$ to D making $MD = AM$.)

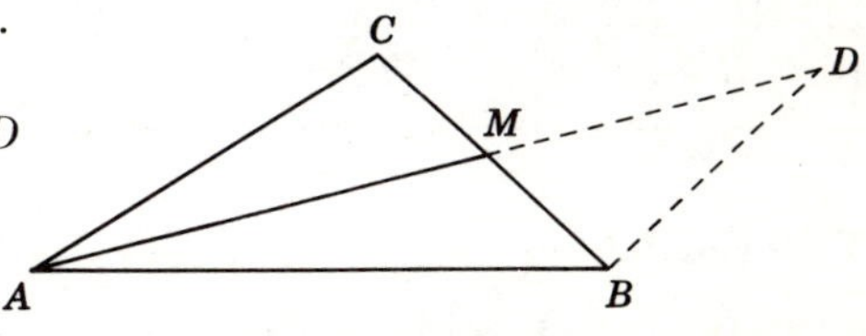

Ex. 12.

13. It is desired in the figure to find the shortest path from point A to line λ and then to point B. Prove that the shortest line is the broken line formed which makes $\angle\alpha \cong \angle\beta$.
    Given: $\angle\alpha \cong \angle\beta$.
    Prove: $AR + RB < AT + TB$.

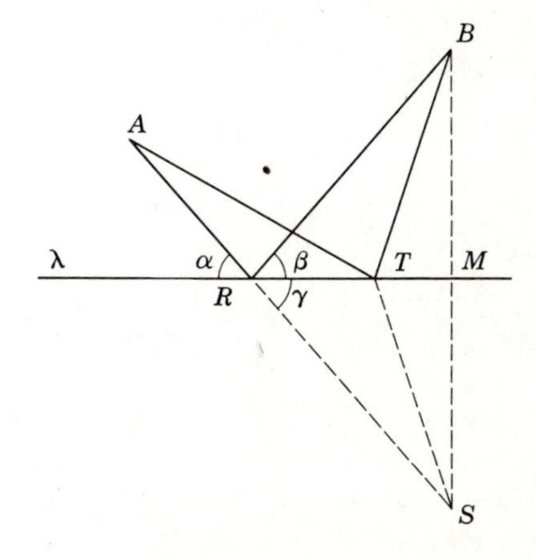

Ex. 13.

Proof:

STATEMENTS	REASONS
1. Draw $\overline{BM} \perp \lambda$.	1. Why?
2. Extend $\overline{BM}$ and $\overline{AR}$ until they intersect at S. Draw $\overline{TS}$.	2. Why?
3. $\angle\alpha \cong \angle\beta$.	3. Why?
4. $\angle\alpha \cong \angle\gamma$.	4. Why?
5. $\angle ? \cong \angle ?$.	5. Why?
6. $RM = RM$.	6. Why?
7. $\triangle RMB \cong \triangle RMS$.	7. Why?
8. $RB = RS$.	8. Why?
9. $BM = SM$.	9. Why?
10. $\underline{\quad ? \quad} = \underline{\quad ? \quad}$.	10. Why?
11. $\triangle TMB \cong \triangle TMS$.	11. Why?
12. $TB = TS$.	12. Why?
13. $AS < AT + TS$.	13. Why?
14. $AR + RS = AS$.	14. Why?
15. $AR + RS < AT + TS$.	15. Why?
16. $AR + RB < AT + TB$.	16. Why?

Theorem 9.6

9.13. In a circle or in congruent circles, if two central angles have unequal measures, the greater central angle has the greater minor arc.

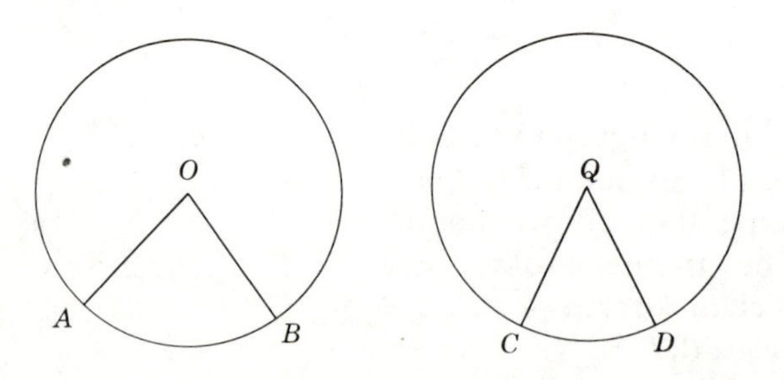

Theorem 9.6.

Given: $\odot O \cong \odot Q$ with $m\angle O > m\angle Q$.
Conclusion: $m\widehat{AB} > m\widehat{CD}$.
Proof:

STATEMENTS	REASONS
1. $\odot O \cong \odot Q$.	1. Given.
2. $m\angle O > m\angle Q$.	2. Given.
3. $m\angle O = m\widehat{AB}$, $m\angle Q = m\widehat{CD}$.	3. §7.9.
4. $\therefore m\widehat{AB} > m\widehat{CD}$.	4. O-7.

Theorem 9.7

9.14. In a circle or in congruent circles, if two minor arcs are not congruent, the greater arc has the greater central angle.
(The proof of this theorem is left to the student.)

Theorem 9.8

9.15. In a circle or in congruent circles, the greater of two noncongruent chords has the greater minor arc.

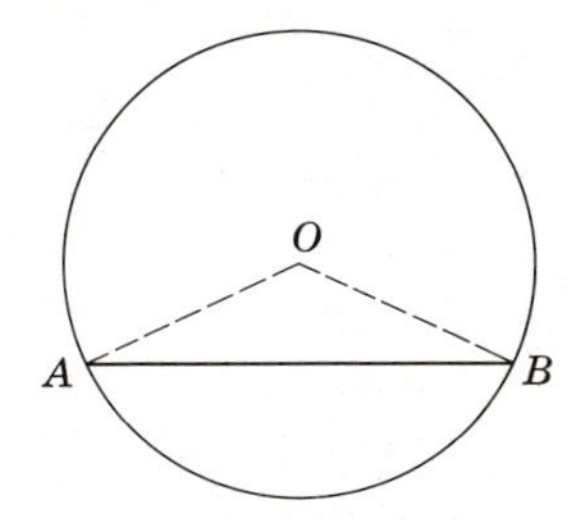

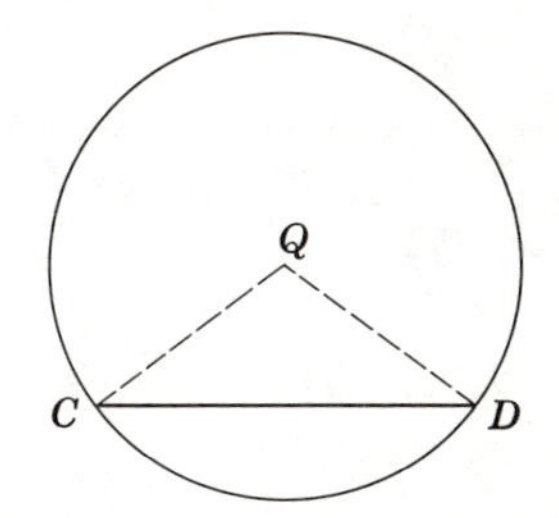

Theorem 9.8.

Given: $\odot O \cong \odot Q$ with chord $AB >$ chord CD.
Conclusion: $m\widehat{AB} > m\widehat{CD}$.

Proof:

STATEMENTS	REASONS
1. $\odot O \cong \odot Q$.	1. Given.
2. Draw radii OA, OB, QC, QD.	2. Postulate 2.
3. $OA = QC$; $OB = QD$.	3. Definition of $\cong$ Ⓢ.
4. Chord $AB >$ chord CD.	4. Given.
5. $m\angle O > m\angle Q$.	5. Theorem 9.5.
6. $m\widehat{AB} > m\widehat{CD}$.	6. Theorem 9.6.

Theorem 9.9

9.16. In a circle or in congruent circles the greater of two noncongruent minor arcs has the greater chord.
(The proof of this theorem is left to the student.)

Theorem 9.10

9.17. In a circle or in congruent circles, if two chords are not congruent, they are unequally distant from the center, the greater chord being nearer the center.

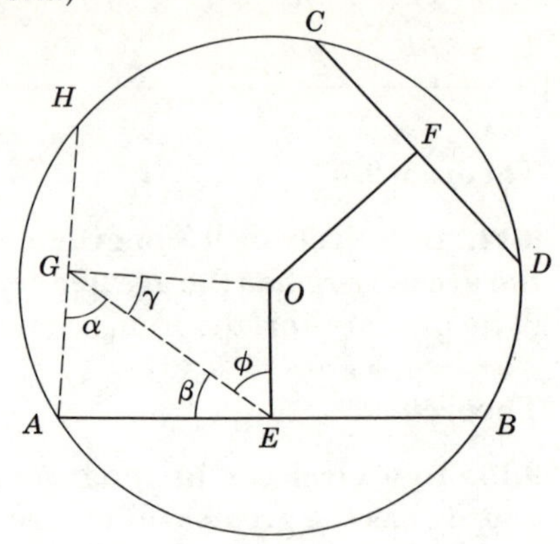

Theorem 9.10.

Given: $\odot O$ with chord $AB >$ chord CD;
$\overline{OE} \perp \overline{AB}$; $\overline{OF} \perp \overline{CD}$.
Conclusion: $OE < OF$.
Proof:

STATEMENTS	REASONS
1. Draw chord $AH \cong$ chord CD.	1. Postulate 11.
2. Draw $\overline{OG} \perp \overline{AH}$.	2. Theorem 5.4.
3. Draw $\overline{GE}$.	3. Postulate 2.
4. $\overline{OE} \perp \overline{AB}$, $\overline{OF} \perp \overline{CD}$.	4. Given.
5. $OG = OF$.	5. Theorem 7.9.
6. $AB > CD$.	6. Given.
7. $AB > AH$.	7. O-7. E-1.
8. G is midpoint of $\overline{AH}$; E is midpoint of $\overline{AB}$.	8. Theorem 7.7.
9. $AE > AG$.	9. O-5.
10. $m\angle\alpha > m\angle\beta$.	10. Theorem 9.1.
11. $m\angle AGO = m\angle AEO$.	11. §1.20; Theorem 3.7.
12. $m\angle\gamma < m\angle\phi$.	12. O-3.
13. $OE < OG$.	13. Theorem 9.2.
14. $\therefore OE < OF$.	14. O-7.

Theorem 9.11

9.18. In a circle or in congruent circles, if two chords are unequally distant from the center, they are not congruent, the chord nearer the center being the greater.
(The proof is left to the student.)

Exercises

In each of the following circles, O is assumed to be the center of a circle.

1. In $\triangle ABC$, $m\overline{AB} = 4$ inches, $m\overline{BC} = 5$ inches, $m\overline{AC} = 6$ inches. Name the angles of the triangle in order of size.
2. $\triangle RST$ is inscribed in a circle. $m\widehat{RS} = 80$ and $m\widehat{ST} = 120$. Name the angles of the triangle in order of size.
3. In $\triangle MNT$, $m\angle N = 60$ and $m\angle M < m\angle T$. Which is the longest side of the triangle?
4. In quadrilateral $LMNT$, $LM = MN$, and $m\angle L > m\angle N$. Which is the longer, $\overline{NT}$ or $\overline{LT}$? Prove your answer.
5. In quadrilateral $QRST$, $QR > RS$ and $\angle Q \cong \angle S$. Which is the longer, $\overline{QT}$ or $\overline{ST}$? Prove your answer.

6. *Given:* $AC > AB$.
   Prove: $AC > AD$.

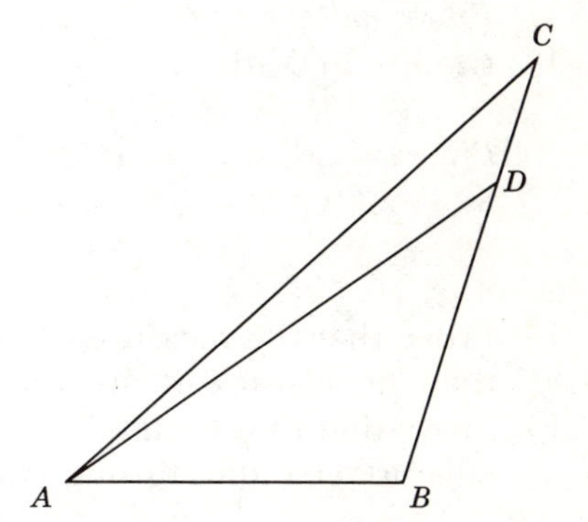

Ex. 6.

7. Prove that, in a circle, if the measure of one minor arc is twice the measure of a second minor arc, the measure of the chord of the first arc is less than twice the measure of the chord of the second arc.
8. Prove that, if a square and an equilateral triangle are inscribed in a circle, the distance from the center of the circle to the side of the square is greater than that to the side of the triangle.

9. *Given:* P a point of diameter CD; chord $AB \perp \overline{CD}$; $\overline{EF}$ any other chord containing P.
   Prove: $AB < EF$.

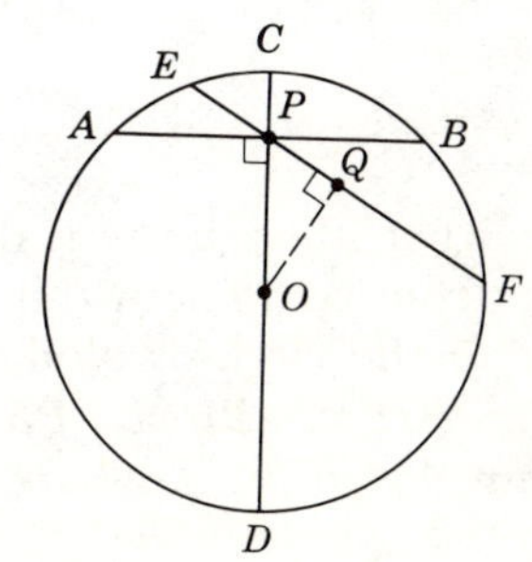

Ex. 9.

10. *Given:* $\odot O$ with $\overline{OA} \perp \overline{PR}$;
$\overline{OB} \perp \overline{SR}$;
$OB > OA$.
Prove: $m\angle POR > m\angle ROS$.

11. *Given:* $\odot O$ with $\overline{OA} \perp \overline{PR}$;
$\overline{OB} \perp \overline{SR}$;
$m\angle POR > m\angle ROS$.
Prove: $OB > OA$.

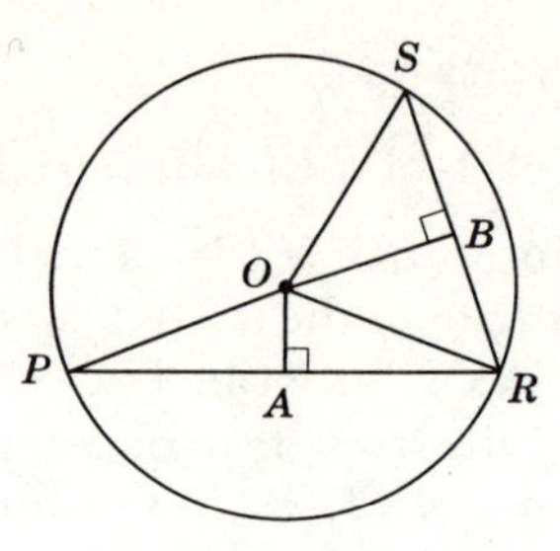

Exs. 10, 11.

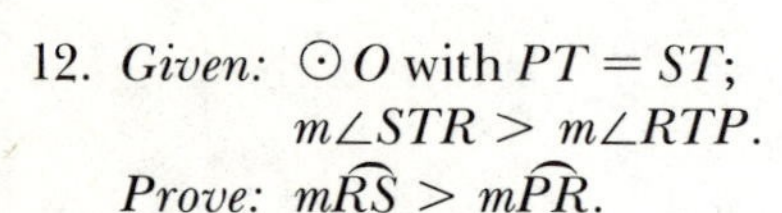

12. *Given:* $\odot O$ with $PT = ST$;
$m\angle STR > m\angle RTP$.
Prove: $m\widehat{RS} > m\widehat{PR}$.

13. *Given:* $\odot O$ with $PT = ST$;
$m\widehat{RS} > m\widehat{PR}$.
Prove: $m\angle STR > m\angle RTP$.

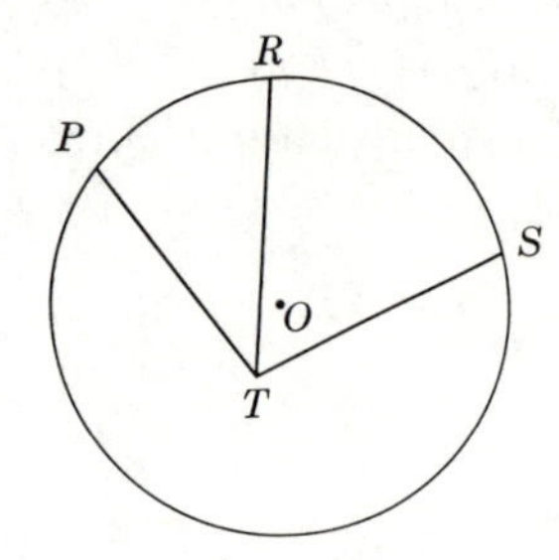

Exs. 12, 13.

14. Prove that the measure of the hypotenuse of a right triangle is greater than the measure of either leg.
15. Prove that the shortest chord through a point inside a circle is perpendicular to the radius through the point.

16. *Given:* $\overline{CM}$ is a median of $\triangle ABC$;
$\overline{CM}$ is not $\perp \overline{AB}$.
Prove: $AC \neq BC$.

17. *Given:* $\overline{CM}$ is a median of $\triangle ABC$;
$AC \neq BC$.
Prove: $\overline{CM}$ cannot be $\perp \overline{AB}$.

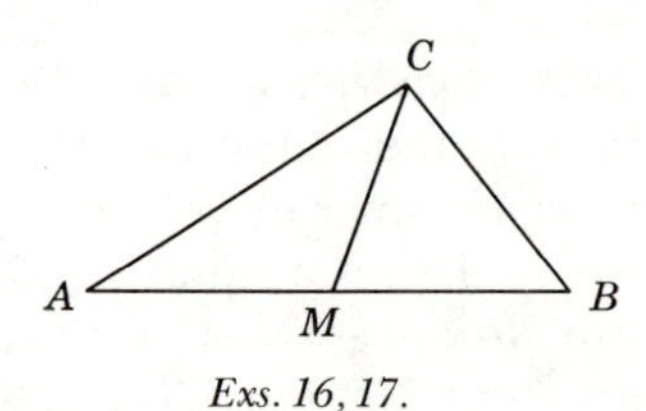

Exs. 16, 17.

18. Prove that the shortest distance from a point within a circle to the circle is along a radius. (*Hint:* Prove $PB < PA$, any $A \neq B$.)

Ex. 18.

19. Prove that the shortest distance from a point outside a circle to the circle is along a radius produced. (*Hint:* $PS + SO > \ldots$; $SO = RO$; $PS > \ldots$.)

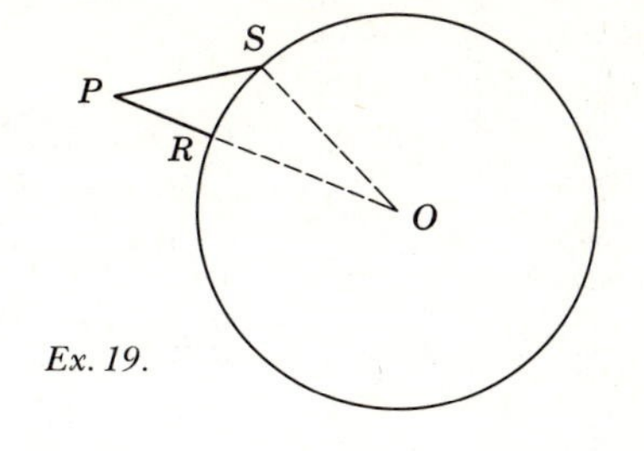

Ex. 19.

20. *Given:* Chord $AC >$ chord BD.
Prove: Chord $AB >$ chord CD.

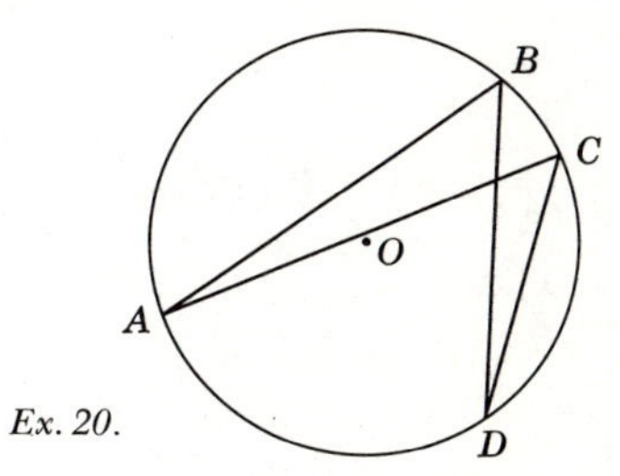

Ex. 20.

Summary Tests

Test 1

COMPLETION STATEMENTS

1. The sum of the measures of any two sides of a triangle is ______ than the measure of the third side.
2. Angle T is the largest angle in triangle RST. The largest side is ______.
3. If $k > h$, then $k+1$ ______ $h+1$.
4. If $m < n$ and $n < p$, then m ______ p.
5. If $l > w$, then $a-l$ ______ $a-w$.
6. If $d < e$, $e < f$, $f = h$, then d ______ h.
7. In $\triangle HJK$, $HJ > JK$, $m\angle J = 80$. Then $m\angle H$ ______ 50.
8. In quadrilateral $LMNP$, $LM = MN$ and $m\angle MLP > m\angle MNP$. Then $m\angle NLP$ ______ $m\angle LNP$.
9. In $\triangle ABC$, $m\angle A = 50$, $m\angle B = 60$, $m\angle C = 70$. Then AB ______ AC.
10. In a circle or in congruent circles, if two central angles are not congruent, the greater central angle has the ______ arc.
11. The measure of an exterior angle of a triangle is equal to the ______ of the measures of the two nonadjacent interior angles.
12. If $x+y = k$, then y ______ $k-x$.
13. If $a < b$ and $c > d$, then $a+d$ ______ $b+c$.
14. If $x < y$, then $x-a$ ______ $y-a$.
15. If $x < y$ and $z > y$, then z ______ x.
16. If $xy < 0$ and $x > 0$, then y ______ 0.
17. In quadrilateral $PQRS$, if $PQ = QR$, and $m\angle P > m\angle R$, then PS ______ RS.
18. In quadrilateral $PQRS$, if $PQ > QR$, and $m\angle P = m\angle R$, then PS ______ RS.

19. In $\triangle RST$, if $m\angle R = 60$ and $m\angle S > m\angle T$, then ______ is the longest side of the triangle.
20. The shortest chord through a point inside a circle is ______ to the radius through the point.

Test 2

TRUE-FALSE STATEMENTS

1. The shortest distance from a point to a circle is along the line joining that point and the center of the circle.
2. The measure of the perpendicular segment from a point to a line is the shortest distance from the point to the line.
3. Either leg of a right triangle is shorter than the hypotenuse.
4. If two triangles have two sides of one equal to two sides of the other, and the third side of the first less than the third side of the second, the measure of the angle included by the two sides of the first triangle is greater than the measure of the angle included by the two sides of the second.
5. No two angles of a scalene triangle can have the same measure.
6. The measure of an exterior angle of a triangle is greater than the measure of any of the interior angles.
7. If two sides of a triangle are unequal, the measure of the angle opposite the greater side is less than the measure of the angle opposite the smaller side.
8. If two chords in the same circle are unequal, the smaller chord is nearer the center.
9. If John is older than Mary, and Alice is younger than Mary, John is older than Alice.
10. Bill has twice as much money as Tom, and Tom has one-third as much as Harry. Then Bill has more money than Harry.
11. Angle Q is the largest angle in $\triangle PQR$. Then the largest side is $\overline{PQ}$.
12. If $k > m$ and $m < t$, then $k > t$.
13. If $x > 0$ and $y > 0$, then $xy < 0$.
14. If $x < y$ and $z < 0$, then $xz < yz$.
15. In a circle or in congruent circles, if two central angles are not congruent, the greater central angle has the greater major arc.
16. $x < y \leftrightarrow y > x$.
17. The difference between the lengths of two sides of a triangle is less than the length of the third side.
18. The perimeter of a quadrilateral is less than the sum of its diagonals.
19. If a triangle is not isosceles, then a median to any side is greater than the altitude to that side.

20. The diagonals of a rhombus that is not a square are unequal.

Test 3

EXERCISES

1. Supply the reasons for the statements in the following proof:

Given: $\triangle ABC$ with $\overline{CD}$ bisecting $\angle ACB$.
Prove: $AC > AD$.

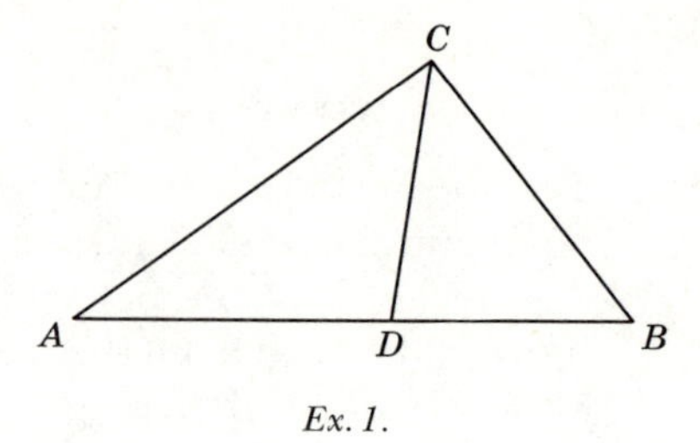

Ex. 1.

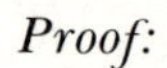
Proof:

STATEMENTS	REASONS
1. $m\angle ACD = m\angle BCD$.	1.
2. $m\angle ADC > m\angle BCD$; $m\angle BDC > m\angle ACD$.	2.
3. $m\angle ADC > m\angle ACD$.	3.
4. $AC > AD$.	4.

2. *Given:* $PR = PT$.
Prove: $m\angle PRS > m\angle S$.

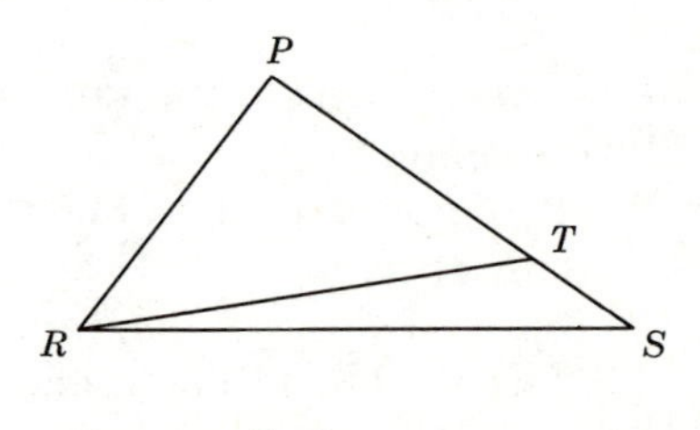

Ex. 2.

3. Prove that the shortest chord through a point within a circle is perpendicular to the radius drawn through that point.

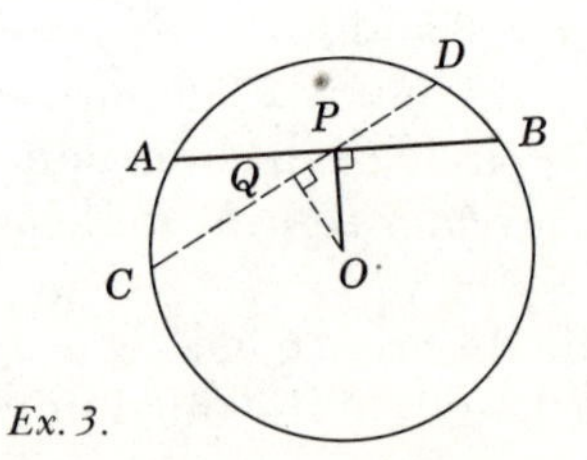

Ex. 3.

10

Geometric Constructions

10.1. Drawing and constructing. In the previous chapters you have been drawing lines with a ruler and measuring angles with the protractor. Mathematicians make a distinction between *drawing* and *constructing* geometric figures. Many instruments are used in drawing. The design engineer and the draftsman in drawing blueprints for airplanes, automobiles, machine parts, and buildings use rulers, compasses, T-squares, parallel rulers, and drafting machines. Any or all of these can be used in drawing geometric figures (Fig. 10.1).

When constructions are made, the only instruments permitted are the straight edge (an unmarked ruler) and a compass. The straight edge is used for constructing straight lines and the compass is used for drawing circles or arcs of circles. It is important that the student distinguish between drawing and constructing. When the student is told to construct a figure, he must not measure the size of angles with a protractor or the length of lines with a ruler. He may use only the compass and the straight edge.

If we are told to construct the bisector of an angle, the method used must be such that we can prove that the figure we have made bisects the given angle.

10.2. Why use only compass and straight edge? The restriction to the use of only a compass and straight edge on the geometry student was first established by the Greeks. It was motivated by their desire to keep geometry simple and aesthetically appealing. To them the introduction of additional instruments would have destroyed the value of geometry as an intellectual exercise. This introduction was considered unworthy of a thinker. The Greeks were not interested in the practical applications of their constructions. They were

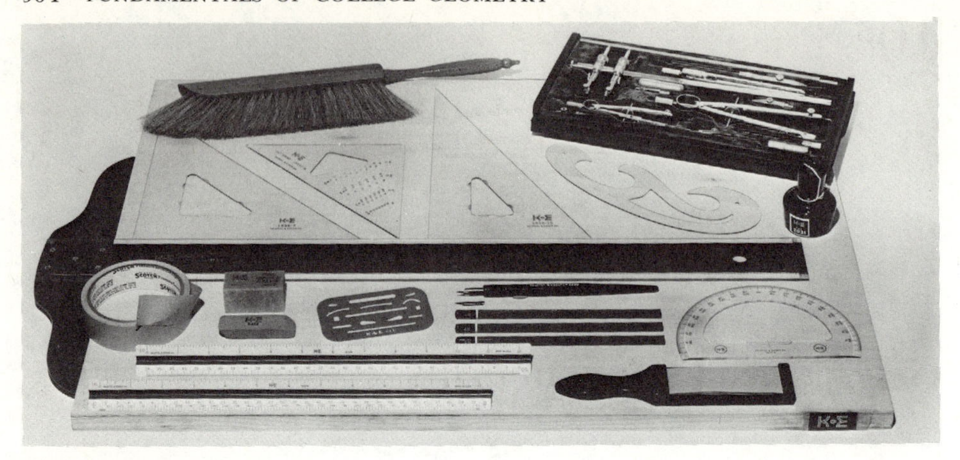

Fig. 10.1.

fascinated by exploring the many constructions possible with the use of only the instruments to which they had limited themselves.

The constructions which we will consider in this chapter should serve objectives similar to those set by the early Greeks. We, too, will restrict ourselves to the use of only the straight edge and the compass.

10.3. Solution of a construction problem. Every construction problem can be solved by steps as follows:

Step I: *A statement of the problem which tells what is to be constructed.*
Step II: *A figure representing the given parts.*
Step III: *A statement of what is given in the representation of Step II.*
Step IV: *A statement of what is to be constructed, that is, the ultimate result to be obtained.*
Step V: *The construction, with a description of each step. An authority for each step in the construction must be given.*
Step VI: *A proof that the construction in Step V gives the desired results.*

Most constructions will involve the intersection properties of two lines, of a line and a circle, or of two circles. In the developments of our constructions, we will assume the following:

1. A straight line can be constructed through any two given points (Postulate 2).

2. It is possible to construct a circle in a plane with a given point P as center and a given segment AB as radius [see Fig. 10.2 (Postulate 19)].

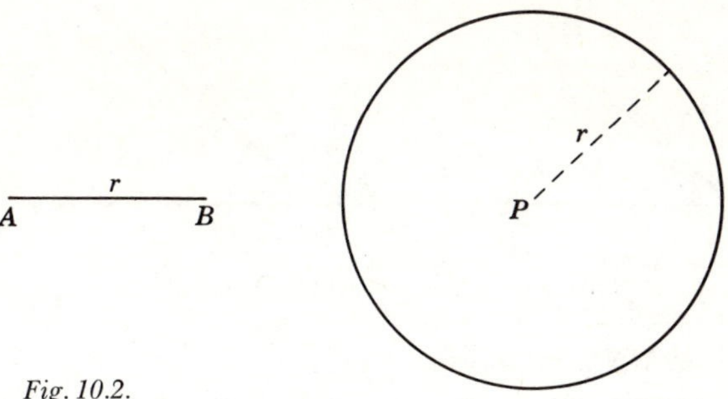

Fig. 10.2.

3. Two coplanar nonparallel lines intersect in a point (Theorem 3.1).
4. Two circles O and P with radii a and b intersect in exactly two points if the distance c between their centers is less than the sum of their radii but greater than the difference of their radii. The intersection points will lie in different half-planes formed by the line of centers (Fig. 10.3).
5. A line and a circle intersect in exactly two points if the line contains a point inside the circle.

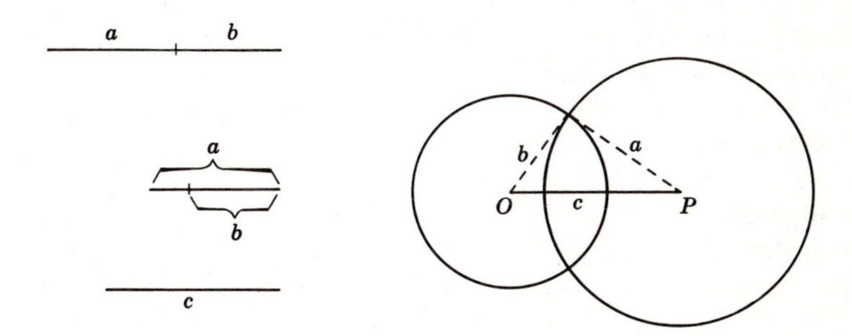

Fig. 10.3. $a+b > c$. $a-b < c$.

The student will find the solution to a construction problem easier to follow if, in the solution, he is able to distinguish three different kinds of lines. We will employ the following distinguishing lines:

(a) *Given lines,* drawn as heavy full black lines.
(b) *Construction lines,* drawn as light (but distinct) lines.
(c) *Lines sought for in the problems,* drawn as heavy dash lines.

Construction 1

10.4. At a point on a line construct an angle congruent to a given angle.

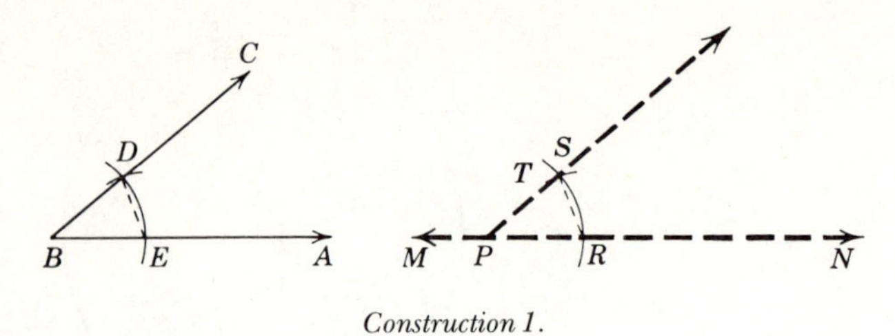

Construction 1.

Given: $\angle ABC$, line MN, and a point P on $\overleftrightarrow{MN}$.
To construct: An angle congruent to $\angle ABC$ having P as vertex and $\overrightarrow{PN}$ as one side.

Construction:

STATEMENTS	REASONS
1. With B as center and any radius, construct an arc intersecting $\overrightarrow{BA}$ at E and $\overrightarrow{BC}$ at D.	1. Postulate 19.
2. With P as center and radius $= BD$, construct $\widehat{RT}$ intersecting $\overleftrightarrow{MN}$ at R.	2. Postulate 19.
3. With R as center and a radius $= DE$, construct an arc intersecting $\widehat{RT}$ at S.	3. Postulate 19.
4. Construct $\overrightarrow{PS}$.	4. Postulate 2.
5. $\angle RPS \cong \angle ABC$.	

Proof:

STATEMENTS	REASONS
1. Draw $\overline{ED}$ and $\overline{RS}$.	1. Postulate 2.
2. $BE = PR$; $BD = PS$.	2. § 7.3.
3. $ED = RS$.	3. § 7.3.
4. $\triangle RPS \cong \triangle EBD$.	4. S.S.S.
5. $\angle RPS \cong \angle ABC$.	5. § 4.28.

Construction 2

10.5. To construct the bisector of an angle.

Given: $\angle ABC$.
To construct: The bisector of $\angle ABC$.

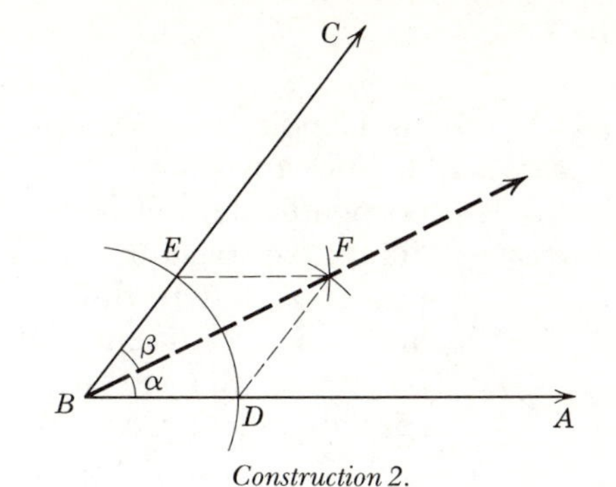

Construction 2.

Construction:

STATEMENTS	REASONS
1. With B as center and any radius, construct an arc intersecting $\overrightarrow{BA}$ at D and $\overrightarrow{BC}$ at E.	1. Postulate 19.
2. With D and E as centers and any radius greater than one-half the distance from D to E, construct arcs intersecting at F.	2. Postulate 19.
3. Construct $\overrightarrow{BF}$.	3. Postulate 2.
4. $\overrightarrow{BF}$ is the bisector of $\angle ABC$.	

Proof:

STATEMENTS	REASONS
1. Draw $\overline{DF}$ and $\overline{EF}$.	1. Postulate 2.
2. $BD = BE$; $DF = EF$.	2. § 7.3.
3. $BF = BF$.	3. Theorem 4.1.
4. $\triangle DBF \cong \triangle EBF$.	4. S.S.S.
5. $\angle\alpha \cong \angle\beta$.	5. § 4.28.
6. $\therefore \overrightarrow{BF}$ bisects $\angle ABC$.	6. § 1.19.

Construction 3

10.6. To construct a perpendicular to a line passing through a point on the line.

Given: Line l and a point P of the line.
To construct: A line containing P and perpendicular to l.
Construction: (The construction and proof are left to the student. The student will recognize this to be a special case of Construction 2.)

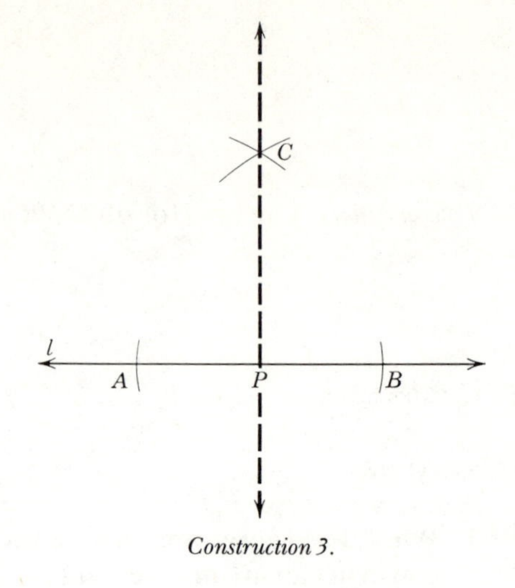

Construction 3.

Exercises

1. Draw an obtuse angle. Then, with a given ray as one side, construct an angle congruent to the obtuse angle.
2. Draw two acute angles. Then construct a third angle whose measure is equal to the sum of the measures of the two given angles.
3. Draw a scalene triangle. Then construct three adjacent angles whose measures equal respectively the measures of the angles of the given triangle. Do the three adjacent angles form a straight angle?

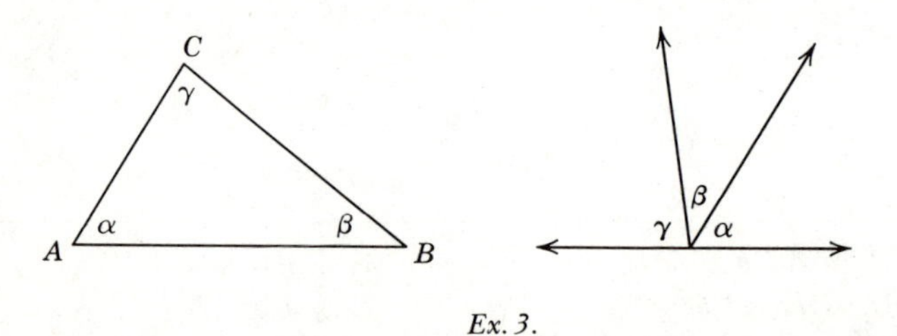

Ex. 3.

4. Draw a quadrilateral. Then construct an angle whose measure is equal to the sum of the measures of the four angles of the given quadrilateral.
5. Draw two angles. Then construct an angle whose measure is the difference of the measures of the given angles. Label the new angle $\angle\alpha$.
6. Draw an obtuse angle. Then construct the bisector of the given angle. Label the bisector $\overrightarrow{RS}$.
7. Construct an angle whose measure is (*a*) 45, (*b*) 135, (*c*) $67\frac{1}{2}$.
8. Draw an acute triangle. Construct the bisectors of the three angles of the acute triangle. What appears to be true of the three angle bisectors?

9. Repeat Ex. 8 with a given right triangle.
10. Repeat Ex. 8 with a given obtuse triangle.
11. Draw a vertical line. At a point on this line construct a perpendicular to the line.
12. Using a protractor, draw $\angle ABC$ whose measure is 45. At any point P on side $\overrightarrow{BA}$ construct a perpendicular to $\overleftrightarrow{AB}$. Label R the point where this perpendicular intersects side BC. At R construct a line perpendicular to $\overleftrightarrow{BC}$. Label S the point where the second perpendicular intersects $\overleftrightarrow{AB}$. What kind of triangle is $\triangle SPR$?

Construction 4

10.7. To construct the perpendicular bisector of a given line segment.

Given: Line segment AB.
To construct: The perpendicular bisector of $\overline{AB}$.

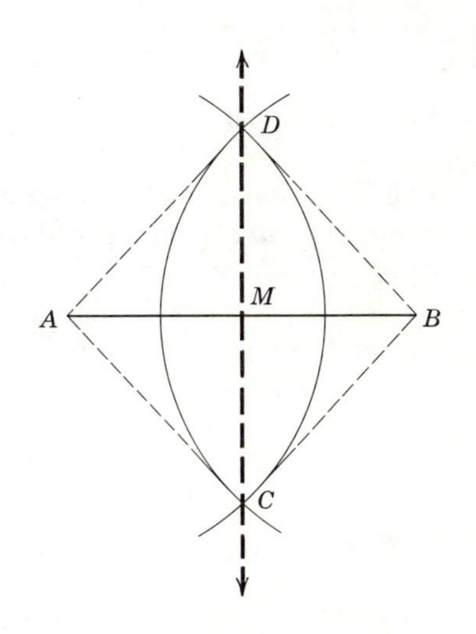

Construction 4.

Construction:

STATEMENTS	REASONS
1. With A and B as centers and with a radius greater than one-half AB, construct arcs intersecting at C and D.	1. Postulate 19.
2. Construct $\overleftrightarrow{CD}$ intersecting $\overleftrightarrow{AB}$ at M.	2. Postulate 2.
3. $\overleftrightarrow{CD}$ is the perpendicular bisector of $\overline{AB}$.	

Proof:

STATEMENTS	REASONS
1. Draw $\overline{AC}$, $\overline{AD}$, $\overline{BC}$, $\overline{BD}$.	1. Postulate 2.
2. $AC = BC$; $AD = BD$.	2. § 7.3.
3. $CD = CD$.	3. Theorem 4.1.
4. $\triangle ACD \cong \triangle BCD$.	4. S.S.S.
5. $\angle ADM \cong \angle BDM$.	5. § 4.28.
6. $DM = DM$.	6. Theorem 4.1.
7. $\triangle ADM \cong \triangle BDM$.	7. S.A.S.
8. $AM = BM$.	8. § 4.28.
9. $\angle AMD \cong \angle BMD$.	9. § 4.28.
10. $\therefore \overleftrightarrow{CD}$ is $\perp$ bisector $\overline{AB}$.	10. Definition of $\perp$ bisector.

Construction 5

10.8. To construct a perpendicular to a given line from a point not on the line.

Given: Line l and point P not on l.
To construct: A line $\perp$ to l from P.

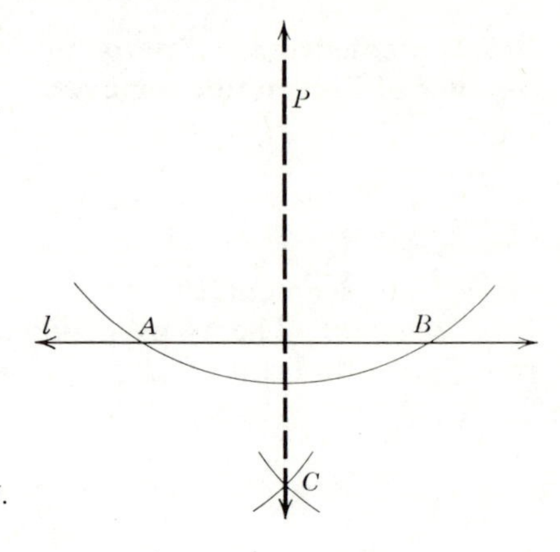

Construction: *Construction 5.*

STATEMENTS	REASONS
1. With P as center and any convenient radius, construct an arc intersecting l at A and B.	1. Postulate 19.
2. With A and B as centers and a radius whose measure is greater than the measure of one-half segment AB, construct arcs intersecting at C.	2. Postulate 19.
3. Construct $\overleftrightarrow{PC}$.	3. Postulate 2.
4. $\overleftrightarrow{PC}$ is $\perp$ to line l.	

Proof: (The proof is left to the student.) *Hint:* See proof of Construction 4.
Discussion: Point C can be either on the same side of l as is P or on the opposite side.

Exercises

1. Draw a line segment. By construction, divide the segment into four congruent segments.
2. Draw an acute scalene triangle. Construct the three altitudes of the triangle.
3. Repeat Ex. 2 for an obtuse scalene triangle.
4. Draw an acute scalene triangle. Construct the perpendicular bisectors of the three sides of the triangle.
5. Repeat Ex. 4 for an obtuse scalene triangle.
6. Draw an acute scalene triangle. Construct the three medians of the triangle.
7. Repeat Ex. 6 for an obtuse scalene triangle.
8. Construct a square.
9. Construct an equilateral triangle ABC. From C, construct the angle bisector, altitude, and median. Are these separate segments? If not, which are the same?

Construction 6

10.9. Through a given point to construct a line parallel to a given line.

Given: Line l and point P not on the line.
To construct: A line through $P \parallel l$.

Construction:

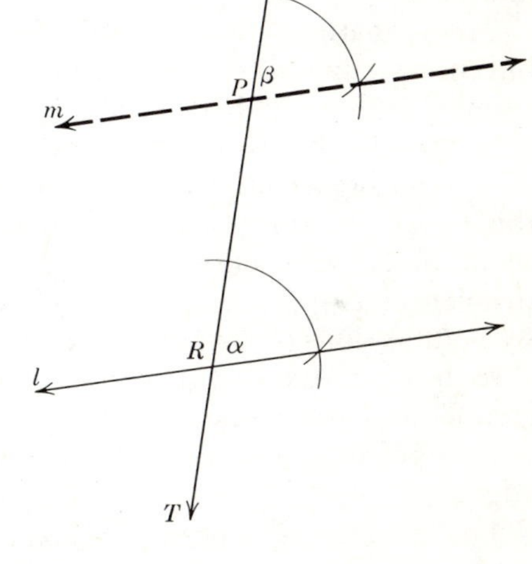

Construction 6.

STATEMENTS	REASONS
1. Through P construct any line ST intersecting l at R. Label $\overleftrightarrow{ST}$ so that P is between S and T.	1. Postulate 2.
2. With P as vertex and $\overrightarrow{PS}$ as a side construct $\angle\beta \cong \angle\alpha$.	2. § 10.4.
3. $m \parallel l$.	

Proof:

STATEMENTS	REASONS
1. $\angle\alpha \cong \angle\beta$.	1. By construction.
2. $m \parallel l$.	2. Theorem 5.12.

10.10. Impossible constructions. Many geometric constructions are impossible if only the unmarked straight edge and compass are used. Among these impossible constructions are three famous ones. These three construction problems were very popular in Greece. Greek mathematicians spent long years of labor in attempting to solve these problems.

These problems are called "trisecting the angle," "squaring the circle," and "doubling the cube." The first problem required the dividing of any angle into three congruent parts. The second required the construction of a square the area of which was equal to that of a given circle. The third problem required constructing a cube the volume of which doubled that of a given cube.

Greek mathematicians made repeated efforts to solve these problems, but none succeeded. Mathematicians for the past 2000 years have persistently attempted theoretical solutions without success.

In the past half century it has been proved that these three constructions never can be accomplished. The proof of this fact is beyond the scope of this book. In spite of this fact, many people are still challenged to attempt theoretical constructions. Occasionally, a suggested solution to one of the problems has appeared, but in each case the solution involved the introduction of modifications of the two instruments permitted.

Each of these constructions can easily be performed by using more complicated instruments. For instance, the angle can be trisected if we are permitted to make just two marks at any two points on the straight edge.

The reader may question the value of the rigor and persistence of these mathematicians. However, it can be shown that the search for solutions to such "impractical" problems has led to a deeper insight and understanding of mathematical concepts as well as to the advanced stage of mathematical science existing today.

Exercises

1. Construct a line passing through a point parallel to a given line by constructing a pair of congruent alternate interior angles.

2. Perform Construction 6 by using Theorem 5.5.
3. Draw a scalene $\triangle ABC$. Through C construct a line $\parallel \overleftrightarrow{AB}$.
4. Draw a scalene $\triangle ABC$. Bisect side AB. Label the point of bisection M. Through M construct a line $\parallel \overleftrightarrow{BC}$.
5. Construct a quadrilateral the opposite sides of which are parallel.
6. Draw $\triangle ABC$. Through each vertex construct a line parallel to the opposite side.
7. Construct a quadrilateral with two sides both congruent and parallel.
8. Inscribe a square in a circle.
9. Circumscribe a square about a circle.
10. Inscribe a regular octagon in a circle.
11. Inscribe a regular hexagon in a circle. (*Hint*: The measure of the central ∠s drawn to the vertices of an inscribed hexagon is equal to 60.)
12. Construct a right $\triangle ABC$ with $m\angle B = 90$. Bisect $\overline{AB}$ at M. Bisect $\overline{BC}$ at N. Through M construct a line $l \parallel$ to $\overleftrightarrow{BC}$. Through N construct a line $m \parallel \overleftrightarrow{AB}$. Where do lines l and m appear to intersect?

Construction 7

10.11. Divide a segment into a given number of congruent segments.

Given: Segment AB.
To construct: Divide $\overline{AB}$ into n congruent segments. (In this figure, we show the case $n = 5$.)

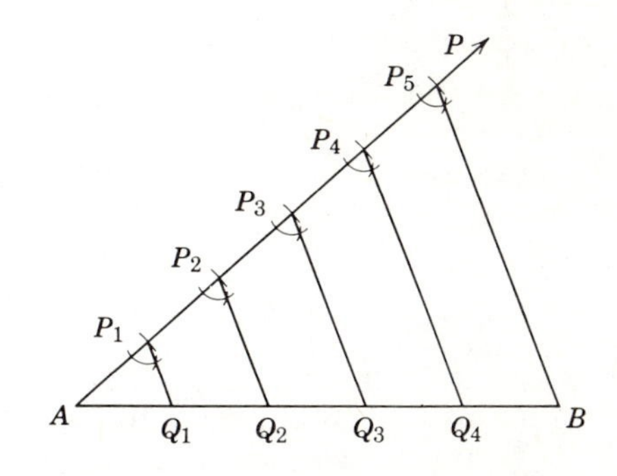

Construction 7.

Construction:

STATEMENTS	REASONS
1. Construct any ray AP not on line AB.	1. Construction.
2. Starting at A, lay off congruent segments $\overline{AP_1}$, $\overline{P_1P_2}$, $\overline{P_2P_3}, \ldots,$ $\overline{P_{n-1}P_n}$ (any length, as long as the segments are of the same length).	2. Construction.
3. Construct $\overleftrightarrow{P_nB}$.	3. Postulate 2.
4. Through points $P_1, P_2, P_3 \ldots, P_{n-1}$ construct lines parallel to $\overleftrightarrow{P_nB}$, intersecting $\overline{AB}$ in points Q_1, Q_2, $Q_3, \ldots, Q_{n-1}$.	4. Construction 6.

5. The points $Q_1, Q_2, Q_3, \ldots, Q_{n-1}$ divide $\overline{AB}$ into n congruent segments.

Proof:

STATEMENTS	REASONS
1. $AP_1 = P_1P_2 = P_2P_3 = \ldots = P_{n-1}P_n$.	1. By construction.
2. $P_1Q_1 \parallel P_2Q_2 \parallel P_3Q_3 \parallel \ldots \parallel P_nB$.	2. By construction.
3. $AQ_1 = Q_1Q_2 = Q_2Q_3 = \ldots = Q_{n-1}B$.	3. Theorem 6.8.

Construction 8

10.12. Circumscribe a circle about a triangle.

Given: $\triangle ABC$.
To construct: Circumscribe a $\odot$ about $\triangle ABC$.

Construction:

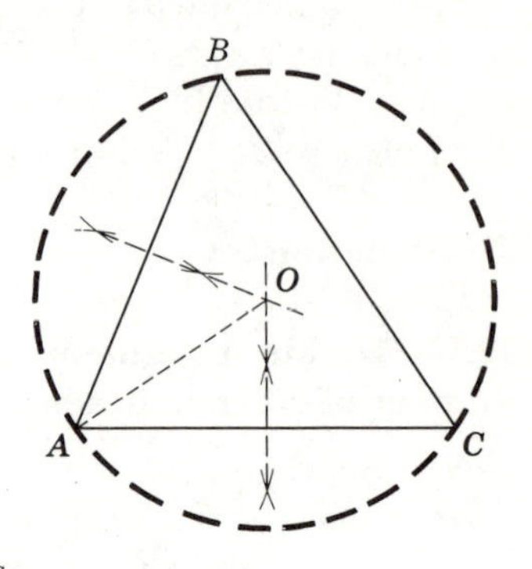

Construction 8.

STATEMENTS	REASONS
1. Construct the perpendicular bisectors of two sides of the $\triangle$.	1. Construction 4.
2. The two lines meet at a point O.	2. Theorem 3.1.
3. With O as the center and $m\overline{OA}$ as the radius construct the circle O.	3. Postulate 19.
4. $\odot O$ is the circumscribed circle.	

(The proof of this construction is left as an exercise.)

Construction 9

10.13. To inscribe a circle in a given triangle.
Given: $\triangle ABC$.
To construct: Inscribe a circle in $\triangle ABC$.

Construction 9.

Construction:

STATEMENTS	REASONS
1. Bisect two angles of $\triangle ABC$.	1. Construction 2.
2. Let O be the point of intersection of the bisectors.	2. Theorem 3.1.
3. Construct $\overline{OD}$ from O perpendicular to $\overline{AC}$.	3. Construction 5.
4. With O as center and $m\overline{OD}$ as as radius, construct $\odot O$.	4. Postulate 19.
5. $\odot O$ is the inscribed $\odot$.	

(The proof of this construction is left as an exercise.)

Exercises (B)

1. Draw two points 3 inches apart. Locate by construction all the points 2 inches from each of these given points.
2. Draw two lines intersecting at 45° and two other parallel lines which are 1 inch apart. Locate by construction all the points equidistant from the intersecting lines and equidistant from the parallel lines.
3. Draw two lines l_1 and l_2 intersecting at a 60° angle. Locate by construction all the points that are 1 inch from l_1 and l_2.
4. Draw a circle O with a radius length of 2 inches. Draw a diameter AB. Locate by construction the points that are 1 inch from the diameter AB and equidistant from A and O.
5. Draw a triangle ABC with measures of the sides equal to 2 inches, $2\frac{1}{2}$ inches, and 3 inches. Locate by construction the points on the altitude from B that are equidistant from B and C.
6. In the triangle of Ex. 5, locate by construction all the points on the altitude from C that are equidistant from sides AB and BC.

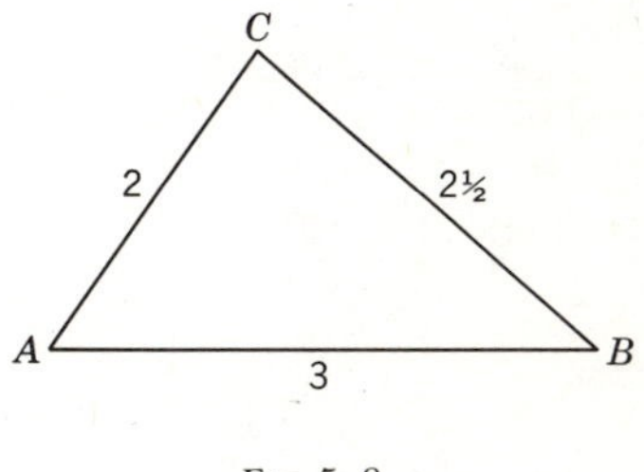

Exs. 5–8.

7. In the triangle of Ex. 5, locate by construction all the points on the median from C that are equidistant from A and C.
8. In the triangle of Ex. 5, locate by construction all the points equidistant from sides AC and BC at a distance of $\frac{1}{2}$ inch from side AB.
9. Draw a triangle ABC. Locate by construction the point P that is equidistant from the vertices of the triangle. With P as center and $m\overline{PB}$ as

radius, construct a circle. (This circle is said to be *circumscribed* about the triangle.)

10. Draw a triangle ABC. Locate by construction the point P equidistant from the sides of the triangle. Construct the segment PM from P perpendicular to $\overleftrightarrow{AB}$. With P as center and $m\overline{PM}$ as radius, construct a circle. (This circle is said to be *inscribed* in the triangle.)

Summary Test

Constructions Test

1–4. With ruler and protractor draw $AB = 3$ inches and $\angle\alpha$ whose measure is 40.

1. Construct an isosceles triangle with base equaling AB and base angle with measure equaling $m\angle\alpha$.
2. Construct an isosceles triangle with leg equaling AB and vertex angle with measure equaling $m\angle\alpha$.
3. Construct an isosceles triangle with altitude to the base equaling AB and vertex angle with measure equaling $m\angle\alpha$.
4. Construct an isosceles triangle with base equaling AB and vertex angle with measure equaling $m\angle\alpha$.

5–9. With ruler and protractor draw segments $AB = 2$ inches, $CD = 3$ inches, $EF = 4$ inches, and $\angle\alpha$ whose measure is 40.

5. Construct $\square PQRS$ with $PS = AB$, $PQ = CD$ and $PR = EF$.
6. Construct $\square PQRS$ with $PS = AB$, $PR = EF$, and $SQ = CD$.
7. Construct $\square PQRS$ with $PS = AB$, $PR = CD$, and $\angle SPR \cong \angle\alpha$.
8. Construct $\square PQRS$ with $PQ = AB$, $PR = EF$, and altitude on $\overleftrightarrow{PQ} = CD$.
9. Construct $\square PQRS$ with $PQ = CD$, $\angle SPQ \cong \angle\alpha$, and altitude on $\overleftrightarrow{PQ} = AB$.
10. Construct an angle whose measure is 75.
11. Draw a line 5 inches long. Then divide it into five congruent segments using only a compass and straight edge.

12. Draw any $\triangle ABC$. Let P be a point outside the triangle. From P construct perpendiculars to the three sides of $\triangle ABC$.
13. Draw an obtuse triangle. Then construct a circle which circumscribes the triangle.
14. Draw a triangle. Then construct a circle which is inscribed in the triangle.
15. Draw an obtuse triangle. Then construct the three altitudes of the triangle.
16. Draw a triangle. Then construct the three medians of the triangle.

11

Geometric Loci

11.1. Loci and sets. The set of all points is *space*. A *geometric figure* is a set of points governed by one or more limiting geometric conditions. Thus, a geometric figure is a subset of space.

In Chapter 7 we defined a circle as a set of points lying in a plane which are equidistant from a fixed point of the plane.

Mathematicians sometimes use the term "locus" to describe a geometric figure.

Definition: A *locus of points* is the set of all the points, and only those points, which satisfy one or more given conditions.

Thus, instead of using the words "the set of points P such that . . . ," we could say "the locus of points P such that . . . ." A circle can be defined as the "locus of points lying in a plane at a given distance from a fixed point of the plane."

Sometimes one will find the locus defined as the path of a point moving according to some given condition or set of conditions.

Consider the path of the hub of a wheel that moves along a level road (Fig. 11.1). A, B, C, D represent positions of the center of the wheel at different instants during the motion of the wheel. It should be evident to the reader that, as the wheel rolls along the road, the set of points which represent the positions of the center of the hub are elements of a line parallel to the road and at a distance from the road equal to the radius of the wheel. We speak of this line as "the locus of the center of the hub of the wheel as the wheel moves along the track." In this text locus lines will be drawn with

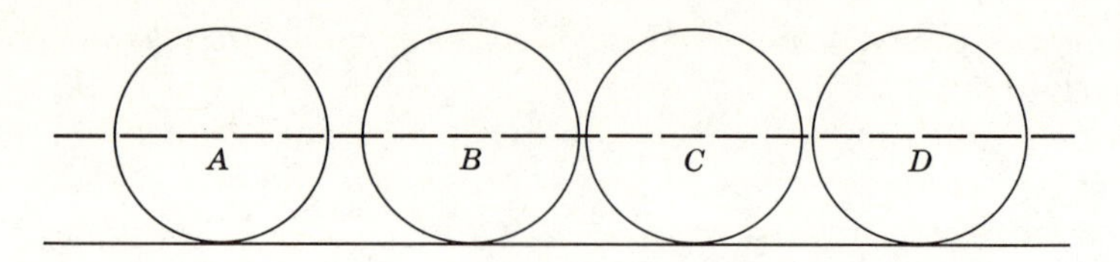

Fig. 11.1.

long dash lines to distinguish them from given and construction lines.

As a second simple illustration of a locus, consider the problem of finding the locus of points in a plane 2 inches from a given point O (Fig. 11.2). Let us first locate several points, such as P_1, P_2, P_3, $P_4, \ldots,$ which are 2 inches from O. Obviously there are an infinite number of such points. Next draw a smooth curve through these points. In this case it appears that the locus is a circle with the center at O and a radius whose measure is 2 inches.

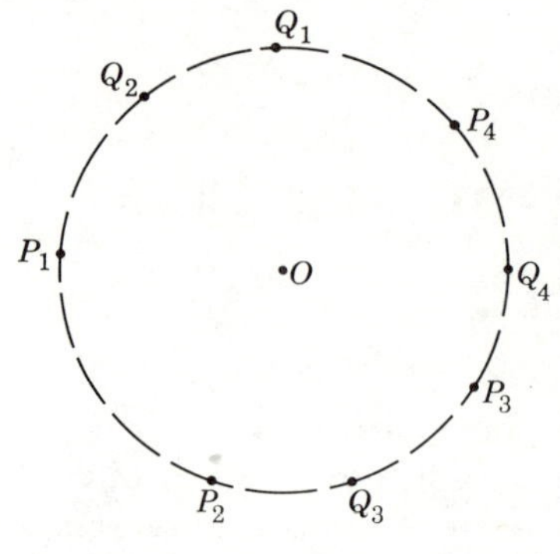

Fig. 11.2.

If now, conversely, we select points such as Q_1, Q_2, Q_3, $Q_4, \ldots,$ each one of which meets the requirement of being 2 inches from O, it is evident that they too will lie on this same circle.

Thus, to prove that a line is a locus, it is necessary to prove the following two characteristics:

1. *Any point on the line satisfies the given condition or set of conditions.*
2. Either (*a*) *any point that satisfies the given condition or set of conditions is on the line* or (*b*) *any point not on the line does not satisfy the condition.*

The word *locus* (plural *loci,* pronounced "lo′-si") is the Latin word meaning "place" or "location." A locus may consist of one or more points, lines, surfaces, or combinations of these.

11.2. Determining a locus. Let us use the example of Fig. 11.2 to illustrate the general method of determining a locus.

Step I: *Locate several points which satisfy the given conditions.*
Step II: *Draw a smooth line or lines (straight or curved) through these points.*
Step III: *Form a conclusion as to the locus, and describe accurately the geometric figure which represents your conclusion.*

Step IV: *Prove your conclusion by proving that the figure meets the two characteristics of loci listed in* § 11.1.

One of the difficulties encountered by the student of geometry is that of describing the geometric figure which represents the locus. These descriptions must be precise and accurate.

11.3. Illustrative Example 1. What is the locus of the center of a circle with radius R_2 that rolls around the outside of a second circle the radius of which is R_1?

Conclusion: The locus is a circle the center of which is the same as that of the second circle and the radius measure of which equals the sum of the measures of radii R_1 and R_2.

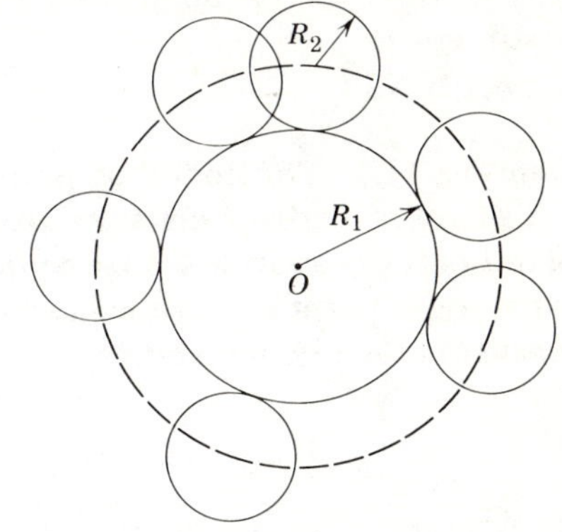

Illustrative Example 1.

Exercises

Using the method outlined in § 11.2, describe the locus for each of the following exercises. No proof is required. Consider only 2-dimensional geometry in these exercises.

1. The locus of points equidistant from two parallel lines l_1 and l_2.
2. The locus of points which are $\frac{3}{8}$ inch from a fixed point P.
3. The locus of points $\frac{3}{4}$ inch from a given straight line l.
4. The locus of the midpoints of the radii of a given circle O.
5. The locus of points equidistant from sides $\overrightarrow{BA}$ and $\overrightarrow{BC}$ of $\angle ABC$.
6. The locus of points equidistant from two fixed points A and B.
7. The locus of points one inch from a circle with center at O and radius equal to 4 inches.
8. The locus of points less than 3 inches from a fixed point P.
9. The locus of the center of a marble as it rolls on a plane surface.
10. The locus of points equidistant from two intersecting straight lines l_1, and l_2.
11. The locus of the midpoints of chords parallel to a diameter of a given circle.
12. The locus of the midpoints of all chords with a given measure of a given circle.
13. The locus of the points equidistant from the ends of a 3-inch chord drawn in a circle with center at O and a radius measure of 2 inches.
14. The locus of the centers of circles that are tangent to a given line at a given point.

15. The locus of the vertex of a right triangle with a given fixed hypotenuse as base.

11.4. Fundamental locus theorems. The following three theorems can easily be discovered and proved by the student. The proofs will be left to the student.

Theorem 11.1. The locus of points in a plane at a given distance from a fixed point is a circle whose center is the given point and whose radius measure is the given distance.

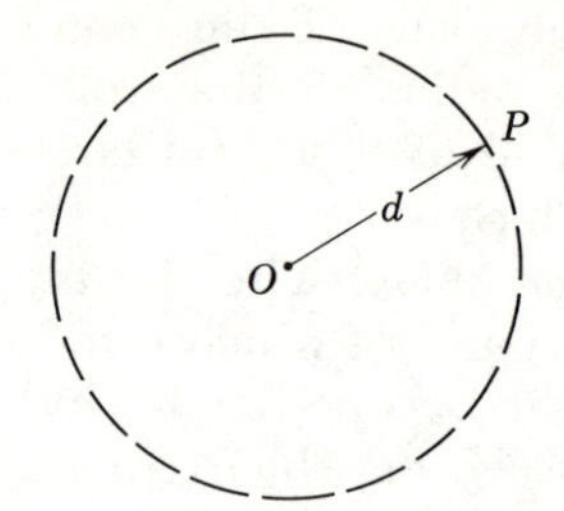

Theorem 11.1.

Theorem 11.2. The locus of points in a plane at a given distance from a given line in the plane is a pair of lines in the plane and parallel to the given line and at the given distance from the given line.

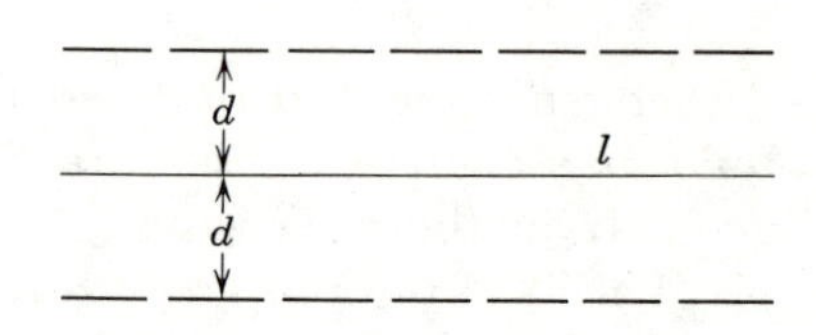

Theorem 11.2.

Theorem 11.3. The locus of points in a plane equidistant from two given parallel lines is a line parallel to the given lines and midway between them.

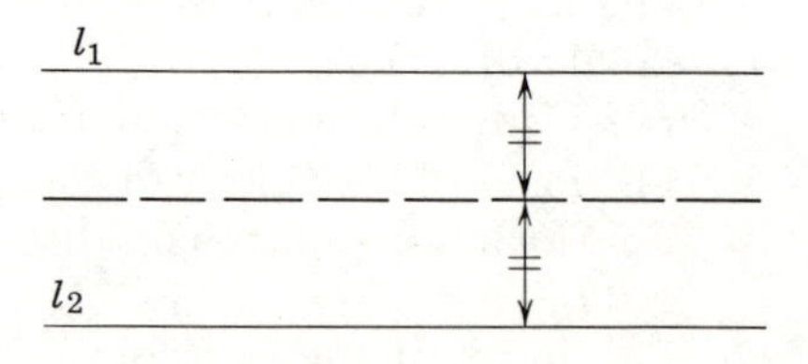

Theorem 11.3.

Theorem 11.4

11.5. The locus of points in a plane which are equidistant from two given points in the plane is the perpendicular bisector of the line segment joining the two points.

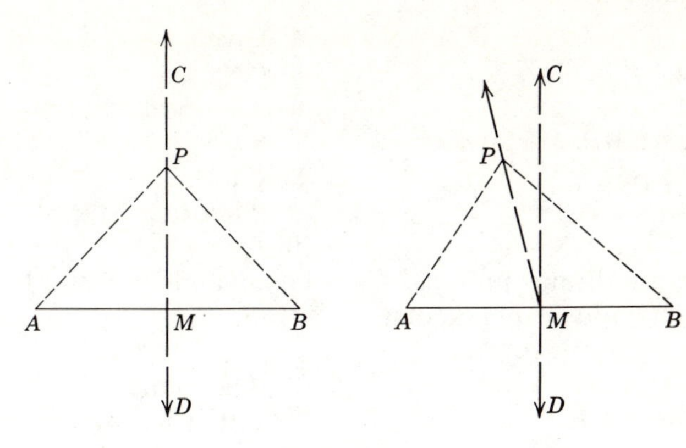

Theorem 11.4.

PART I: *Any point on the perpendicular bisector of the line segment joining two points is equidistant from the two points.*

Given: $\overleftrightarrow{CD} \perp \overline{AB}$; $AM = MB$; P is any point on $\overleftrightarrow{CD}$, $P \neq M$.
Conclusion: $AP = BP$.
Proof:

STATEMENTS	REASONS
1. $\overleftrightarrow{CD} \perp \overline{AB}$; $AM = MB$.	1. Given.
2. $\angle AMP$ and $\angle BMP$ are right $\angle$s.	2. § 1.20.
3. Draw $\overline{PA}$ and $\overline{PB}$.	3. Postulate 2.
4. $PM = PM$.	4. Theorem 4.1.
5. $\triangle AMP \cong \triangle BMP$.	5. Theorem 4.13.
6. $\therefore AP = BP$.	6. § 4.28.

PART II: *Any point equidistant from two points lies on the perpendicular bisector of the line segment joining those two points.*

Given: P any point such that $AP = BP$; $\overleftrightarrow{CM} \perp \overline{AB}$; $AM = BM$.
Conclusion: P lies on $\overleftrightarrow{CM}$.
Proof:

STATEMENTS	REASONS
1. P lies on $\overleftrightarrow{CM}$ or P does not lie on $\overleftrightarrow{CM}$.	1. Law of excluded middle.
2. Assume P does not lie on $\overleftrightarrow{CM}$.	2. Temporary assumption.
3. Draw $\overline{PM}$.	3. Postulate 2.

4. $AP = BP$.	4. Given.
5. $AM = BM$.	5. Given.
6. $PM = PM$.	6. Theorem 4.1.
7. $\triangle AMP \cong \triangle BMP$.	7. S.S.S.
8. $\angle AMP \cong \angle BMP$.	8. § 4.28.
9. $\overrightarrow{MP} \perp \overline{AB}$.	9. Theorem 3.14.
10. $\overleftrightarrow{CM} \perp \overline{AB}$.	10. Given.
11. There are two distinct lines passing through P and perpendicular to $\overline{AB}$.	11. Statements 9 and 10.
12. This is impossible.	12. Theorem 5.2.
13. $\therefore P$ must lie on $\overleftrightarrow{CM}$.	13. Either p or not-p; [not(not-p)] $\leftrightarrow$ p.

Theorem 11.5

11.6. The locus of points in the interior of an angle which are equidistant from the sides of the angle is the bisector of the angle minus its endpoint.

PART I: *Any point on the bisector of the angle is equidistant from the sides of the angle.*

Given: $\overrightarrow{BF}$ bisects $\angle ABC$; point $P \neq B$ on $\overrightarrow{BF}$; $\overline{PE} \perp \overrightarrow{BA}$; $\overline{PD} \perp \overrightarrow{BC}$.
Conclusion: $PE = PD$.

Proof:

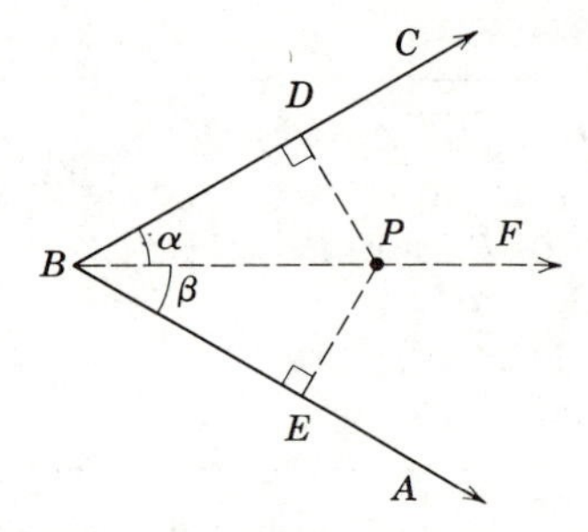

Theorem 11.5.

STATEMENTS	REASONS
1. $\overrightarrow{BF}$ bisects $\angle ABC$.	1. Given.
2. $\angle\alpha \cong \angle\beta$.	2. § 1.19.
3. $\overline{PE} \perp \overrightarrow{BA}$, $\overline{PD} \perp \overrightarrow{BC}$.	3. Given.
4. $\angle BEP$ and $\angle BDP$ are right $\angle$s.	4. § 1.20.
5. $BP = BP$.	5. Theorem 4.1.
6. Right $\triangle BEP \cong$ right $\triangle BDP$.	6. §5.27 and A.S.A.
7. $PE = PD$.	7. § 4.28.

PART II: *Any point equidistant from the sides of an angle is on the bisector of the angle.*

Given: $\overrightarrow{BF}$ bisects $\angle ABC$; $\overline{PE} \perp \overrightarrow{BA}$; $\overline{PD} \perp \overrightarrow{BC}$; $PE = PD$, $\overrightarrow{BP}$
Conclusion: P lies on $\overrightarrow{BF}$.

Proof:

STATEMENTS	REASONS
1. $\overline{PE} \perp \overrightarrow{BA}$, $\overline{PD} \perp \overrightarrow{BC}$.	1. Given.
2. $\angle BEP$ and $\angle BDP$ are right ∠s.	2. § 1.20.
3. $PE = PD$.	3. Given.
4. $BP = BP$.	4. Theorem 4.1.
5. $\triangle BEP \cong \triangle BDP$.	5. Theorem 5.20.
6. $\angle\alpha \cong \angle\beta$.	6. § 4.28.
7. $\therefore \overrightarrow{BP}$ bisects $\angle ABC$.	7. § 1.19.

11.7. Corollary: The locus of points equidistant from two given intersecting lines is the pair of perpendicular lines which bisects the vertical angles formed by the given lines.

Theorem 11.6

11.8. The locus of all points such that $\triangle APB$ is a triangle having $\overline{AB}$ a fixed line segment as hypotenuse is a circle having $\overline{AB}$ as a diameter, except for points A and B themselves.

11.9. Intersection of loci. In our study of loci thus far we have limited ourselves to finding points which satisfy only one condition. Sometimes a point must satisfy each of two or more given conditions. In such cases each condition will determine a locus. The required points will then be the intersection of the loci, since only those points will lie on each of the lines which, in turn, represent the given conditions.

Thus *to locate the point (or set of points) which satisfies two or more conditions, determine the locus for each condition. The point (or set of points) at which these loci intersect will be the required point (or set of points).*

In solving a problem involving intersecting loci, it is customary to place the given parts in the most general positions in order to determine the most general solution for the problem. Then, in a discussion that follows the general solution, consideration is given to special positions of the given parts and to the solutions for these special cases.

11.10. Illustrative Example 1. Find the set of all the points that are a given distance d from a fixed point A and which are also equidistant from two points B and C.

Given: Points $A, B,$ and C.
Find: All points a distance d from A and equidistant from B and C.

Solution:

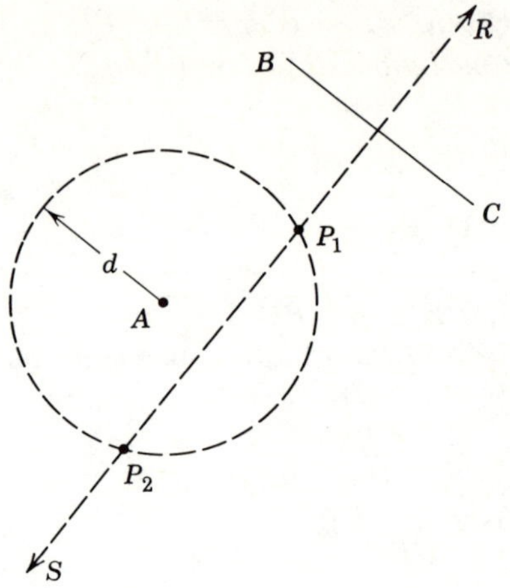

Illustrative Example 1.

STATEMENTS	REASONS
1. $\overleftrightarrow{RS}$, the $\perp$ bisector of $\overline{BC}$, is the locus of points equidistant from B and C.	1. Theorem 11.4.
2. Circle A with center at A and radius equal to d is the locus of all points a distance d from A.	2. Theorem 11.1.
3. The required points are P_1 and P_2, the intersection of $\overleftrightarrow{RS}$, and $\odot A$.	3. § 11.9.

Discussion:
1. If $\overleftrightarrow{RS}$ is a tangent to $\odot A$, the required solution set will be only one point.
2. If $\overleftrightarrow{RS}$ falls outside $\odot A$ (i.e., if distance from A to $\overleftrightarrow{RS}$ is more than d), there are no points which will satisfy the required conditions. The solution set is a null set.
3. There can never be more than two points in the solution set.

11.11. Illustrative Example 2. Find all the points that are equidistant from two fixed points and equidistant from two intersecting lines.

Given: Points A and B; lines CD and EF intersecting at O.
Find: All points equidistant from A and B and also equidistant from intersecting lines CD and EF.

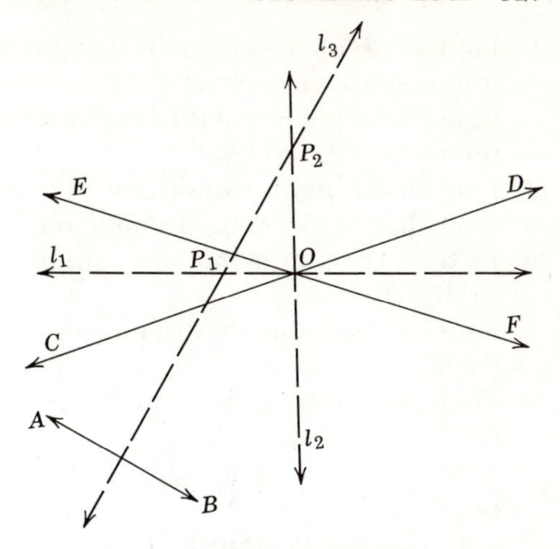

Illustrative Example 2.

Solution:

STATEMENTS	REASONS
1. The locus of points equidistant from A and B is line l_3 the $\perp$ bisector of $\overline{AB}$.	1. Theorem 11.4.
2. The loci of points equidistant from $\overleftrightarrow{CD}$ and $\overleftrightarrow{EF}$ are l_1, which bisects $\angle COE$, and l_2, which bisects $\angle EOD$.	2. § 11.7. Corollary.
3. The solution set is P_1 and P_2, the intersections of line l_3 with lines l_1 and l_2.	3. § 11.9.

Discussion:

1. If $l_3 \parallel l_2$ (or if $l_3 \parallel l_1$), the solution set will consist of only one point.
2. If l_3 coincides with either l_1 or l_2, the solution set will consist of an infinite number of points.
3. If l_3 passes through O, there is one point in the solution set.
4. In all other cases there will be two points in the solution set.

Exercises

State and prove the solutions for each of the following locus problems. Discuss each.

1. Find all points a given distance from a fixed point and equidistant from two parallel lines.

2. Find all points within a given angle equidistant from the sides and a given distance from the vertex.
3. Find all the points equidistant from two parallel lines and equidistant from the sides of an angle.
4. Find all the points equidistant from the three vertices of $\triangle ABC$.
5. Find all the points equidistant from the three sides of $\triangle ABC$.
6. Find all the points that are equidistant from two parallel lines and lie on a third line.
7. Find all the points a distance d_1 from a given line and d_2 from a given circle.
8. Find all points equidistant from two parallel lines and a given distance from a third line.
9. Find all points equidistant from two parallel lines and equidistant from two points.
10. Find all points equidistant from two intersecting lines and also at a given distance from a given point.

Additional Loci (Optional)

11.12. Loci other than straight lines and circles. Euclidean geometry confines itself to figures formed by straight lines and circles. Our loci have thus far all resolved into straight lines and circles or combinations of them. The locus of points equidistant from two straight lines, for example, is another straight line. The locus of points equidistant from two points is a straight line. The locus of points a given distance from a fixed point is a circle.

Let us briefly consider three other loci configurations. The early Greeks were familiar with these curves but were not able to relate them to our physical world. By the seventeenth century the advances of science and technology had produced a need for a clearer understanding of the properties of these curves. By that time mathematicians had developed powerful techniques of algebra and analysis which aided them in the study of these curves. We shall leave the algebraic analysis of these curves to the student when he studies analytic geometry in his future mathematical pursuits and limit ourselves in this text to a general discussion of these curves.

8.13. The parabola. As mathematicians studied loci of points equidistant from two points and loci of points equidistant from two straight lines, it was only natural that they should consider the locus of points equidistant from a point and a line. Such a locus is the parabola.

Definition: A *parabola* is the locus of the points whose distances from a fixed line and a fixed point are equal.

Thus, in Fig. 11.3, if $P_1D_1 = P_1F$, $P_2D_2 = P_2F, \ldots$, the curve is a parabola. Conversely, if the curve is a parabola, $P_3D_3 = P_3F, \ldots$. With a little study the student should discover that the shape of the parabola varies as the distance from the fixed point to the fixed line varies.

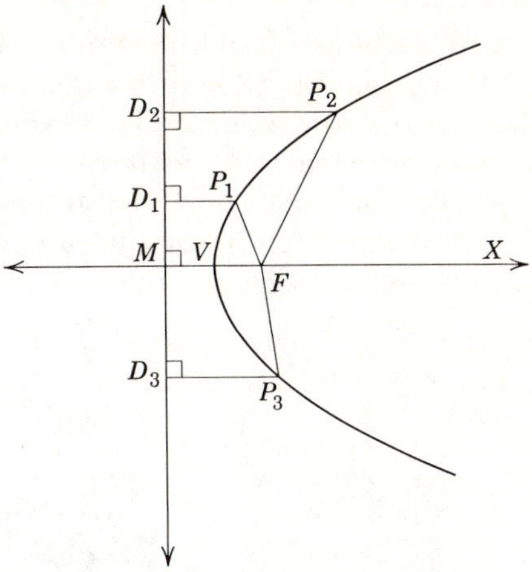

Fig. 11.3.

We know that many moving objects travel in parabolic paths. A ball thrown in the air, the projectile fired from a cannon (see Fig. 11.4), the bomb released from an airplane, a stream of water from a garden hose would all follow a parabolic path if the resistance of air could be neglected.

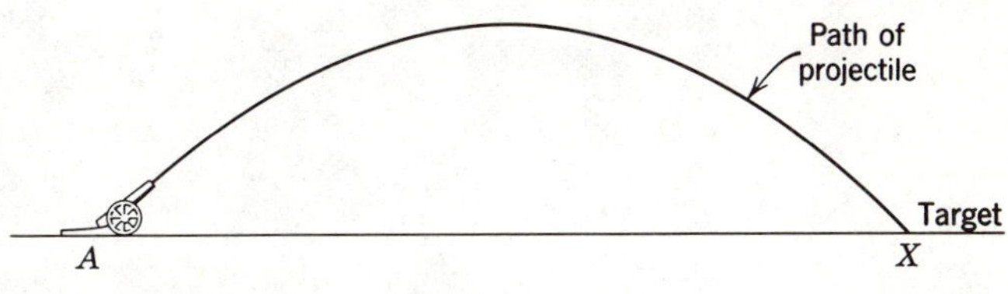

Fig. 11.4.

Thus, in firing an artillery shell, if the angle of elevation of the gun and its muzzle velocity are known, it is possible to calculate the equation of the path of flight. It is then possible to calculate in advance how far the projectile will go and how long it will take to go that distance. By varying the angle of elevation of the gun, the path of flight can be varied.

In like manner the equation of the path of the bomb released from an airplane can be determined (Fig. 11.5). From the equation, the speed of the airplane, the height of plane and the position of the target, it can be deter-

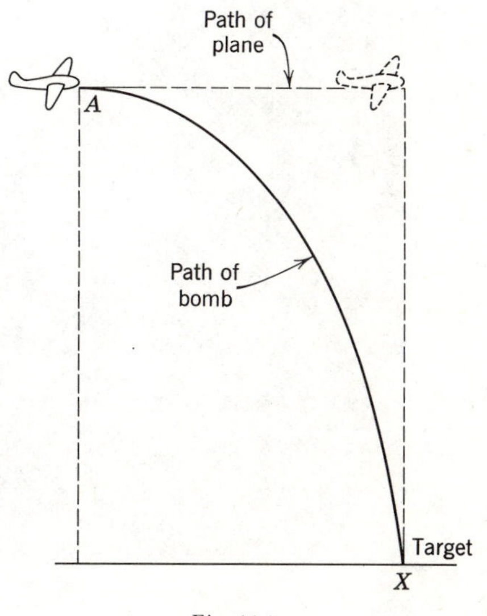

Fig. 11.5.

mined when to drop the bomb. Today the whole procedure is so mechanized that the bombardier does not need to consider the equation, nor even be aware of its existence. However, the persons responsible for the mechanized procedure had to use extensively these equations.

The parabolic curve is also useful in construction of many physical objects. A parabolic arch is often used in constructing bridges since it is stronger than any other. (See Fig. 11.6.)

Fig. 11.6.

Parabolic curves are used extensively in constructing reflectors of light, sound, and heat. In order to understand why the parabola is used, consider Fig. 11.7. Let $\overleftrightarrow{D_1D_2}$ be the fixed line and F the fixed point. Let P be any

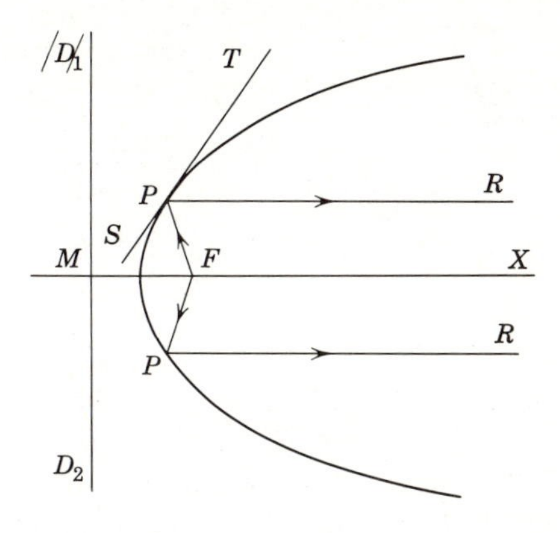

Fig. 11.7.

point on the curve. Through P draw $\overleftrightarrow{ST}$ tangent to the curve. Draw $\overline{FP}$. Draw $\overrightarrow{PR}$ so that $\angle TPR \cong \angle FPS$. It is known that, if a ray of light strikes a smooth surface at a point such as P, it will be reflected and that the receding reflected ray PR will make the same angle with the tangent as does the oncoming incident ray FP. It can also be shown in analytic geometry that, if the incident ray passes through F, the reflected ray is parallel to $\overleftrightarrow{MX}$. Because of this property, if a small source of light is placed at the focus F of a polished parabolic surface, all the rays will be reflected parallel to the line $\overleftrightarrow{MX}$ passing through the focus and perpendicular to the fixed line. $\overleftrightarrow{MX}$ is called the *axis* of the parabola. Searchlights, spotlights, headlights and radar antennae are examples of parabolic reflectors which are obtained by rotating a parabola about its axis. The surface thus formed is called a *paraboloid of revolution.* Since all the reflected rays travel in the same direction, they form a stronger beam.

Conversely, if parallel rays strike a parabolic surface, the reflected rays will converge at the focus of the parabola. This property is used in some of our reflecting telescopes. The rays coming from the heavenly bodies are very nearly parallel when traveling through the telescope. These rays are concentrated at the focus of the parabolic mirror of the telescope, thus forming a relatively bright and clear image.

Figure 11.8 illustrates how the parabolic reflector can be used to concentrate heat rays from the sun to the focal point of the reflector. This principle is used in some solar-radiant water heaters today.

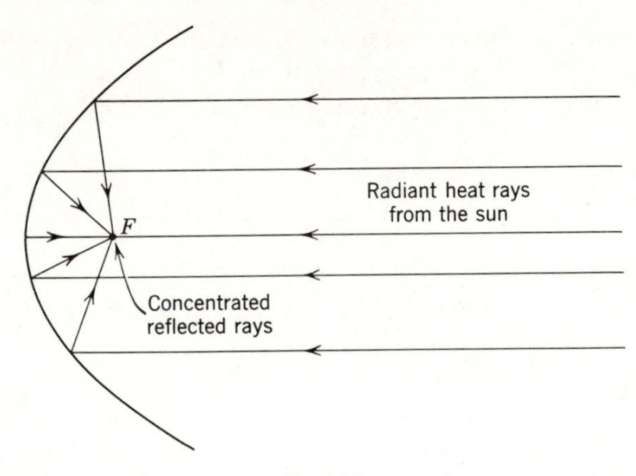

Fig. 11.8.

11.14. The ellipse. We recall that the circle is the locus of points that are a given distance from a fixed point. We shall now consider two loci associated with distances from two fixed points.

Definition: The *ellipse* is the locus of all points the sum of whose distances from two fixed points is a constant.

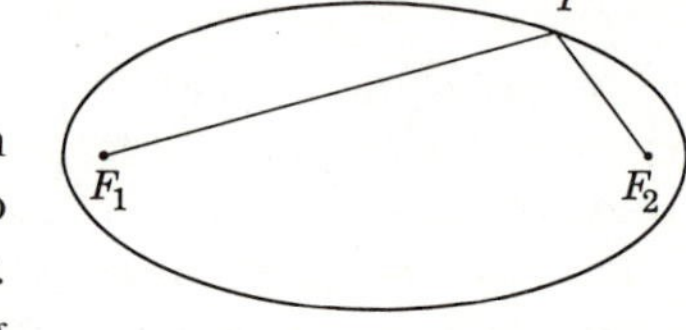

Fig. 11.9.

The fixed points are called *foci* of the ellipse.

A simple mechanical construction of an ellipse can be obtained with the aid of two thumbtacks and a piece of string (see Fig. 11.9). Place two thumbtacks in a drawing board at points F_1 and F_2. These points will be the foci of the ellipse. Then take a piece of string the length of which is larger than the distance F_1F_2. Fix the ends of the string to the tacks. Place a pencil in the loop F_1PF_2 and let it move, keeping the string taut. Then $F_1P + F_2P$ is constant and the curve traced will be an ellipse. Again, the student should discover the varying shapes of the ellipses when F_1F_2 remains fixed and the length of the string is varied.

The ellipse can be constructed by ruler and compass, but we are primarily interested in this discussion in the properties and applications of the curve.

The ellipse is frequently used for artistic effects. Flower gardens, curved walks, swimming pools, and pieces of furniture and chinaware are often seen in elliptic shapes. The elliptic arch is used in construction work where beauty is desirable and strength is not critical. The elliptic arch is considered more beautiful than the parabolic arch (Fig. 11.10).

Elliptic gears are used on machines where a slow drive and a quick return are desirable.

Fig. 11.10.

Astronomers have proved that the orbits of planets are ellipses with the sun at one focus. Our earth, for example, in its yearly journey around the sun travels in an elliptical path with the sun at one focus. Thus, at different seasons of the year, the distance from the sun to the earth will vary. In like manner, the path of the moon with respect to the earth is an ellipse with the earth at one focus. Orbits of the satellites of other planets are also elliptical. A knowledge of the paths (loci) along which these planets and their satellites move is essential in the study of astronomy. Astronomers are able to express these motions by equations and, from these equations, predict with a high degree of accuracy such things as lunar and solar eclipses.

The ellipse, like the parabola, has a geometric property which makes it useful for reflecting light and sound (see Fig. 11.11). Two lines drawn from foci F_1 and F_2 to any point P on the ellipse will make congruent angles with the tangent to the curve at P. Thus $\angle RPF_1 \cong \angle SPF_2$. Because of this geometric property and the property of reflecting surfaces mentioned in the discussion on parabolic reflectors, if a source of light is placed at either focus of an elliptical surface, the rays on striking the surface will be reflected to the other focus. Actually, when such elliptical reflectors are used, the reflecting surface is that obtained by rotating the ellipse about the line $\overleftrightarrow{F_1F_2}$. This surface is termed an *ellipsoid of revolution*.

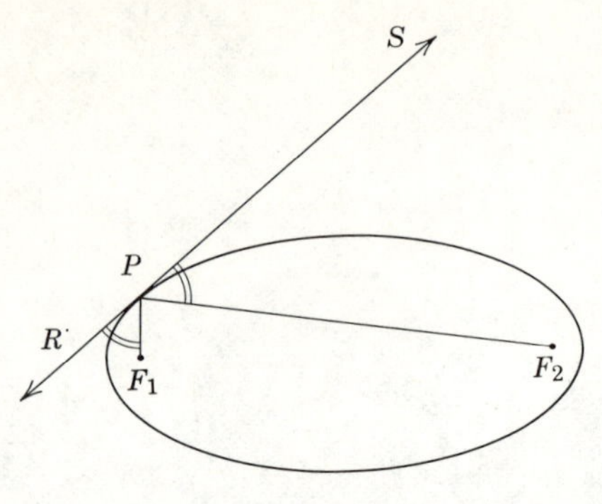

Fig. 11.11.

Sound waves are reflected in the same way as light waves. Thus, if the ceiling of a room is in the shape of half an ellipsoid (Fig. 11.12), a faint sound produced at one focus is clearly heard at the other focus which may be a considerable distance away. The sound is usually not heard at other points. Such rooms are known as *whispering galleries*. Examples of such rooms are found in the dome of the Mormon Tabernacle in Salt Lake City and in Statuary Hall in the capitol at Washington, D.C.

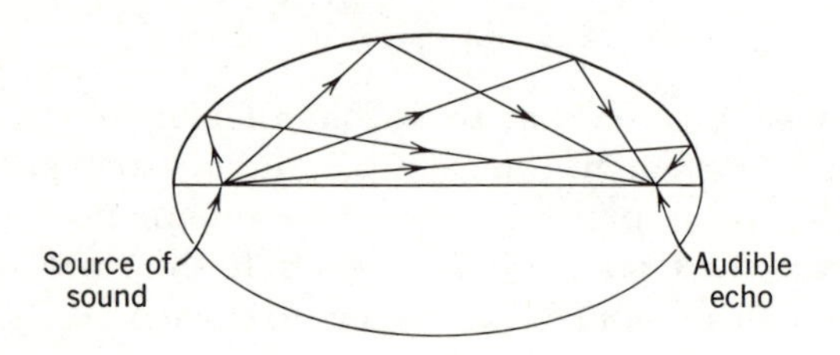

Fig. 11.12.

11.15. The hyperbola. Another locus associated with distances from two fixed points is the hyperbola.

Definition: The *hyperbola* is the locus of all points the difference of whose distances from two fixed points is a constant. (The fixed points are called the *foci* of the hyperbola.)

A mechanical construction of the hyperbola is as follows (see Fig. 11.13). Place two thumbtacks in a drawing board at points F_1 and F_2. These points will be the foci of the hyperbola. Tie a pencil to a string at P so that the string does not slip on the pencil. Let one end of the string be carried under tack F_1 and then join the other end over the tack F_2. If the two ends R_1 and R_2 coincide as both string ends are pulled in or let out the same length, then $PF_2 - PF_1$ will be a constant (if the string is kept taut). The resultant path will

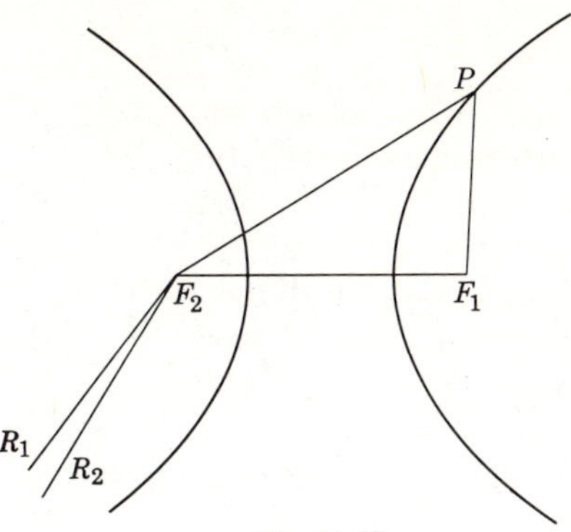

Fig. 11.13.

describe a hyperbola. The left branch of the hyperbola can be obtained by reversing the role of F_1 and F_2.

Simple applications of the hyperbola are not as common as those for the parabola and ellipse. Equations for various hyperbolas are studied in analytic geometry courses. Many laws of nature are represented by hyperbolic equations. The physics student learns that the relation between the volume and pressure of a gas is an equation of a hyperbola. Other relationships which graphically are represented by hyperbolic curves are (*a*) distance, velocity, time; (*b*) area of square, its length, its width; (*c*) total cost, cost per article, number of articles; (*d*) current, voltage, and resistance in electricity.

Hyperbolas are utilized in warfare for locating hidden enemy artillery emplacements. The navigator of an airplane often uses the hyperbola in determining his position. The system involves reception of radio signals from several radio stations at known fixed positions. By noting the times of arrivals of the signals and the finding of the point of intersection of two hyperbolas derived from plotting these times of arrivals, the navigator can determine the plane's position.

If the hyperbola of Fig. 11.13 is rotated around the line $\overleftrightarrow{F_2F_1}$, a surface called a *hyperboloid of revolution* is formed. This type of surface is sometimes used as a sound reflector in the form of a band shell for large outdoor amphitheaters. If the sound of the speaker or of a musical ensemble originates near the focus of the shell, the sound will be directed toward the audience in front of the shell. The sound is spread more uniformly throughout the audience than would be true if the shell were in the shape of a paraboloid of revolution.

Exercises

1. Define a parabola.
2. Determine from the accompanying figure

how to construct a parabola. Then construct a parabola with the focus one inch from the fixed line.
(*Hint:* $MN = FP = DP$.)

3. List five of the applications of parabolas.
4. What is the advantage of using a parabolic reflector in the searchlight?
5. Define an ellipse.
6. Name some objects that have elliptical shapes.
7. What property of the elliptic curve is used in the "whispering galleries"?
8. What heavenly body is at one focus of the earth's elliptic path of movement?
9. Define a hyperbola.
10. List three applications of the hyperbola.

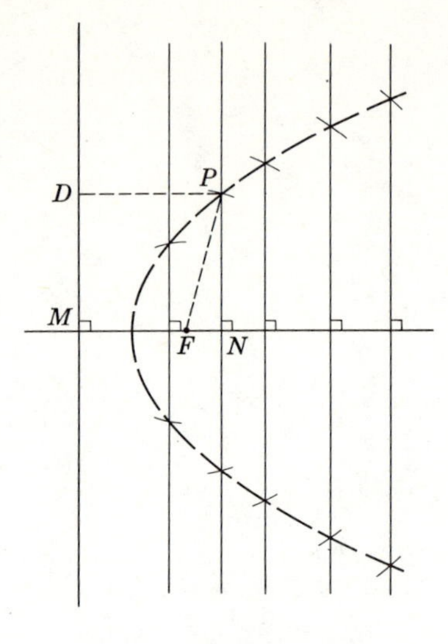

Ex. 2.

Summary Tests

Test 1

COMPLETION STATEMENTS

In each problem, the figures are assumed to lie in a single plane.

1. A point equally distant from the ends of a line segment lies on the ______ of that segment.
2. The distance from a point to a line is the length of the ______ line segment from the point to the line.
3. The locus of all points at a given distance from a fixed point is a ______.
4. The maximum number of points that are a given distance from two intersecting lines is ______.
5. The number of points on a circle equidistant from the endpoints of its diameter is ______.
6. The perpendicular bisectors of the legs of a right triangle intersect on the ______ of the triangle.
7. The maximum number of points taken at random through which it is possible to draw a circle is ______.
8. The locus of a point which is a given distance from a fixed point is a ______.
9. The locus of a point equidistant from two sides of a triangle is the ______ of an angle of the triangle.
10. The locus of the middle of a chord of a given length in a given circle is a ______.
11. Two circles have the same center. The locus of the center of a circle tangent to both is a ______.

12. To inscribe a circle in a triangle, it is necessary to construct two of the ______ of the triangle.
13. The locus of a point equidistant from two concentric circles and also equidistant from two given points, in general, is ______.
14. The locus of the vertex of a right triangle with a given hypotenuse is a ______.

Test 2

MULTIPLE-CHOICE STATEMENTS

In each problem, the figures are assumed to lie in a plane.

1. The locus of a point equidistant from two intersecting lines is (*a*) one line; (*b*) a circle; (*c*) two parallel lines; (*d*) two intersecting lines; (*e*) none of these.
2. The locus of a point equidistant from two points and a fixed distance from a line, in general, is (*a*) one line; (*b*) two points; (*c*) a circle; (*d*) two intersecting lines; (*e*) none of these.
3. The locus of a point a given distance from a fixed point and equidistant from two parallel lines is, in general, (*a*) one point; (*b*) one line; (*c*) a circle; (*d*) two points; (*e*) none of these.
4. The locus of a point a given distance from a fixed line and a second given distance from a fixed point, in general, is (*a*) two intersecting lines; (*b*) one point; (*c*) four points; (*d*) two points; (*e*) none of these.
5. The locus of a point equidistant from two fixed points A and B and also equidistant from two other points C and D, in general, is (*a*) two points; (*b*) two intersecting lines; (*c*) four points; (*d*) one point; (*e*) none of these.
6. The locus of a point equidistant from two intersecting lines and a given distance from a fixed line, in general, is (*a*) four points; (*b*) two intersecting lines; (*c*) two points; (*d*) one point; (*e*) none of these.
7. The locus of the vertex of a right triangle with AB as base is (*a*) a line parallel to AB; (*b*) a line perpendicular to AB; (*c*) a circle; (*d*) a point; (*e*) none of these.
8. The locus of a point equidistant from the three sides of a triangle is (*a*) three intersecting lines; (*b*) three points; (*c*) three parallel lines; (*d*) one point; (*e*) none of these.
9. The locus of a point equidistant from two intersecting lines and a given distance from a fixed point, in general, is (*a*) three lines; (*b*) three points; (*c*) one point; (*d*) four points; (*e*) none of these.
10. The locus of a point equidistant from two parallel lines and a given distance from a third line is (*a*) one point; (*b*) three lines; (*c*) three points; (*d*) four points; (*e*) none of these.

11. The locus of a point a given distance from a circle and a given distance from a line, in general, is (*a*) one point; (*b*) four points; (*c*) three points; (*d*) two points; (*e*) none of these.
12. The locus of the midpoint of chords parallel to a fixed diameter of a circle and a given distance from the circle, in general, is (*a*) two circles; (*b*) two points; (*c*) one point; (*d*) two lines; (*e*) none of these.
13. The locus of a point equidistant from two concentric circles and also equidistant from two given points, in general, is (*a*) one point; (*b*) two points; (*c*) four points; (*d*) six points; (*e*) none of these.
14. Point A is on line BC. The locus of a point at a given distance from A is a line parallel to $\overleftrightarrow{BC}$.

12

Areas of Polygons

12.1. Need for determining areas. From earliest times the measurement of surface areas has been an important and necessary practice. The early Egyptians were assigned plots of land by their rulers. The size of these plots had to be determined. Frequently the Nile river would overflow and sweep away the boundaries of these parcels of land. To re-establish these boundaries, the Egyptians developed a crude system of measuring the land.

Surveyors today have developed the art of measuring land boundaries to a precise science. The history of the growth and expansion of the United States involves the determining of areas. As this nation expanded in a westerly direction, new frontiers opened which eventually had to be measured and the areas of which had to be determined.

Anyone wishing to buy a farm or a tract of land is interested in the area of the land. When a person builds a new house, he is concerned with the "area of his floor plan." The paint required to cover a given surface depends upon the area of that surface. The engineer, architect, tinsmith, carpenter, and artist find that an exact knowledge of the subject of areas is fundamental to their vocations.

12.2. Polygonal regions. In previous chapters we have defined various polygons. We are now going to be concerned with the "regions" of these polygons and their areas.

Definitions: A *triangular region* is a set of points which is the union of a triangle and its interior (see Fig. 12.1). A *polygonal region* (see Fig. 12.2) is a set of points which is the union of a finite number of triangular regions lying in a given plane,

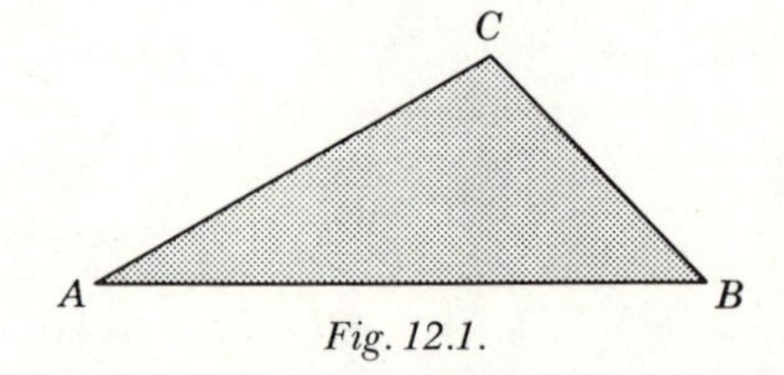

Fig. 12.1.

such that, if any two of them intersect, the intersection is either a segment or a point.

The shaded portions represent the regions of the figures. A polygonal region can be "cut up" into triangular regions in many different ways. The triangular regions of any such decomposition are called *component triangular regions* of the polygonal region. Fig. 12.3 shows three ways of cutting into triangular regions the region of a parallelogram and its interior.

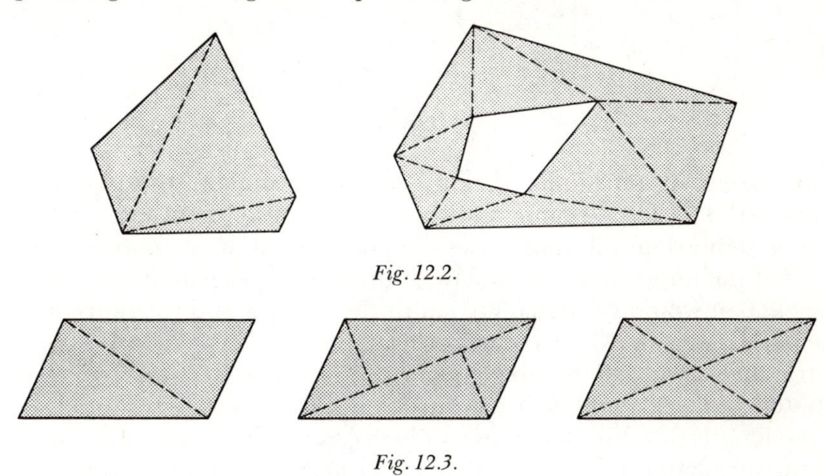

Fig. 12.2.

Fig. 12.3.

12.3. Areas of polygonal regions. We have discussed measures of segments and measures of angles. Let us now consider the measures of areas. After we have discussed units of area measures, we will give as a postulate the basis for the computation of area.

The "unit of area" is closely associated with the unit of distance and can be considered as the region formed by a square of unit length and its interior points. Thus, if *ABCD* in Fig. 12.4 is a square the side of which is one inch long, the measure of the region enclosed is called a *square inch*. Other common units of area are the *square foot, square yard, square mile,* and the *square centimeter*.

The *area of a polygonal region* is the number which tells how many times a given unit of area is contained in the polygonal region. Thus, if, in Fig. 12.5, *AEFG* is a square unit, we can count the number of such units in the total area of *ABCD*. We state then that the area of *ABCD* is 12 such units. If the area of *AEFG* is 1 square inch, the area of *ABCD* is 12 square inches.

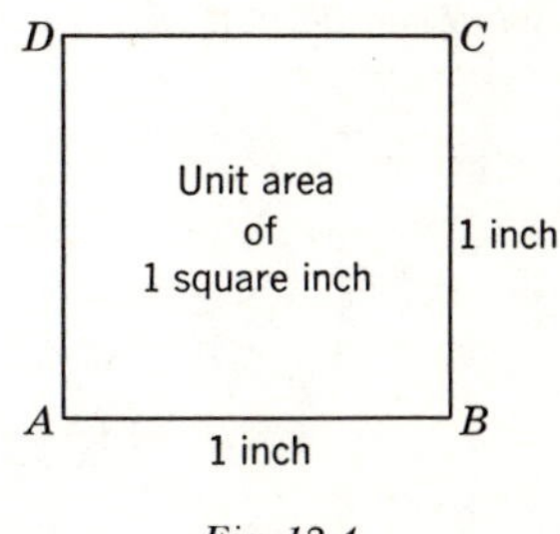

Fig. 12.4.

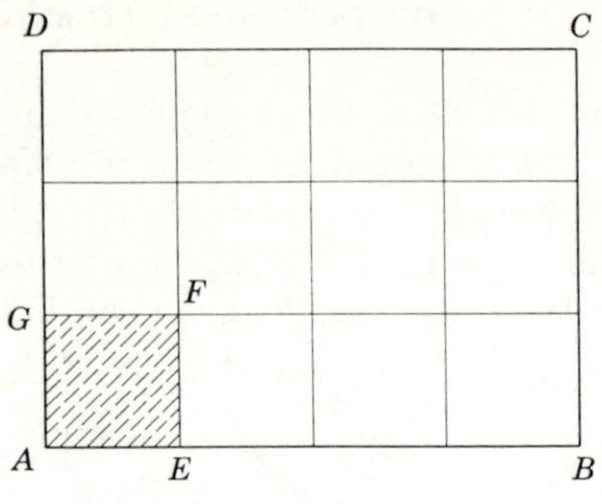

Fig. 12.5.

Thus, areas of polygonal regions can be found by drawing small unit squares in the enclosed region and counting the number of such units. This would be tedious and in most cases inaccurate. In many figures it would be difficult, if not impossible, to count the number of square units. For example, in Fig. 12.6, it would be difficult to count the squares and fractions of squares in the parallelogram $ABCD$ and in circle O.

Fortunately, we will be able to derive formulas by which areas can be computed when certain linear measurements are known. It should be noted that the length of a line segment can be *measured* directly by using a ruler or tape measure, but the area of a region is *computed by formula.* Formulas have been developed for areas of the triangle, parallelogram, trapezoid, and circle. The areas of other shapes can often be found by splitting them up into triangles, rectangles, and trapezoids, and then summing up the areas of these figures.

Postulate 21. (area postulate). *Given a unit of area, to each polygonal region there corresponds a unique number, which is called the area of the region.*

Postulate 22. *The area of a polygonal region is the sum of the area measures of any set of component regions into which it can be cut.*

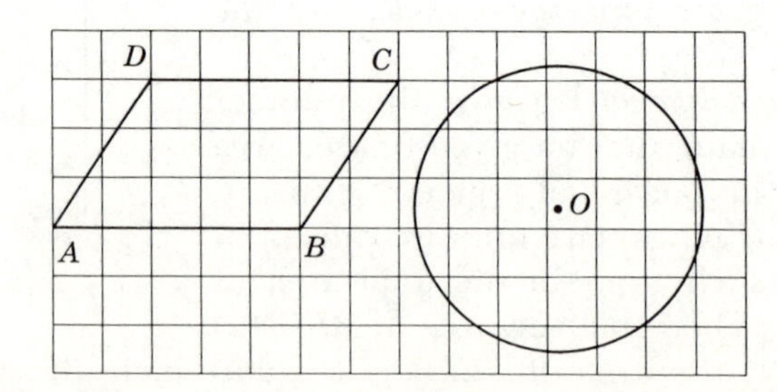

Fig. 12.6.

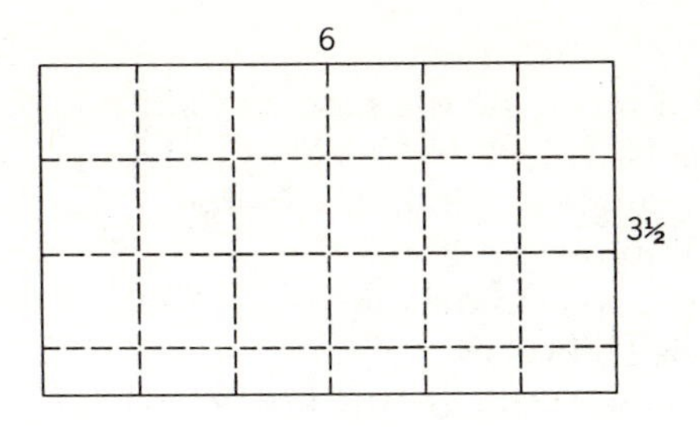

Fig. 12.7.

Postulate 23. *If two polygons are congruent, then the corresponding polygonal regions have the same area.*

Hereafter we will denote the area measure of a region R simply by "area R." We will also use "region" to stand for "polygonal region."

12.4. Area of a rectangle. If the base of a rectangle (see Fig. 12.5) measures 4 linear units and its altitude 3, its area is 4×3, or 12 corresponding square units of surface. If the base of a rectangle (see Fig. 12.7) measures 6 linear units and its altitude $3\frac{1}{2}$ units, it is possible to count 18 whole square units and 6 one-half square units. These 6 one-half square units are equivalent to 3 square units, making a total of 21 square units of surface. This number could also be obtained by multiplying $6 \times 3\frac{1}{2}$.

These examples suggest the following postulate.

Postulate 24. *The area of a rectangular region is equal to the product of the length of its base and the length of its altitude* ($A = bh$).

Since a square is an equilateral rectangle, we can state: The area of a square region is equal to the square of the length of a side ($A = s^2$).

The student will note in Postulate 24 the use of the words "length of its base" and "length of its altitude." Hereafter unless confusion will result, we will let the words "base" and "altitude" mean their lengths if the context of the sentence implies measurements. Thus, Postulate 24 will be stated, "The area of a rectangle is equal to the product of its base and its altitude."

The context of a given statement will usually make clear if the words "rectangle," "base," "altitude," and "side" refer to sets of points or to their measures. Thus, we will speak of "squaring the hypotenuse" and of "bisecting the hypotenuse." The meaning of the word "hypotenuse" in each case should be evident.

Exercises

1. Find the area of a rectangle the base and altitude of which are: (*a*) 7 feet by 4 feet; (*b*) 5 feet by $2\frac{1}{2}$ feet; (*c*) $2\frac{3}{4}$ feet by $3\frac{1}{2}$ feet.
2. The area of a rectangle is 288 square inches. The base is 12 inches long. How long is the altitude?
3. Find the area of a rectangle the base of which is 12.3 yards and the altitude of which is 11.4 yards.
4. Find the altitude of a rectangle the area of which is 63.8 square feet and the base of which is 7.6 feet.
5. How many acres are in a rectangular plot of ground 60 rods wide and 80 rods long, if 160 square rods = 1 acre?
6. What is the cost of laying a cement patio 18 feet wide and 24 feet long at 40 cents a square foot?
7. How many square yards of carpeting will be required for a room that is 15 feet wide and 18 feet long?
8. Compute the area of the cross section of the L-beam in the figure.
9. Determine the cross-sectional area of the T-section.

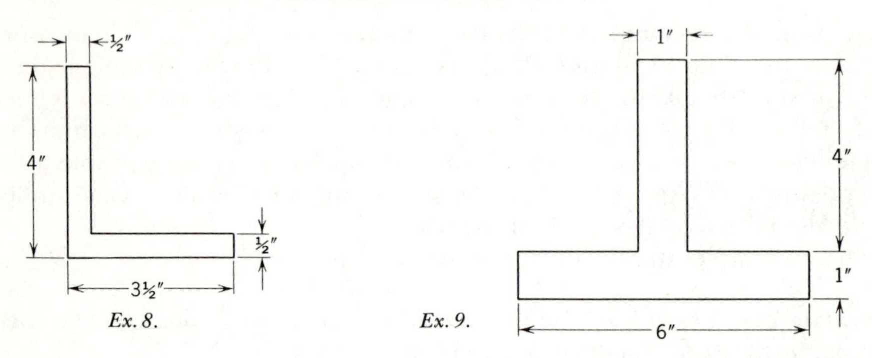

Ex. 8. *Ex. 9.*

10. What is the area of the cross section of the I-beam shown in the figure?
11. Compute the area of the cross section of the accompanying H-beam.

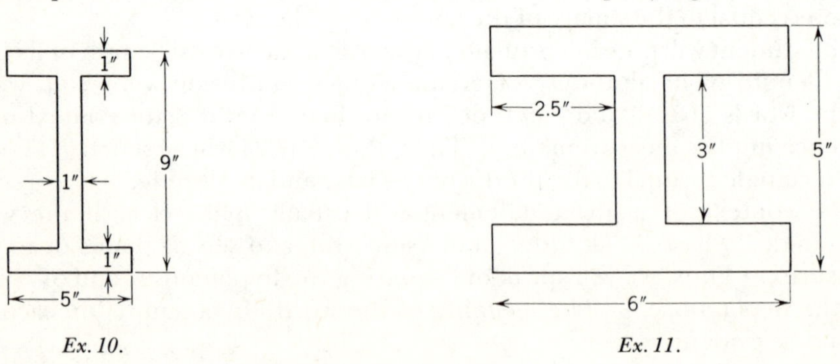

Ex. 10. *Ex. 11.*

12. Compute the cross-sectional area of the Z-bar.

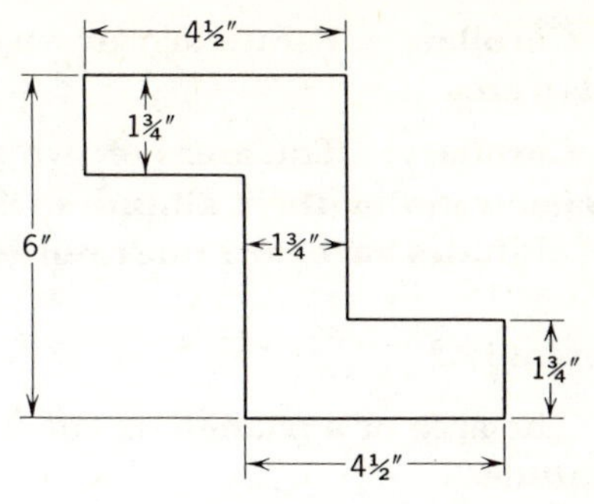

Ex. 12.

Theorem 12.1

12.5. The area of a parallelogram is equal to the product of its base and its altitude.

Given: ▱ $ABCD$ with base $AB = b$ units; altitude $DE = h$ units.

Conclusion: Area of ▱ $ABCD = bh$ square units.

Proof:

Theorem 12.1.

STATEMENTS	REASONS
1. $ABCD$ is a ▱.	1. Given.
2. $\overleftrightarrow{DC} \parallel \overleftrightarrow{AB}$.	2. The opposite sides of ▱ are $\parallel$.
3. $\overline{DE} \perp \overleftrightarrow{AB}$.	3. §6.3.
4. Draw $\overline{CF} \perp \overleftrightarrow{AB}$.	4. Theorem 5.3.
5. $\overleftrightarrow{DE} \parallel \overleftrightarrow{CF}$.	5. Theorem 5.5.
6. $EFCD$ is a ▱.	6. Definition of a ▱.
7. $\angle DEF$ is a right $\angle$.	7. Definition of $\perp$ lines.
8. $EFCD$ is a ▭.	8. Definition of a ▭.
9. $DE = CF$; $AD = BC$.	9. Theorem 6.2.
10. $\triangle AED \cong \triangle BFC$.	10. Theorem 5.20.
11. Area $EBCD$ = area $EBCD$.	11. Reflexive property.
12. Area $EBCD$ + area AED = area $EBCD$ + area BFC.	12. Additive property.
13. Area $ABCD$ = area $EBCD$ + area AED; area $EFCD$ = area $EBCD$ + area BFC.	13. Postulate 22.
14. Area $ABCD$ = area $EFCD$.	14. Substitution property.
15. $AB = b$; $DE = h$.	15. Given.
16. Area $EFCD = bh$.	16. Postulate 24.
17. Area $ABCD = bh$.	17. Substitution property.

12.6. Corollary: Parallelograms with equal bases and equal altitudes are equal in area.

12.7. Corollary: The areas of two parallelograms having equal bases have the same ratio as their altitudes; the areas of two parallelograms having equal altitudes have the same ratio as their bases.

Theorem 12.2

12.8. The area of a triangle is equal to one-half the product of its base and its altitude.

Given: $\triangle ABC$ with base $AB = b$ and altitude $CE = h$.

Conclusion: Area of $\triangle ABC = \frac{1}{2}bh$.

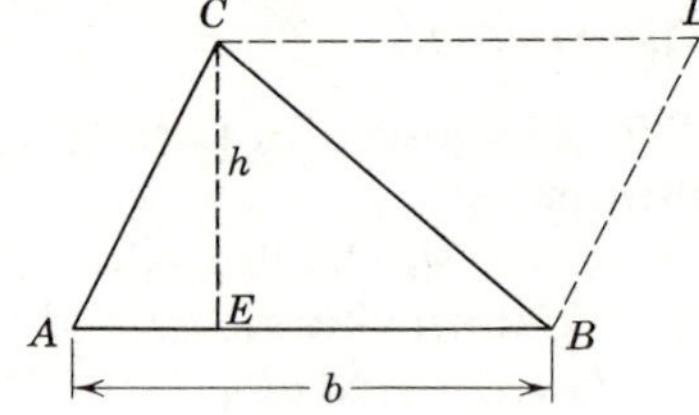

Theorem 12.2.

Proof:

STATEMENTS	REASONS
1. $\triangle ABC$ has base $= b$, altitude $CE = h$.	1. Given.
2. Draw $\overleftrightarrow{CD} \parallel \overleftrightarrow{AB}$ and $\overleftrightarrow{BD} \parallel \overleftrightarrow{AC}$, meeting at D.	2. Postulate 18; Theorem 5.7.
3. $ABDC$ is a $\square$.	3. Definition of a $\square$.
4. $\triangle ABC \cong \triangle DCB$.	4. §6.6.
5. Area $ABDC$ = area ABC + area DCB.	5. Postulate 22.
6. Area $ABDC$ = area ABC + area ABC.	6. Substitution property.
7. Area $ABDC = bh$.	7. Theorem 12.1.
8. 2 Area $ABC = bh$.	8. Theorem 3.5.
9. Area of $\triangle ABC = \frac{1}{2}bh$.	9. E-7.

12.9. Corollary: Triangles with equal bases and equal altitudes are equal in area.

12.10. Corollary: The areas of two triangles having equal bases have the same ratio as their altitudes; the areas of two triangles having equal altitudes have the same ratio as their bases.

12.11. Corollary: The area of a rhombus is equal to one-half the product of its diagonals.

Exercises

1. Find the area of a parallelogram the base of which is 16 inches and the altitude of which is 10 inches.
2. Find the area of a parallelogram the base of which is 16.4 feet and the altitude of which is 11.6 feet.
3. Find the altitude of the parallelogram the area of which is 204 square inches and the base of which is 26 inches.
4. Find the area of $\square ABCD$ if $AB = 24$ yards, $AD = 18$ yards, and $m\angle A = 30$.
5. Find the area of $\square ABCD$ if $AD = 12$ inches, $AB = 18$ inches, $m\angle A = 60$.

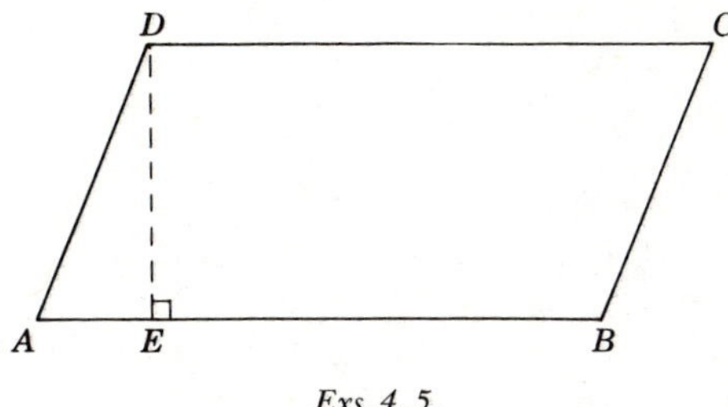

Exs. 4, 5.

6. Find the area of the rhombus, the diagonals of which are 35 inches and 24 inches.
7. *Given:* $\overline{GF} \parallel \overline{AE}$; $\overline{CF} \parallel \overline{AG}$; $\overline{CG} \parallel \overline{EF}$; $GF = 14$ inches; $BG = 12$ inches; $AB = DE = 7$ inches. Find the area of (*a*) $\square ACFG$; (*b*) $\square CEFG$; (*c*) $\square BDFG$; (*d*) $\triangle ABG$; (*e*) $\triangle GCF$; (*f*) $\triangle CEF$.
8. Find the area of $\triangle RTK$ if $RK = 15$, $KS = 12$, $ST = 18$.

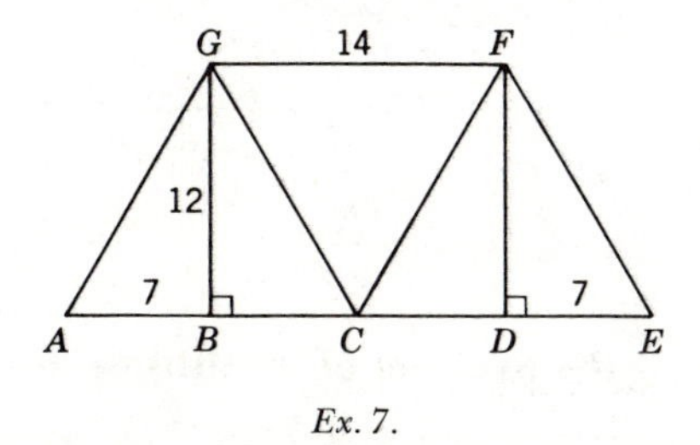

Ex. 7.

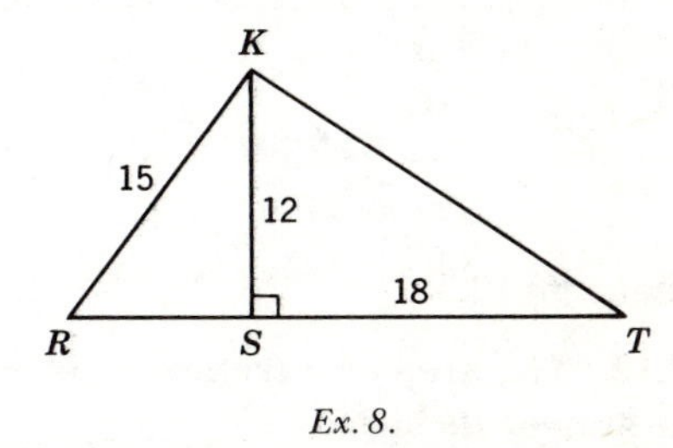

Ex. 8.

9. *Given:* $\triangle ABC$ with $\overline{CD} \perp \overline{AB}$, $\overline{AE} \perp \overline{BC}$, $AB = 24$ inches, $CD = 15$ inches, $BC = 21$ inches. Find AE.
10. How long is the leg of an isosceles right triangle the area of which is 64 square feet?
11. Find the area of trapezoid $ABCD$.
12. Find the area of trapezoid $RSTQ$.

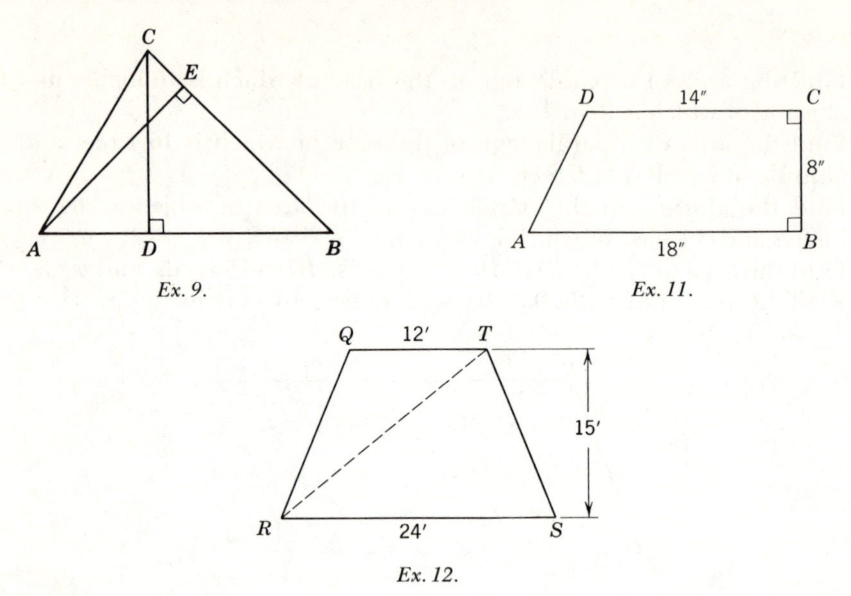

Ex. 9.

Ex. 11.

Ex. 12.

13. *Given:* D is the midpoint of $\overline{AF}$, area of $\square ABCD = 36$ square inches. Find the area of $\triangle ABF$.
14. K is the midpoint of $\overline{RS}$. Area of $\triangle RST = 30$ square inches. Area of $\triangle RKL = 7$ square inches. Find area of $\triangle LST$.

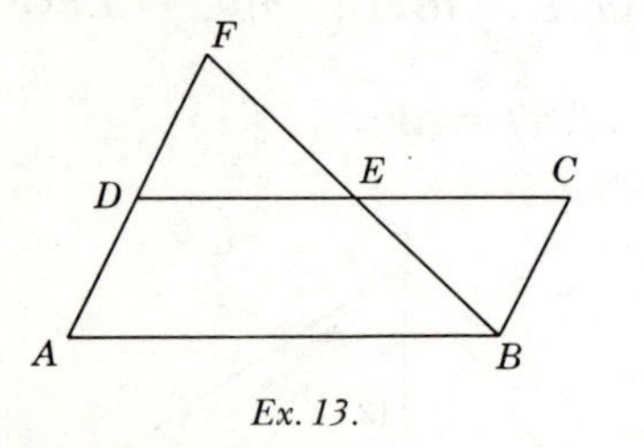

Ex. 13.

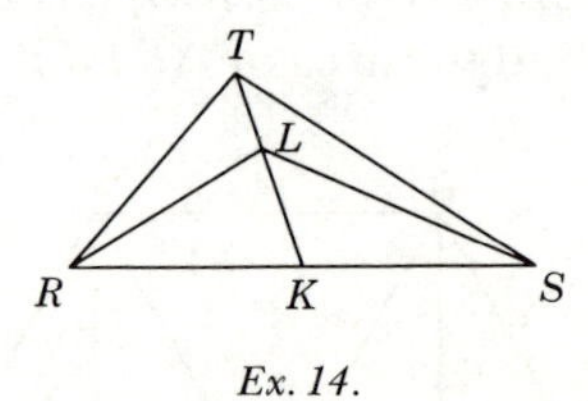

Ex. 14.

Theorem 12.3

12.12. The area of a trapezoid is equal to half the product of its altitude and the sum of its bases.

Given: Trapezoid $ABCD$ with altitude $CE = h$, base $AB = b_1$, and base $DC = b_2$.

Conclusion: Area of trapezoid $ABCD = \frac{1}{2}h(b_1 + b_2)$.

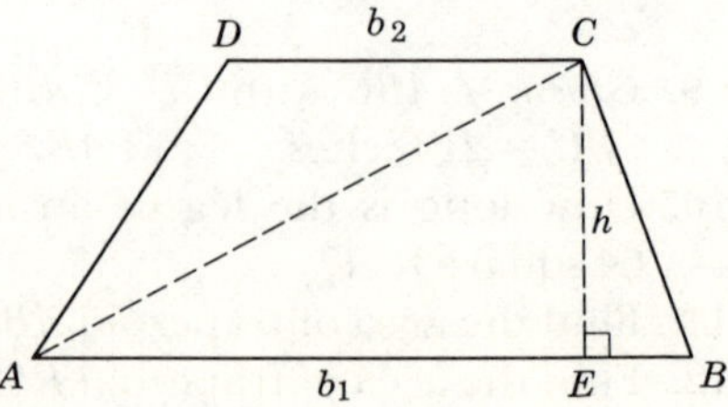

Theorem 12.3.

Proof:

STATEMENTS	REASONS
1. Draw diagonal AC dividing the trapezoid into $\triangle ABC$ and ACD.	1. Postulate 2.
2. Area of $\triangle ABC = \frac{1}{2}b_1h$.	2. Theorem 12.2.
3. Area of $\triangle DAC = \frac{1}{2}b_2h$.	3. Theorem 12.2.
4. Area of $\triangle ABC +$ area of $\triangle DAC = \frac{1}{2}b_1h + \frac{1}{2}b_2h = \frac{1}{2}h(b_1+b_2)$.	4. E-4.
5. Area of $\triangle ABC +$ area of $\triangle DAC =$ area of trapezoid $ABCD$.	5. Postulate 22.
6. $\therefore$ Area of trapezoid $ABCD = \frac{1}{2}h(b_1+b_2)$.	6. Theorem 3.5.

12.13. Other formulas for triangles. In this section we shall develop some important formulas which are used quite extensively in solutions of geometric problems. It is assumed in this discussion that the student has a basic knowledge of square roots and radicals. A table of square roots can be found on page 421 of this text.

Formula 1: *Relate the diagonal d and the side s of a square.*

$$d^2 = s^2 + s^2$$
$$= 2s^2$$
$$\therefore d = s\sqrt{2}$$

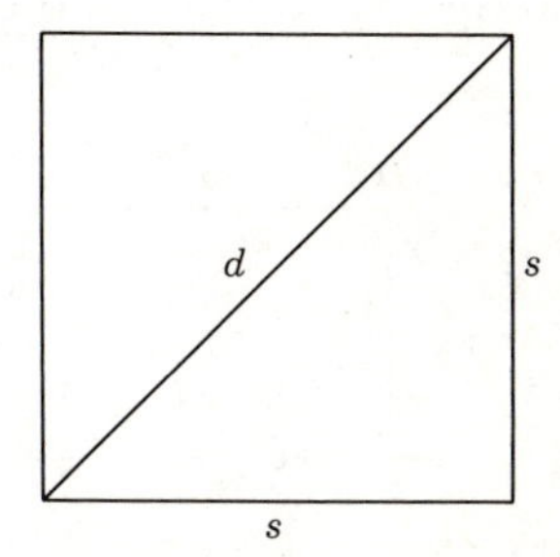

Formula 1.

Formula 2: *Relate the side s and the diagonal d of a square.*

$$s\sqrt{2} = d \text{ (Formula 1)}$$

$$s = \frac{d}{\sqrt{2}}$$

$$= \frac{d}{\sqrt{2}} \times \frac{\sqrt{2}}{\sqrt{2}} = \frac{d\sqrt{2}}{2}$$

Formula 3: *Relate altitude h and side s of an equilateral triangle.*

$m\angle RTM = 30$; $\overline{TM} \perp \overline{RS}$.

$$h^2 = s^2 - \left(\frac{s}{2}\right)^2$$

$$= \frac{3s^2}{4}$$

$$h = \frac{s\sqrt{3}}{2}$$

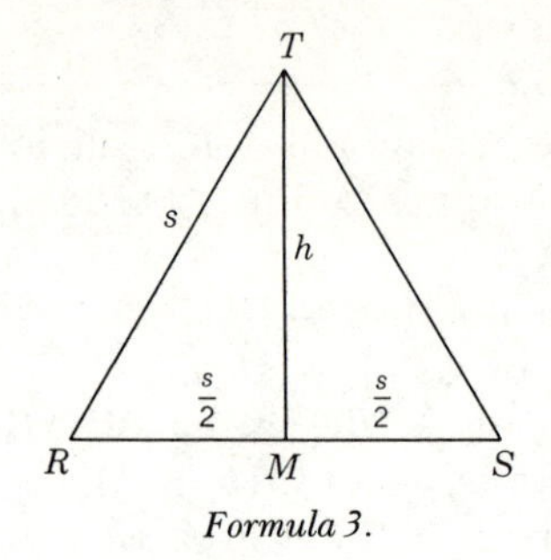

Formula 3.

Formula 4: *Relate the area A and the side s of an equilateral triangle.*

$$A = \tfrac{1}{2}RS \times TM$$

$$= \tfrac{1}{2}s \times \frac{s\sqrt{3}}{2}$$

$$= \frac{s^2\sqrt{3}}{4}$$

Exercises

1–8. Find the area of each of the following trapezoids.

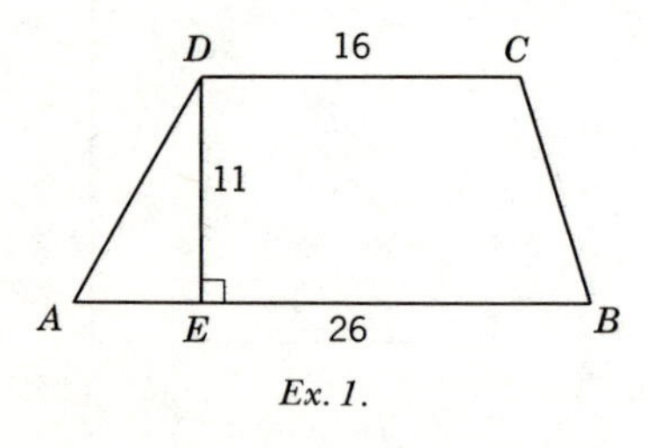

Ex. 1.

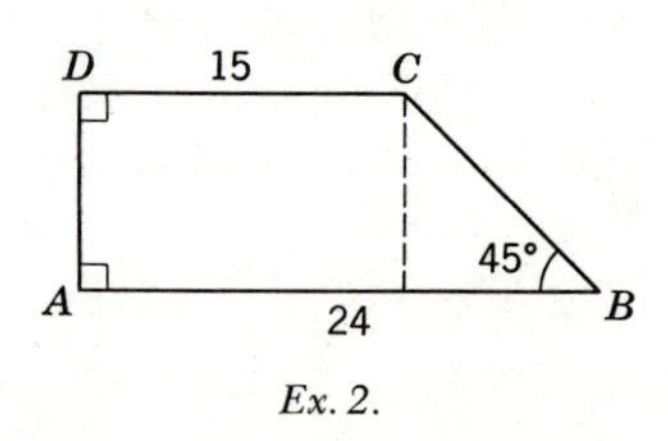

Ex. 2.

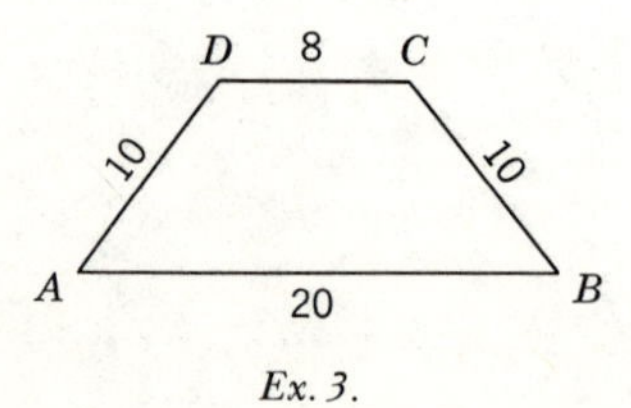

Ex. 3.

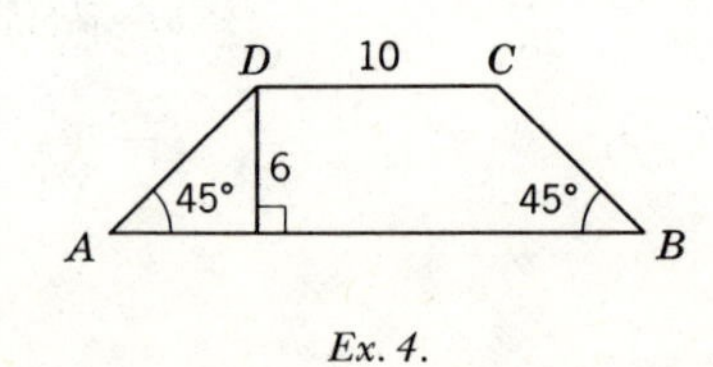

Ex. 4.

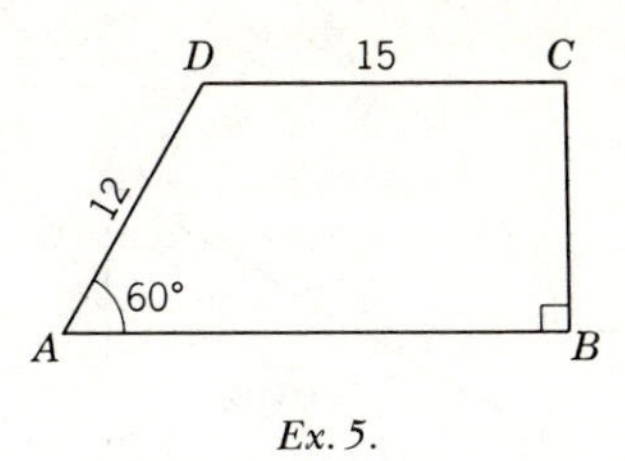

Ex. 5.

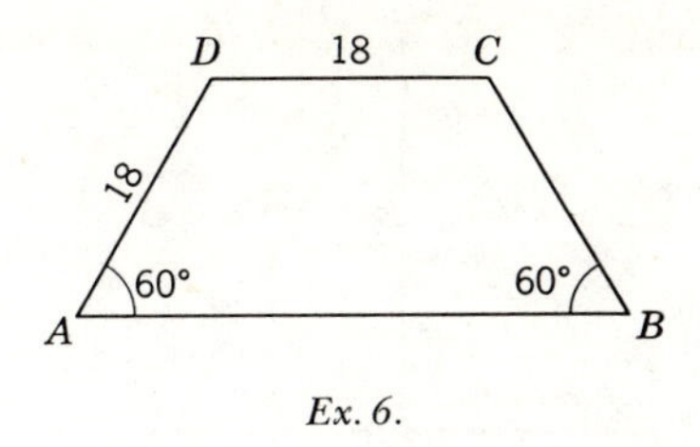

Ex. 6.

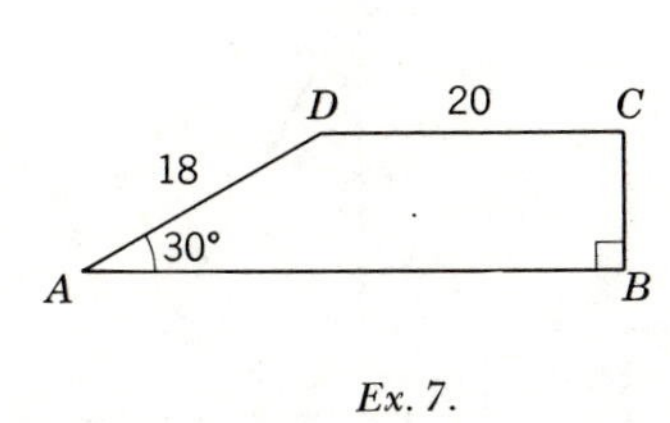

Ex. 7.

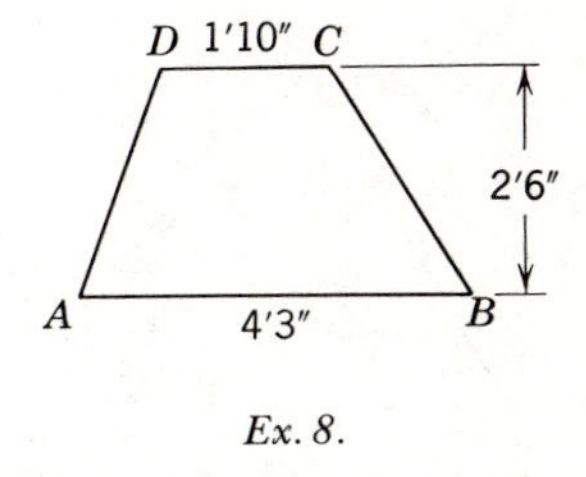

Ex. 8.

9–14. Solve for x in each of the following figures.

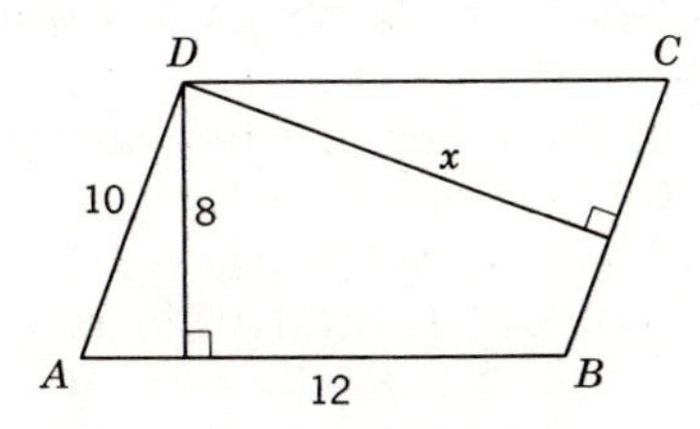

Ex. 9. Given: ABCD is a ▱.

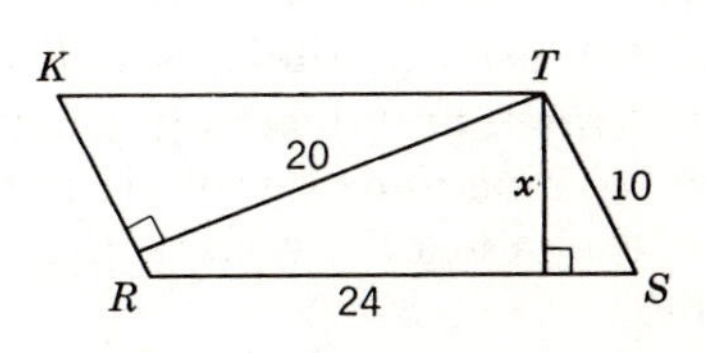

Ex. 10. Given: KRST is a ▱.

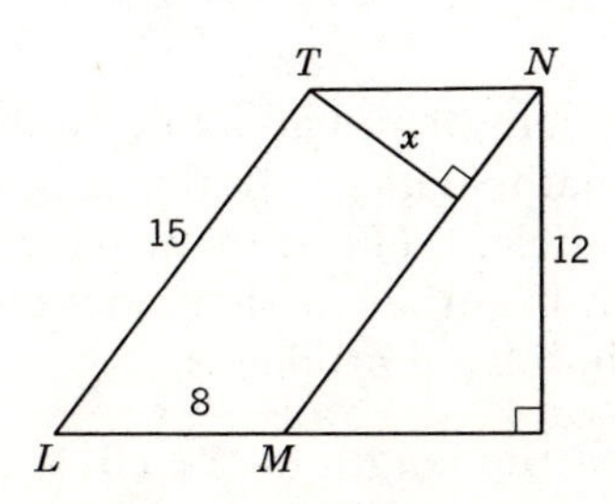

Ex. 11. Given: LMNT is a ▱.

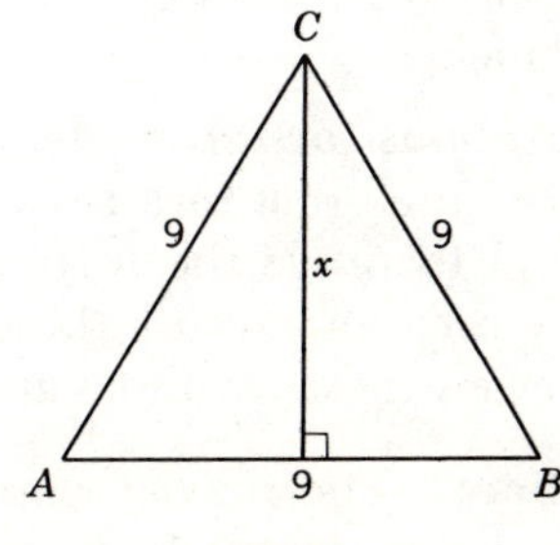

Ex. 12.

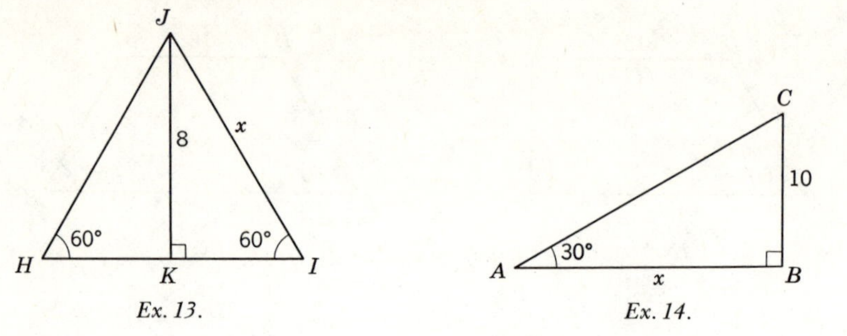

Ex. 13. *Ex. 14.*

15–16. Find the areas of the following equilateral triangles.

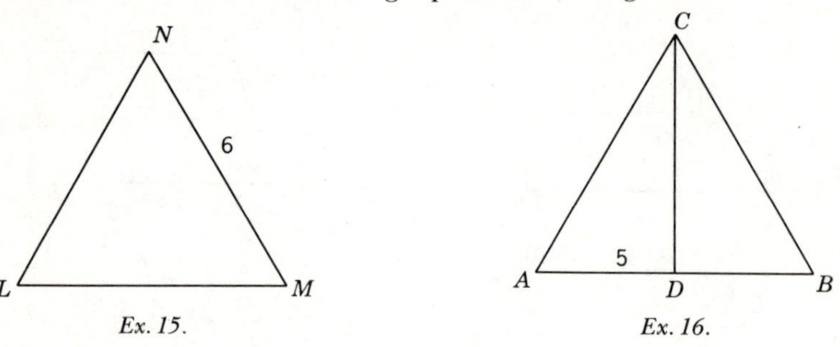

Ex. 15. *Ex. 16.*

17. The area of a trapezoid is 76 square inches. The bases are 11 inches and 8 inches. Find the altitude.
18. Find the area of the isosceles trapezoid the bases of which are 9 yards and 23 yards and the diagonal of which is 17 yards.
19. Find the diagonal of a square of side (*a*) 6 inches; (*b*) 15 inches.
20. Find the side of a square the diagonal of which is (*a*) 16 inches; (*b*) 32 inches.
21. Find the area of a square the diagonal of which is (*a*) 12 inches; (*b*) 52 inches.
22. Find the altitude of an equilateral triangle the side of which is (*a*) 8 inches; (*b*) 17 inches.
23. Find the area of an equilateral triangle the side of which is (*a*) 7 inches; (*b*) 52 inches.

12.14. Formulas for the circle. Finding the length and area of a circle have been two of the great historic problems in mathematics. In this text we will not attempt to prove the formulas for the circle. The student has already had many occasions to use these formulas in his other mathematics courses. We will review these formulas and use them in solving problems.

Definition: The *circumference* of a circle is the length of the circle (sometimes called its *perimeter*).

It can be shown that *the ratio of the circumference of a circle to its diameter is a constant.* This constant is represented by the Greek letter π (*pi*). Thus, in Fig. 12.8, $C_1/D_1 = C_2/D_2 = \pi$.

12.15. Historical note on π. The above fact was known in antiquity. Various values, astoundingly accurate, were found by the ancients for this constant.

Perhaps the first record of an attempt to evaluate π is credited to an Egyptian named Ahmes, about 1600 B.C. His evaluation of π was 3.1605. Archimedes (287–212 B.C.) estimated the value of π by inscribing in and circumscribing about a circle regular polygons of 96 sides. He then calculated the perimeters of the regular polygons and reasoned that the circumference of the circle would lie between the two calculated values. From these results he proved that π lies between $3\frac{1}{7}$ and $3\frac{10}{71}$. This would place π between 3.1429 and 3.1408. Ptolemy (?100–168 A.D.) evaluated π as 3.14166. Vieta (1540–1603) gave 3.141592653 as the value of π.

Students of calculus can prove that π is what is termed an *irrational number*; i.e., no matter to what degree of accuracy the constant is carried, it will never be exact. It can be shown in advanced mathematics work that $\pi = 4(1 - \frac{1}{3} + \frac{1}{5} - \frac{1}{7} + \frac{1}{9} - \frac{1}{11} \ldots)$. The right-hand expression is what is termed an *infinite series*.

By using the modern calculating machines of today, the value of π has been found accurate to more than 100,000 digits. This is a degree of accuracy which has no practical value. The value of π accurate to 10 decimal places is 3.1415926536.

12.16. Circumference of a circle. Since $C/D = \pi$, we can now derive a formula for the circumference. If we multiply each side of the equation by D, we obtain $C = \pi D$. Since the diameter D equals two times the radius, we can substitute in the equation and get $C = 2\pi R$.

Thus, *the circumference of a circle is expressed by the formula* $C = \pi D$, or $C = 2\pi R$.

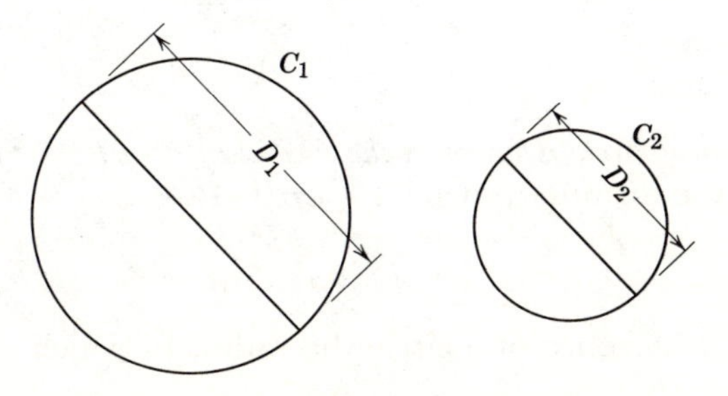

Fig. 12.8.

In using this formula, it is best to use π rounded off to one more digit than the data, D or R, it is used with. Then round off the answer to the degree of accuracy of the given data.

12.17. Illustrative Example 1. Find the circumference of a circle the diameter of which is 5.7 inches.

Solution: $C = \pi D$

$$C = 3.14(5.7)$$
$$= 17.898$$

Answer: 17.9 inches.

12.18. Illustrative Example 2. Find the radius of a circle the circumference of which is 8.25 feet.

Solution: $C = 2\pi R$

$$R = \frac{C}{2\pi}$$

$$= \frac{8.25}{2(3.142)}$$

$$= 1.313$$

Answer: 1.31 feet.

12.19. Area of a circle. It can be shown that the ratio of the area of a circle to the square of its radius is a constant. This is the same constant π, which equals the ratio of the circumference and the diameter of a circle. In Fig. 12.9.

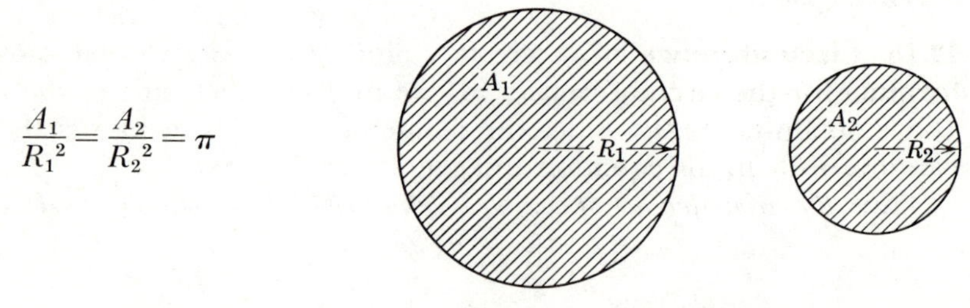

Fig. 12.9.

Thus, *the area of a circle is given by the formula* $A = \pi R^2$. Since $R = D/2$, we can substitute in the formula and get $A = (\pi D^2/4)$.

Exercises

1. Find the circumference of a circle the radius of which is (*a*) 5.2; (*b*) 2.54; (*c*) 32.58.
2. Find the area of a circle the radius of which is (*a*) 7.0; (*b*) 8.34; (*c*) 25.63.

3. Find the diameter of a circle the circumference of which is (*a*) 280; (*b*) 87.54; (*c*) 68.3562.
4. Find the radius of a circle the circumference of which is (*a*) 140; (b) 26.38; (*c*) 86.6512.
5. Find the radius of a circle the area of which is (*a*) 24.5; (*b*) 37.843; (c) 913.254.
6. Find the diameter of a circle the area of which is (*a*) 376; (*b*) 62.348; (*c*) 101.307.
7. If the radii of two circles are 2 and 3 inches respectively, what is the ratio of their areas?
8. If the radii of two circles are 3 and 5 inches respectively, what is the ratio of their areas?
9. Find the area of a circle the circumference of which is 28.7 feet.
10. Find the perimeter of the track *ABCDEF* in the figure.
11. Find the area enclosed by the track *ABCDEF* in the figure.

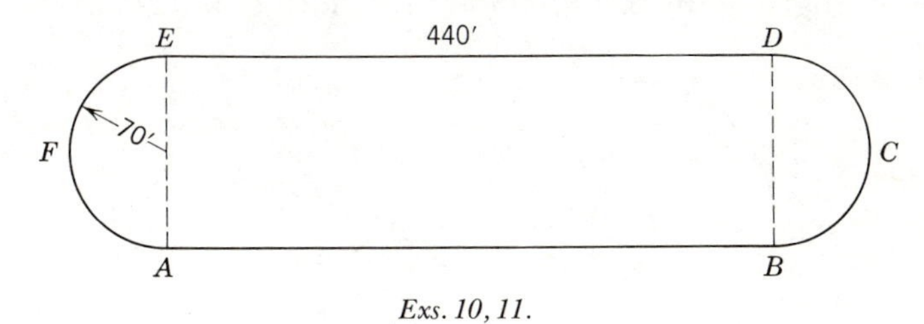

Exs. 10, 11.

12. Find the area of a semicircle the perimeter of which is 36.43.
13. Find the shaded area.
14. Find the area of the shaded portion of the semicircle.

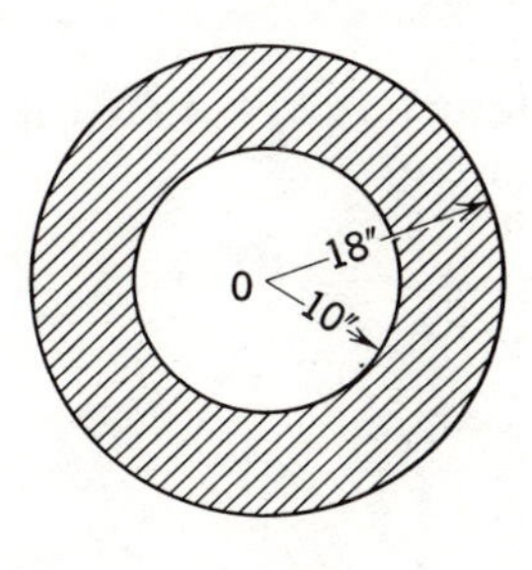

Ex. 13.

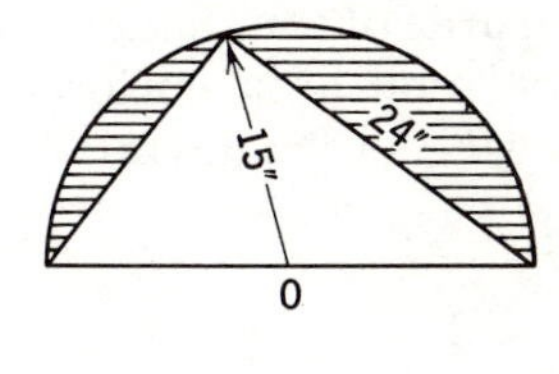

Ex. 14.

15. Find the area of the shaded portion in the figure.
16. Find the perimeter of the figure.
17. Find the total area enclosed in the figure.

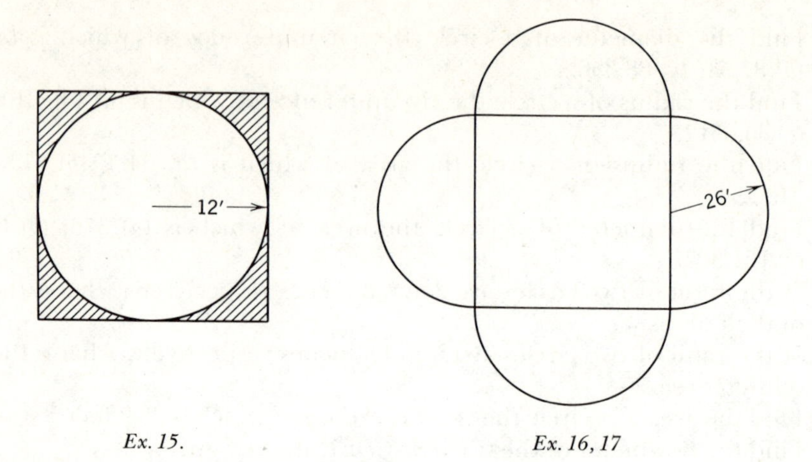

Ex. 15. *Ex. 16, 17*

18. Find the length of the belt used joining wheels O and O'.

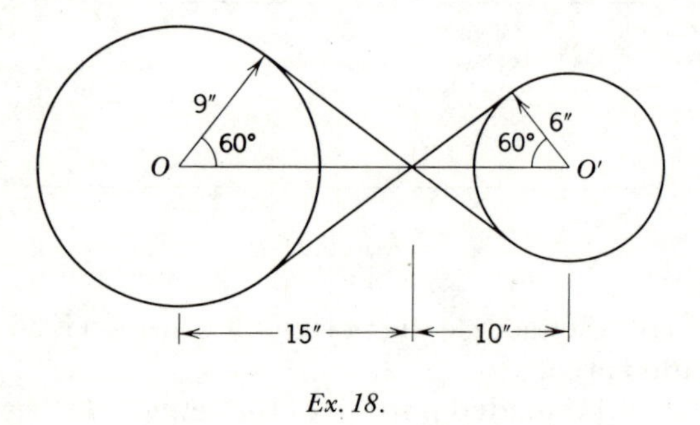

Ex. 18.

19. Find the area of the shaded portion.
20. Find the total area enclosed in the figure, given that the sides of the equilateral $\triangle$ are diameters of the semicircles.

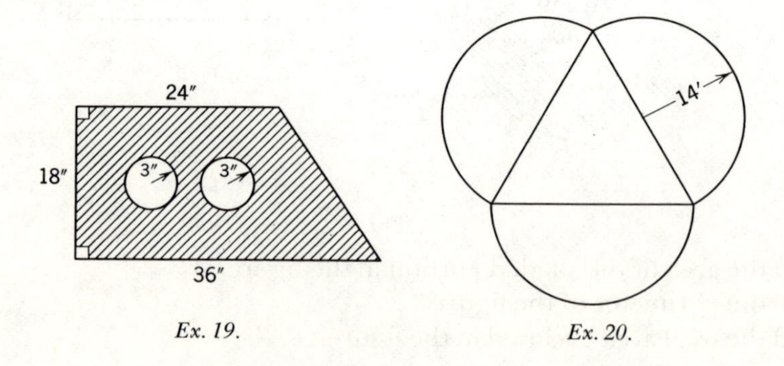

Ex. 19. *Ex. 20.*

Summary Tests

Test 1

COMPLETION STATEMENTS

1. The ratio of the circumference to the diameter of a circle is ______.
2. The area of a rhombus is equal to one-half the product of ______ of the rhombus.
3. The area of an equilateral triangle with sides equal to 8 inches is ______ square inches.
4. The area of the square with the diagonal equal to $\sqrt{32}$ inches is ______ square inches.
5. The number π represents the ratio of the area of a circle to its ______.
6. If the altitude of a triangle is doubled while the area remains constant, the base is multiplied by ______.
7. Doubling the diameter of a circle will multiply the circumference by ______.
8. The ratio of the circumference of a circle to the perimeter of its inscribed square is ______.

Test 2

TRUE-FALSE STATEMENTS

1. The median of a triangle divides it into two triangles with equal areas.
2. Two rectangles with equal areas have equal perimeters.
3. The area of a circle is equal to $2\pi R^2$.
4. If the radius of a circle is doubled, its area is doubled.

5. The area of a trapezoid is equal to the product of its altitude and its median.
6. Triangles with equal altitudes and equal bases have the same areas.
7. The area of a triangle is equal to half the product of the base and one of the sides.
8. If the base of a rectangle is doubled while the altitude remains unchanged, the area is doubled.
9. A square with a perimeter equal to the circumference of a circle has an area equal to that of the circle.
10. Doubling the radius of a circle will double its circumference.
11. If a triangle and a parallelogram have the same base and the same area, their altitudes are the same.
12. The line joining the midpoints of two adjacent sides of a parallelogram cuts off a triangle the area of which is equal to one-eighth that of the parallelogram.
13. The sum of the lengths of two perpendiculars drawn from any point on the base to the legs of an isosceles triangle is equal to the altitude on a leg.

Test 3

PROBLEMS

1. Find the area of the I-beam.
2. Find the area of the trapezoid.

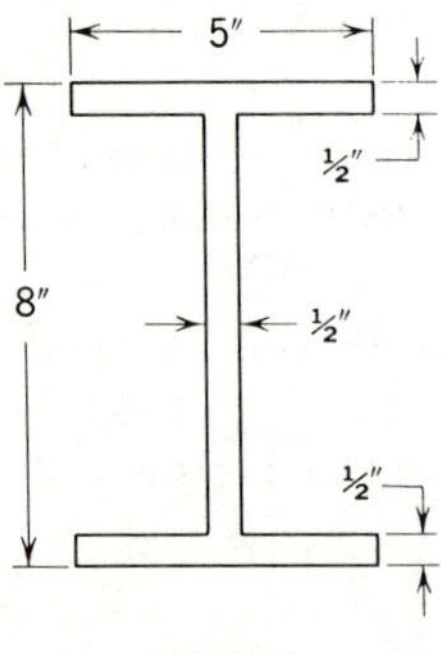

Prob. 1.

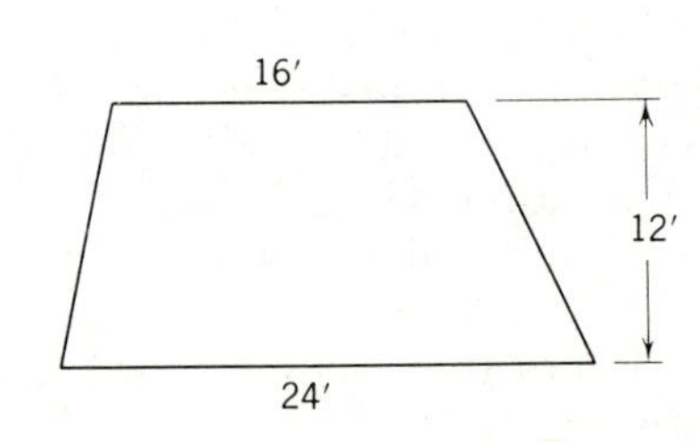

Prob. 2.

3. Find the area of the triangle.

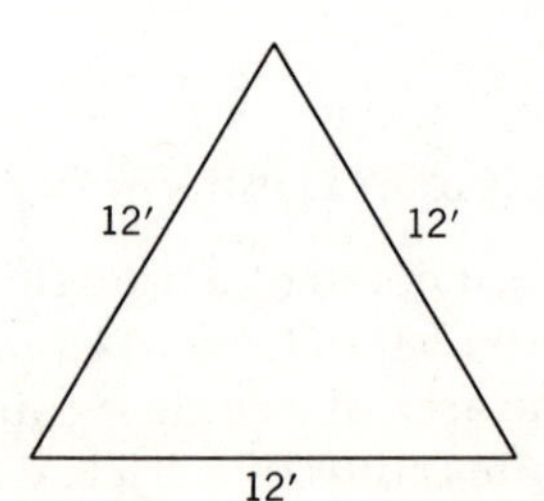

Prob. 3.

4. Find the area of the trapezoid.

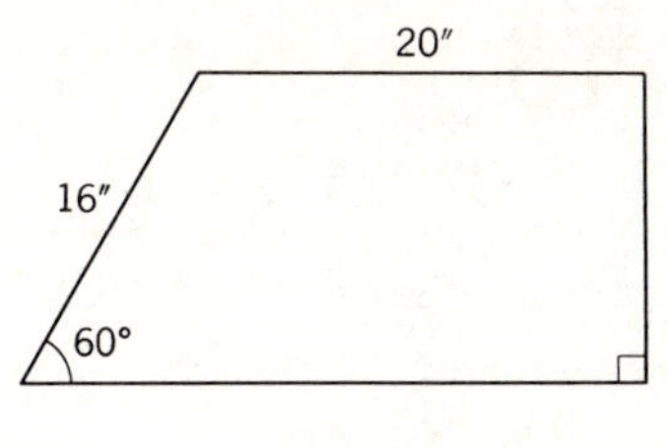

Prob. 4.

5. Find the shaded area.
6. Find the shaded area.

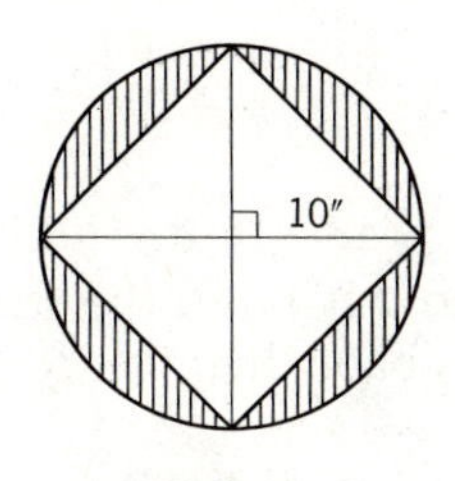

Prob. 5.

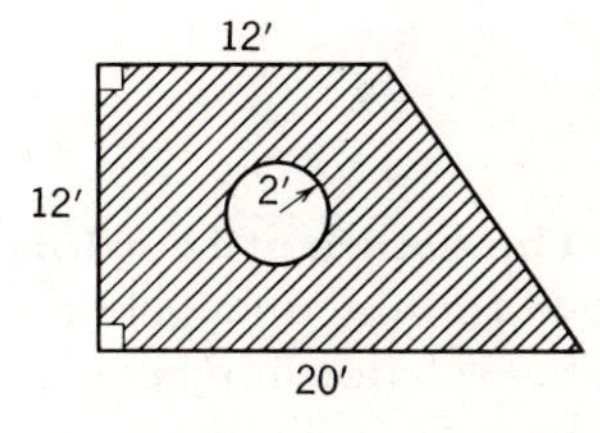

Prob. 6.

7. Find the perimeter of the figure.
8. Find the ratio of the circumferences of $\odot O$ and Q.

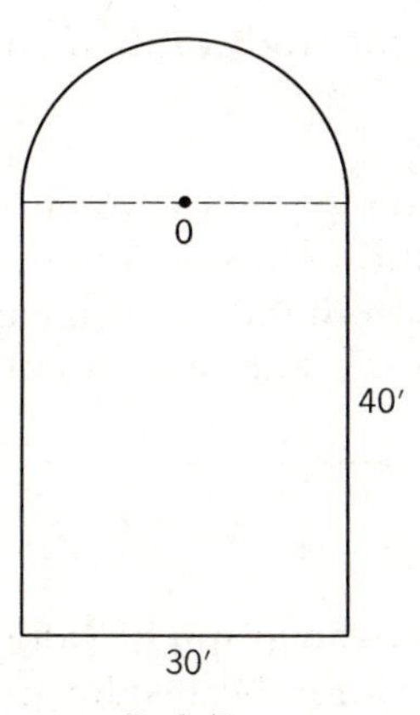

Prob. 7.

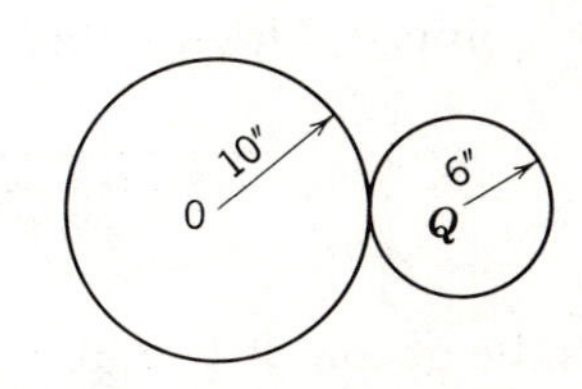

Prob. 8.

13

Coordinate Geometry

13.1. The nature of coordinate geometry. Up to this point the student has received training in algebra and geometry, but probably had little occasion to study the relationship between the two.

In the year 1637 the French philosopher and mathematician René Descartes (1596–1650) established a landmark in the field of mathematics when his book *La Geométrie* was published. In this book he showed the connection between algebra and geometry. By establishing this relationship, he was able to study geometric figures by examining various equations which represented the figures. The geometric properties, in turn, were used to study equations. The study of geometric properties of figures by the study of their equations is called *coordinate* or *analytic* geometry.

13.2. Plotting on one axis. In Chapter 1 we discussed the one-to-one correspondence that exists between real numbers and points on a line. A given set of points on the line is the *graph* of the set of numbers which correspond to the points. The number line is called the *axis*. The diagram of Fig. 13.1

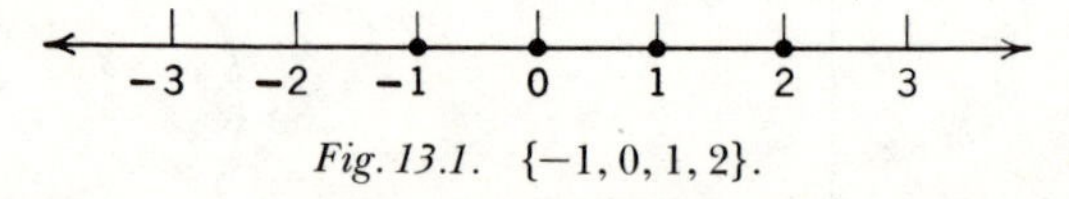

Fig. 13.1. $\{-1, 0, 1, 2\}$.

shows the graph of $\{-1, 0, 1, 2\}$. Fig. 13.2 shows the graph of {all real numbers x such that $x > -1$}. Notice that the open dot indicates the graph to be

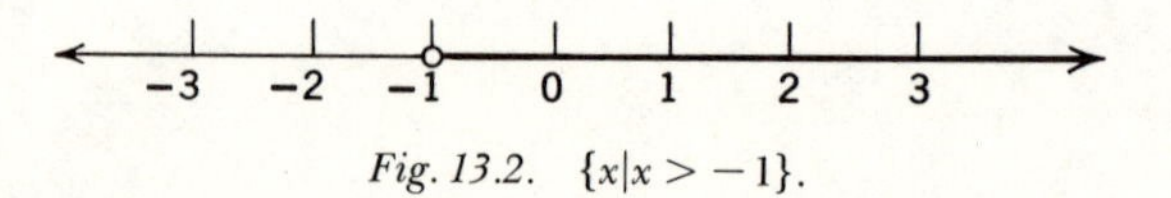

Fig. 13.2. $\{x|x > -1\}$.

a half-line which does not include the number -1 as a member of the set it represents.

In Fig. 13.3, we see the graph of the set of all real numbers from -1 to 2 inclusive. Mathematicians have developed a concise way to describe sets such as the one shown in Fig. 13.3. It is written $\{x \mid -1 \leqslant x \leqslant 2\}$ and is read "the set of all real numbers x such that $x \geqslant -1$ and $x \leqslant 2$."

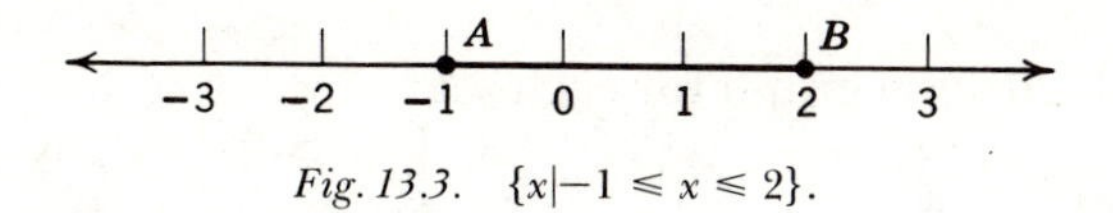

Fig. 13.3. $\{x \mid -1 \leqslant x \leqslant 2\}$.

The intersection of the set of all real numbers x such that x is less than 2 and the set of all real numbers x such that x is greater than or equal to -1 can be written as $\{x \mid x < 2\} \cap \{x \mid x \geqslant -1\}$. The graph of such a set is shown in Fig. 13.4.

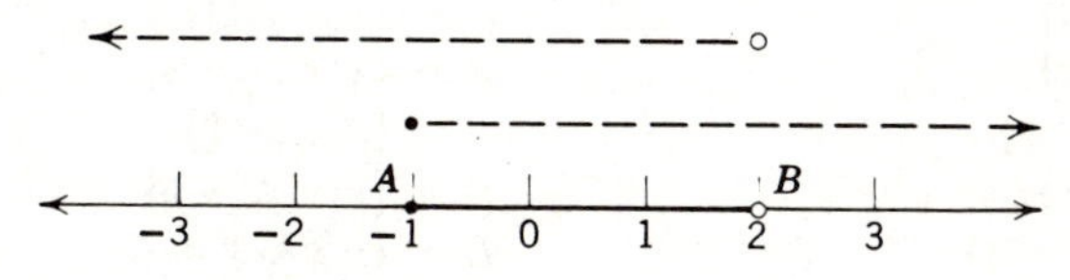

Fig. 13.4. $\{x \mid x < 2\} \cap \{x \mid x \geqslant 1\}$.

Exercises

In Exercises 1–6, use set notation to describe the set that has been graphed.

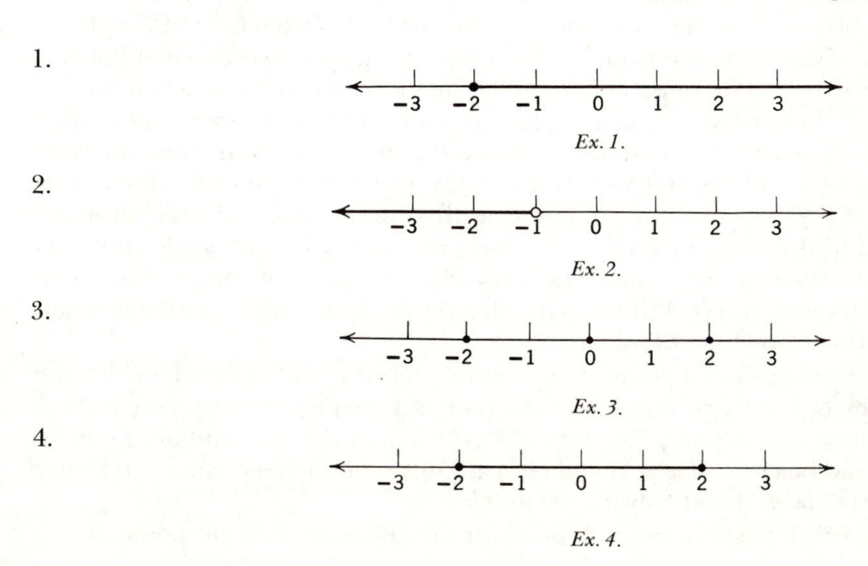

Ex. 1.

Ex. 2.

Ex. 3.

Ex. 4.

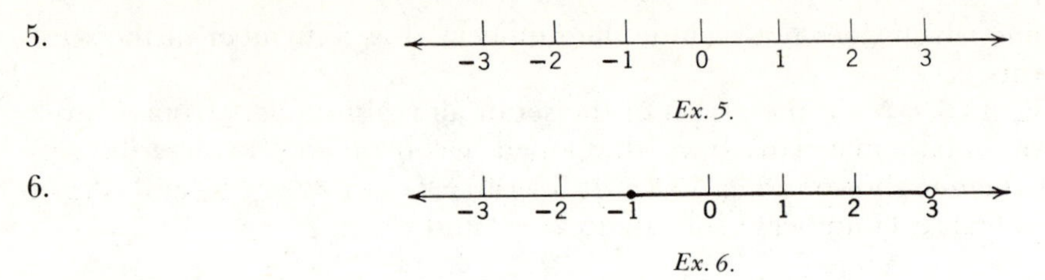

Ex. 5.

Ex. 6.

In Exercises 7-12, draw a graph for each set.

7. $\{x|x < -2\}$.
8. $\{x|x \geqslant 2\}$.
9. $\{x|5 < x < 10\}$.
10. $\{x|-4 < x < 2\}$.
11. $\{x|-2 \leqslant x \leqslant 3\}$.
12. $\{x|-1 \leqslant x < 7\}$.

In Exercises 13–20, draw graphs of $A \cap B$ and $A \cup B$.

13. $A = \{x|x > 1\}$ $B = \{x|x \geqslant 2\}$.
14. $A = \{x|x > 2\}$ $B = \{x|x \leqslant 4\}$.
15. $A = \{x|x < 1\}$ $B = \{x|x > -2\}$.
16. $A = \{x|x < -1\}$ $B = \{x|x > 1\}$.
17. $A = \{x|x < 1\}$ $B = \{x|x < -1\}$.
18. $A = \{x|x-1 = 3\}$ $B = \{x|x+1 = 5\}$.
19. $A = \{x|x-2 = 1\}$ $B = \{x|x < 3\}$.
20. $A = \{x|x < 1\}$ $B = \{x|x > -1\}$.

13.3. Rectangular coordinate system. We will now develop a method of representing points in a plane by pairs of numbers. Consider any two perpendicular lines x and y which intersect at the point 0 (see Fig. 13.5). Let U and V be points on line x and line y, respectively, such that $OU = OV = 1$.

Given any point in the plane of line x and line y, let L be the foot of the perpendicular from P to line x and let M be the foot of the perpendicular from P to line x. The coordinate of L on line x is called the *x-coordinate* or *abscissa* of point P. The coordinate of M on line y is called the *y-coordinate* or *ordinate* of P. The x-coordinate and y-coordinate taken together are called the *coordinates* of P. The matching of points with ordered pairs of real numbers obtained in this manner is called a *rectangular* (or Cartesian) *coordinate system.* Line x is called the *x-axis* and line y is called the *y-axis* of the system. The point of intersection O of the axes is called the *origin* and the plane determined by the axes is called the *XY*-plane.

The coordinates of a point is an ordered pair of real numbers in which the x-coordinate is the first number of the pair and the y-coordinate is the second. The coordinates of point P of Fig. 13.5 are (3,2), of L is (3,0) and of M is (0,2).

It should be clear that (3,2) and (2,3) are different ordered pairs. It is true that $(a, b) = (c, d)$ if and only if $a = c$ and $b = d$.

The coordinates of a point depend on the choice of the unit point. If you

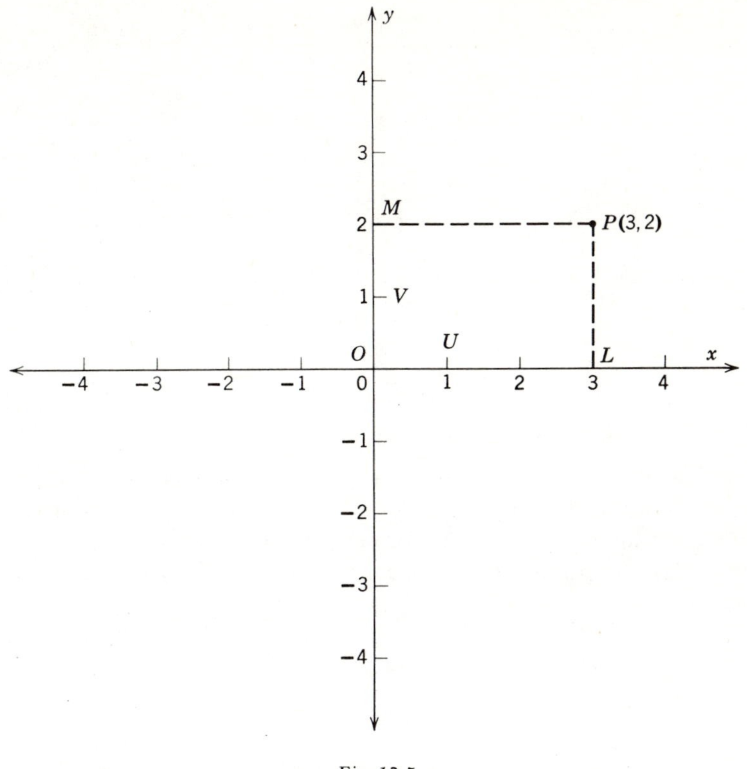

Fig. 13.5.

halve the length of a unit segment in a given coordinate system, you double the coordinates of each point in the system. If you double the length of the unit segment, the coordinate of each point will be halved.

Postulate 25. *There is exactly one pair of real numbers assigned to each point in a given coordinate system. Conversely, if* (a, b) *is any ordered pair of real numbers, there is exactly one point* P *in a given system which has* (a, b) *as its coordinates.*

13.4. Quadrants. The rectangular coordinate axes divide a plane into four regions, called quadrants. These quadrants are numbered I, II, III, IV as shown in Fig. 13.6. The *first quadrant* or *quadrant* I is the set of all points whose x- and y-coordinates are both positive. The *second quadrant* or *quadrant* II is the set of all points whose x-coordinate is negative and whose y-coordinate is positive. The *third quadrant* or *quadrant* III is the set of all points whose x- and y-coordinates are both negative. The *fourth quadrant* or *quadrant* IV is the set of all points whose x-coordinate is positive and whose y-coordinate is negative.

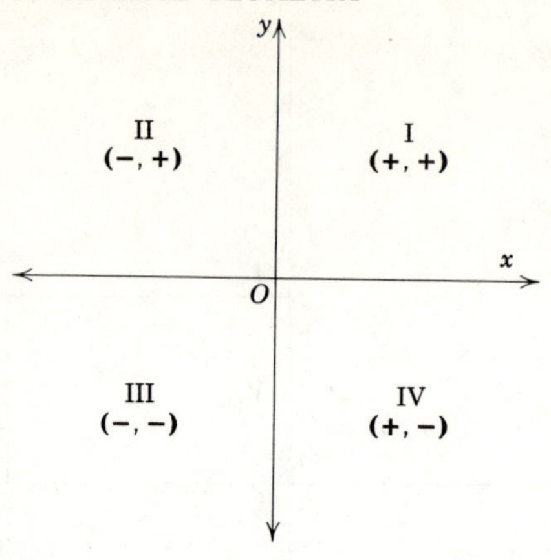

Fig. 13.6.

The process of locating a point when its coordinates are given is called *plotting* or *graphing*. Plotting is facilitated by the use of coordinate or graph paper whose parallel lines divide the paper into small squares.

Example 1. Plot the points $A(-3, 1)$, $B(2, -1\frac{1}{2})$, $C(-1, -2\frac{1}{2})$.

Solution:

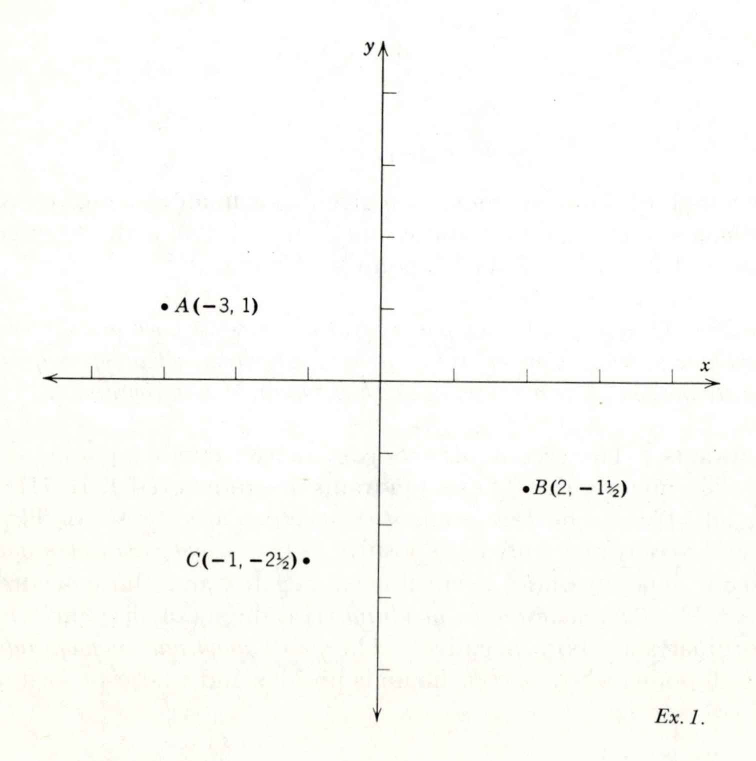

Ex. 1.

Example 2. Draw a line segment with endpoints of $(2, -3)$ and $(3, 1)$.

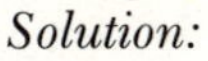

Solution:

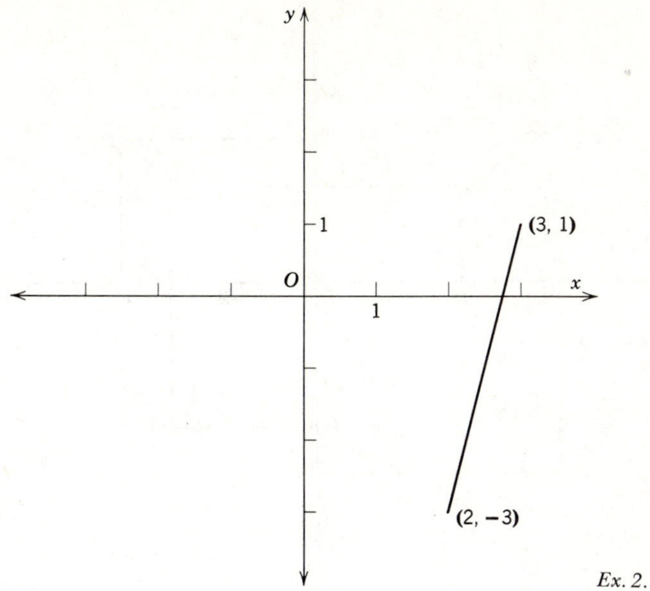

Ex. 2.

Exercises

1. Plot the points $A(2, 0)$, $B(0, -1)$, $C(-2, -2)$, $D(3, 1\frac{1}{2})$, $E(-\frac{1}{2}, 1\frac{1}{2})$, $F(3, \pi)$.
2. In the accompanying figure, $OU = 1$. Name the coordinates of the points shown.

	A	B	C	D	E	P_1	P_2	P_3	P_4
x-coordinate									
y-coordinate									

3. Using the figure of Ex. 2, but letting $OU = \frac{1}{2}$, complete the table.

	A	B	C	D	E	P_1	P_2	P_3	P_4
x-coordinate									
y-coordinate									

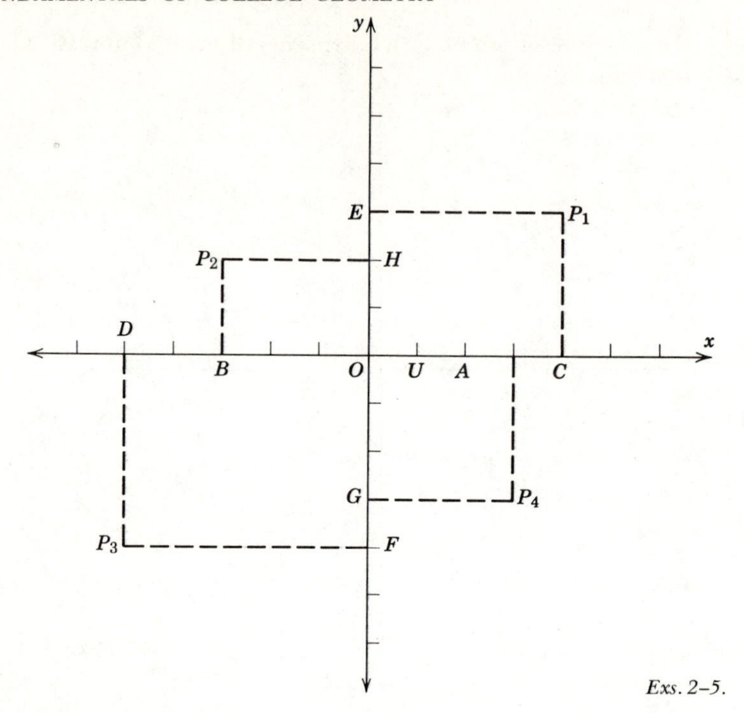

Exs. 2–5.

4. Using the figure of Ex. 2, but letting $OU = 5$, complete the table.

	A	B	C	D	E	P_1	P_2	P_3	P_4
x-coordinate									
y-coordinate									

5. Using the figure for Ex. 2, if $OU = 1$, find (*a*) OA; (*b*) UC; (*c*) BU; (*d*) OD; (*e*) AB; (*f*) CD; (*g*) CP_1; (*h*) EP_1; (*i*) HP_2; (*j*) BP_2; (*k*) FP_3; (*l*) GP_4.
6. If $A = (0, 0)$, $B = (3, 0)$, $C = (4, 5)$, $D = (1, 2)$, plot the set of all points which belong to polygon $ABCD$.
7. Plot the points $R(1, 2)$, $S(-1, 1)$, $T(-1, 3)$. Indicate the points which lie in the interior of $\triangle RST$.
8. Describe the set of all points for which the x-coordinate is 2.
9. Describe the set of all points for which the y-coordinate is -3.
10. Plot the set of all points (x, y) for which x and y are integers which satisfy the following conditions.

(a) $x = 3$, $-2 \leq y \leq 4$. (Read "y is greater than or equal to -2 and y is also less than or equal to 4.")
(b) $y = -2, -1 < x < 3$.
(c) $-3 < x \leq 0, -2 \leq y < 3$.
(d) $x = 3, y = 1$.

11. Plot the set of points (x, y) which satisfies the following conditions.
(a) $x \geq 1, y \geq 2$.
(b) $x \leq 2, y \leq 4$.
(c) $x \geq 3, -3 \leq y \leq -1$.
12. If $A = (2, 5)$ and $B = (2, 9)$, find AB.
13. If $R = (-3, 6)$ and $S = (2, 6)$, find RS.
14. If $C = (8, -7)$ and $D = (-2, -7)$, find CD.
15. If $A = (x_1, y_1)$ and $B = (x_1, y_2)$, find AB.
16. If $K = (x_1, y_1)$ and $L = (x_2, y_1)$, find KL.
17. If $E = (0, 1)$ and $F = (0, 7)$, find the coordinates of the midpoint of $\overline{EF}$.
18. If $G = (4, 6)$ and $H = (4, -2)$, find the coordinates of the midpoint of $\overline{GH}$.
19. If $H = (x_1, y_1)$ and $K = (x_2, y_1)$, find the coordinates of the midpoint of $\overline{HK}$.
20. If $S = (x_1, y_1)$ and $T = (x_1, y_2)$, find the coordinates of the midpoint of $\overline{ST}$.

13.5. The distance formula. Suppose P and Q are two points with coordinates (x_P, y_P) and (x_Q, y_Q). If $\overleftrightarrow{PQ}$ is not perpendicular to either the x-axis, or the y-axis, then perpendiculars from P and Q to the coordinate axes will intersect at R and S (Fig. 13.7). It can be readily proved that $\triangle PRQ$ is a right triangle. (Why?) $\overline{PQ}$ is the hypotenuse of right $\triangle PRQ$.

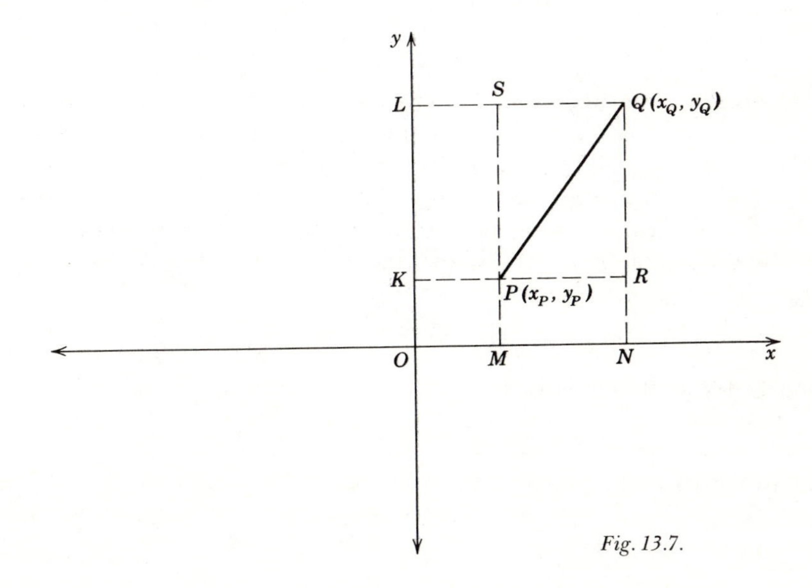

Fig. 13.7.

By the Pythagorean theorem,

$$(PQ)^2 = (PR)^2 + (RQ)^2$$

or

$$PQ = \sqrt{(PR)^2 + (RQ)^2}$$

But, $PR = MN = x_Q - x_P$ and $RQ = KL = y_Q - y_P$. Using the substitution property of equality, we get

$$PQ = \sqrt{(x_Q - x_P)^2 + (y_Q - y_P)^2}$$

Since $(x_Q - x_P)^2 = (x_P - x_Q)^2$ and $(y_Q - y_P)^2 = (y_P - y_Q)^2$, we can state the following.

Theorem 13.1 (The distance formula). *For any two points P and Q*

$$PQ = \sqrt{(x_Q - x_P)^2 + (y_Q - y_P)^2} = \sqrt{(x_P - x_Q)^2 (y_P - y_Q)^2}.$$

The student should convince himself that the distance formula will also work if the segment is perpendicular to either axis.

Example: Find the distance between $P_1(-3, -4)$ and $P_2(2, -7)$.

Solution: Using Theorem 13.1, we get

$$\begin{aligned} P_1P_2 &= \sqrt{[2-(-3)]^2 + [-7-(-4)]^2} \\ &= \sqrt{25+9} \\ &= \sqrt{34} \end{aligned}$$

13.6. Midpoint of a segment. Quite frequently we will wish to find the coordinates of a point which bisects the segment between two given points. In Fig. 13.8, let $M(x_M, y_M)$ be the midpoint of the line segment joining $A(x_A, y_A)$ and $B(x_B, y_B)$. If through each point A, M, and B we draw lines parallel to the coordinate axes, two congruent right triangles will be formed. (Can you prove them congruent?). We know that $AP = MQ$. But $AP = x_M - x_A$, and $MQ = x_B - x_M$. Therefore,

$$x_M - x_A = x_B - x_M$$

Adding $x_M + x_A$ to both terms,

$$x_M + x_M = x_A + x_B$$

Then, dividing by 2,

$$x_M = \frac{x_A + x_B}{2}$$

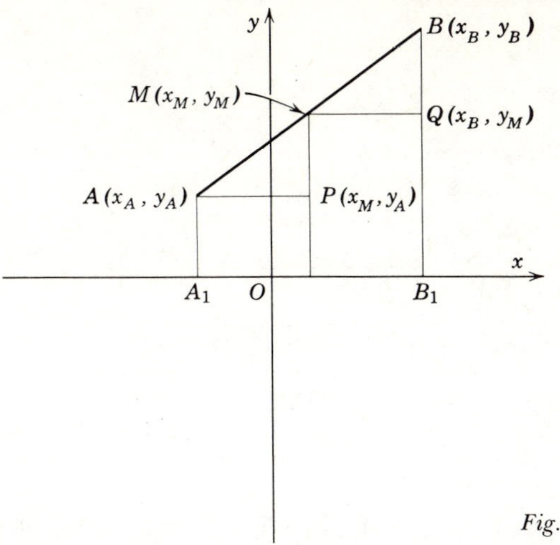

Fig. 13.8.

Similarly, by setting $PM = QB$, we get

$$y_M = \frac{y_A + y_B}{2}$$

Theorem 13.2 (the midpoint formula). *M is the midpoint of* $\overline{AB}$ *if and only if* $x_M = \frac{1}{2}(x_A + x_B)$ *and* $y_M = \frac{1}{2}(y_A + y_B)$.

Example. Find the length of the median from vertex B of $\triangle ABC$ with the following vertices: $A(-1, 1)$, $B(3, 4)$, $C(5, -7)$.

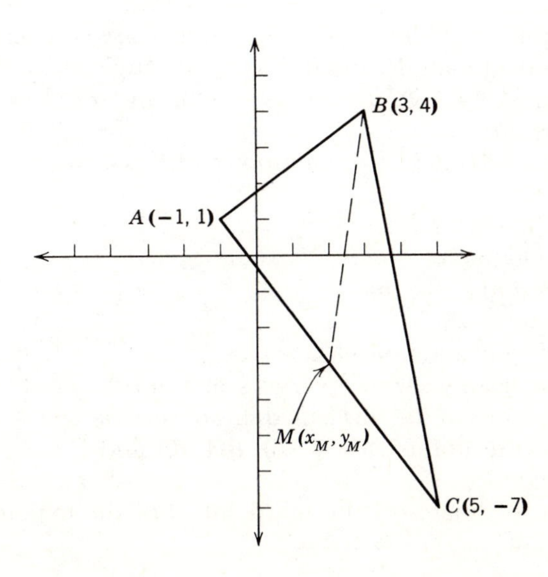

Solution:

$$x_M = \frac{x_A + x_C}{2} = \frac{-1+5}{2} = 2$$

$$y_M = \frac{y_A + y_C}{2} = \frac{1-7}{1} = -3$$

$$\begin{aligned} BM &= \sqrt{(x_M - x_B)^2 + (y_M - y_B)^2} \\ &= \sqrt{(2-3)^2 + (-3-4)^2} \\ &= \sqrt{50} \text{ or } 5\sqrt{2} \end{aligned}$$

Exercises

1. Given points $O(0, 0)$, $A(3, 4)$, $B(-5, 12)$, $C(15, -8)$, $D(11, -3)$, $E(-9, -4)$. Determine the lengths of the following segments:
   (*a*) $\overline{OA}$; (*b*) $\overline{OB}$; (*c*) $\overline{OC}$; (*d*) $\overline{AD}$; (*e*) $\overline{AB}$; (*f*) $\overline{AC}$; (*g*) $\overline{BE}$.
2. Plot points $A(-1, 3)$, $B(-4, 7)$, $C(0, 4)$. Prove $\triangle ABC$ isosceles.
3. Find the perimeter of $\triangle ABC$ of Ex. 2.
4. Prove that points $A(-2, -2)$, $B(4, -2)$, and $C(4, 6)$ are vertices of a right triangle. (Assume the converse of the Pythagorean theorem.)
5. Find the area of $\triangle ABC$ of Ex. 4.
6. Prove that $A(-4, -3)$, $B(1, 4)$, and $C(6, 11)$ are collinear.
7. Prove that points $R(-3, 1)$, $S(5, 6)$, and $T(7, -15)$ are vertices of a right triangle.
8. Prove that $A(-2, 0)$, $B(6, 0)$, $C(6, 6)$, $D(-2, 6)$ are vertices of a rectangle.
9. Prove that diagonals AC and BD of the rectangle of Ex. 8 are congruent.
10. Prove that $A(-4, -1)$, $B(1, 0)$, $C(7, -3)$, and $D(2, -4)$ are vertices of a parallelogram.
11. Find the coordinates of the midpoint of the segment with the given endpoints.
    (*a*) $(8, -5)$ and $(-2, 9)$.
    (*b*) $(7, 6)$ and $(3, 2)$.
    (*c*) $(-2, 3)$ and $(-9, -6)$.
    (*d*) (a, b) and (c, d).
    (*e*) $(a+b, a-b)$ and $(-a, b)$, $a \neq 0$.
12. Graph the triangle whose vertices are $A(5, 2)$, $B(-3, -4)$ and $C(8, -6)$. Find the lengths of the three medians of the triangle.
13. Prove that the points $A(-8, -6)$, $B(4, 10)$ and $C(4, -6)$ are vertices of a right triangle.
14. Find the coordinates of the midpoint M of the hypotenuse of $\triangle ABC$ of Ex. 3. Show that $MA = MB = MC$.

15. Given points $R(-6, 0)$, $S(6, 0)$, $T(0, 8)$; M the midpoint of $\overline{RT}$; N the midpoint of $\overline{ST}$. Show that $\triangle RST$ is isosceles and that $RN = SM$.

16. *Given:* Right $\triangle ABC$ with vertices at $A(0, 0)$, $B(a, 0)$, and $C(0, b)$; M is the midpoint of the hypotenuse BC.

 Prove: $AM = BM = CM$.

17. *Given:* $\triangle ABC$ with vertices at $A(-a, 0)$, $B(a, 0)$, and $C(0, b)$.

 Prove: $\triangle ABC$ is isosceles; median $AN =$ median BM.

18. Show that $ABCD$ is a parallelogram if its vertices are at $A(0, 0)$, $B(a, 0)$, $C(a+b, c)$, $D(b, c)$.

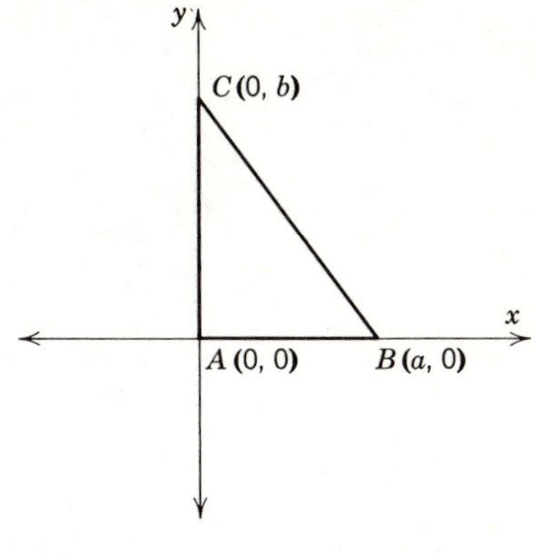

Ex. 16.

19. Find the coordinates of the midpoints of the diagonals of ▱ $ABCD$ of Ex. 18 and then show analytically that the diagonals bisect each other.

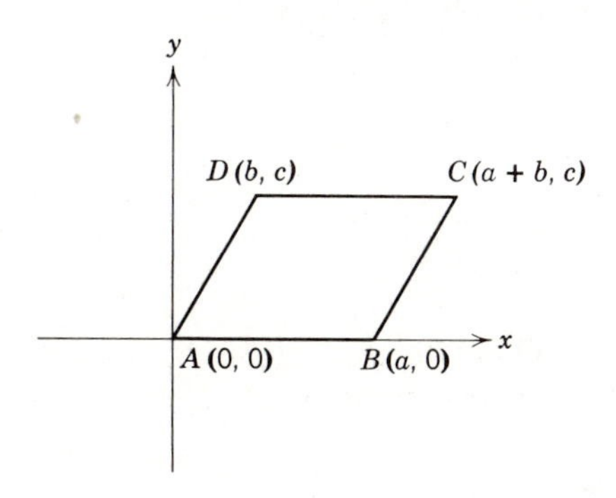

Ex. 18.

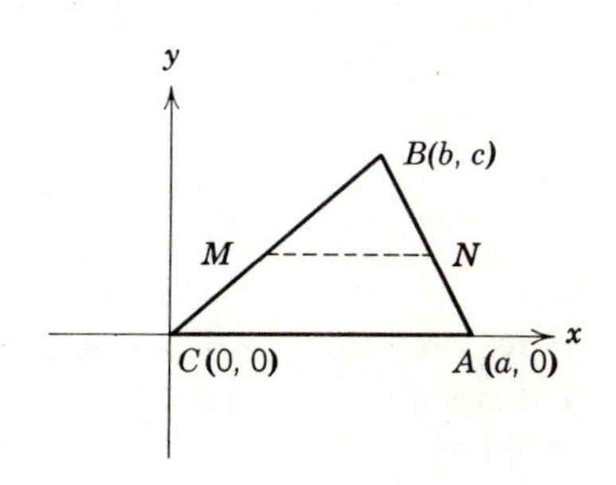

Ex. 20.

20. Prove analytically that the measure of the segment joining the midpoints of the two sides of a triangle is equal to half the measure of the third side.
21. Prove analytically that $O(4, 3)$ is the center of a circle passing through $P(5, 10)$, $Q(11, 4)$, and $R(-1, -2)$.
22. Draw a circle with center at $O(2, 3)$ and passing through $P(-8, 9)$. Does the circle pass through: (*a*) $A(8, 13)$? (*b*) $B(12, -3)$? (*c*) $C(5, -7)$?

13.7. Slope and inclination. Let l be a straight line that is not parallel to the y-axis. Let $P(x_P, y_P)$ and $Q(x_Q, y_Q)$ be any two distinct points on l (see Fig. 13.9) and consider the number m defined by $m = (y_Q - y_P)/(x_Q - x_P)$. The

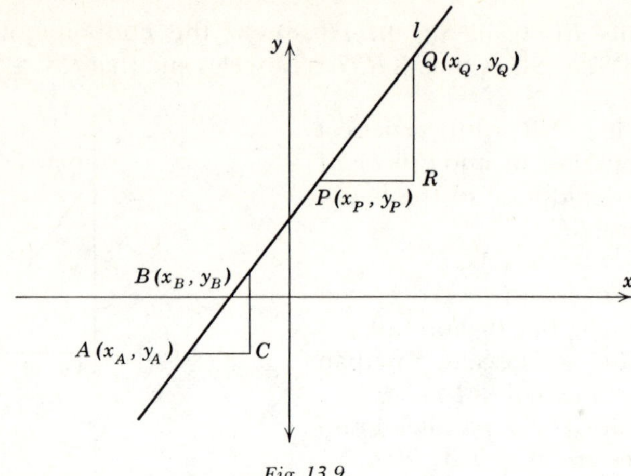

Fig. 13.9.

question naturally arises, "Will different pairs of points on the line lead to different values for m?"

Consider any other two points of the line, $A(x_A, y_A)$ and $B(x_B, y_B)$. It can be shown that $\triangle ACB \sim \triangle PRQ$.
Then

$$\frac{RQ}{PR} = \frac{CB}{AC}$$

or

$$\frac{y_Q - y_P}{x_Q - x_P} = \frac{y_B - y_A}{x_B - x_A}$$

Theorem 13.3. (The slope formula). If $P \neq Q$ *are any pair of points on a line not parallel to the y-axis of a rectangular coordinate system, then there is a unique real number*

$$m = \frac{y_Q - y_P}{x_Q - x_P}$$

This number is called the *slope* of the line.

In layman's language, the slope is a measure of the steepness of a line, i.e., it is the ratio of the "rise" to "run" of the line. Thus a 10 percent slope rises 10 feet for every 100 feet of horizontal run.

Let line l intersect the x-axis at B (Fig. 13.10). If we take any point P on l which lies on the positive y-side of the x-axis (quadrant I or II), then $\angle\alpha$ is called the *angle of inclination* and $m\angle\alpha$ is called the *inclination* of l. It should be evident that $0 \leq m\angle\alpha \leq 180$.

If a line is parallel to the x-axis, or coincident with it, then $m\angle\alpha = 0$ and slope $m = 0$. We say such a line is a *horizontal line*. If a line is parallel to the

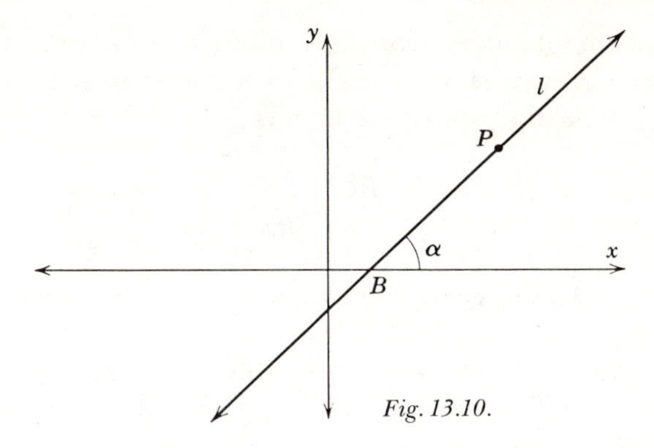

Fig. 13.10.

y-axis, or is coincident with it, $m\angle\alpha = 90$ and slope m is not defined. Such lines are called *vertical lines*. We also speak of horizontal and vertical segments and rays.

A review of the slope function will reveal that:

1. If $0 < m\angle\alpha < 90$, the slope m is positive and the line slopes upward to the right.
2. If $90 < m\angle\alpha < 180$, the slope m is negative and the line slopes downward to the right.

13.8. Parallel lines. When two lines are both parallel to the x-axis each has a slope of zero. Hence their slopes are equal. Next consider two nonhorizontal, nonvertical lines l_1 and l_2 that are parallel (see Fig. 13.11). Let the angles of inclination for lines l_1 and l_2 be α_1 and α_2, respectively. Since $l_1 \parallel l_2$, we know $\angle\alpha_1 \cong \angle\alpha_2$.

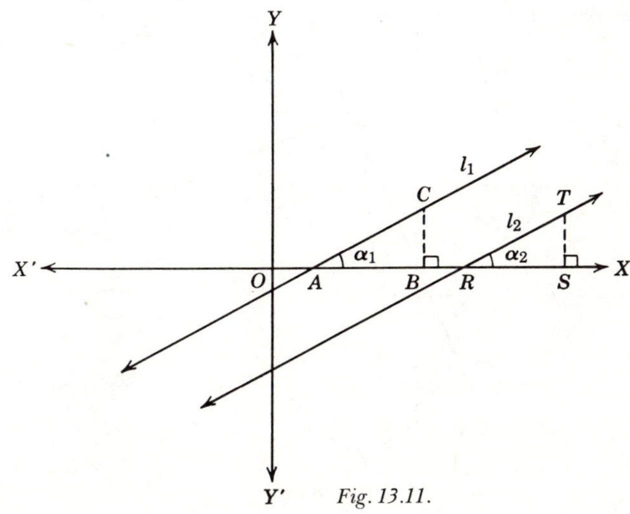

Fig. 13.11.

If we drop perpendiculars from any point C on l_1 to $\overleftrightarrow{X'X}$ and from any point T on l_2 to $\overleftrightarrow{X'X}$, two similar triangles will be formed. Hence the corresponding sides will be in proportion. That is,

$$\frac{BC}{AB} = \frac{ST}{RS}.$$

But $BC/AB = m_1$, and $ST/RS = m_2$. Then, $m_1 = m_2$ by the substitution property. Thus, $l_1 \parallel l_2 \rightarrow m_1 = m_2$.

Conversely, when $m_1 = m_2$, we can readily show that $l_1 \parallel l_2$. For if $m_1 = m_2$, then $BC/AB = RT/RS$. Since $\angle ABC \cong \angle RST$, we know by Theorem 8.13 that $\triangle ABF \sim \triangle RST$. Hence, $\angle \alpha_1 \cong \angle \alpha_2$ and $l_1 \parallel l_2$ by Theorem 5.12.

Theorem 13.4. *Two nonvertical lines l_1 and l_2 are parallel if and only if their slopes m_1 and m_2 are equal:*

$$l_1 \parallel l_2 \leftrightarrow m_1 = m_2$$

13.9. Perpendicular lines. Consider the two perpendicular lines l_1 and l_2 neither of which is vertical and with respective slopes m_1 and m_2 (see Fig. 13.12). Let P be their point of intersection. Next select any point R

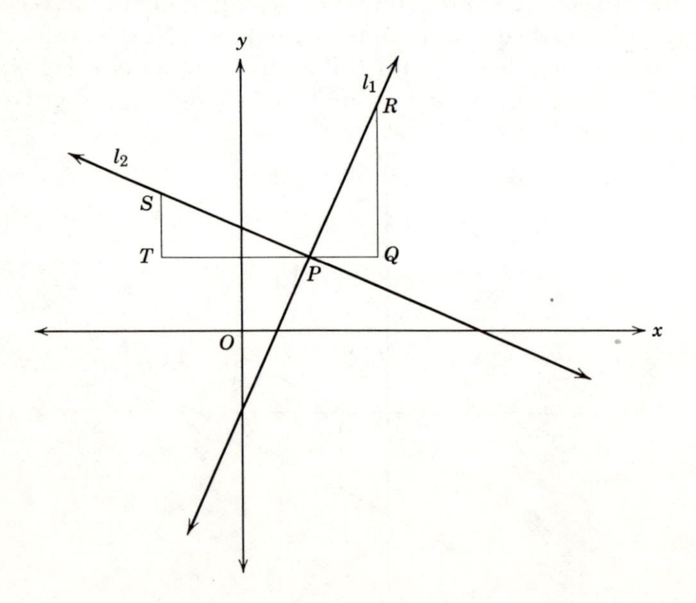

Fig. 13.12.

on l_1 lying above and to the right of P and a point S on l_2 above and to the left of P such that $PS = PR$. By drawing lines through S and R perpendicular to the x-axis and a line through P parallel to the x-axis, we form two right triangles $\triangle PQR$ and $\triangle PTS$.

By Theorem 5.19 we can prove $\triangle PQR \cong \triangle STP$. Therefore,

$$QR = TP \quad \text{and} \quad PQ = ST$$

and

$$\frac{QR}{PQ} = \frac{TP}{ST}$$

But

$$\frac{QR}{PQ} = m_1 \quad \text{and} \quad m_2 = -\frac{ST}{TP} = -\frac{PQ}{QR}$$

Therefore

$$m_2 = -\frac{1}{m_1}$$

Thus, we have shown that the slopes of two nonvertical perpendicular lines are negative reciprocals of each other. The converse of this fact can be proved by retracing our steps.

Theorem 13.5. *Two nonvertical lines l_1 and l_2 are perpendicular if and only if their slopes are negative reciprocals of each other*:

$$l_1 \perp l_2 \leftrightarrow m_1 = -\frac{1}{m_2}$$

Exercises

In Exs. 1–6 find the slope of the line passing through the points.

1. $(7, 3), (4, -3)$.
2. $(-3, 3), (3, -4)$.
3. $(4, 1), (-1, 6)$.
4. $(-3, 0), (1, 2)$.
5. $(-3, 2), (2, 1)$.
6. $(6, -4), (2, -3)$.

In Exs. 7–12 check to see if $\overline{AB} \parallel \overline{CD}$ or if $\overline{AB} \perp \overline{CD}$.

7. $A(-4, 2), B(4, -1), C(3, 2), D(11, -1)$.
8. $A(2, -2), B(-3, 4), C(5, 0), D(-1, -5)$.
9. $A(2, 4), B(0, 0), C(-2, 1), D(-1, 3)$.
10. $A(-1, -2), B(-2, -4), C(3, -2), D(1, -1)$.
11. $A(0, 4), B(0, 11), C(3, -3), D(4, -3)$.
12. $A(-4, -5), B(-3, 4), C(3, 1), D(-6, 2)$.

In Ex. 13–15 prove the points are collinear.

13. $A(3,4), B(4,6), C(2,2)$.
14. $A(-3,0), B(-1,1), C(3,3)$.
15. $A(-4,-6), B(0,-7), C(-8,-5)$.

Prove the following theorems analytically (figures are drawn as an aid for your proofs):

16. The segment joining the midpoints of two sides of a triangle is parallel to the third side.
17. The median of a trapezoid is parallel to the bases and has a measure equal to half the sum of the measure of the bases.
18. The diagonals of a square intersect at right angles.

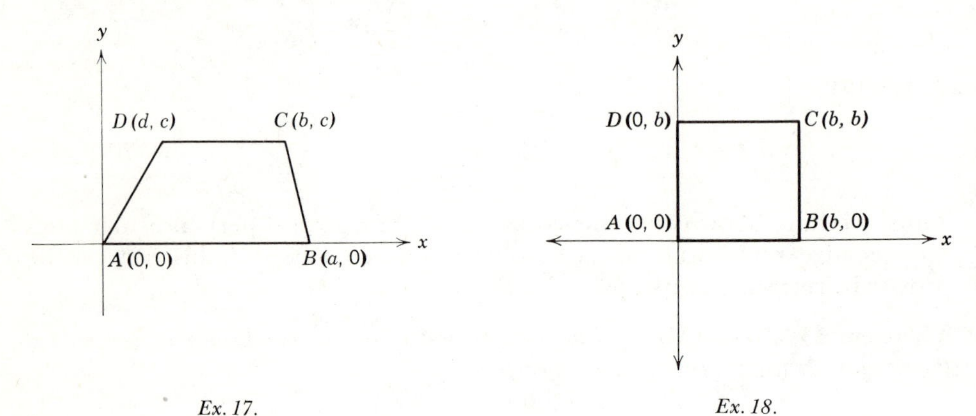

Ex. 17. *Ex. 18.*

19. The lines joining the midpoints of the opposite sides of a quadrilateral bisect each other.
20. The diagonals of a rhombus intersect at right angles.

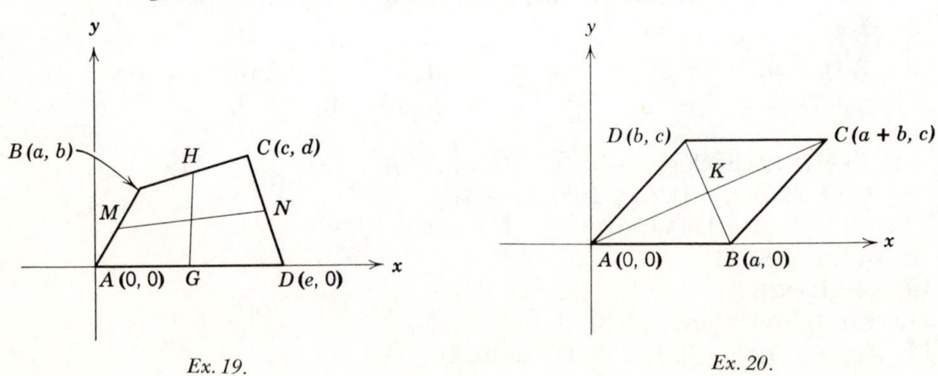

Ex. 19. *Ex. 20.*

13.10. The graph of a condition. The *graph of a condition* imposed on two variables is the set of points whose coordinates (x, y) satisfy the condition. Often the condition is given in equation or inequation form. *To graph an equation* (or inequation) or condition in x and y means to draw its graph. To obtain it, we draw the figure which represents *all* the points whose coordinates satisfy the condition. The graph might be lines, rays, segments, triangles, circles, half-planes, or subsets of each. There follow examples which illustrate the relationship between a condition and its graph.

Example. Draw the graphs of the point $P(x, y)$ which satisfy the following conditions: (*a*) $x = 3$; (*b*) $x = 0$; (*c*) $y = -5$; (*d*) $y \geq 0$; (*e*) $1 < x \leq 4$; (*f*) $OP \leq 2$.

Solution:

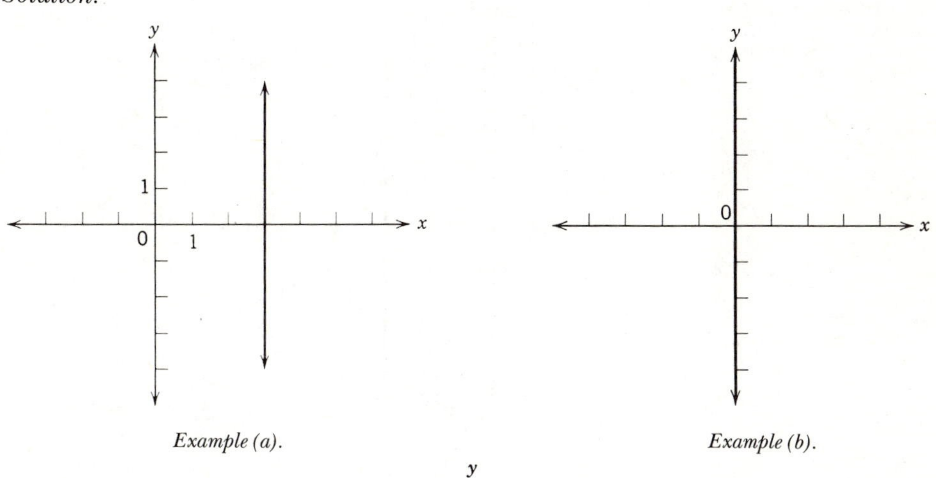

Example (a).

Example (b).

Example (c).

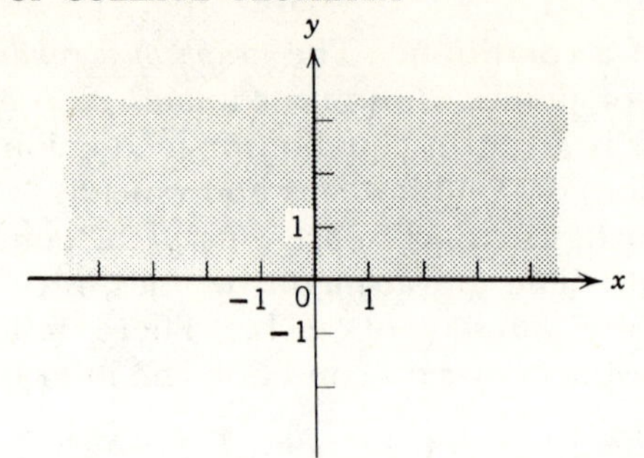

Example (d).

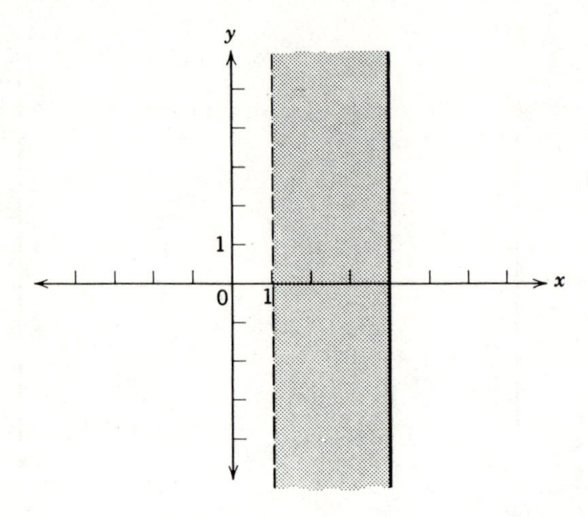

Example (e).

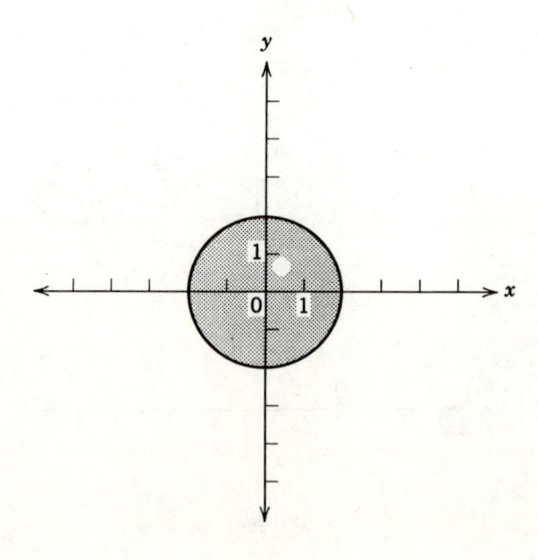

Example (f).

Exercises

Draw and describe the graphs which satisfy the following conditions.

1. $x \geq 0$.
2. $y < 0$.
3. $x > 0$ and $y > 0$.
4. $x < 0$ and $y > 0$.
5. $x > 0$ and $y < 0$.
6. $-1 < x < 2$.
7. $x > 0$ or $x < -2$.
8. $y < 1$ or $y > 4$.
9. $-1 < x < 3$ and $2 < y < 4$.
10. $2 \leq x \leq 4$ and $3 \leq y \leq 5$.
11. $x = 1$ and $y \geq 0$.
12. $|x| = 3$.
13. $|y| < 2$.
14. $|y| > 2$.
15. $x > 2$ and $y < -1$.
16. $x > 0, y > 0$, and $y = x$.

13.11. Equation of a line. The equation of a line in a plane is an equation in two variables, such as x and y, which is satisfied by every point on the line and is not satisfied by any point not on the line. The form of the equation will depend upon the data used in determining the line. A straight line is determined geometrically in several ways. If two points are used to determine the line, the equation of the line will have a different form than if one point and a direction were used. We will consider some of the more common forms of the equation for a straight line.

13.12. Horizontal and vertical lines. If a line l_1 is parallel to the y-axis, then every point on l_1 has the same x-coordinate (see Fig. 13.13). If this x-coordinate is a, then the point $P(x, y)$ is on l_1 *if and only if* $x = a$.

In like manner, $y = b$ is the equation of l_2, a line through $(0, b)$ parallel to the x-axis.

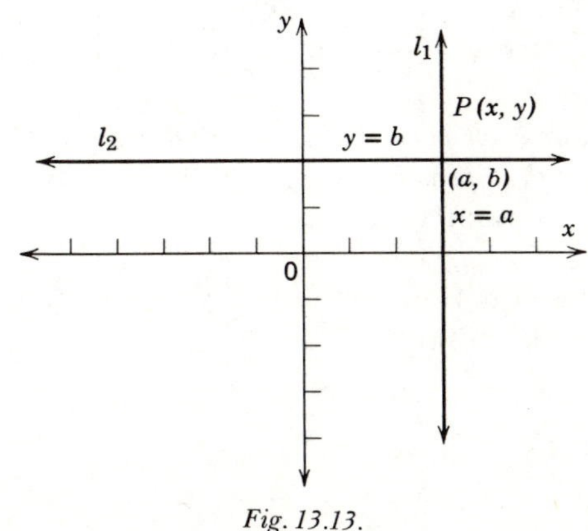

Fig. 13.13.

13.13. Point-slope form of equation of a line. One of the simplest ways in which a line is determined is to know the coordinates of a point through which it passes and the slope of the line.

Consider a nonvertical line l passing through $P_1(x_1, y_1)$ with a slope m (see Fig. 13.14). Let $P(x, y)$ be *any* point other than P_1 on the given line. P

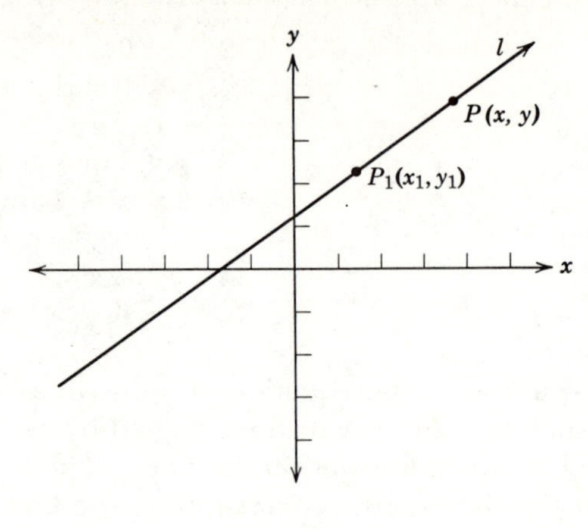

Fig. 13.14.

will lie on l *if and only if* the slope of $\overleftrightarrow{PP_1}$ is m; that is $P(x, y)$ is on $l \leftrightarrow$ slope of $\overleftrightarrow{PP_1}$ is m, or

$$P(x, y) \text{ is on } l \leftrightarrow \frac{y-y_1}{x-x_1} = m$$

and

$$P(x, y) \text{ is on } l \leftrightarrow y-y_1 = m(x-x_1).$$

Theorem 13.6. *For each point $P_1(x_1, y_1)$ and for each number m, the equation of the line through P_1 with slope m is $y-y_1 = m(x-x_1)$.* This equation is called the *point-slope form* of an equation of a line.

Example. Find the equation of the line which contains the point with coordinates $(2, -3)$ and has a slope of 5.

Solution:

$$\begin{aligned} y-y_1 &= m(x-x_1) \\ y-(-3) &= 5(x-2) \\ y+3 &= 5x-10 \\ 5x-y-13 &= 0 \end{aligned}$$

Example. Find the slope of the line whose equation is $5x-2y = 11$.

Solution: We must first reduce our equation to the point-slope form, as follows:

$$5x-2y=11$$
$$5x-11=2y \quad \text{(Why?)}$$
$$2y=5x-11 \quad \text{(Why?)}$$
$$=5(x-\tfrac{11}{5})$$

$$y=\tfrac{5}{2}(x-\tfrac{11}{5}) \quad \text{(Why?)}$$

or

$$y-0=\tfrac{5}{2}(x-\tfrac{11}{5})$$

Comparing this equation with the standard point-slope form equation, we find that the line passes through $(\frac{11}{5}, 0)$ and has a slope of $\frac{5}{2}$.

13.14. Two-point form of equation of a line. The equation of a straight line that passes through two points can be obtained by use of the point-slope form and the equation for the slope of a line through two points. Thus, if $P_1(x_1, y_1)$ and $P_2(x_2, y_2)$ are coordinates of two points through which the line passes, the slope of the line is $m=(y_1-y_2)/(x_1-x_2)$ and substituting this value for m in the point-slope form, we get the equation

$$y-y_1=\frac{y_1-y_2}{x_1-x_2}(x-x_1)$$

This is called the *two-point form* of the equation of a straight line.

Exercises

Find an equation for each of the lines described.

1. The line contains the point with coordinates (7, 3) and its slope is 4.
2. The line contains the point with coordinates (−2, −5) and has a slope of 3.
3. The line has a slope of −2 and passes through (−6, 8).
4. The line has an inclination of 45 and passes through (3, 5).
5. The line passes through (−9, −3) and is parallel to the x-axis.
6. The line contains the point (5, −7) and is perpendicular to the x-axis.
7. The line contains the points with coordinates (4, 7) and (6, 11).
8. The line contains the point whose coordinates are (−1, 1) and has an inclination of 90.

In Exs. 9–16 find the slope of each of the lines with the following equations.

9. $3x-y=7$.
10. $2x+y=8$.
11. $5x+3y=9$.
12. $y=x$.

13. $2x = y$.
14. $y = 2x - 7$.
15. $y = -3$.
16. $x = 6$.

In Exs. 17–22, points $A(-2, 4)$, $B(2, -4)$, $C(6, 6)$ are vertices of $\triangle ABC$.

17. Find the equation of $\overleftrightarrow{AB}$.
18. Find the equation of $\overleftrightarrow{BC}$.
19. Find the equation of the median drawn from C.
20. Find the equation of the median drawn from A.
21. Find the equation of the perpendicular bisector of $\overline{AB}$.
22. Find the equation of the perpendicular bisector of $\overline{BC}$.

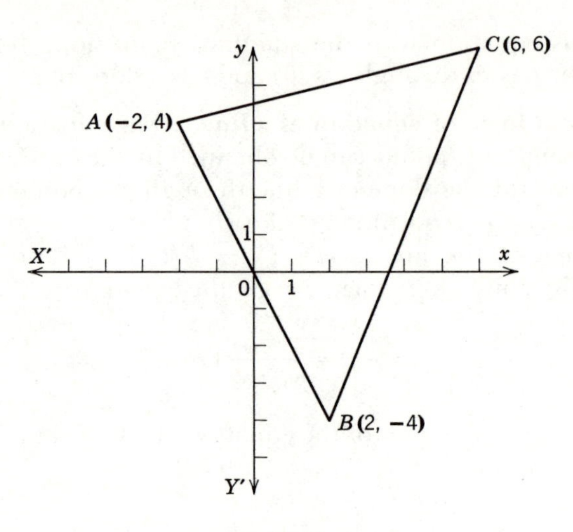

Exs. 17–22.

13.15. Intercept form of equation of a line. The *x-intercept* and *y-intercept* of a line are defined as the coordinates of the points where the line crosses the x-axis and the y-axis respectively. The terms are also used for the distances these points are from the origin. The context of the statement will make clear if a coordinate or distance is meant.

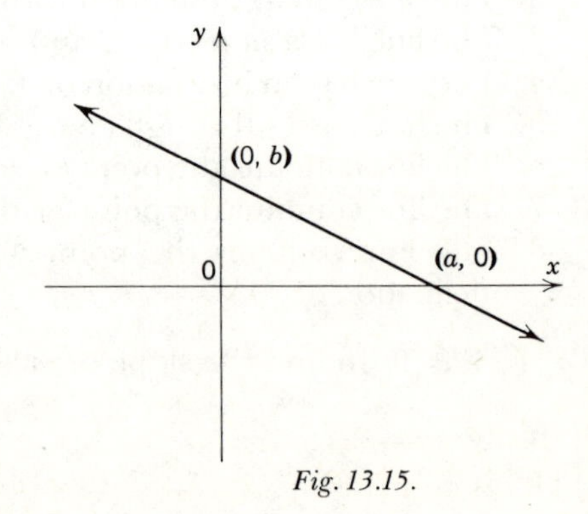

Fig. 13.15.

If the x-intercept and y-intercept of a line are respectively a and b (Fig. 13.15), the coordinates of the points of intersection of the line and axes are $(a, 0)$ and $(0, b)$. Using the two-point form,

$$y - b = \frac{b - 0}{0 - a}(x - 0)$$

or

$$y - b = -\frac{b}{a}x$$

This may be reduced to

$$bx + ay = ab$$

and by dividing both sides of the equation by ab, we get

$$\frac{x}{a} + \frac{y}{b} = 1$$

which is the *intercept form* of the equation of a line.

Example. Draw the graph of the line L whose equation is $3x - 4y - 12 = 0$.
Solution: $3x - 4y - 12 = 0$.

Adding 12 to both sides,

$$3x - 4y = 12.$$

Dividing both sides by 12,

$$\frac{x}{4} - \frac{y}{3} = 1$$

or

$$\frac{x}{4} + \frac{y}{(-3)} = 1$$

Hence the x-intercept is 4 and the y-intercept is -3.

13.16. Slope and y-intercept form of the equation of a line. If the y-intercept of a line is b and the slope of the line is m, we can determine the equation of the line by using the point-slope form. Thus,

$$y - b = m(x - 0)$$

or

$$y = mx + b$$

This is called the *slope y-intercept* form.

Theorem 13.7. *The graph of the equation $y = mx + b$ is the line with slope m and y-intercept b.*

Example. What is the slope of the line whose equation is $2x - 5y - 17 = 0$.
Solution: Reduce the equation to $y = mx + b$ form, as follows

$$2x - 5y - 17 = 0$$
$$2x - 17 = 5y \quad \text{(Why?)}$$
$$5y = 2x - 17 \quad \text{(Why?)}$$
$$y = \tfrac{2}{5}x - \tfrac{17}{5} \quad \text{(Why?)}$$

The slope of the line is $\frac{2}{5}$.

13.17. The general form of the equation of a line. The most general form of an equation of the first degree in x and y is

$$Ax + By + C = 0$$

where A and B are not both zero.

If $B \neq 0$, we can solve for y to get

$$y = -\frac{A}{B}x - \frac{C}{B}$$

This equation has the form $y = mx + b$ and, hence, must be the equation of a straight line with

$$\text{slope } m = -\frac{A}{B} \quad \text{and} \quad y\text{-intercept } b = -\frac{C}{B}$$

If $B = 0$, we can solve the equation for x to get $x = -C/A$, which is the equation of a line parallel to the y-axis. If $A = 0$, $y = -C/B$, which is the equation for a line parallel to the x-axis.

Thus, we have proved the following:

Theorem 13.8. *The graph of every linear equation in x and y is always a straight line (in the XY-plane).*

The converse of Theorem 13.8 can be proved by noting that every straight line must either intersect the y-axis or be parallel to it and, hence, can be expressed either by the equation $x = a$, where a is a constant, or by $y = mx + b$. The equation $y = mx + b$ can be converted to $Ax + By + C = 0$ form.

Theorem 13.9. *Every straight line in the XY-plane is the graph of a linear equation in x and y.*

Exercises

What are the x- and y-intercepts of the graphs of the following equations? What are the slopes of each equation? Draw graphs of the equations.

1. $3x+4y=12$.
2. $2x-3y-6=0$.
3. $x+5=0$.
4. $y-7=0$.

Determine the equations, in $Ax+By+C=0$ form, of which the following lines are graphs.

5. The line through (2, 3) with slope 5.
6. The line through (−5, 1) with slope 7.
7. The line through (4, −3) with slope −2.
8. The line through (1, 1) and (4, 6).
9. The line through (2, −3) and (0, −9).
10. The line through (−10, −7) and (−6, −2).
11. The line with y-intercept 5 and slope 2.
12. The line with y-intercept −3 and slope 1.
13. The line with y-intercept −4 and slope −3.
14. The x-axis.
15. The y-axis.
16. The vertical line through (−5, 7).
17. The vertical line through (3, −8).
18. The line through (2, 5) and parallel to the line passing through (−2, −4) and (6, 8).
19. The line through (4, 4) and perpendicular to the line passing through (1, 1) and (7, 7).
20. The line with x-intercept 5 and y-intercept 1.
21. The line through (−1, 3) and parallel to the line whose equation is $6x-2y-15=0$.
22. The line through (5, −2) and parallel to the line whose equation is $2x+5y+20=0$.
23. The line through (0, 0) and perpendicular to the line whose equation is $5x-y-9=0$.
24. The line through (−4, −7) and perpendicular to the line whose equation is $6x+5y-18=0$.

Summary Tests

Test 1

COMPLETION STATEMENTS

1. The coordinates of the origin are ______ .
2. The line through $(2, -3)$ perpendicular to the x-axis intersects the x-axis at the point whose coordinates are ______ .
3. The point $(8, -5)$ lies in the ______ quadrant.
4. If k is a negative number and h is a positive number, the point $(-k, -h)$ lies in the ______ quadrant.
5. The coordinates of the point which is the intersection of the x-axis and the y-axis is ______ .
6. The line through $(-2, -4)$ and (______, -8) is vertical.
7. A road with a 6% slope will rise ______ feet for every horizontal run of 500 feet.
8. The slope of the line through $A(a-b, b)$ and $B(a+b, a)$ is ______ .
9. The line through $(2, -4)$ and (-3, ______) has a slope of $\frac{1}{2}$.
10. The midpoint of the segment whose endpoints are $(-7, 6)$ and $(3, -4)$ has the coordinates ______ .
11. A point lies in the ______ quadrants if its abscissa is numerically equal to its ordinate but opposite in sign.
12. The points $(3, -2)$, $(4, 3)$, and $(-6, 5)$ are vertices of a ______ triangle.
13. Three vertices of a rectangle are the points whose coordinates are $(-3, 4)$, $(-3, -2)$, and $(1, 4)$. The coordinates of the fourth vertex are ______ .
14. The points $(1, 6)$, $(4, -5)$, and $(0, 17)$ are coordinates of a (an) ______ triangle.

15. $3x - 4y - 12 = 0$ and $6x + (______) - 6 = 0$ are equations of two parallel lines.
16. $4y - 10x - 30 = 0$ and $2x + (______) - 15 = 0$ are equations of two perpendicular lines.

Test 2

PROBLEMS

1–7. Given $\triangle ABC$ with vertices at $A(-4, 7)$, $B(10, 5)$, and $C(-6, -8)$.

1. Find the length of $\overline{BC}$.
2. Find the coordinates of the midpoint of $\overline{AB}$.
3. Find the slope of $\overleftrightarrow{AC}$.
4. Find the equation of $\overleftrightarrow{AB}$.
5. Find the equation of the median from C.
6. Find the equation of the altitude from B.
7. Find the equation of the line through B and parallel to $\overleftrightarrow{AC}$.
8. What are the coordinates of the center of a circle with a diameter whose endpoints are at $(3, -7)$ and $(-5, 2)$.
9. Find the area of $\triangle RST$ if its vertices are at $R(7, 5)$, $S(-4, 2)$, and $T(0, -2)$.
10. Find the area of the circle circumscribed about $\triangle RST$ of Ex. 9.
11. Find the equation of the line whose x-intercept is -2 and whose y-intercept is 6.
12. Prove that the points $(5, 2)$, $(2, 6)$, $(-6, 0)$, and $(-3, -4)$ are vertices of a rectangle.
13. Determine the equation of the line through $(-2, 5)$ and parallel to the line whose equation is $6x - 3y - 24 = 0$.
14. Find the equation of the line through $(1, -4)$ and perpendicular to the line whose equation is $3x - 4y = 12$.
15. Draw and describe the graph which satisfies the conditions: $-1 \leq x < 3$ and $2 < y \leq 4$.
16. Prove that the segments joining the midpoints of the opposite sides of a quadrilateral bisect each other.

|14|

Areas and Volumes of Solids

14.1. Space geometry. The set of all points is called space. *Space geometry* (often called *solid geometry*) treats, primarily, figures the parts of which do not lie in the same plane (see Fig. 14.1). Examples of space figures (also called *solids*) are the cube, sphere, cylinder, cone, and pyramid (see Fig. 1.9).

A space figure is a combination of points, lines, and surfaces. The eight corners of the solid in Fig. 14.2 are points; the twelve edges, such as $\overline{AB}$ and $\overline{CF}$, are line segments; the six faces, such as $ABCD$, are flat surfaces or portions of planes.

14.2. Proofs of theorems in space geometry. All the definitions, postulates, theorems, and corollaries we have studied in the first 13 chapters of this text are applicable to the study of space geometry.

Most of the postulates and definitions used in plane geometry apply without reference to a plane. Many theorems and corollaries of plane geometry hold without reference to any plane. Among these are propositions on congruence and similarity of triangles. If a proposition or postulate applies only to figures lying in one plane, it cannot be used in space geometry.

Several simple theorems on space figures have been proved in other chapters of this text. In this chapter, we will not be concerned with formal proofs of theorems of figures in space. We will list the more important theorems, without proof, and apply them to solutions of problems dealing with the more common space figures. The reader can find the proofs of each of the propositions in any standard text on solid geometry.

14.3. Dihedral and polyhedral angles. As defined in §1.18, a *dihedral angle* is the union of a line and two noncoplanar half-planes having the line as their common edge. The line is called the *edge* of the dihedral angle and the

Fig. 14.1.

union of the edge and either half-plane is called a *face*, or *side*, of the dihedral angle. In Fig. 14.3, BD and BF are faces and $\overleftrightarrow{AB}$ is the edge of dihedral $\angle C\text{-}BA\text{-}F$.

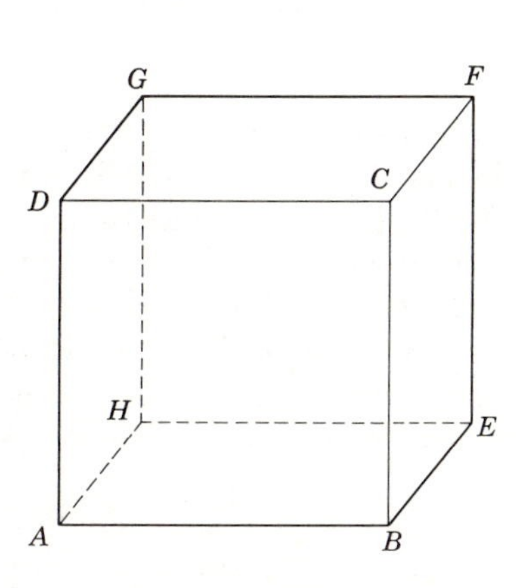

Fig. 14.2.

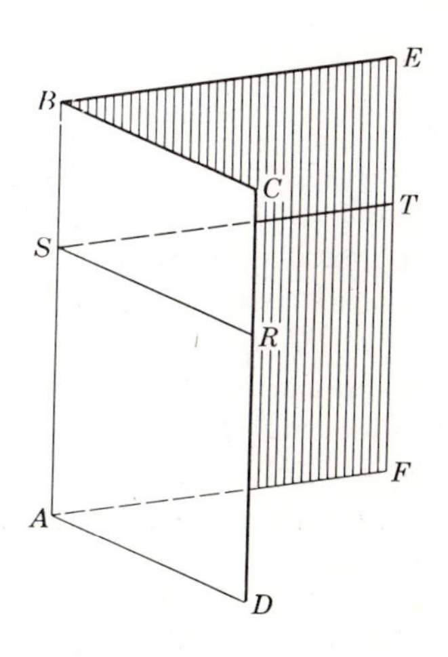

Fig. 14.3.

The *plane angle of a dihedral angle* is the angle formed by two rays, one in each face, with a common vertex and both perpendicular to the edge at the vertex. Thus, $\angle RST$ is a plane angle of dihedral $\angle C\text{-}BA\text{-}F$ if $\overrightarrow{SR} \perp \overleftrightarrow{AB}$ and $\overrightarrow{ST} \perp \overleftrightarrow{AB}$. All plane angles of a given dihedral angle are congruent. The measure of the plane angle of a dihedral angle is the *measure of the dihedral angle.*

A dihedral angle is *acute, right,* or *obtuse* if its plane angle is acute, right, or obtuse, respectively. Two planes are *perpendicular,* iff they intersect to form right dihedral angles.

Two dihedral angles are congruent iff their plane angles are congruent.

Let *ABCDE*. . . be a simple closed polygon lying in one plane and let *V* be a point not in the plane of the polygon. The set of points *Q* on the rays from *V* through all the points *P* of the polygon is called a *polyhedral angle* (Fig. 14.4). The point *V* is the *vertex* of the polyhedral angle.

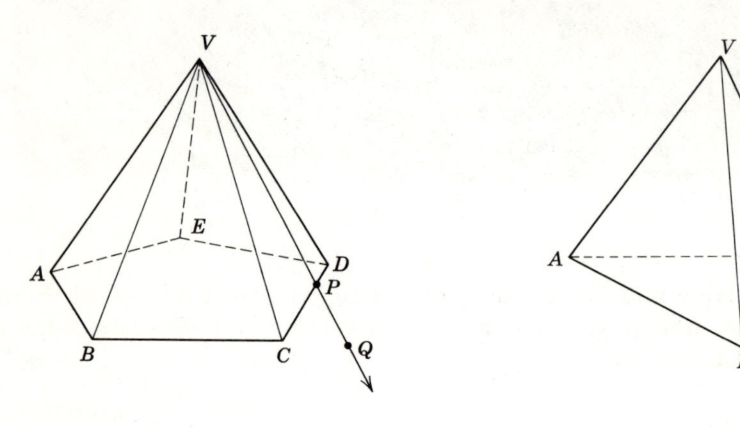

Fig. 14.4.

Fig. 14.5.

A polyhedral angle can be named by the vertex or by the vertex and a point of each edge of the angle. Thus, in Fig. 14.4, the polyhedral angle may be read "polyhedral $\angle V$" or "polyhedral $\angle V\text{-}ABCDE$." The parts of the planes which form the polyhedral angle are called the *faces* of the angle. Such faces are *VAB*, *VBC*, and *VCD* of the figure. The intersections of adjacent faces are the *edges* of the polyhedral angle. Angle *BVC* is a *face angle* of the polyhedral angle. The dihedral angles formed by the planes, at the edges, such as $\angle A\text{-}VB\text{-}C$ and $\angle B\text{-}VC\text{-}D$, are the dihedral angles of the polyhedral angle.

A *trihedral angle* is a polyhedral angle having three faces; as $\angle V\text{-}ABC$ in Fig. 14.5.

14.4. Theorems on planes and polyhedral angles. There follows a list of fundamental theorems on space geometry. We will not attempt to prove the statements. The student is advised to study them carefully and consider their implications. Many of these propositions will be analogous to theorems we have proved in plane geometry.

Theorem 14.1. *If a line intersects a plane not containing it, then the intersection is a single point.* If there were two points of intersection common to the plane and the line, the line would lie in the plane.

Theorem 14.2. *All the perpendiculars drawn through a point on a given line lie in a plane perpendicular to the given line at that point.* Thus, in Fig. 14.6, if $\overleftrightarrow{TA}$, $\overleftrightarrow{TB}$, and $\overleftrightarrow{TC}$ are all perpendicular to $\overleftrightarrow{PT}$ at T, then $\overleftrightarrow{PT}$ is perpendicular to plane MN.

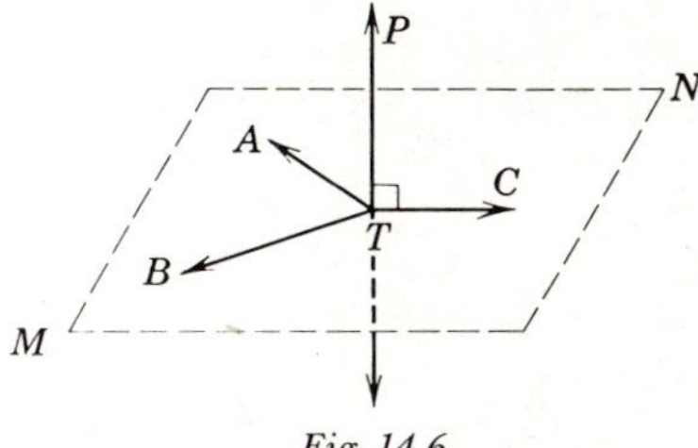

Fig. 14.6.

Theorem 14.3. *Through a given point there passes one and only one plane perpendicular to a given line.*

Theorem 14.4. *One and only one perpendicular line can be drawn to a plane from a point not on the plane.*

Theorem 14.5. *The perpendicular from a point not on a plane to the plane is the shortest line segment from the point to the plane.* The *distance from a point to a plane* is the perpendicular distance from the point to the plane.

Theorem 14.6. *If two lines are perpendicular to a plane, they are parallel to each other.* Thus, in Fig. 14.7, if $\overleftrightarrow{AB}$ and $\overleftrightarrow{CD}$ are each perpendicular to plane MN, $\overleftrightarrow{AB}$ is parallel to $\overleftrightarrow{CD}$.

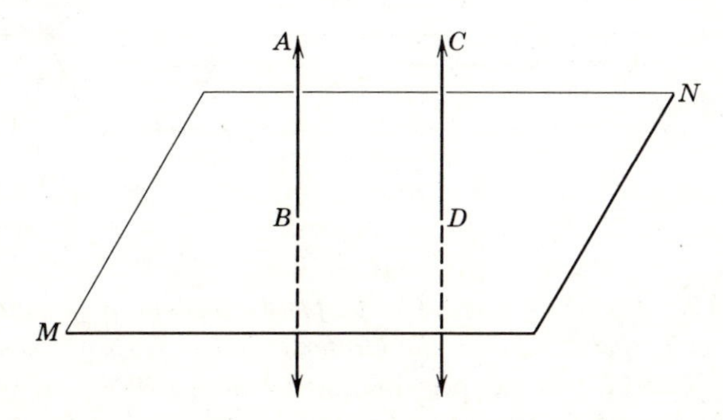

Fig. 14.7.

Theorem 14.7. *If one of two parallel lines is perpendicular to a plane, the other is also perpendicular to the plane.* In Fig. 14.7, if $\overleftrightarrow{AB}$ is parallel to $\overleftrightarrow{CD}$, and $\overleftrightarrow{AB}$ is perpendicular to plane MN, then $\overleftrightarrow{CD}$ is also perpendicular to plane MN.

Theorem 14.8. *If each of three non-collinear points of a plane is equidistant from two points, then every point of the plane is equidistant from these points.* Thus, if in Fig. 14.8, $PA = QA$, $PB = QB$, and $PC = QC$, then every point of the plane determined by A, B, and C is equidistant from P and Q.

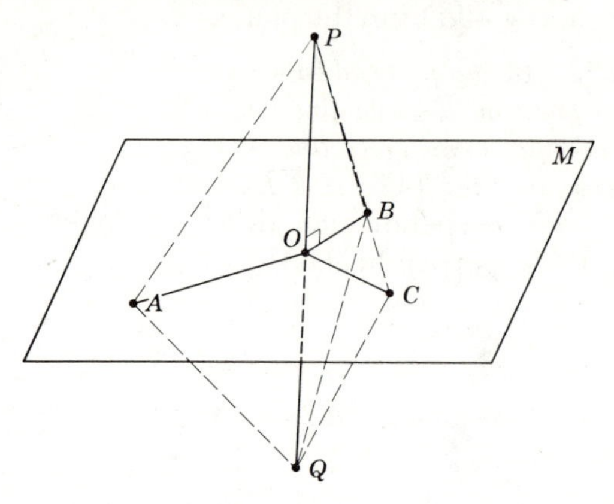

Fig. 14.8.

Theorem 14.9. *The distance between two parallel planes is the perpendicular distance between them. Two parallel planes are everywhere equidistant.*

Theorem 14.10. *Through one straight line any number of planes may be passed* (see Fig. 14.9).

Fig. 14.9.

Theorem 14.11. *If two planes are perpendicular to each other, a straight line in one of them perpendicular to their intersection is perpendicular to the other.* If in Fig. 14.10, plane RS is perpendicular to plane MN and intersects it in $\overleftrightarrow{QS}$ and if $\overleftrightarrow{AB}$ in plane RS is perpendicular to $\overleftrightarrow{QS}$, then $\overleftrightarrow{AB}$ is perpendicular to plane MN.

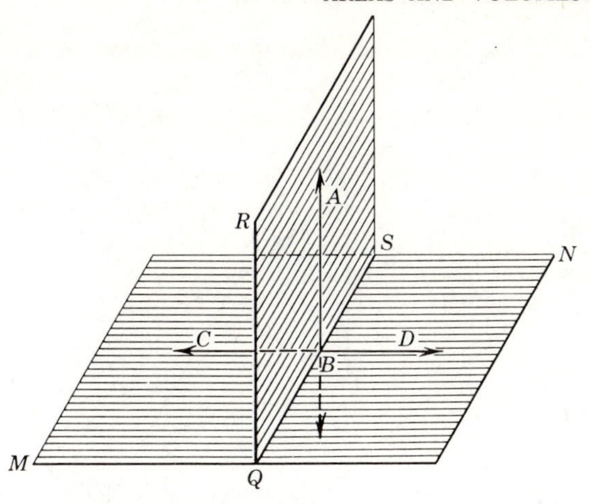

Fig. 14.10.

Theorem 14.12. *If two planes are perpendicular to each other, a perpendicular to one of them at a point of their intersection lies in the other.* Thus, in Fig. 14.10, if plane RS is perpendicular to plane MN, and $\overleftrightarrow{AB}$ is perpendicular to plane MN at a point B on the line of intersection of the planes, then $\overleftrightarrow{AB}$ must lie in plane RS.

Theorem 14.13. *If two intersecting planes are perpendicular to a third plane, their intersection is also perpendicular to that plane.* If in Fig. 14.11, planes RS and PQ are each perpendicular to plane MN, and if planes RS and PQ intersect at $\overleftrightarrow{AB}$, then $\overleftrightarrow{AB}$ is perpendicular to plane MN.

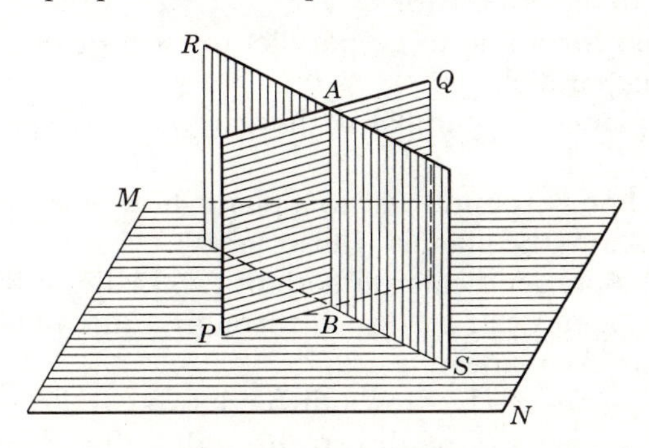

Fig. 14.11.

Theorem 14.14. *Every point in a plane bisecting a dihedral angle is equidistant from the faces of the angle.* Thus, in Fig. 14.12 if $\angle\alpha \cong \angle\beta$, then $PA = PB$.

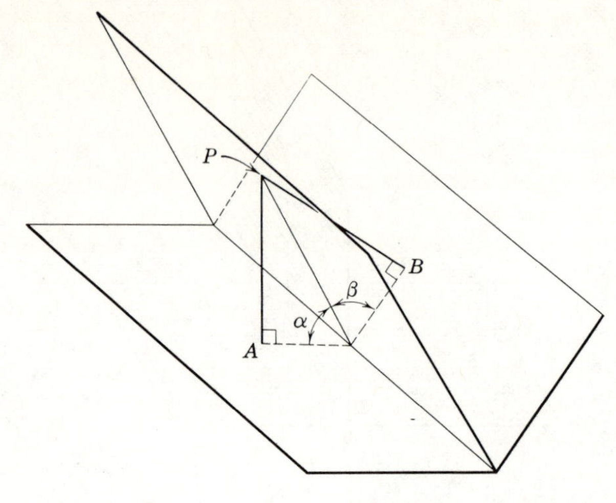

Fig. 14.12.

Theorem 14.15. *The set of points equidistant from the faces of a dihedral angle is the plane bisecting the dihedral angle* (see Fig. 14.12).

Exercises

1. At a point on a line how many lines can be drawn perpendicular to the line?
2. At a point on a line how many planes can be perpendicular to the line?
3. Is it possible for three lines to intersect in a point so that each is perpendicular to the other two?
4. Is it possible for a line to be parallel to each of two planes without the planes being parallel?
5. Must two planes be parallel if they are each perpendicular to a third plane?
6. Through a line perpendicular to a plane, how many planes can be drawn perpendicular to the plane? Illustrate.
7. Can a line be perpendicular to both of two planes if they are not parallel?
8. How many planes can be drawn through a line oblique to a plane perpendicular to the plane?
9. How many planes can be drawn through two parallel lines?
10. How many planes can be drawn through a line parallel to a plane and also be perpendicular to the plane?
11. Must two parallel lines be in a common plane?
12. Must lines that are parallel to the same plane be parallel to each other?
13. Are two lines parallel if they are perpendicular to the same plane?

14. A straight line, not in either of two given planes, is parallel to the intersection of the planes. Is the line parallel to each of the planes?
15. How many lines can be drawn parallel to a plane through a point not in the plane?
16. If two planes are parallel to a third plane, are they parallel to each other?
17. If one of two parallel lines is perpendicular to a plane, must the other also be perpendicular to the plane?
18. Is it possible for a line to be perpendicular to each of two lines that are not parallel? Explain.
19. Is it possible for a line to be perpendicular to two lines in the same plane? Illustrate.
20. Describe the shortest distance from a point to a plane.
21. How many planes can there be parallel to a given plane through a given point outside the given plane?
22. Is it possible to have a plane pass through two lines which are perpendicular to a given plane? Illustrate.
23. If a line is perpendicular to a line in a plane, is it perpendicular to the plane?
24. If a line and a plane never meet, must the line be parallel to the plane?
25. In the figure, $\overline{PA} \perp$ plane MN at A; $AB = AC$; $\overline{AB}$ and $\overline{AC}$ lie in plane MN.

    Prove $PB = PC$.

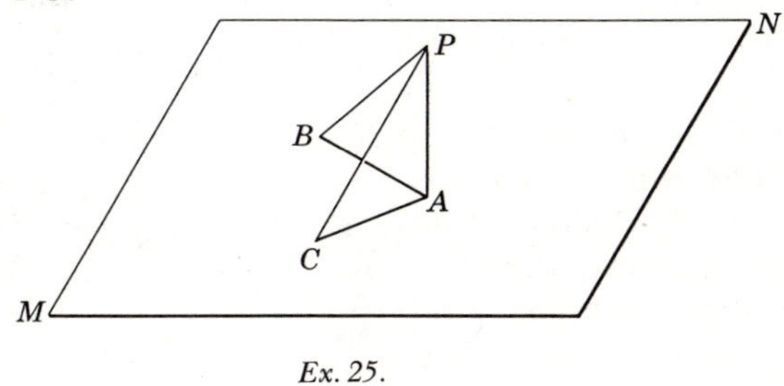

Ex. 25.

26. In the figure circle O lies in plane RS; $\overline{PO} \perp$ plane RS. Prove $\angle PAO \cong \angle PBO$.

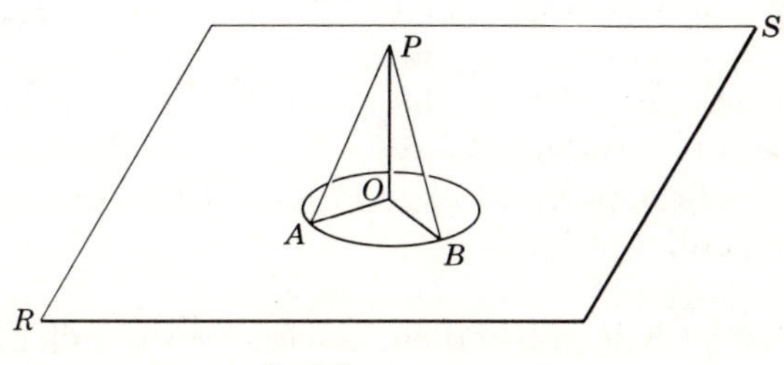

Ex. 26.

Study the facts given in the following exercises, draw figures if necessary, and then state a conclusion reached from the facts.

27. Plane XY cuts plane RS in $\overleftrightarrow{GH}$, and plane XY cuts plane MN in $\overleftrightarrow{KL}$; $\overleftrightarrow{GH} \parallel \overleftrightarrow{KL}$.
28. $\overleftrightarrow{PB} \perp \overleftrightarrow{AB}$; $\overleftrightarrow{PB} \perp \overleftrightarrow{CB}$; $\overleftrightarrow{AB}$ lies in plane MN; $A \neq C$.
29. Points E and F lie on both planes MN and RS; $E \neq F$.
30. Plane $XY \perp$ line PQ; plane RS is $\perp$ line PQ.
31. Plane RS cuts plane XY in $\overleftrightarrow{EJ}$; plane XY cuts plane GH in $\overleftrightarrow{WZ}$; plane $RS \parallel$ plane GH.
32. Line AB passes through points P and Q; P and Q are points in plane MN.
33. Line CD lies in plane RS; point Q is on $\overleftrightarrow{CD}$.
34. Plane $XY \perp$ plane WZ; plane WZ intersects plane XY in $\overleftrightarrow{GH}$; $\overleftrightarrow{EF}$ lies in plane WZ and is $\perp$ $\overleftrightarrow{GH}$; line AB lies in plane XY.
35. Plane $RS \parallel$ plane MN; points A and B lie in plane RS; points E and K lie in plane MN.
36. Plane $AB \parallel$ plane CD; plane $EF \perp$ plane CD.
37. Line $AB \perp$ plane RS; plane $RS \perp$ line KL.
38. Plane $KL \perp$ plane MN; plane KL intersects plane MN at $\overleftrightarrow{PS}$; H lies on $\overleftrightarrow{PS}$; $\overleftrightarrow{RH}$ is in plane MN.
39. Plane $XY \perp$ plane KL; plane HG is $\perp$ plane KL; plane XY intersects plane HG in line PQ.
40. Line PT is $\perp$ plane MN; Q and T are points in plane MN; $m\angle QPT = 60$; $PQ = 30$ feet. Find projection of $\overline{PQ}$ on plane MN.

14.5. Polyhedron. A *polyhedron* is the union of a finite number of polygonal regions, each of which contains a polygon and its interior, such that (1) the interior of any two of the regions do not intersect and (2) every side of any of the polygons is also a side of exactly one of the other polygons. Each of the polygonal regions is called a *face* of the polyhedron. The intersection of any two faces of the polyhedron is called an *edge* of the polyhedron. The intersection of any two edges is a *vertex* of the polyhedron. Figure 14.13 represents a polyhedron of 8 faces, 16 edges, and 9 vertices.

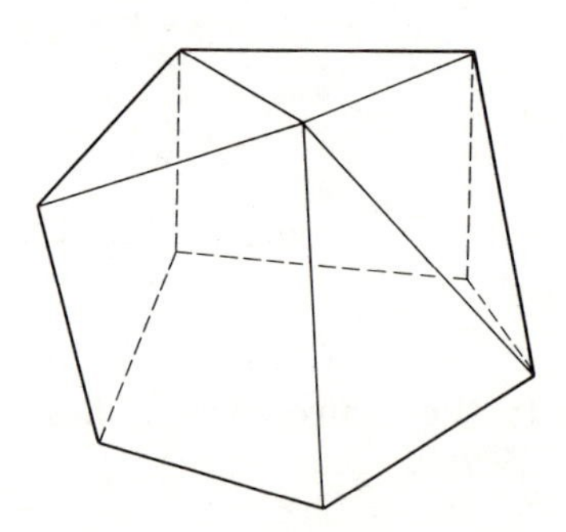

Fig. 14.13. A polyhedron.

14.6. Prism. A *prism* is a polyhedron having two parallel faces, called the *bases*, with the remaining faces being parallelograms (Fig. 14.14). In prism AI base $ABCDE$ is congruent to base $FGHIJ$. The faces which are parallelograms are called the *lateral faces*, and their intersections are called *lateral edges*. The lateral edges are equal and parallel. The *lateral area* is the sum of the areas of the lateral faces. The *total area* is the sum of the lateral areas and the areas of the two bases. The *altitude* h of a prism is the perpendicular distance between the planes of the bases.

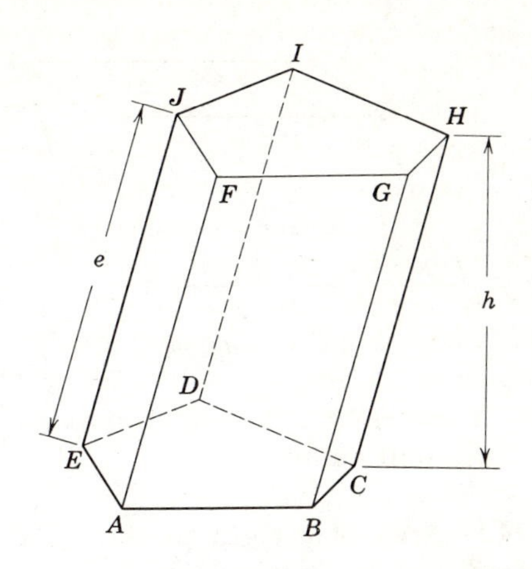

Fig. 14.14. A prism.

A *right prism* is a prism the lateral edges of which are perpendicular to the bases (Fig. 14.15). It can be shown that, in a right prism, the lateral faces are rectangles and the lateral edges equal the altitude.

A *regular prism* is a right prism the bases of which are regular polygons.

14.7. Parallelepiped. A *parallelepiped* is a prism the bases of which are

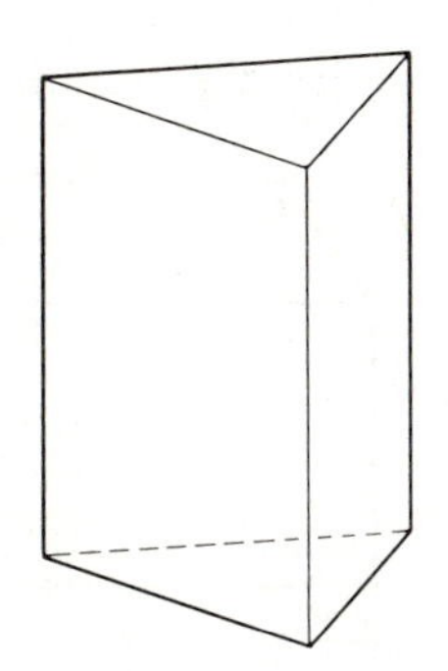
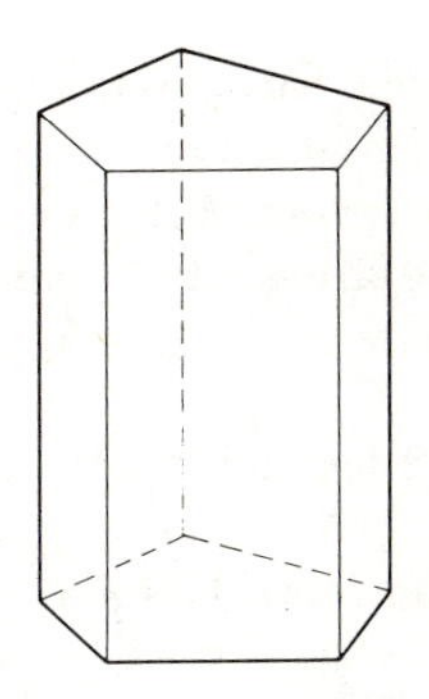

Fig. 14.15. Right prisms.

parallelograms (Fig. 14.16). It can be shown that the opposite faces of a parallelepiped are parallel and congruent.

A *rectangular parallelepiped* is a parallelepiped the faces of which are rectangles (Fig. 14.17). All the lateral edges of a rectangular parallelepiped are perpendicular to the planes of the parallel bases.

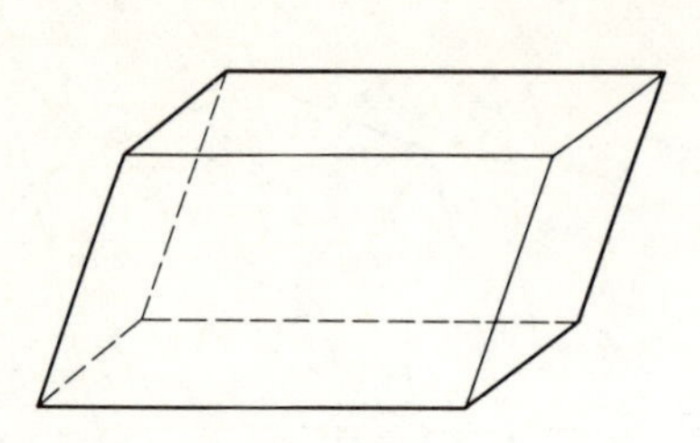

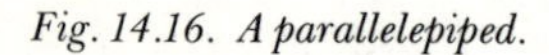

Fig. 14.16. *A parallelepiped.*

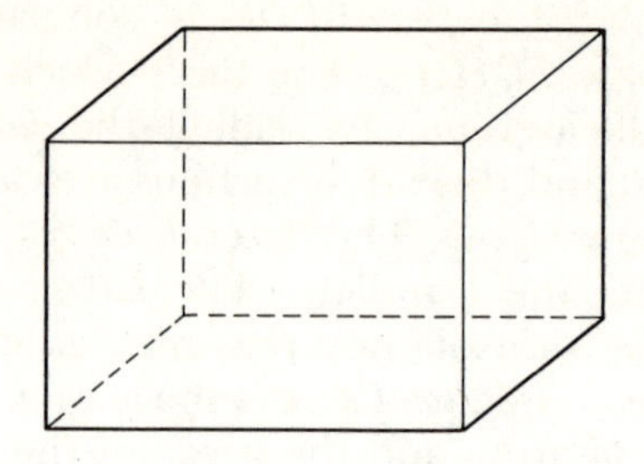

Fig. 14.17. *A rectangular parallelepiped.*

A *cube* is a rectangular parallelepiped the bases and faces of which are congruent squares.

14.8. Area of a right prism. *The lateral area of a right prism is equal to the product of its altitude and the perimeter of its base.* Thus, if we denote the lateral area by S, the perimeter of the base by P, and the altitude by h, we get the formula

$$S = hP$$

If we denote the total area by T and the area of a base by A, we get the formula

$$T = S + 2A$$

The total area for a rectangular parallelepiped (Fig. 14.18), the length, width, and height of which are denoted by l, w, and h respectively is equal to the sum of the areas of the six faces, $2lw + 2wh + 2lh$ or

$$T = 2(lw + wh + lh)$$

For a cube with a lateral edge e,

$$T = 6e^2$$

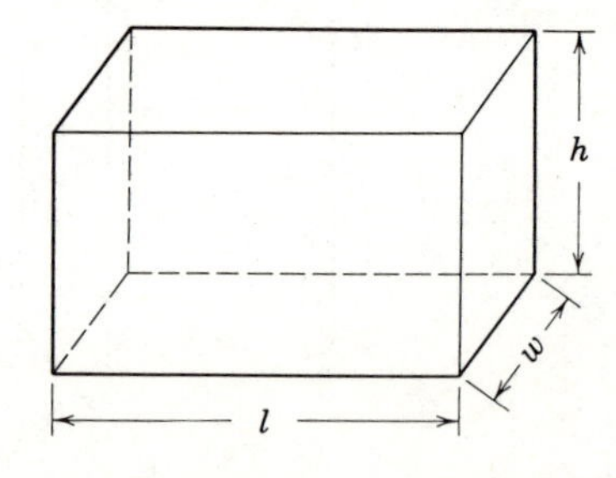

Fig. 14.18.

14.9. Volume of a prism. The *volume* of a solid is defined as the number of units of space measured in the solid. This unit of space, called a cubic unit, is that of a cube the edges of which are equal to some unit for measuring length. Consider the rectangular parallelepiped, shown in Fig. 14.19, which is 4 units long, 3 units wide, and 2 units high. If planes are passed parallel to the faces of the solid as shown, the solid will consist of two layers, each layer containing 4×3, or 12 cubic units.

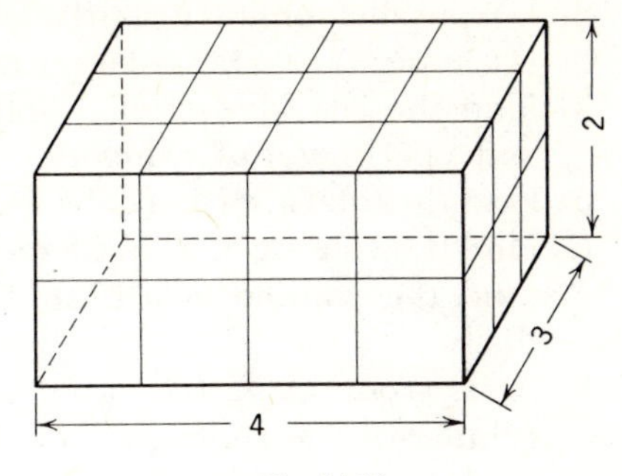

Fig. 14.19.

The two layers contain 2×12, or 24 cubic units. Thus, the volume of the solid equals 24 cubic units. This number may be obtained by multiplying together the three dimensions or by multiplying the area of the base by the altitude.

In Fig. 14.18, if we denote the volume by V and the area of the base by A, then

$$V = lwh \qquad \text{or} \qquad V = Ah$$

It can be shown that the *volume of any prism is the product of the area of its base and its altitude.*

14.10. Illustrative Example 1. A storage vault has a rectangular floor 72 feet by 48 feet. The walls are vertical and 15 feet high. (*a*) Find the total area of walls, floor, and ceiling; (*b*) find the storage space (volume) of the room.

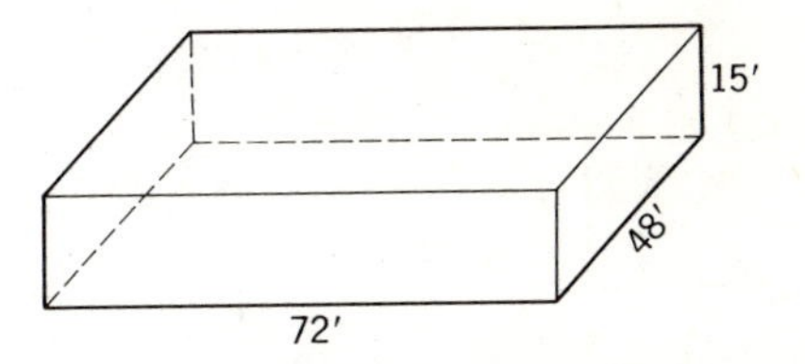

Illustrative Example 1.

Solution:

(*a*) The total surface is found by using the formula

$$\begin{aligned} T &= 2(lw + wh + lh) \\ &= 2[(72)(48) + (48)(15) + (72)(15)] \\ &= 10{,}512 \end{aligned}$$

Answer: 10,512 square feet.

(*b*) The storage space is found by using the formula

$$\begin{aligned} V &= lwh \\ &= (72)(48)(15) \\ &= 51{,}840 \end{aligned}$$

Answer: 51,840 cubic feet.

Exercises

1. In a rectangular parallelepiped, there are how many faces? How many edges? How many vertices?

2. Is a parallelepiped (*a*) a prism? (*b*) a polyhedron? (*c*) a cube?
3. Is a cube (*a*) a rectangular parallelepiped? (*b*) a prism? (*c*) a polyhedron?
4. Find the lateral area of a right prism which has an altitude of 18 inches and a perimeter of 30 inches.
5. Find the total area of a cube 8 inches on an edge.
6. Find the volume of a cube 8 inches on an edge.
7. Find the volume of a beam 12 feet long, 12 inches wide, and 2 inches thick.
8. A classroom is 42 feet long, 30 feet wide and 12 feet high. What is the volume of the room in cubic yards? What is the lateral area in square yards?
9. Find the lateral area of a right prism which has an altitude 5 feet and a base of a regular hexagon with a side of 2 feet.

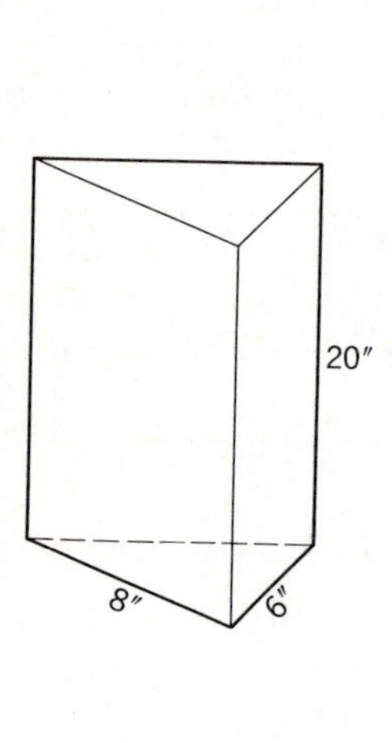

Exs. 10–12.

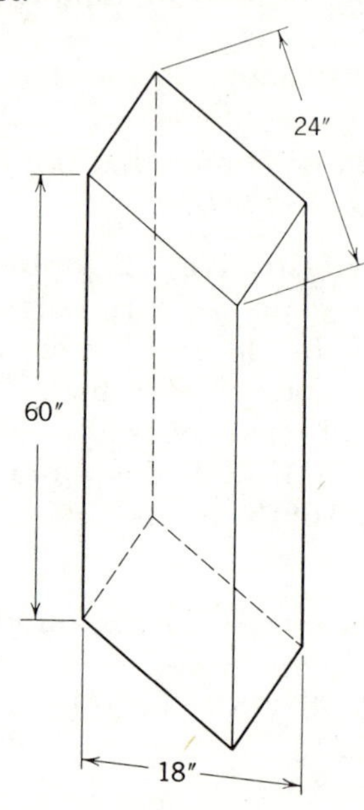

Exs. 13–15.

10. Find the volume of a right prism the base of which is a right triangle with legs of 6 inches and 8 inches and the altitude of which is 20 inches.
11. Find the lateral area of the figure in Ex. 10.
12. Find the total area of the figure in Ex. 10.
13. Find the volume of a right prism the base of which is a rhombus having diagonals 18 inches and 24 inches long and the altitude of which is 60 inches.

14. Find the lateral area of the prism in Ex. 13.
15. Find the total area of the prism in Ex. 13.
16. How many packages $5 \times 8 \times 12$ inches can be placed in a box that has the dimensions of $24 \times 30 \times 60$ inches?
17. How many cubic yards of concrete are needed to build a retaining wall 120 feet long, 8 inches thick, and 5 feet high?
18. How many gallons of water will be required to fill a pool 45 feet long, 30 feet wide, and 6 feet deep? (*Note:* 231 cubic inches = 1 gallon.)
19. How many gallons of paint will be needed to paint the exterior walls of a building 60 feet long, 30 feet wide, and 15 feet high if 1 gallon of paint will cover 500 square feet?
20. Find the weight of a steel plate 12 feet long, 5 feet wide, and $\frac{3}{8}$ inch thick if steel weighs 490 pounds per cubic foot.

14.11. Pyramid. A *pyramid* is a polyhedron with one face, called the *base*, a polygon of any number of sides, and the other faces are triangles that meet in a common point called the *vertex.* The triangular faces are called the *lateral faces*, and the meeting of the lateral faces are *lateral edges*. The *altitude* of the pyramid is the length of the perpendicular dropped from the vertex to the plane of the base. The *lateral area* of a pyramid is equal to the sum of the areas of the lateral faces of the pyramid. The *total area* of a pyramid is equal to the sum of the lateral area and the area of the base. (See Fig. 14.20.)

A *regular pyramid* is one having a base which is a regular polygon and the altitude from the vertex perpendicular to the base at its center. The lateral

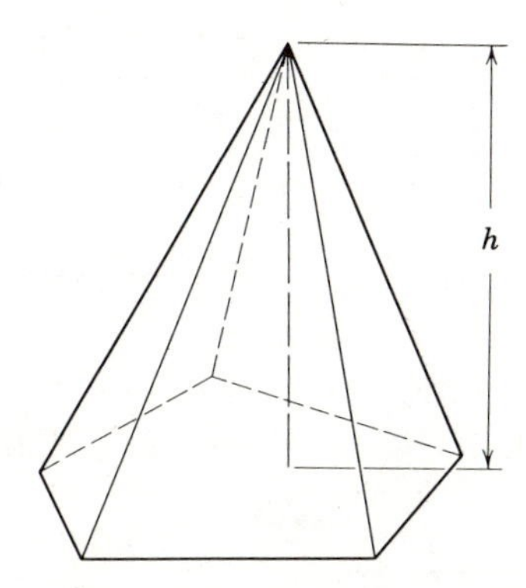

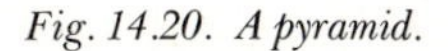
Fig. 14.20. A pyramid.

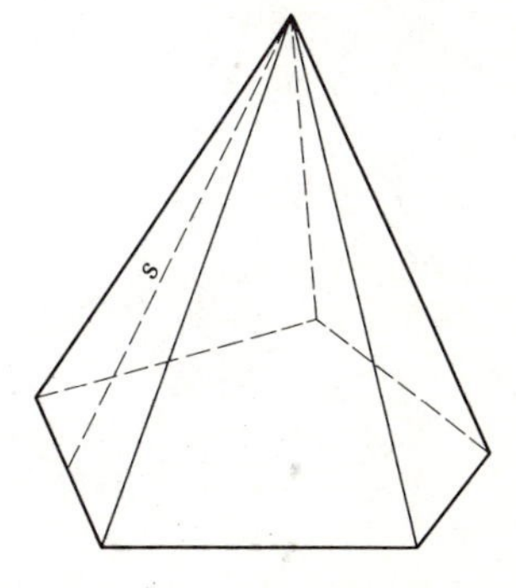

Fig. 14.21. A regular pyramid.

edges of a regular pyramid are congruent. The lateral faces of a regular pyramid are congruent isosceles triangles. The *slant height* of a regular pyramid is the altitude of any of its lateral faces. (See Fig. 14.21.)

14.12. Volume of a pyramid. The volume of any pyramid is equal to one-third the product of the area of its base and altitude.

$$\text{Volume} = \tfrac{1}{3}\,\text{area of base} \times \text{altitude}$$
$$V = \tfrac{1}{3} Ah$$

14.13. Area of a regular pyramid. The lateral area of a regular pyramid is equal to half the product of its slant height s and the perimeter p of its base.

$$\text{Lateral area} = \tfrac{1}{2}\,\text{slant height} \times \text{perimeter of base}$$
$$S = \tfrac{1}{2} sp$$

14.14. Illustrative Example 1. A regular pyramid has a base of a regular hexagon with sides equal to 8 inches and an altitude of 15 inches. Find (*a*) the volume and (*b*) the lateral area of the pyramid.

Solution: Draw sketches of the pyramid, its base, and a side view of one of its lateral faces.

(*a*) To determine the volume, we must first find the area of the base. Remembering our plane geometry, we know that the total area can be found by dividing it into twelve 30°–60° triangles, as shown in (2) of the figure.

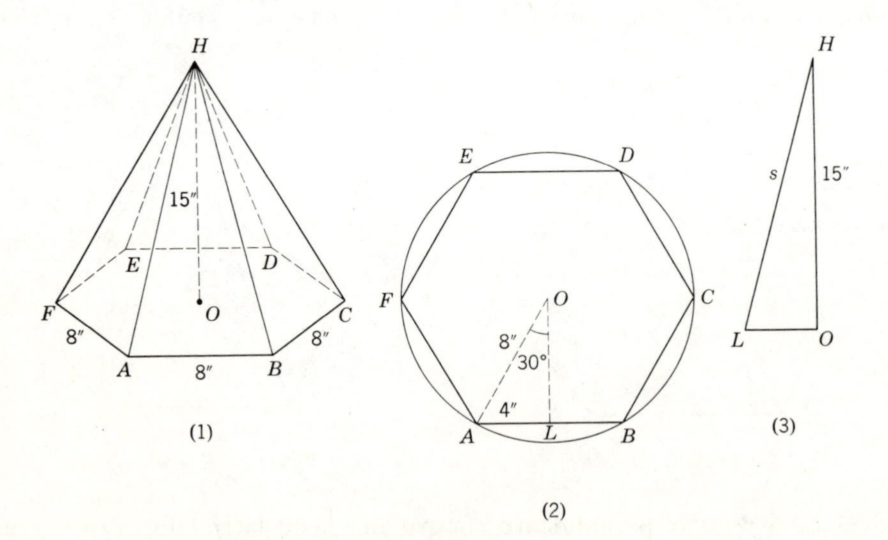

Illustrative Example 1.

$$
\begin{aligned}
OA &= 2AL = 8 \text{ inches} \\
(OL)^2 &= (8)^2 - (4)^2 \\
OL &= \sqrt{48} \\
&= 6.928 \text{ inches (by Table I)} \\
\text{Area of } \triangle AOL &= \tfrac{1}{2}(4)(6.928) \\
&= 13.856 \text{ square inches.} \\
A &= 12(13.856) \\
\therefore V &= \tfrac{1}{3}Ah \\
&= \tfrac{1}{3}(12)(13.856)(15) \\
&= 831.36
\end{aligned}
$$

Answer: 831.36 cubic inches.

(*b*) We need to determine the slant height s before we can compute the lateral area. In figure (3) we know that $OL = \sqrt{48}$ inches and $OH = 15$ inches. Using the pythagorean theorem,

$$
\begin{aligned}
s^2 &= (OL)^2 + (OH)^2 \\
&= (\sqrt{48})^2 + (15)^2 \\
&= 48 + 225 \\
s^2 &= 273 \\
s &= 16.523 \text{ inches (by Table I)} \\
S &= \tfrac{1}{2}sp \\
&= \tfrac{1}{2}(16.523)(6 \times 8) \\
&= 396.53
\end{aligned}
$$

Answer: 396.53 square inches.

Exercises

1. What is the volume of a pyramid the base of which has an area of 42 square inches and the altitude of which equals 15 inches?
2. Find the lateral area of a pyramid the base of which has a perimeter of 36 inches and the slant height of which equals 23 inches.
3. What is the volume of a pyramid having a square base with a 10-inch side and an altitude of 12 inches?
4. Find the lateral area of the pyramid having a square base with a 10-inch side and an altitude of 12 inches.
5. *ACGE* is a cube with an 18-inch side. Find the volume of pyramid *ABEC*.
6. Find the volume of a pyramid having a base a right $\triangle$, with hypotenuse $RT = 18$ feet, leg $RS = 15$ feet, and altitude $PO = 24$ feet.

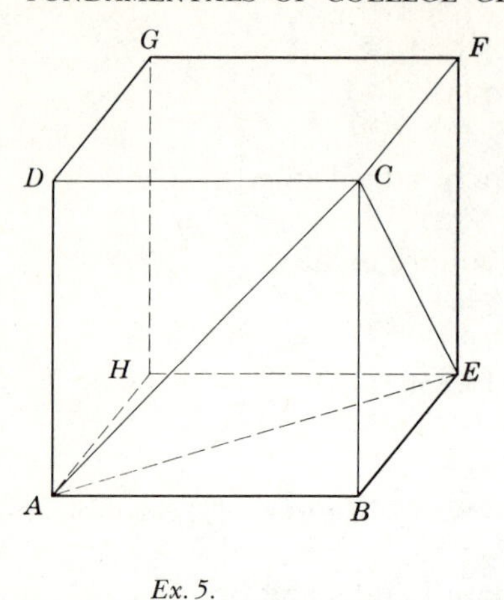

Ex. 5.

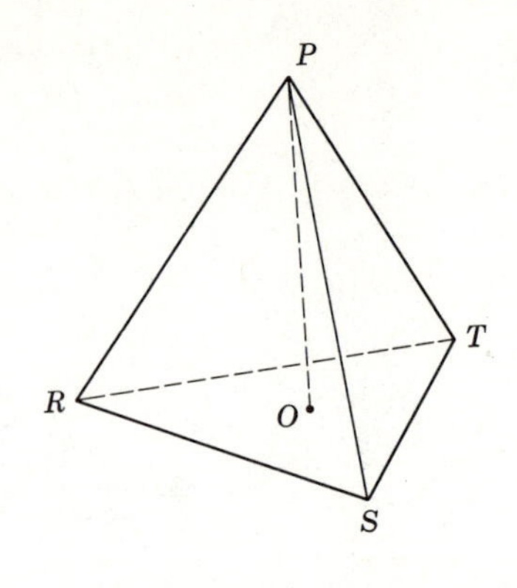

Ex. 6.

7. Find the volume of a regular pyramid having as a base a regular hexagon with a 6-inch side an an altitude of 10 inches.
8. Find the lateral area of a regular pyramid having as a base a regular hexagon with a 6-inch side and an altitude of 10 inches.
9. Find the lateral area of a regular pyramid having as a base a regular octagon (8 sides) with a 12-inch side and a lateral edge equal to 25 inches.

14.15. Conical surface. A *conical surface* is a surface generated by a moving straight line which turns around one of its points and intersects a given plane in a curve. The moving line QS is called the *generatrix*. In Fig. 14.22, point P on $\overleftrightarrow{QS}$ follows the given curve PRT called the *directrix*. $\overleftrightarrow{QS}$ in any of its positions is called an *element* of the surface. B is the *vertex* of the surface.

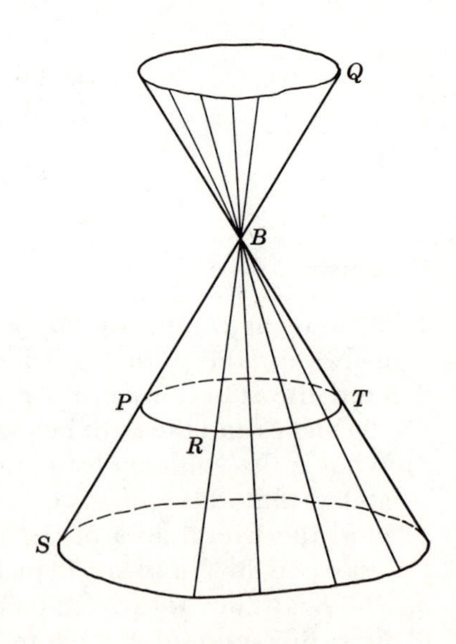

Fig. 14.22

14.16. Cone. A *cone* is that part of a conical surface which is bounded by the vertex and a plane cutting all the elements on one side of the vertex. The cone in Fig. 14.23 is labeled cone B-RST. The *base* RST of the cone is the curve cut from the conical surface by the plane. The *altitude* of a cone is the perpendicular distance from the vertex to the plane of the base. The *lateral area* of a cone is the area of the lateral surface.

A *circular cone* is one the base of which is a circle. A *right circular cone* is one in which the line through the vertex and center of the base is perpendicular to the base. This perpendicular is often termed the *axis* of the cone. The elements of a right circular cone are congruent.

The *slant height* of a right circular cone is the length of an element of the cone.

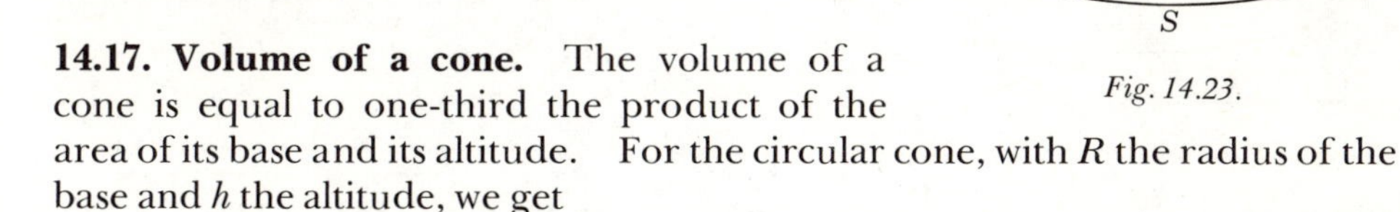

Fig. 14.23.

14.17. Volume of a cone. The volume of a cone is equal to one-third the product of the area of its base and its altitude. For the circular cone, with R the radius of the base and h the altitude, we get

$$\text{Volume} = \tfrac{1}{3}\text{area of base} \times \text{altitude}$$
$$V = \tfrac{1}{3}\pi R^2 h$$

14.18. Lateral area of a right circular cone. The lateral area of a right circular cone is equal to half the product of its slant height and the circumference of its base. Using $2\pi R$ for the circumference of the base and s for the slant height, we get

$$\text{Lateral area} = \tfrac{1}{2}\text{circumference of base} \times \text{slant height}$$
$$S = \tfrac{1}{2}(2\pi R)(s)$$
$$\text{or } S = \pi R s$$

The total area T of a cone is equal to the sum of the lateral area and the area of the base.

$$\text{Total area} = \text{lateral area} + \text{area of base}$$
$$T = \pi R s + \pi R^2$$

Exercises

1. What is the volume of a circular cone having an altitude of 18 inches and a radius of 5 inches?
2. Find the lateral area of a right circular cone having a slant height of 24 inches and a radius of 8 inches.
3. Find the total surface area of a right circular cone having a slant height of 40 feet and a radius of 10 feet.
4. Find the lateral area of a right circular cone having an altitude of 12 inches and a radius of 5 inches.

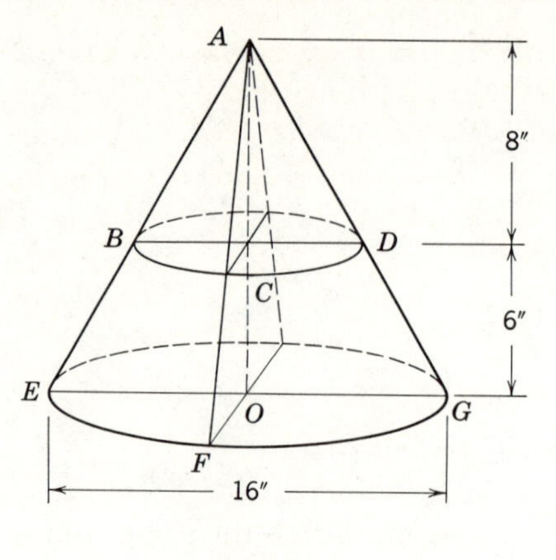

Exs. 5–8.

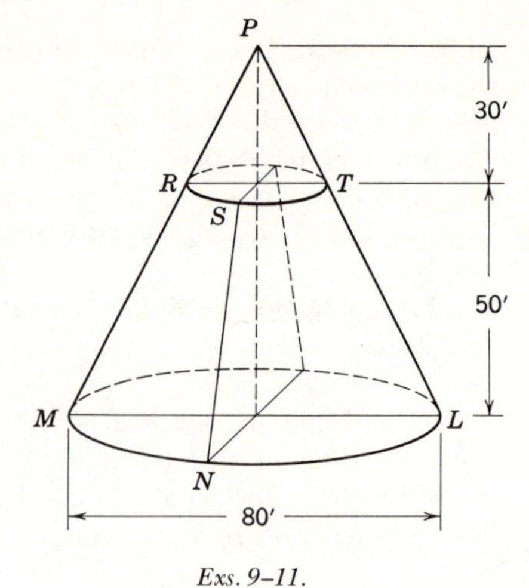

Exs. 9–11.

5. Find the volume of cone *A-EFG*.
6. Find the volume of cone *A-BCD*.
7. Find the lateral area of cone *A-EFG*.
8. Find the total area of cone *A-EFG*.
9. Find the volume of the solid *RTLM* cut from cone *P-MNL* by the plane *RST* ∥ plane *MNL*.
10. Find the lateral area of the solid *RTLM*.
11. Find the total surface area of the solid *RTLM*.

14.19. Cylindrical surface. A surface generated by a straight line which moves parallel to itself and intersects a given plane curve is called a *cylindrical surface*. The moving line *QS* is called the *generatrix*. In Fig. 14.24, point *R* on $\overleftrightarrow{QS}$ follows the given curve *PRT*, called the *directrix*. $\overleftrightarrow{QS}$ in any of its positions is called an *element* of the surface.

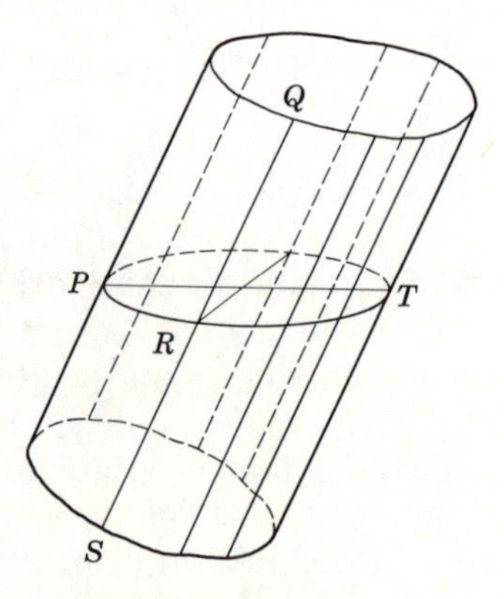

Fig. 14.24.

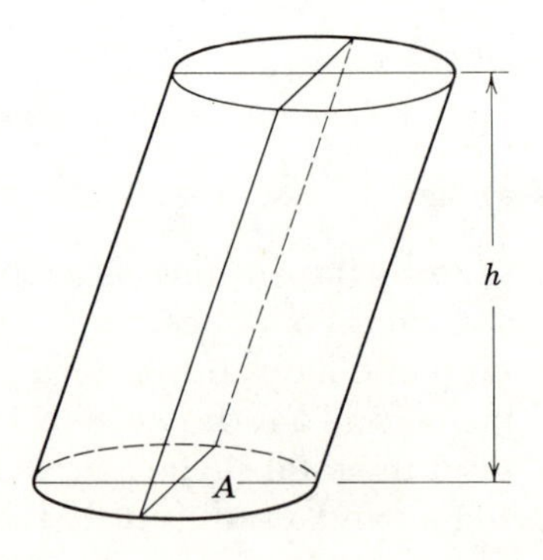

Fig. 14.25.

14.20. Cylinder. A *cylinder* is that part of a cylindrical surface bounded between two parallel planes cutting all the elements. The intersection of the cylindrical surface and one of the parallel planes is a *base* of the cylinder. The bases of a cylinder are congruent. The bounding cylindrical surface is the *lateral surface* of the cylinder. The *altitude* of a cylinder is the perpendicular distance between the bases. (See Fig. 14.25.)

A *circular cylinder* is one the bases of which are circles. A *right circular cylinder* is a circular cylinder in which the elements are perpendicular to the

Fig. 14.26. Municipal water tank of Topeka, Kansas. A vertical cylinder on the inside, this water tower on the outside seems to consist of vertically bisected cylinders with the concave outward.

bases. The elements of a right circular cylinder are parallel and congruent. Fig. 14.26 shows an interesting use made of right cylindrical surfaces which is functional and yet attractive.

14.21. Volume of a circular cylinder. The volume of a circular cylinder is equal to the product of the area of its base and the length of its altitude. (See Fig. 14.25.) The formula is

$$\textit{Volume} = \textit{area of base} \times \textit{altitude}$$
$$V = \pi R^2 h$$

14.22. Lateral area of a right circular cylinder. The lateral area of a right circular cylinder is equal to the product of the circumference of its base and the length of its altitude. The formula is

$$\textit{Lateral area} = \textit{circumference of base} \times \textit{altitude}$$
$$S = 2\pi R h$$

The total area T of a cylinder is equal to the sum of the lateral area and the area of its two bases. The formula is

$$\textit{Total area} = \textit{lateral area} + \textit{area of bases}$$
$$T = 2\pi R h + 2\pi R^2$$

12″

22″

Exs. 2–4.

Exercises

1. Is a rectangular parallelepiped a right cylinder?
2. What is the volume of the right circular cylinder?
3. Find the lateral area of the right circular cylinder.
4. Find the total area of the right circular cylinder.
5. How many gallons of gasoline will a cylindrical tank hold that is 6 feet in diameter and 25 feet long? (*Note:* 1 cubic foot is equivalent to 7.5 gallons.)
6. How high is a right cylindrical tank that holds 100 gallons if its diameter is 30 inches? (*Note:* 1 gallon is equivalent to 231 cubic inches.)
7. A steel roller is 4 feet long and 30 inches in diameter. What area will it cover in rolling through 250 revolutions?
8. How much will 1000 cylindrical steel rods $\frac{5}{8}$ inch in diameter and 15 feet long weigh if 1 cubic foot of iron weighs 490 pounds?

9. A right circular rod is made on a lathe from a solid steel bar 4 by 4 by 60 inches. How much waste will result in making the largest possible cylindrical rod from the bar?

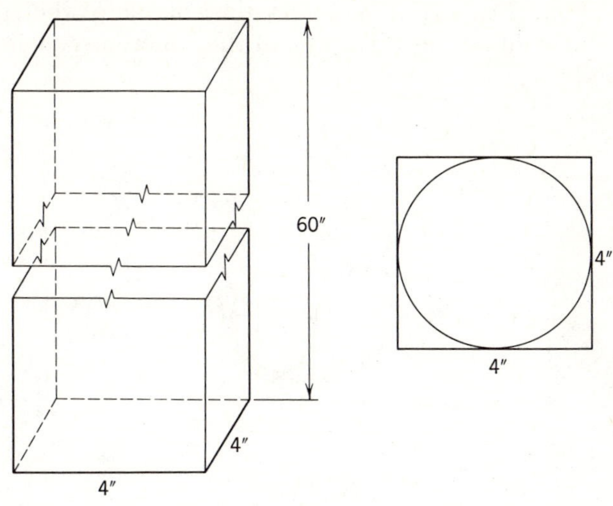

Ex. 9.

10. Find the amount of steel in 480 feet of pipe having an inside diameter of 1 inch and an outside diameter of $1\frac{1}{4}$ inches.

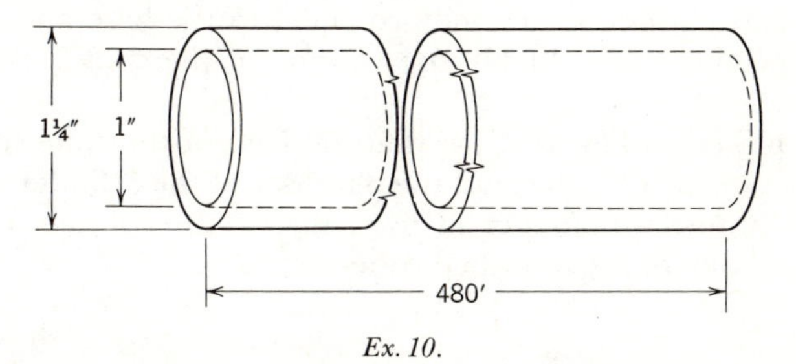

Ex. 10.

14.23. Volume of a sphere. The volume of a sphere is equal to $\frac{4}{3}\pi$ times the cube of its radius. The formula is

$$Volume = \tfrac{4}{3}\pi\,(radius)^3$$
$$V = \tfrac{4}{3}\pi R^3$$

14.24. Surface area of a sphere. The area of the surface of a sphere is equal to the area of four great circles. The formula is

$$Area = 4 \text{ times area of great circle}$$
$$A = 4\pi R^2$$

14.25. (optional). Relations of the sphere, cone, cylinder, and cube. Some interesting relations can be proved to exist between the solids we have thus far studied. We will list several of them. If a right circular cylinder is circumscribed about a sphere (Fig. 14.27), the volume of the sphere is two-thirds that of the cylinder, and the area of the sphere is equal to the lateral area of the cylinder.

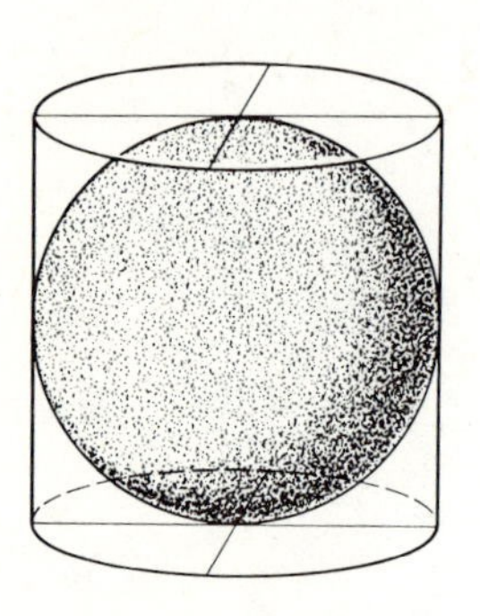

Fig. 14.27.

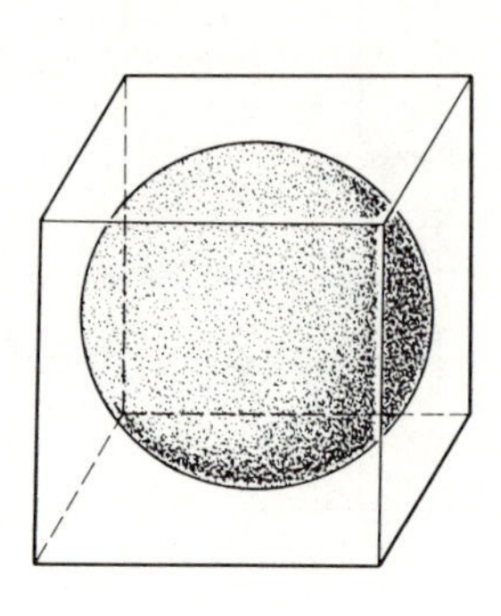

Fig. 14.28.

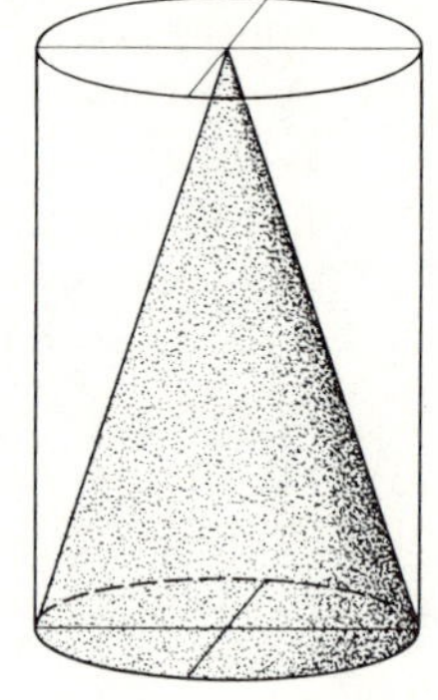

Fig. 14.29.

If a cube is circumscribed about a sphere (Fig. 14.28), the volume of the sphere is very nearly half the volume of the cube.

If the base and altitude of a cone are equal to the base and altitude of a right circular cylinder (Fig. 14.29), the volume of the cone is one-third that of the cylinder.

If a cone is placed in Fig. 14.27 so that the base of the cone coincides with one of the bases of the cylinder and the altitudes of the cylinder and cone are equal (also equal to the diameter of the sphere), the volume of the cone is equal to one-half that of the inscribed sphere.

Exercises

1. Find the volume of a sphere the radius of which is 6 inches.
2. Find the surface area of a sphere the radius of which is 8 inches.
3. The area of a great circle of a sphere is 231 square inches. Find the area of the sphere.
4. What is the weight of an iron ball having a diameter of 30 inches if 1 cubic foot of iron weighs 490 pounds?
5. A storage tank for gas is in the shape of a sphere. Its inside diameter is 20 feet. How many cubic feet of gas is stored in the tank?
6. Find the volume enclosed in the figure.

7. Find the total surface area of the figure. Include the lateral surface area of the cylinder, the area of the base, and the area of the hemisphere.
8. A sphere just fits into a cube having a 15-inch side. Find the volume of the sphere.
9. A cylindrical vessel 12 inches in diameter is filled with water. Then a rock is immersed in it, causing some of the water to overflow. When the rock is removed, it is found that the water level in the cylinder drops 10 inches. What is the volume of the rock?
10. A sphere 8 inches in diameter is inscribed in a right circular cylinder. What is (*a*) the volume? (*b*) the lateral area of the cylinder?
11. A hollow spherical metal ball has an inside diameter of 11 inches and is $\frac{1}{2}$ inch thick. Find the volume of metal in the ball.

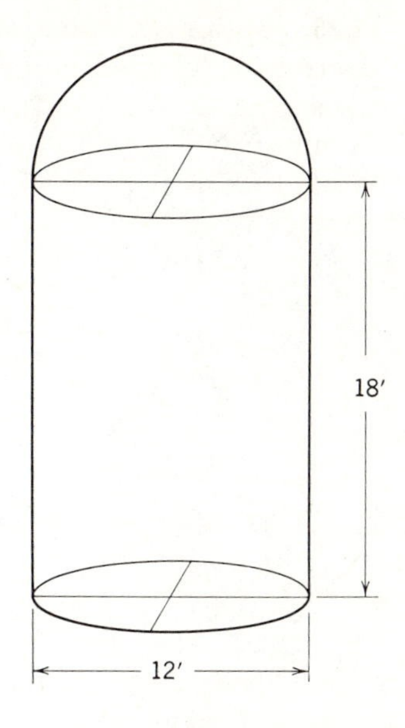

Exs. 6, 7.

14.26. (optional). Conic section. The curve formed by the intersection of a plane and a right circular conical surface is called a *conic section.* If the plane is perpendicular to the axis, the conic is a circle (Fig. 14.30*a*). If the cutting plane is oblique to the axis, and cuts all the elements, the conic is an ellipse (Fig. 14.30*b*). If the plane is parallel to one, and only one, element of the cone, the conic is a parabola (Fig. 14.30*c*). If the plane is parallel to the axis, it will cut both parts of the surface, forming a conic which is a hyperbola (Fig. 14.30*d*).

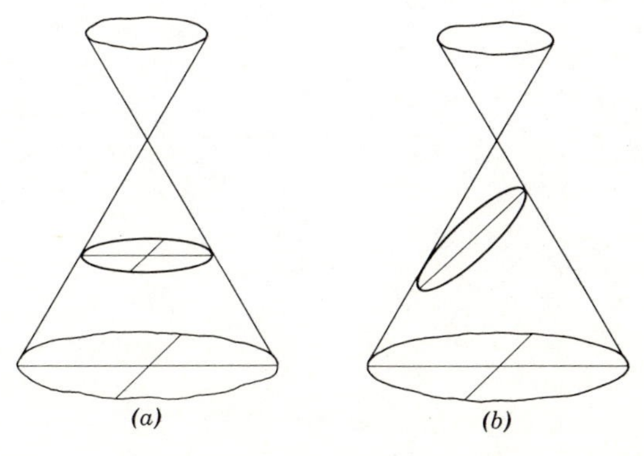

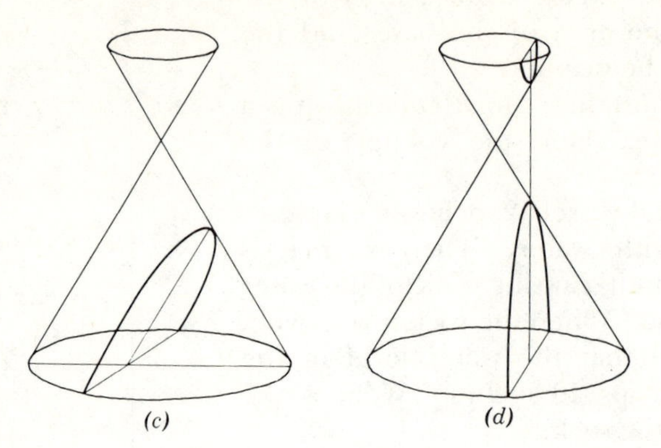

Fig. 14.30.

The ancient Greek mathematicians were familiar with the conics and discovered many properties of the conics. They were able to express these properties by means of mathematical equations. The content of these defining equations and their properties are included in the study of analytic geometry.

Summary Tests

Test 1

COMPLETION STATEMENTS

1. The locus of points in a room equidistant from two adjacent walls and the ceiling is a ______ .
2. The number of planes determined by four points not all in one plane is ______ .
3. The formula for the volume of a sphere is ______ .
4. The intersection of a plane and a sphere is a ______ .
5. The number of tangents that can be drawn from an external point to a sphere is ______ .
6. The number of lines that can be drawn through a point on a sphere tangent to the sphere is ______ .
7. The number of tangent planes to a sphere that can be drawn at a point on the sphere is ______ .
8. Two lines may intersect in a ______ .
9. Two planes may intersect in a ______ .
10. Given two intersecting lines, there is (are) exactly ______ plane(s) containing them.
11. The number of edges in a tetrahedron is ______ .
12. Two planes parallel to the same plane are ______ to each other.
13. The locus of points equidistant from a circle is a ______ .
14. The locus of points in a plane at a given distance from a given point without the plane is a ______ .
15. At a given point in a given line there can be only one ______ perpendicular to the given line.

Test 2

TRUE-FALSE STATEMENTS

1. A line perpendicular to one of two parallel planes must be parallel to the other.
2. A line intersecting one of two parallel lines must also intersect the other.
3. It is possible to have two lines parallel to the same plane and be perpendicular to each other.
4. It is possible for two planes to be perpendicular to the same line and intersect each other.
5. If two planes are parallel, then any line in one of them is parallel to any line in the other.
6. Two lines parallel to the same plane are parallel to each other.
7. Given plane MN; line l not in plane MN; plane $MN \parallel$ line m; $l \parallel m$. Then plane $MN \parallel$ line l.
8. Given plane MN; line l not in plane MN; line $m \perp$ plane MN; line $l \perp$ line m. Then line $l \parallel$ plane MN.
9. Given plane $RS \parallel$ plane MN; plane GH cuts plane RS and plane MN in lines l and m respectively. Then $l \parallel m$.
10. Given line $l \parallel$ plane GH; line $m \parallel$ plane GH. Then $l \parallel m$.
11. Given plane $AB \perp$ plane MN; plane $GH \perp$ plane MN. Then plane $AB \parallel$ plane GH.
12. The projection of a segment on a plane is always another segment.
13. The projections of congruent segments on a plane will also be congruent.
14. It is possible for the projection of a segment on a plane to be greater than the length of the segment.
15. Given line l lies in plane MN; line $l \perp$ line m. Then line $m \perp$ plane MN.
16. Given $\overleftrightarrow{AB} \perp \overleftrightarrow{BC}$; $\overleftrightarrow{BD} \perp \overleftrightarrow{BC}$. Then $\overleftrightarrow{AB} \perp \overleftrightarrow{BD}$.
17. If two intersecting planes are each perpendicular to a third plane, then their line of intersection is perpendicular to the third plane.
18. Given plane MN bisects $\overline{AB}$. Then every point of plane MN is equidistant from A and B.
19. The lateral area of a pyramid equals half the sum of its slant height and the perimeter of the base.
20. If a line intersects a plane in only one point, there is at least one line in the plane perpendicular to the line.
21. It is always possible to have intersecting lines both perpendicular to a given plane.
22. It is possible to have two planes perpendicular to a given line.
23. It is possible to have two lines perpendicular to a given plane at a point on the plane.

24. It is possible to have two lines perpendicular to a given line at a point on the line.
25. If a line intersects two parallel lines, all three lines lie in the same plane.

Problems Test

1. Find the volume of the solid in cubic feet.
2. Find the total surface area in square feet.

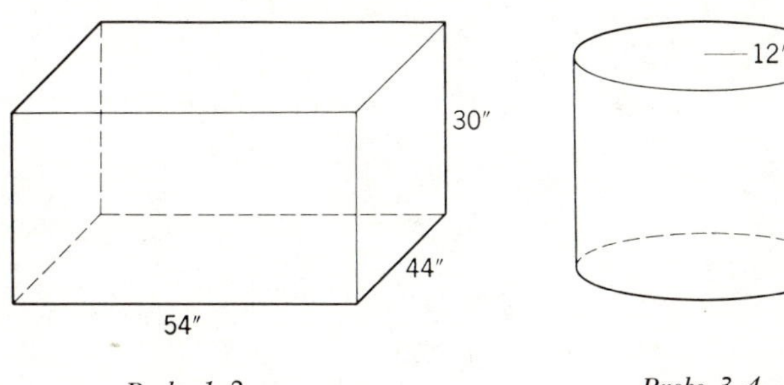

Probs. 1, 2. *Probs. 3, 4.*

3. Find the volume of the tank.
4. Find the total surface area (include top and bottom) of the tank.
5. Find the lateral area of the right circular cone.
6. Find the volume of the cone.

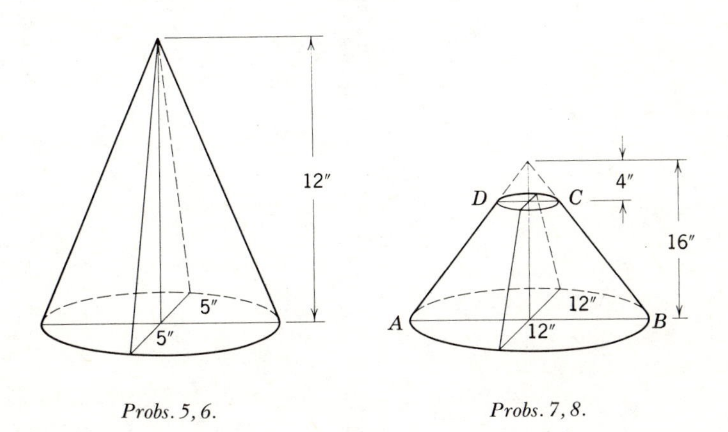

Probs. 5, 6. *Probs. 7, 8.*

7. Find the volume of the solid *ABCD*.
8. Find the lateral area of the solid *ABCD*.

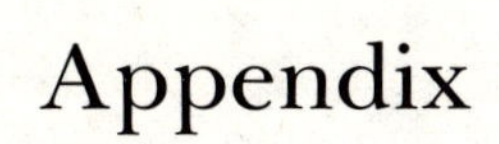

Appendix

GREEK ALPHABET

LETTERS		NAMES	LETTERS		NAMES	LETTERS		NAMES
A	α	Alpha	I	ι	Iota	P	ρ	Rho
B	β	Beta	K	κ	Kappa	Σ	σ	Sigma
Γ	γ	Gamma	Λ	λ	Lambda	T	τ	Tau
Δ	δ	Delta	M	μ	Mu	Υ	υ	Upsilon
E	ϵ	Epsilon	N	ν	Nu	Φ	ϕ	Phi
Z	ζ	Zeta	Ξ	ξ	Xi	X	χ	Chi
H	η	Eta	O	o	Omicron	Ψ	ψ	Psi
Θ	θ	Theta	Π	π	Pi	Ω	ω	Omega

SYMBOLS AND ABBREVIATIONS

Symbol	*Meaning*
$A \cup B$	the union of sets A and B
$A \cap B$	the intersection of sets A and B
A'	the complement of set A
$A \subset B$	A is a subset of B
$x \in B$	x is a member of set B
$x \notin B$	x is not a member of set B
$p \wedge q$	p and q
$p \vee q$	p or q (inclusive)
$p \veebar q$	p or q (exclusive)
$p \rightarrow q$	p implies q; if p, then q
$p \leftrightarrow q$	p is equivalent to q; p if, and only if, q
iff	if, and only if
{ }	notation for set
$\emptyset$	null set
$\overset{\circ\quad\circ}{AB}$	interval AB
$\overline{AB}$	segment AB

$\overleftrightarrow{AB}$	line AB
$\overset{\circ}{\overrightarrow{AB}}$	half-line AB
$\overrightarrow{AB}$	ray AB
$m\overline{AB}$	the measure of segment AB
AB	the measure of segment AB
$\angle$	angle
$\angle$s	angles
$m\angle ABC$	the measure of angle ABC
$\widehat{AB}$	arc AB
$m\widehat{AB}$	the degree measure of arc AB
$=$	is equal to; equals
$\neq$	is not equal to
$\sqrt{x}$	the nonnegative square root of x
$\|x\|$	the absolute value of x.
$\sim$	is similar to
$\perp$	is perpendicular to
$\parallel$	is parallel to
$>$	is greater than
$<$	is less than
$\geqslant$	is greater than or equal to
$\leqslant$	is less than or equal to
$\therefore$	therefore
$\ldots$	and so on
§	section
$\triangle$	triangle
$\triangle$s	triangles
▱	parallelogram
▱s	parallelograms
▭	rectangle
▭s	rectangles
$\odot$	circle
ⓢ	circles
$^\circ$	degree

S.A.S. If two triangles have two sides and the included angle of one congruent respectively to two sides and the included angle of the other, the triangles are congruent.

A.S.A. If two triangles have two angles and the included side of one congruent respectively to two angles and the included side of the other, the triangles are congruent.

S.S.S. If two triangles have the three sides of one congruent respectively to the three sides of the other, the triangles are congruent.

TABLE I. SQUARE ROOTS

N	√N	N	√N	N	√N	N	√N	N	√N	N	√N
1	1.000	51	7.141	101	10.050	151	12.288	201	14.177	251	15.843
2	1.414	52	7.211	102	10.100	152	12.329	202	14.213	252	15.875
3	1.732	53	7.280	103	10.149	153	12.369	203	14.248	253	15.906
4	2.000	54	7.348	104	10.198	154	12.410	204	14.283	254	15.937
5	2.236	55	7.416	105	10.247	155	12.450	205	14.318	255	15.969
6	2.449	56	7.483	106	10.296	156	12.490	206	14.353	256	16.000
7	2.646	57	7.550	107	10.344	157	12.530	207	14.387	257	16.031
8	2.828	58	7.616	108	10.392	158	12.570	208	14.422	258	16.062
9	3.000	59	7.681	109	10.440	159	12.610	209	14.457	259	16.093
10	3.162	60	7.746	110	10.488	160	12.649	210	14.491	260	16.125
11	3.317	61	7.810	111	10.536	161	12.689	211	14.526	261	16.155
12	3.464	62	7.874	112	10.583	162	12.728	212	14.560	262	16.186
13	3.606	63	7.937	113	10.630	163	12.767	213	14.595	263	16.217
14	3.742	64	8.000	114	10.677	164	12.806	214	14.629	264	16.248
15	3.873	65	8.062	115	10.724	165	12.845	215	14.663	265	16.279
16	4.000	66	8.124	116	10.770	166	12.884	216	14.697	266	16.310
17	4.123	67	8.185	117	10.817	167	12.923	217	14.731	267	16.340
18	4.243	68	8.246	118	10.863	168	12.961	218	14.765	268	16.371
19	4.359	69	8.307	119	10.909	169	13.000	219	14.799	269	16.401
20	4.472	70	8.367	120	10.954	170	13.038	220	14.832	270	16.432
21	4.583	71	8.426	121	11.000	171	13.077	221	14.866	271	16.462
22	4.690	72	8.485	122	11.045	172	13.115	222	14.900	272	16.492
23	4.796	73	8.544	123	11.091	173	13.153	223	14.933	273	16.523
24	4.899	74	8.602	124	11.136	174	13.191	224	14.967	274	16.553
25	5.000	75	8.660	125	11.180	175	13.229	225	15.000	275	16.583
26	5.099	76	8.718	126	11.225	176	13.267	226	15.033	276	16.613
27	5.196	77	8.775	127	11.269	177	13.304	227	15.067	277	16.643
28	5.292	78	8.832	128	11.314	178	13.342	228	15.100	278	16.673
29	5.385	79	8.888	129	11.358	179	13.379	229	15.133	279	16.703
30	5.477	80	8.944	130	11.402	180	13.416	230	15.166	280	16.733
31	5.568	81	9.000	131	11.446	181	13.454	231	15.199	281	16.763
32	5.657	82	9.055	132	11.489	182	13.491	232	15.232	282	16.793
33	5.745	83	9.110	133	11.533	183	13.528	233	15.264	283	16.823
34	5.831	84	9.165	134	11.576	184	13.565	234	15.297	284	16.852
35	5.916	85	9.220	135	11.619	185	13.601	235	15.330	285	16.882
36	6.000	86	9.274	136	11.662	186	13.638	236	15.362	286	16.912
37	6.083	87	9.327	137	11.705	187	13.675	237	15.395	287	16.941
38	6.164	88	9.381	138	11.747	188	13.711	238	15.427	288	16.971
39	6.245	89	9.434	139	11.790	189	13.748	239	15.460	289	17.000
40	6.325	90	9.487	140	11.832	190	13.784	240	15.492	290	17.029
41	6.403	91	9.539	141	11.874	191	13.820	241	15.524	291	17.059
42	6.481	92	9.592	142	11.916	192	13.856	242	15.556	292	17.088
43	6.557	93	9.644	143	11.958	193	13.892	243	15.588	293	17.117
44	6.633	94	9.695	144	12.000	194	13.928	244	15.620	294	17.146
45	6.708	95	9.747	145	12.042	195	13.964	245	15.652	295	17.176
46	6.782	96	9.798	146	12.083	196	14.000	246	15.684	296	17.205
47	6.856	97	9.849	147	12.124	197	14.036	247	15.716	297	17.234
48	6.928	98	9.899	148	12.166	198	14.071	248	15.748	298	17.263
49	7.000	99	9.950	149	12.207	199	14.107	249	15.780	299	17.292
50	7.071	100	10.000	150	12.247	200	14.142	250	15.811	300	17.321

PROPERTIES OF REAL NUMBER SYSTEM

Equality Properties

E-1 (reflexive property): $a = a$ (p. 72).

E-2 (symmetric property): $a = b \rightarrow b = a$ (p. 72).

E-3 (transitive property): $(a = b) \wedge (b = c) \rightarrow a = c$ (p. 72).

E-4 (addition property): $(a = b) \wedge (c = d) \rightarrow (a+c) = (b+d)$ (p. 72).

E-5 (subtraction property): $(a = b) \wedge (c = d) \rightarrow (a-c) = (b-d)$ (p. 72).

E-6 (multiplication property): $(a = b) \wedge (c = d) \rightarrow ac = bd$ (p. 72).

E-7 (division property): $(a = b) \wedge (c = d \neq 0) \rightarrow \frac{a}{c} = \frac{b}{d}$ (p. 72).

E-8 (substitution property): Any expression may be replaced by an equivalent expression in an equation without destroying the truth value of the equation (p. 73).

Order Properties

O-1 (trichotomy property): For every pair of real numbers, a and b, exactly one of the following is true: $a < b$, $a = b$, $a > b$ (p. 73).

O-2 (addition property): $(a < b) \wedge (c \leqslant d) \rightarrow (a+c) < (b+d)$ (p. 73).

O-3 (subtraction property): $(a < b) \wedge (c \geqslant 0) \rightarrow (a-c) < (b-c)$ (p. 73).
$(a < b) \wedge (c \geqslant 0) \rightarrow (c-a) > (c-b)$ (p. 73).

O-4 (multiplication property): $(a < b) \wedge (c > 0) \rightarrow ac < bc$ (p. 73).
$(a < b) \wedge (c < 0) \rightarrow ac > bc$ (p. 73).

O-5 (division property): $(a < b) \wedge (c > 0) \rightarrow a/c < b/c \wedge c/a > c/b$ (p. 73).
$(a < b) \wedge (c < 0) \rightarrow a/c > b/c \wedge c/a < c/b$ (p. 73).

O-6 (transitive property): $(a < b) \wedge (b < c) \rightarrow a < c$ (p. 73).

O-7 (substitution property): Any expression may be substituted for an equivalent expression in an inequality without changing the truth value of the inequality (p. 73).

O-8 (partition property): $(c = a+b) \wedge (b > 0) \rightarrow c > a$ (p. 73).

Properties of a Field

F-1 (closure property for addition): $a+b$ is a unique real number (p. 73).

F-2 (associative property for addition): $(a+b)+c = a+(b+c)$ (p. 73).

F-3 (commutative property for addition): $a+b = b+a$ (p. 73).

F-4 (additive property of zero): There is a unique number 0, the additive identity element, such that $a+0 = 0+a = a$ (p. 73).

F-5 (additive inverse property): For every real number a there exists a real number $-a$, the additive inverse of a, such that $a+(-a) = (-a)+a = 0$ (p. 73).

F-6 (closure property for multiplication): $a \cdot b$ is a unique real number (p. 73).

F-7 (associative property for multiplication): $(a \cdot b) \cdot c = a \cdot (b \cdot c)$ (p. 73).

F-8 (commutative property for multiplication): $a \cdot b = b \cdot a$ (p. 74).

F-9 (multiplicative property of 1): There is a unique real number 1, the multiplicative identity element, such that $a \cdot 1 = 1 \cdot a = a$ (p. 74).

F-10 (multiplicative inverse property): For every real number a $(a \neq 0)$, there is a unique real number $1/a$, the multiplicative inverse of a, such that $a \cdot (1/a) = (1/a) \cdot a = 1$ (p. 74).

F-11 (distributive property): $a(b+c) = a \cdot b + a \cdot c$ (p. 74).

LIST OF POSTULATES

1. A line contains at least two points; a plane contains at least three points not all collinear; and space contains at least four points not all coplanar (p. 76).
2. For every two distinct points, there is exactly one line that contains both points (p. 76).
3. For every three distinct noncollinear points, there is exactly one plane that contains the three points (p. 76).
4. If a plane contains two points of a straight line, then all points of the line are points of the plane (p. 76).
5. If two distinct planes intersect, their intersection is one and only one line (p. 76).

6. (The ruler postulate): The points on a line can be placed in a one-to-one correspondence with real numbers in such a way that
   (1) for every point of the line, there corresponds exactly one real number;
   (2) for every real number, there corresponds exactly one point of the line; and
   (3) the distance between two points on a line is the absolute value of the difference between the corresponding numbers (p. 79).
7. To each pair of distinct points there corresponds a unique positive number, which is called the distance between the two points (p. 79).
8. For every three collinear points, one and only one is between the other two (p. 80).
9. If A and B are two distinct points, then there is at least one point C such that $C \in \overline{AB}$. (p. 80).
10. If A and B are two distinct points, there is at least one point D such that $\overline{AB} \subset \overline{AD}$ (p. 80).
11. (Point plotting postulate): For every $\overrightarrow{AB}$ and every positive number n, there is one and only one point of $\overrightarrow{AB}$ such that $m\overline{AB} = n$ (p. 80).
12. (Angle construction postulate): If $\overrightarrow{AB}$ is a ray on the edge of the half-plane h, then for every n between 0 and 180 there is exactly one ray AP, with P in h, such that $m\angle PAB = n$ (p. 80).
13. (Segment addition postulate): A set of points lying between the endpoints of a line segment divides the segment into a set of consecutive segments the sum of whose lengths equals the length of the given segment (p. 80).
14. (Angle addition postulate): In a given plane, rays from the vertex of an angle through a set of points in the interior of the angle divides the angle into consecutive angles the sum of whose measures equals the measure of the given angle (p. 80).
15. A segment has one and only one midpoint (p. 81).
16. An angle has one and only one bisector (p. 81).
17. (The S.A.S. postulate): Two triangles are congruent if two sides and the included angle of one are, respectively, congruent to the two sides and the included angle of the other (p. 113).
18. (The parallel postulate): Through a given point not on a given line, there is at most one line which can be drawn parallel to the given line (p. 155).
19. In a plane one, and only one, circle can be drawn with a given point as center and a given line segment as radius (p. 210).
20. (Arc addition postulate): If the intersection of $\overset{\frown}{AB}$ and $\overset{\frown}{BC}$ of a circle is the single point B, then $m\overset{\frown}{AB} + m\overset{\frown}{BC} = m\overset{\frown}{AC}$ (p. 214).
21. (Area postulate): Given a unit of area, to each polygonal region there

corresponds a unique positive number, which is called the area of the region (p. 342).

22. The area of a polygonal region is the sum of the area measures of any set of component regions into which it can be cut (p. 342).
23. If two polygons are congruent, their corresponding polygonal regions have the same area (p. 343).
24. The area of a rectangular region is equal to the product of the length of its base and the length of its altitude (p. 343).
25. There is exactly one pair of real numbers assigned to each point in a given coordinate system. Conversely, if (a, b) is any ordered pair of real numbers, there is exactly one point in a given system which has (a, b) as its coordinates (p. 363).

LISTS OF THEOREMS AND COROLLARIES

3-1. If two distinct lines in a plane intersect in a point, then their intersection is at most one point (p. 77).

3-2. If a point P lies outside a line l, exactly one plane contains the line and the point (p. 77).

3-3. If two distinct lines intersect, exactly one plane contains both lines (p. 77).

3-4. For any real number, a, b, and c, if $a = c$, and $b = c$, then $a = b$ (p. 82).

3-5. For any real numbers a, b, and c, if $c = a$ and $c = b$, then $a = b$ (p. 83).

3-6. For any real numbers a, b, c, and d, if $a = c$, $b = d$, and $c = d$, then $a = b$ (p. 83).

3-7. All right angles are congruent (p. 83).

3-8. Complements of the same angle are congruent (p. 84).
Corollary: Complements of congruent angles are congruent (p. 85).

3-9. All straight angles are congruent. (p. 84).

3-10. Supplements of the same angle are congruent (p. 84).
Corollary: Supplements of congruent angles are congruent (p. 85).

3-11. Two adjacent angles whose noncommon sides form a straight angle are supplementary (p. 90).

3-12. Vertical angles are congruent (p. 91).

3-13. Perpendicular lines form four right angles (p. 91).

3-14. If two lines meet to form congruent adjacent angles, they are perpendicular (p. 92).

4-1. (Reflexive property): Every segment is congruent to itself (p. 102).

4-2. (Symmetric property): If $\overline{AB} \cong \overline{CD}$, then $\overline{CD} \cong \overline{AB}$. (p. 102).

4-3. (Transitive property): If $\overline{AB} \cong \overline{CD}$ and $\overline{CD} \cong \overline{EF}$, then $\overline{AB} \cong \overline{EF}$ (p. 102).

4-4. (Addition property): If B is between A and C, E between D and F, and if $\overline{AB} \cong \overline{DE}$ and $\overline{BC} \cong \overline{EF}$, then $\overline{AC} \cong \overline{DF}$ (p. 102).

4-5. (Subtractive property): If B is between A and C, E is between D and F, $\overline{AC} \cong \overline{DE}$, and $\overline{BC} \cong \overline{EF}$, then $\overline{AB} \cong \overline{DE}$ (p. 103).

4-6. (Reflexive property): Every angle is congruent to itself (p. 103).

4-7. (Symmetric property): If $\angle A \cong \angle B$, then $\angle B \cong \angle A$ (p. 103).

4-8. (Transitive property): If $\angle A \cong \angle B$ and $\angle B \cong \angle C$, then $\angle A \cong \angle C$ (p. 103).

4-9. (Angle addition property): If D is in the interior of $\angle ABC$, P is in the interior of $\angle RST$, $\angle ABD \cong \angle RSP$, and $\angle DBC \cong \angle PST$, then $\angle ABC \cong \angle RST$ (p. 103).

4-10. (Angle subtraction property): If D is in the interior of $\angle ABC$, P is in the interior of $\angle RST$, $\angle ABC \cong \angle RST$, and $\angle ABD \cong \angle RSP$, then $\angle DBC \cong \angle PST$ (p. 103).

4-11. If $\overline{AC} \cong \overline{DF}$, B bisects $\overline{AC}$, E bisects $\overline{DF}$, then $\overline{AB} \cong \overline{DE}$ (p. 104).

4-12. If $\angle ABC \cong \angle RST$, $\overrightarrow{BD}$ bisects $\angle ABC$, $\overrightarrow{SP}$ bisects $\angle RST$, then $\angle ABD \cong \angle RSP$ (p. 105).

4-13. If the two legs of one right triangle are congruent, respectively, to the two legs of another right triangle, the triangles are congruent (p. 114).

4-14. If two triangles have two angles and the included side of one congruent to the corresponding two angles and the included side of the other, the triangles are congruent (p. 119).

4-15. If a leg and the adjacent acute angle of one right triangle are congruent, respectively, to a leg and the adjacent acute angle of another, the right triangles are congruent (p. 120).

4-16. The base angles of an isosceles triangle are congruent (p. 126).
Corollary: An equilateral triangle is also equiangular (p. 126).

4-17. The measure of an exterior angle of a triangle is greater than the measure of either of the two nonadjacent interior angles (p. 131).

4-18. If two triangles have the three sides of one congruent, respectively, to the three sides of the other, the triangles are congruent to each other (p. 131).

5-1. If two parallel planes are cut by a third plane, the lines of intersection are parallel (p. 141).

5-2. In a given plane, through any point of a straight line, there can pass one and only one line perpendicular to the given line (p. 147).

5-3. Through a point not on a given line, there is at least one line perpendicular to that given line (p. 149).

5-4. Through a given external point, there is at most one perpendicular to a given line (p. 150).

5-5. If two lines are perpendicular to the same line, they are parallel to each other (p. 153).

5-6. Two planes perpendicular to the same line are parallel (p. 153).

5-7. In a plane containing a line and a point not on the line, there is at least one line parallel to the given line (p. 154).

5-8. Two lines parallel to the same line are parallel to each other (p. 155).

5-9. In a plane containing two parallel lines, if a line is perpendicular to one of the two parallel lines, it is perpendicular to the other also (p.156).

5-10. A line perpendicular to one of two parallel planes is perpendicular to the other (p. 157).

5-11. If two straight lines form congruent alternate interior angles when they are cut by a transversal, they are parallel (p. 158).

5-12. If two straight lines are cut by a transversal so as to form a pair of congruent corresponding angles, the lines are parallel (p. 159).

Corollary: If two lines are cut by a transversal so as to form interior supplementary angles in the same closed half-plane of the transversal, the lines are parallel (p. 159).

5-13. If two parallel lines are cut by a transversal, the alternate interior angles are congruent (p. 163).

5-14. If two parallel lines are cut by a transversal, the corresponding angles are congruent (p. 163).

5-15. If two parallel lines are cut by a transversal, the interior angles on the same side of the transversal are supplementary (p. 163).

5-16. The measure of an exterior angle of a triangle is equal to the sum of the measures of the two nonadjacent interior angles (p. 167).

5-17. The sum of the measures of the angles of a triangle is 180 (p. 167).

Corollary: Only one angle of a triangle can be a right angle or an obtuse angle (p. 168).

Corollary: If two angles of one triangle are congruent, respectively, to two angles of another triangle, the third angles are congruent (p. 168).

Corollary: The acute angles of a right triangle are complementary (p. 168).

5-18. If two angles of a triangle are congruent, the sides opposite them are congruent (p. 172).
Corollary: An equiangular triangle is equilateral (p. 173).

5-19. If two right triangles have a hypotenuse and an acute angle of one congruent, respectively, to the hypotenuse and an acute angle of the other, the triangles are congruent (p. 173).

5-20. If two right triangles have the hypotenuse and a leg of one congruent to the hypotenuse and a leg of the other, the triangles are congruent (p. 173).

5-21. If the measure of one acute angle of a right triangle equals 30, the length of the side opposite this angle is one-half the length of the hypotenuse (p. 175).

6-1. All angles of a rectangle are right angles (p. 186).

6-2. The opposite sides and the opposite angles of a parallelogram are congruent (p. 188).
Corollary: Either diagonal divides a parallelogram into two congruent triangles (p. 188).
Corollary: Any two adjacent angles of a parallelogram are supplementary (p. 188).
Corollary: Segments of a pair of parallel lines cut off by a second pair of parallel lines are congruent (p. 188).
Corollary: Two parallel lines are everywhere equidistant (p. 188).
Corollary: The diagonals of a rectangle are congruent (p. 188).

6-3. The diagonals of a parallelogram bisect each other (p. 189).
Corollary: The diagonals of a rhombus are perpendicular to each other (p. 189).

6-4. If the opposite sides of a quadrilateral are congruent, the quadrilateral is a parallelogram (p. 191).

6-5. If two sides of a quadrilateral are congruent and parallel, the quadrilateral is a parallelogram (p. 192).

6-6. If the diagonals of a quadrilateral bisect each other, the quadrilateral is a parallelogram (p. 192).

6-7. If three or more parallel lines cut off congruent segments on one transversal, they cut off congruent segments on every transversal (p. 193).

6-8. If two angles have their sides so matched that corresponding sides have the same directions, the angles are congruent (p. 196).

6-9. If two angles have their sides so matched that two corresponding sides have the same direction and the other two corresponding sides are oppositely directed, the angles are supplementary (p. 197).

6-10. The segment joining the midpoints of two sides of a triangle is parallel to the third side and its measure is one-half the measure of the third side (p. 198).

6-11. A line that bisects one side of a triangle and is parallel to a second side bisects the third side (p. 199).

6-12. The midpoint of the hypotenuse of a right triangle is equidistant from its vertices (p. 199).

7-1. If two central angles of the same or congruent circles are congruent, then their intercepted arcs are congruent (p. 214).

7-2. If two arcs of a circle or congruent circles are congruent, then the central angles intercepted by these arcs are congruent (p. 215).

7-3. The measure of an inscribed angle is equal to half the measure of its intercepted arc (p. 217).
Corollary: An angle inscribed in a semicircle is a right angle (p. 218).
Corollary: Angles inscribed in the same arc are congruent (p. 218).
Corollary: Parallel lines cut off congruent arcs on a circle (p. 218).

7-4. In the same circle, or in congruent circles, congruent chords have congruent arcs (p. 222).

7-5. In the same circle, or in congruent circles, congruent arcs have congruent chords (p. 223).

7-6. In the same circle, or in congruent circles, chords are congruent iff they have congruent central angles (p. 223).

7-7. A line through the center of a circle and perpendicular to a chord bisects the chord and its arc (p. 227).

7-8. If a line thorugh the center of a circle bisects a chord that is not a diameter, it is perpendicular to the chord (p. 227).
Corollary: The perpendicular bisector of a chord of a circle passes through the center of the circle (p. 227).

7-9. In a circle, or in congruent circles, congruent chords are equidistant from the center (p. 227).

7-10. In a circle, or in congruent circles, chords equidistant from the center of the circle are congruent (p. 228).

7-11. If a line is tangent to a circle, it is perpendicular to the radius drawn to the point of tangency (p. 230).
Corollary: if a line lying in the plane of a circle is perpendicular to a tangent at the point of tangency, it passes through the center of the circle (p. 231).

7-12. If a line lying in the plane of a circle is perpendicular to a radius at at its point on the circle, it is tangent to the circle (p. 231).

7-13. Tangent segments from an external point to a circle are congruent and make congruent angles with the line passing through the point and the center of the circle (p. 232).

7-14. The measure of the angle formed by a tangent and a secant drawn from the point of tangency is half the measure of its intercepted arc (p. 236).

7-15. The measure of an angle formed by two chords intersecting within a circle is half the sum of the measures of the arcs intercepted by it and its vertical angle (p. 236).

7-16. The measure of the angle formed by two secants intersecting outside a circle is half the difference of the measures of the intercepted arcs (p. 237).

Corollary: The measure of the angle formed by a secant and a tangent intersecting outside a circle is half the difference of the measures of the intercepted arcs (p. 238).

Corollary: The measure of the angle formed by two tangents drawn from an external point to a circle is half the difference of the measures of the intercepted arcs (p. 238).

8-1. In a proportion, the product of the extremes is equal to the product of the means (p. 249).

8-2. In a proportion, the second and third terms may be interchanged to obtain another valid proportion (p. 249).

8-3. In a proportion, the ratios may be inverted to obtain another valid proportion (p. 250).

8-4. If the product of two quantities is equal to the product of two other quantities, either pair of quantities can be used as the means and the other as the extremes of a proportion (p. 250).

8-5. If the numerators of a proportion are equal, the denominators are equal and conversely (p. 250).

8-6. If three terms of one proportion are equal to the corresponding three terms of another proportion, the remaining terms are equal (p. 250).

8-7. In a series of equal ratios the sum of the numerators is to the sum of the denominators as the numerator of any one of the ratios is to the denominator of that ratio (p. 250).

8-8. If four quantities are in proportion, the terms are in proportion by by addition or subtraction; that is, the sum (or difference) of the first and second terms is to the second term as the sum (or difference) of the third and fourth terms is to the fourth term (p. 251).

8-9. If a line parallel to one side of a triangle cuts a second side into segments which have a ratio with interger terms, the line will cut the third side into segments which have the same ratio (p. 253).

8-10. A line parallel to one side of a triangle and intersecting the other two sides divides these sides into proportional segments (p. 254).
Corollary: If a line is parallel to one side of a triangle and intersects the other two sides, it divides these sides so that either side is to one of its segments as the other is to its corresponding segment (p. 254).
Corollary: Parallel lines cut off proportional segments on two transversals (p. 254).

8-11. If a line divides two sides of a triangle proportionally, it is parallel to the third side (p. 255).
Corollary: If a line divides two sides of a triangle so that either side is to one of its segments as the other side is to its corresponding segment, the line is parallel to the third side (p. 255).

8-12. If two triangles have the three angles of one congruent, respectively, to the three angles of the other, the triangles are similar (p. 259).
Corollary: If two triangles have two angles of one congruent to two angles of the other, the triangles are similar (p. 260).
Corollary: If two right triangles have an acute angle of one congruent to an acute angle of the other, they are similar (p. 260).
Corollary: Two triangles which are similar to the same triangle or two similar triangles are similar to each other (p. 260).
Corollary: Corresponding altitudes of two similar triangles have the same ratio as any two corresponding sides (p. 260).

8-13. If two triangles have an angle of one congruent to an angle of the other and the sides including these angles proportional, the triangles are similar (p. 266).

8-14. If two triangles have their corresponding sides proportional, they are similar (p. 266).

8-15. The altitude on the hypotenuse of a right triangle forms two right triangles which are similar to the given triangle and similar to each other (p. 269).
Corollary: The altitude on the hypotenuse of a right triangle is the mean proportional between the measures of the segments of the hypotenuse (p. 270).
Corollary: Either leg of a right triangle is the mean proportional between the measure of the hypotenuse and the measure of the segment of the hypotenuse cut off by the altitude which is adjacent to that leg (p. 270).

8-16. The square of the measure of the hypotenuse of a right triangle is equal to the sum of the squares of the measures of the legs (p. 270).
Corollary: The square of the measure of the leg of a right triangle is

equal to the square of the measure of the hypotenuse minus the square of the measure of the other leg (p. 271).

8-17. If two chords intersect within a circle, the product of the measures of the segments of one chord is equal to the product of the measures of the segments of the other (p. 274).

8-18. If a tangent and a secant are drawn from the same point outside a circle, the measure of the tangent is the mean proportional between the measures of the secant and its external segment (p. 274).

8-19. If two secants are drawn from the same point outside a circle, the product of the measures of one secant and its external segment is equal to the product of the measures of the other secant and its external segment (p. 275).

9-1. If two sides of a triangle are not congruent, the angle opposite the longer of the two sides has a greater measure than does the angle opposite the shorter side (p. 287).

9-2. If two angles of a triangle are not congruent, the side opposite the larger of the two angles is greater than the side opposite the smaller of the two angles (p. 288).
Corollary: The shortest segment joining a point to a line is the perpendicular segment (p. 288).
Corollary: The measure of the hypotenuse of a right triangle is greater than the measure of either leg (p. 288).

9-3. The sum of the measures of two sides of a triangle is greater than the the measure of the third side (p. 289).

9-4. If two triangles have two sides of one congruent, respectively, to two sides of the other and the measure of the included angle of the first greater than the measure of the included angle of the second triangle, the third side of the first is greater than the third side of the second (p. 289).

9-5. If two triangles have two sides of one congruent, respectively, to two sides of the other and the third side of the first greater than the third side of the second, the measure of the angle opposite the third side of the first is greater than the measure of the angle opposite the third side of the second (p. 290).

9-6. In a circle or in congruent circles, if two central angles have unequal measures, the greater central angle has the greater minor arc (p. 294).

9-7. In a circle or in congruent circles, if two minor arcs are not congruent, the greater arc has the greater central angle (p. 295).

9-8. In a circle or in congruent circles, the greater of two noncongruent chords has the greater minor arc (p. 295).

9-9. In a circle or in congruent circles, the greater of two noncongruent minor arcs has the greater chord (p. 296).

9-10. In a circle or in congruent circles, if two chords are not congruent, they are unequally distant from the center, the greater chord being nearer the center (p. 296).

9-11. In a circle or in congruent circles, if two chords are unequally distant from the center, they are not congruent, the chord nearer the center being the greater (p. 296).

11-1. The locus of points in a plane at a given distance from a fixed point is a circle whose center is the given point and whose radius measure is the given distance (p. 322).

11-2. The locus of points in a plane at a given distance from a given line in the plane is a pair of lines parallel to the given line and at the given distance from the given line (p. 322).

11-3. The locus of points in a plane equidistant from two given parallel lines is a line parallel to the given lines and midway between them (p. 322).

11-4. The locus of points in a plane which are equidistant from two given points in the plane is the perpendicular bisector of the line segment joining the two points (p. 322).

11-5. The locus of points in the interior of an angle which are equidistant from the sides of the angle is the bisector of the angle minus its endpoint. (p. 324).
Corollary: the locus of points equidistant from two given intersecting lines is the pair of perpendicular lines which bisects the vertical angles formed by the given lines.(p. 325).

11-6. The locus of all points such that $\triangle APB$ is a right triangle having $\overline{AB}$ a fixed line segment as hypotenuse is a circle having $\overline{AB}$ as a diameter, for points A and B themselves (p. 325).

12-1. The area of a parallelogram is equal to the product of its base and its altitude (p. 345).
Corollary: Parallelograms with equal bases and equal altitudes are equal in area (p. 346).
Corollary: The areas of two parallelograms having equal bases have the same ratio as their altitudes; the areas of two parallelograms having equal altitudes have the same ratio as their bases (p. 346).

12-2. The area of a triangle is equal to one-half the product of its base and its altitude (p. 346).
Corollary: Triangles with equal bases and equal altitudes are equal in area (p. 346).
Corollary: The areas of two triangles having equal bases have the

same ratios as their altitudes; the areas of two triangles having equal altitudes have the same ratio as their bases (p. 346).
Corollary: The area of a rhombus is equal to one-half the product of its diagonals (p. 346).

12-3. The area of a trapezoid is equal to half the product of its altitude and the sum of its bases (p. 348).

13-1. (The distance formula). For any two points P and Q.

$$PQ = \sqrt{(x_Q - x_P)^2 + (y_Q - y_P)^2} = \sqrt{(x_P - x_Q)^2 + (y_P - y_Q)^2}. \qquad \text{(p. 368)}.$$

13-2. (The midpoint formula). M is the midpoint of $\overline{AB}$ if and only if $x_M = \frac{1}{2}(x_A + x_B)$ and $y_M = \frac{1}{2}(y_A + y_B)$ (p. 369).

13-3. (Slope formula). If $P \neq Q$ are any pair of points on a line not parallel to the y-axis of a rectangular coordinate system, then there is a unique real number m, called slope, such that

$$m = \frac{y_Q - y_P}{x_Q - x_P}. \qquad \text{(p. 372)}.$$

13-4. Two nonvertical lines l_1 and l_2 are parallel if and only if their slopes m_1 and m_2 are equal (p. 374).

13-5. Two nonvertical lines l_1 and l_2 are perpendicular if and only if their slopes are negative reciprocals of each other (p. 375).

13-6. (Point-slope formula). For each point $P_1(x_1, y_1)$ and for each number m, the equation of the line through P with slope m is $y - y_1 = m(x - x_1)$ (p. 380).

13-7. (Slope y-intercept formula). The graph of the equation $y = mx + b$ is the line with slope m and y-intercept b (p. 384).

13-8. The graph of every linear equation in x and y is always a straight line in the XY-plane (p. 384).

13-9. Every straight line in a plane is the graph of a linear equation in x and y (p. 384).

14-1. If a line intersects a plane not containing it, then the intersection is a single point (p. 391).

14-2. All the perpendiculars drawn through a point on a given line lie in a plane perpendicular to the given line at that point (p. 391).

14-3. Through a given point, there passes one and only one plane perpendicular to a given line (p. 391).

14-4. One and only one perpendicular line can be drawn to a plane from a point not on the line (p. 391).

14-5. The perpendicular from a point not on a plane to the plane is the shortest line segment from the point to the plane (p. 391).

14-6. If two lines are perpendicular to a plane, they are parallel to each other (p. 391).

14-7. If one of two parallel lines is perpendicular to a plane, the other is also perpendicular to the plane (p. 392).

14-8. If each of three noncollinear points of a plane is equidistant from two points, then every point of the plane is equidistant from these points (p. 392).

14-9. Two parallel planes are everywhere equidistant (p. 392).

14-10. Through one straight line any number of planes may be passed (p. 392).

14-11. If two planes are perpendicular to each other, a straight line in one of them perpendicular to their intersection is perpendicular to the other (p. 392).

14-12. If two planes are perpendicular to each other, a perpendicular to one of them at their intersection lies in the other (p. 393).

14-13. If two intersecting planes are perpendicular to a third plane, their intersection is also perpendicular to that plane (p. 393).

14-14. Every point in a plane bisecting a dihedral angle is equidistant from the faces of the angle (p. 393).

14-15. The set of points equidistant from the faces of a dihedral angle is the plane bisecting the dihedral angle (p. 394).

Answers to Exercises

Pages 4–5

1. Ten. None.
3. No. E is not a set; F is a set with one element.
5. 1, 2, 3, 4, 5
7. A, B, C, E, F, G
9. There are none.
11. a. $\in$ b. $\notin$ c. $\in$ d. $\notin$ e. $\notin$ f. $\in$
13. {Tuesday, Thursday}
15. $\{0\}$
17. $\{10, 11, 12, \ldots\}$
19. {January, June, July}
21. {vowels of the alphabet}
23. {colors of the spectrum}
25. {even numbers greater than 1 and less than 11}
27. {negative even integers}

Pages 8–9

1. a. $\{3, 6, 9\}$ b. $\{2, 3, 4, 5, 6, 7, 8, 9, 10\}$
3. a. Q b. P
5. $\{2, 4, 6, \ldots\}$
6. a. B c. $\varnothing$ e. A
7. a. true c. true e. false g. false i. false k. false
9. the null set

11.

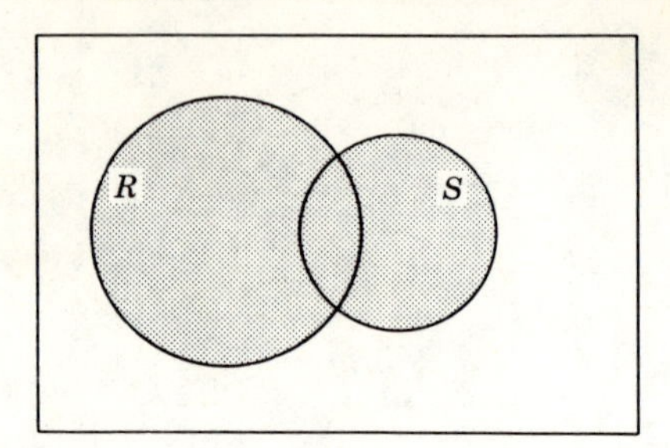

13.

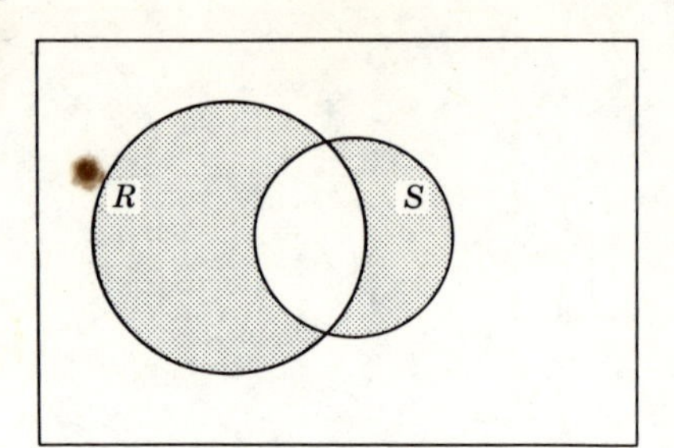

15.

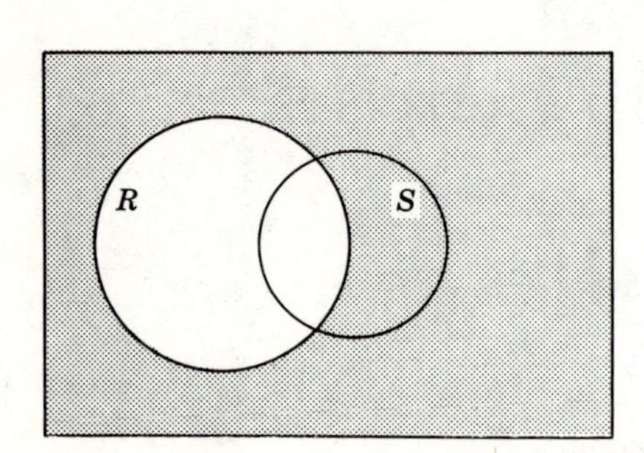

17.

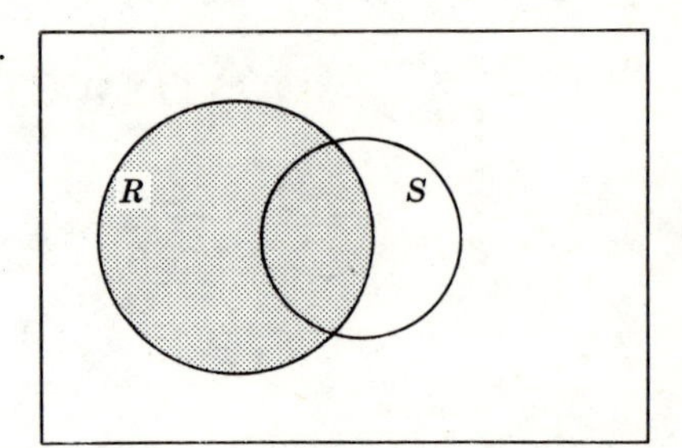

19.

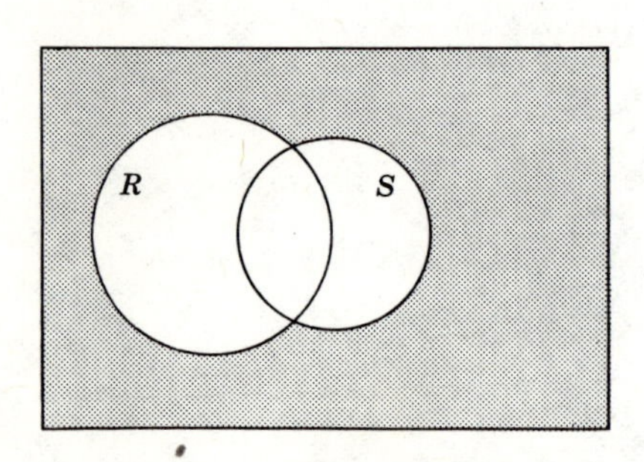

21.

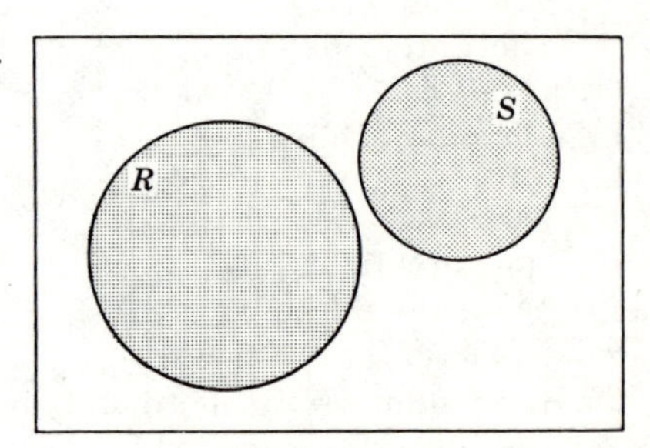

23.

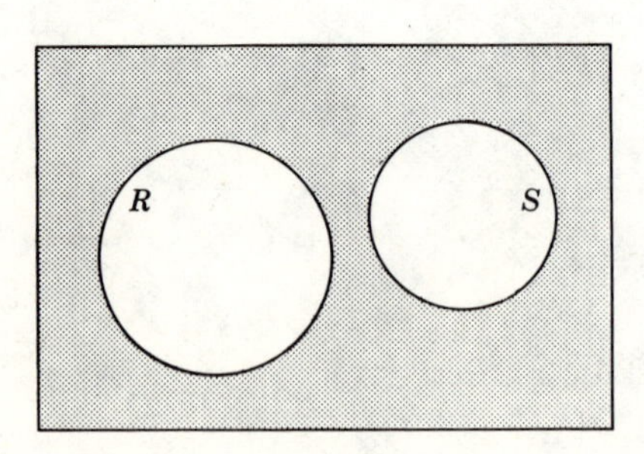

25.

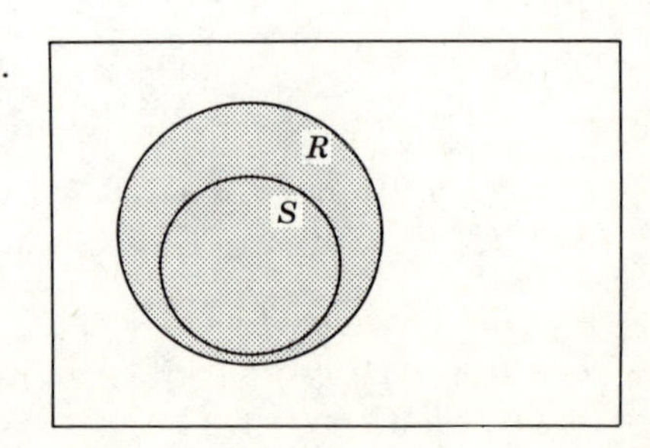

27. 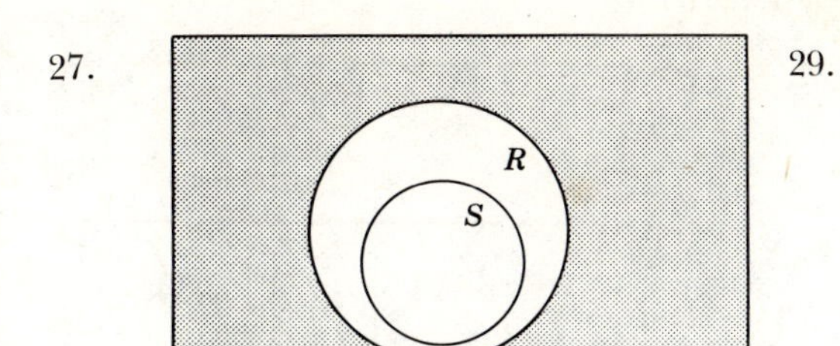

29. 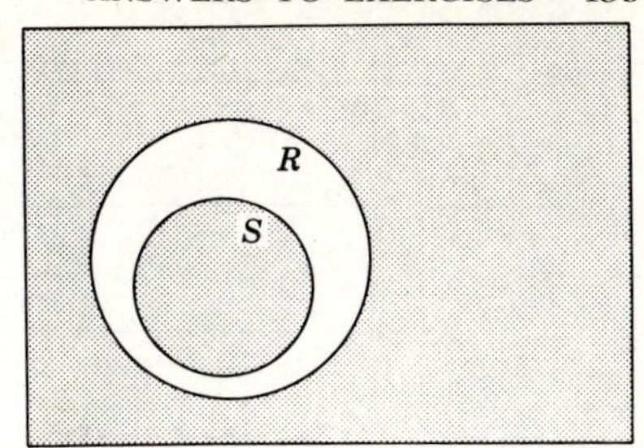

Pages 14–16

1. an infinite number 3. one 5. no 7. no 9. false
11. true 13. false 15. true 17. true

19. 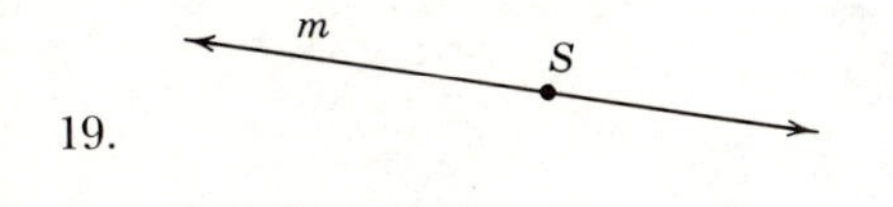

21. 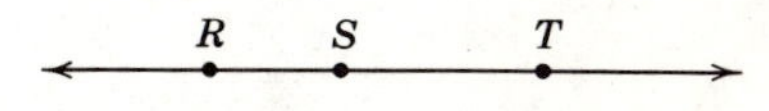

Draw points R, S, T of a line in any order.

23. 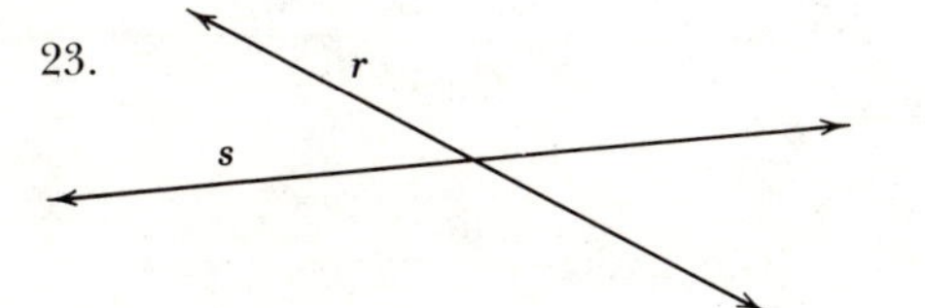

25. 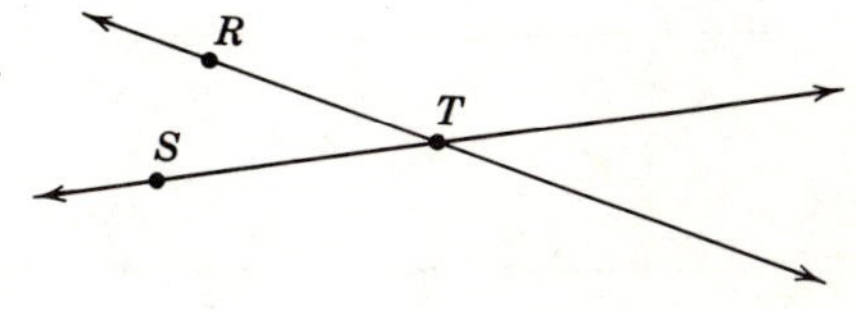

27. P Q R S

Draw a line; label points P, Q, R, S (any order) on that line.

29. Not possible.

31. Not possible.

33. 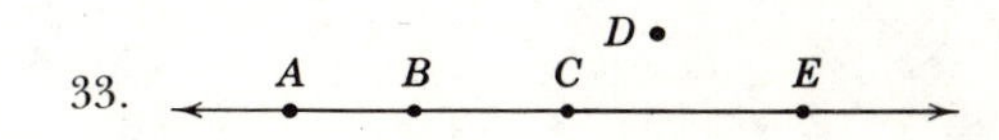

A, B, C, E are collinear (any order); D does not lie on the line.

35. A B C D

37.

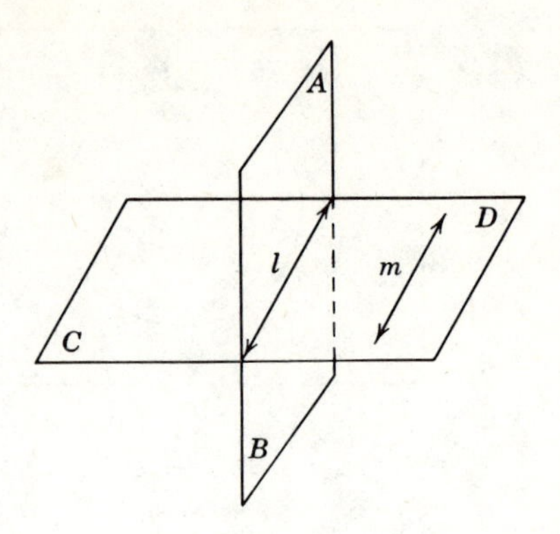

Pages 19–20

1. -4; 2 3. -5 5. 2 7. $1\frac{1}{2}$ 9. 4 11. -3 13. 2 15. 3
17. 9 19. 7 21. -2 23. 8 25. 8

Page 23

1. yes 3. yes 5. no 7. yes 9. { } 11. $\overrightarrow{AB}$ (or $\overrightarrow{AC}$ or $\overrightarrow{AD}$)

13. A B C D

15. R S T

17. R Q P G

19. Not possible.

21. Q P R

(*P* between *Q* and *R*)

23.

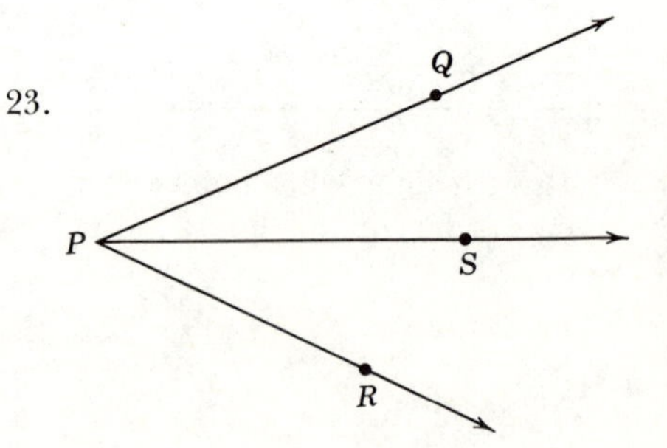

25. R P Q

(*P* between *Q* and *R*)

27.

Q P R

(Q between P and R)

29.

l

m

n

(l, m, n are 3 ∥ lines taken in any order)

31.

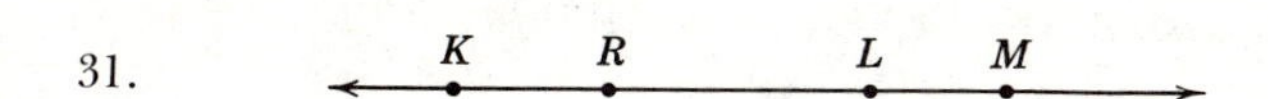

Pages 28–30

1. $\angle DMC$; $\angle CMD$; $\angle\beta$ 3. $\overleftrightarrow{MD}$; $\overleftrightarrow{BD}$; $\overleftrightarrow{DB}$; (also $\overleftrightarrow{BM}$ or $\overleftrightarrow{MB}$) 5. $\overrightarrow{BD}$
7. $\angle ABF$; $\angle AMC$; $\angle BMD$ 9. $\angle AMD$

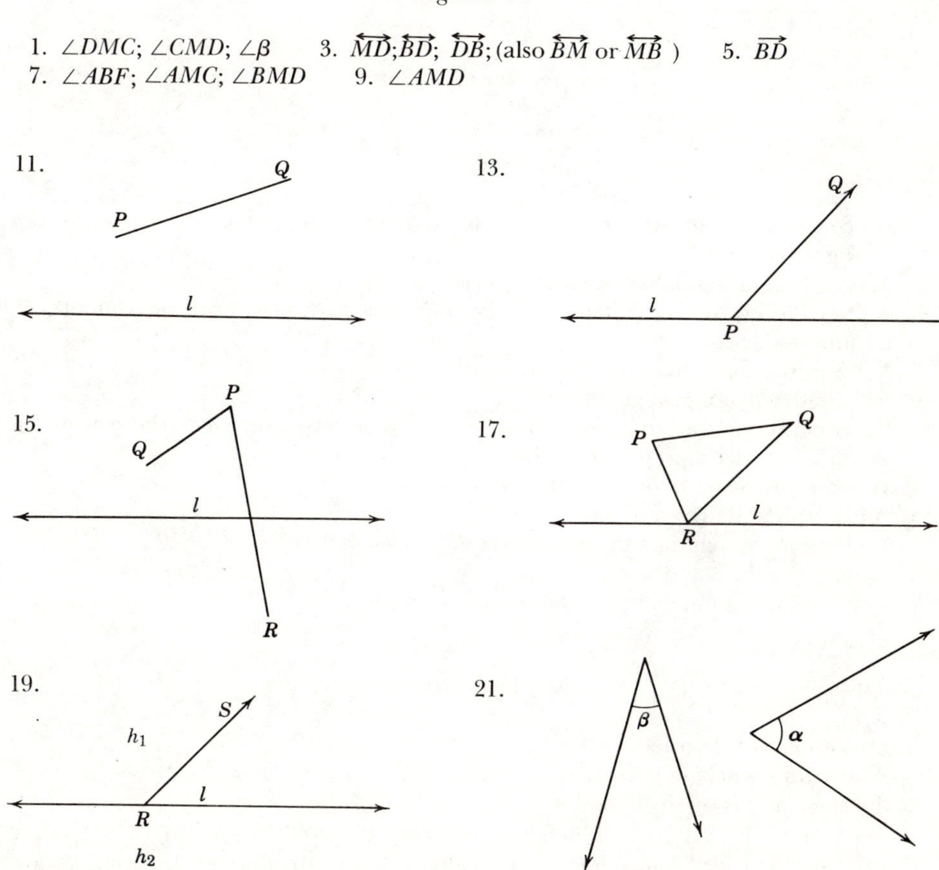

25. 69 27. 45

Pages 38–41

1. scalene; obtuse 3. isosceles; acute
5. scalene; acute 7. equilateral; equiangular
9. scalene; right 11. isosceles; right
13. $AB, CD; AC, BC$ 15. $\angle A$ and $\angle B$
17. $\angle RST = 180$; $m\angle USW = 90$. Subtracting the second equation from the first, we get $m\angle\alpha + m\angle\beta = 90$. Hence, $\angle\alpha$ and $\angle\beta$ are complementary angles.
19. $\angle AOB$ and $\angle BOC$; $\angle AOD$ and $\angle DOC$; $\angle AOE$ and $\angle EOC$
21. $\angle RPT$ and $\angle TPS$; $\angle TPS$ and $\angle SPW$; $\angle SPW$ and $\angle WPR$; $\angle WPR$ and $\angle RPT$
23. $\angle ABE$ and $\angle EBD$; $\angle ACE$ and $\angle DCE$
25. (1) 150 (b) 135 (c) 90 $180 - x$
27. $\angle ABD \cong \angle CBD$
29. $\overline{AE} \cong \overline{CE}$; $\overline{BE} \cong \overline{DE}$
31. $\overline{CD} \cong \overline{BD}$

Pages 44–46

Exercises (A)

1. no conclusion; dog may be barking for a reason other than the presence of a stranger.
3. Mary Smith must take an orientation class.
5. no conclusion; the given statement does not indicate that *only* college students will be admitted free.
7. Mr. Smith is a citizen of the United States.
9. Bill Smith will not pass geometry.
11. no conclusion; the given statement does not indicate that *only* those who eat Zeppo cereal are alert on the diamond.
13. It is not customary to bury living persons.
15. A five-cent and fifty-cent piece.
17. It is not stated that the men played five games *against each other*.
19. Two.
21. Coins are not stamped in advance of an uncertain date.

Exercises (B)

1. Bob is heavier than Jack.
3. Misfortune will befall Mr. Grimes.
5. I will get a wart on my hand.
7. (a) yes; (b) yes; (c) doesn't logically follow; (d) not true
9. (a) yes; (b) yes; (c) doesn't logically follow; (d) doesn't logically follow.

Pages 47–50

Test 1

1. perpendicular
3. obtuse
5. scalene
7. isosceles
9. bisector
11. 90
13. 360
15. acute
17. straight line
19. midpoint
21. {1, 2, 3, 4, 5}
23. {1, 2, 3, 4, 5, 6, 7, 8, 9}
25. {1, 2, 3, 4, 5, 6, 7, 8, 9}
27. {5}
29. {1, 2, 3, 4, 5, 6, 7, 8, 9}
31. $>$
33. $=$
35. $<$
37. 10
39. 2

Test 2

1. T 3. T 5. T 7. T 9. F 11. F 13. F 15. T
17. F 19. F 21. T 23. F 25. F 27. T 29. F 31. F
33. F 35. F 37. F 39. T 41. T 43. T 45. T 47. T
49. F

Test 3

1. 3 3. $\frac{5}{3}$ 5. $\frac{5}{2}$ 7. 1 9. 50 11. 20 13. 55
15. 110 17. 95; 95 19. 55

Pages 52–53

Exercises (A)

1. no 3. yes 5. yes 7. yes 9. yes 11. yes 15. yes

Exercises (B)

1. It is hot. I am tired.
3. His action was deliberate. His action was careless.
5. The figure is not a square. The figure is not a rectangle.
7. He is clever. I am not clever.
9. Sue dislikes Kay. Kay dislikes Sue.
11. Two lines intersect. Two lines are parallel.

13. The animal is a male. The animal is a female.
15. I would buy the car. The car costs too much.

Page 55

1. true; true 3. false; true 5. false; false 7. false; true 9. false; false
11. false; true 13. false; true 15. true; true

Pages 56–57

1. Gold is heavy.
3. Not everyone who wants a good grade in this course needs to study hard.
5. A hexagon does not have seven sides.
7. Every banker is rich.
9. Two plus 4 does not equal 8.
11. Not all equilateral triangles are equiangular.
13. No blind men carry white canes.
15. Not all these cookies are delicious.
17. Not every European lives in Europe.
19. There are no girls in the class.
21. Every question can be answered.
23. Not every ZEP is a ZOP.
25. It is not true that a null set is a subset of itself.

Pages 58–59

1. An apricot is not a fruit or a carrot is not a vegetable. False.
3. No men like to hunt or no men like to fish. False.
5. Some numbers are odd or not every number is even. True.
7. The sides of a right angle are not perpendicular or not all right angles are congruent. False.
9. Not every triangle has a right angle or not every triangle has an acute angle. True.
11. Not every triangle has a right angle and not every triangle has an obtuse angle. True.
13. No triangles have three acute angles or none have only two acute angles. False.
15. A ray does not have one endpoint and a segment does not have two endpoints. False.

Page 60

1. Premise: It is snowing. Conclusion: The rain will be late.
3. He is a citizen; he has the right to vote.
5. He is a student; he must take a physical examination.

7. The two lines are not parallel; the two lines intersect.
9. The numbers are natural numbers; the numbers are either even or odd.
11. It is a parallelogram; it is a quadrilateral.
13. It is a bird; it does not have four feet.
15. He studies; he will pass this course.
17. The person steals; the person will be caught.
19. He is a worker; he will be a success.
21. I have your looks; I will be a movie star.

Page 63

1. Bob is heavier than Jack.
3. My dog does not bite.
5. Figure *ABCD* is a quadrilateral.
7. $a+c=b+c$.
9. No conclusion.
11. I will get warts on my hand.
13. Jones lives in Houston.
15. $a \neq b$.
17. No conclusion.
19. $y=4$.
21. $a \neq b$.
23. $S \notin \overrightarrow{RT}$.
25. If l is not parallel to m, then $l \cap m \neq \emptyset$.

Pages 64–65

1. True. Vegetables are carrots. False.
3. True. Cars are Fords. Flase.
5. False. If he is not a poor speller, then he is a journalist. False.
7. Don't know. If he is a moron, then he will accept your offer. Dont' know.
9. Don't know. If a person studies, then he will succeed in school. False.
11. True. If it is hard, then it is a diamond. False.
13. True. If it has three congruent sides, then it is an equilateral triangle. True, if you are talking about triangles; otherwise it is false.
15. True. If x is larger than y, then $x-y=1$. False.
17. True (?). If he lives in California, then he lives in Los Angeles. False.
19. True. If $x^2=25$, then $x=5$. False.

Pages 66–67

1. yes 3. yes 5. yes 7. no 9. yes 11. yes 13. no 15. yes
17. no 19. yes 21. no (in space geometry) 23. yes

Pages 68–69

1. (a) If $T \in \overrightarrow{RX}$, then $T \in \mathring{\overrightarrow{RX}}$.
   (b) If $T \notin \overrightarrow{RX}$, then $T \notin \mathring{\overrightarrow{RX}}$.
   (c) If $T \notin \mathring{\overrightarrow{RX}}$, then $T \notin \overrightarrow{RX}$.
3. (a) If $C \in \overline{AB}$, then $C \in \overleftrightarrow{AB}$.
   (b) If $C \notin \overline{AB}$, then $C \notin \overleftrightarrow{AB}$.
   (c) If $C \notin \overleftrightarrow{AB}$, then $C \notin \overline{AB}$.
5. (a) If $a = -b$, then $a + b = 0$.
   (b) If $a \neq -b$, then $a + b \neq 0$.
   (c) If $a + b \neq 0$, then $a \neq -b$.
7. (a) If I pass this course, then I have studied.
   (b) If I do not pass this course, then I have not studied.
   (c) If I do not study, then I will not pass this course.
9. (a) If lines do not meet, then they are parallel.
   (b) If lines meet, then they are not parallel.
   (c) If lines are not parallel, then they will meet.
11. (a) If this is not a square, then it is not a rectangle.
    (b) If this is a square, then it is a rectangle.
    (c) If this is a rectangle, then it is a square.
13. (a) If the triangle is equiangular, it is equilateral.
    (b) If the triangle is not equiangular, then it is not equilateral.
    (c) If the triangle is not equilateral, then it is not equiangular.

15. not valid   17. not valid
19. valid   21. valid
23. valid   25. not valid

Pages 70–71

1. T   3. F   5. T   7. T   9. F   11. T   13. F   15. F
17. T   19. F   21. T   23. T   25. F   27. F   29. F   31. F
33. T   35. T   37. F   39. F   41. F

Pages 74–75

1. commutative property under addition
3. additive property of zero
5. distributive property
7. addition property of equality
9. symmetric property of equality
11. subtraction property of equality
13. multiplication property of equality
15. subtraction property of order
17. transitive property of order
19. division property of order
21. associative property of multiplication

23. multiplicative property of order
25. given; addition property of equality; subtraction property of equality; division property of equality
27. given; distributive property; addition property of equality; subtraction property of equality; division property of equality
29. given; addition property of order; subtraction property of order; division property of order

Pages 78–79

1. (a) any natural number (b) one
3. not necessarily
5. one
7. four
9. Line AB lies entirely in one plane.
11. yes
13. any nonnegative whole number
15. six
17. yes
19. collinear: d
    coplanar but not collinear: a, b, c
    not coplanar: e

Page 81

1. B 3. 5 5. 1 7. 8 9. no 11. C 13. C
15. AEC 17. AED 19. 78 21. 42

Pages 98–100

Test 1

1. T 3. T 5. F 7. F 9. T 11. T 13. T 15. T 17. F
19. F 21. T 23. T (except if one of the angles has a measure of zero) 25. T

Test 2

1. postulate 3. perpendicular 5. obtuse
7. 132 9. congruent 11. right
13. 60 15. plane 17. complementary
19. line

Pages 105–107

Exercises (A)

1. F 3. F 5. T 7. F 9. F 11. T 13. T 15. T
17. T 19. F 21. T

Exercises (B)

1. $\overline{AC} \cong \overline{FD}$. Segment addition property.
3. $\overline{AG} \cong \overline{GE} \cong \overline{BG} \cong \overline{FG}$. Definition of segment bisector and Theorem 4.11.
5. $\angle DAG \cong \angle CBE$. Angle subtraction property.
7. $AE \cong BG$. Transitive property of congruence.

Pages 109–110

1. $GHJ \leftrightarrow KLM$
   $GHJ \leftrightarrow KML$
   $GHJ \leftrightarrow LKM$
   $GHJ \leftrightarrow MLK$
   $GHJ \leftrightarrow MKL$
3. $ABC \leftrightarrow ABC$
   $ABC \leftrightarrow ACB$
   $ABC \leftrightarrow BAC$
   $ABC \leftrightarrow BCA$
   $ABC \leftrightarrow CAB$
   $ABC \leftrightarrow CBA$
5. yes; no
7. $\angle A \leftrightarrow \angle B$
   $\angle ADC \leftrightarrow \angle BDC$
   $\angle ACD \leftrightarrow \angle BCD$
   $\overline{AD} \leftrightarrow \overline{BD}$
   $\overline{AC} \leftrightarrow \overline{BC}$
   $\overline{CD} \leftrightarrow \overline{CD}$
9. $\angle FAC \leftrightarrow \angle EBD$
   $\angle ACF \leftrightarrow \angle BDE$
   $\angle F \leftrightarrow \angle E$
   $\overline{AF} \leftrightarrow \overline{BE}$
   $\overline{AC} \leftrightarrow \overline{BD}$
   $\overline{FC} \leftrightarrow \overline{ED}$
11. $\angle DAB \leftrightarrow \angle CBA$
    $\angle ADB \leftrightarrow \angle BCA$
    $\angle ABD \leftrightarrow \angle BAC$
    $\overline{AD} \leftrightarrow \overline{BC}$
    $\overline{BD} \leftrightarrow \overline{AC}$
    $\overline{AB} \leftrightarrow \overline{AB}$

Pages 116–117

1. yes 3. no 5. no 7. no 9. yes 11. no

Pages 122–123

1. yes 3. no 5. yes 7. no 9. yes

Pages 136–137

Test 1

1. exterior 3. corresponding 5. corresponding 7. base 9. right

Test 2

1. F 3. T 5. T 7. F 9. F 11. T 13. F 15. F
17. T 19. F 21. T 23. T 25. T

Page 146

11. q and w are true 13. (a) β; ψ (b) no

Pages 151–153

1. T 3. T 5. F 7. T 9. T 11. F 13. F 15. F
17. F 19. T 21. T 23. F

Pages 168–170

1. 90 3. 150 5. 45 7. 80

Pages 178–181

Test 1

1. 180 3. parallel 5. indirect 7. perpendicular
9. isosceles 11. complementary 13. right 15. parallel
17. parallel 19. obtuse 21. vertical; congruent

Test 2

1. F 3. F 5. F 7. F 9. T 11. F 13. T 15. T
17. F 19. F 21. T 23. F 25. F 27. T

Test 3

1. 55 3. 18 5. 50 7. 50

Pages 187–188

Exercises (A)

1. T 3. T 5. T 7. T 9. T 11. T 13. F 15. F
17. T 19. F 21. T 23. F 25. F

Exercises (B)

1. 360 3. (a) four (b) 720 5. 1800
7. 120 9. 144 11.

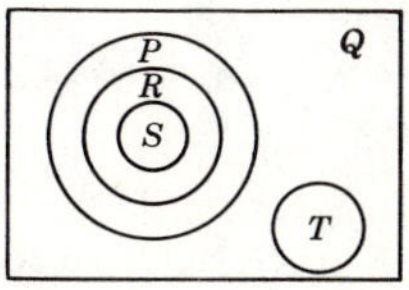

Pages 203–205

Test 2

1. T 3. T 5. F 7. T 9. F 11. F 13. T 15. F

17. T 19. F 21. F 23. T 25. T 27. F 29. T

Test 3

1. 30 3. 120 5. 14 in. 7. 13 in. 9. 5 11. 4 13. 5
15. 108 17. 36

Pages 215–216

1. F 3. T 5. T 7. F 9. F 11. F 13. T 15. F
17. F 19. T 21. 110 23. 60 25. (a) 90 (b) 120

Pages 219–221

1. 100; 140; 66; 54 3. (a) 50 (b) 50 5. 40 7. 50 9. 60 11. (a) 65 (b) 65

Pages 238–241

1. 60 3. 25 5. 160 7. 50 9. 30 11. 40 13. $m\angle\alpha = 25$; $m\angle\beta = 90$
15. $m\angle\alpha = 30$; $m\angle\beta = 60$ 17. $m\angle\alpha = 45$; $m\angle\beta = 70$; $s = 70$ 19. $m\angle\alpha = 80$; $m\angle\beta = 35$

Pages 242–244

Test 1

1. perpendicular 3. chords 5. one 7. diameter
9. supplementary 11. perpendicular 13. congruent

Test 2

1. T 3. T 5. F 7. F 9. F 11. F 13. F 15. T
17. F 19. F 21. F 23. F 25. T

Test 3

1. $m\angle\alpha = 70$; $m\angle\beta = 80$; $s = 60$ 3. $m\angle\alpha = 88$; $m\angle\beta = 65$; $s = 46$.
5. $m\angle\alpha = 72$; $m\angle\beta = 55.5$; $s = 111$ 7. $m\angle\alpha = 17.5$; $m\angle\beta = 72.5$; $s = 110$

Pages 246–248

1. (a) 2:3 (b) 5:3 (c) 3:5 (d) 2:3 (e) 4:5 3. 64:345 5. 72; 18 7. 3:1
9. 440:21 11. $68/83 \approx 0.819$ 13. $\pi:1$ 15. $AB:AC = 1.25:1$; $BC:CD = 2.1:1$
17. $DE:BE = AE:CE =$ a constant

Pages 251–252

1. (a) 16/5 (b) 6/5 (c) 3/2 (d) 4.8 (e) 10 2/7 (f) 5/9 3. (a) 5/2 (b) 4/9

(c) 5/3 (d) 1/3 (e) b/a (f) r/s 5. (a) 12 (b) $\sqrt{126}$ 9. 30 gal 11. $11\frac{1}{4}$ in.

Pages 255–256

1. 7.2 3. 10 5. 20 7. 16 9. yes 11. 15 13. 15 15. 24

Page 265

13. 36.5 ft

Pages 267–269

1. $DF:AB = EF:AE$ 3. $DC:AE = BC:AB$ 5. $CE:BE = AC:BD$
7. $RP:PT = PT:PS$ 9. $PJ:HP = RJ:SH$ 11. $8\frac{1}{3}$ 13. 18 15. 16

Pages 272–273

1. 6 3. $11\frac{1}{4}$ 5. 20 7. 15.3 9. 8 11. 10 13. 9 15. 18
17. 8 19. 24 21. 8 23. 30.6 25. 16

Pages 276–277

1. 10 3. $10\frac{5}{7}$ 5. 17 7. 12.8 9. $13\frac{1}{2}$ 11. 20 13. 6 15. 8
17. 9.6 19. 5.5 21. 7

Pages 278–281

Test 1

1. PT 3. similar 5. $EC \times DC$ 7. 20 9. $\frac{a+b}{b}$ 11. 6 13. 3:7
15. 8:5

Test 2

1. F 3. F 5. T 7. F 9. F 11. F 13. F 15. T 17. F
19. F

Test 3

1. 10 3. 13.856 5. 9.798 7. 8 9. 30 11. 20 13. 13

Pages 285–287

11. $a+c > b+d$ 13. $x < r$ 15. $z > x$ 17. $m\angle ABC > m\angle DEF$
19. $AD > BE$ 21. $BD < AC$ 23. $m\angle\alpha > m\angle A$

Page 297

1. $m\angle B > m\angle A > m\angle C$ 3. $\overline{NM}$ 5. $\overline{ST}$

Pages 300–302

Test 1

1. greater 3. $>$ 5. $<$ 7. $<$ 9. $>$ 11. sum 13. $<$ 15. $>$
17. $<$ 19. $\overline{RT}$

Test 2

1. T 3. T 5. T 7. F 9. T 11. F 13. F 15. F
17. T 19. T

Pages 337–339

Test 1

1. perpendicular bisector 3. circle 5. 2 7. 3
9. bisector 11. circle 13. two points

Test 2

1. (d) 3. (d) 5. (d) 7. (c) 9. (d) 11. (e) 13. (b)

Page 344

1. (a) 28 ft^2; (b) 12.5 ft^2; (c) 77/8 ft^2 3. 140 yd^2 5. 30 7. 30
9. 10 in.^{2} 11. 15 in.^{2}

Pages 347–348

1. 160 in.^{2} 3. $7\frac{11}{13}$ in. $\approx$ 7.84 in. 5. 187 in.^{2} (approx.) 7. (a) 168 in.^{2} (b) 168 in.^{2}
(c) 168 in.^{2} (d) 42 in.^{2} (e) 84 in.^{2} (f) 84 in.^{2} 9. $17\frac{1}{7}$ in. 11. 128 in.^{2} 13. 36 in.^{2}

Pages 350–352

1. 231 3. 112 5. 187 7. 250.2 9. 9.6 11. 6.4 13. 9.24
15. 15.6 17. 8 in. 19. 8.48 in; (b) 67.1 in. 21. (a) 72 in.^{2}; (b) 1352 in.^{2}
23. (a) 21.1 in.^{2}; (b) 1171 in.^{2}

Pages 354–356

1. (a) 33; (b) 16.0; (c) 204.6 3. (a) 89; (b) 27.88; (c) 21.7585 5. (a) 2.79;

(b) 3.4248; (c) 17.0499 7. 4:9 9. 65.6 ft^2 11. 76,986 ft^2 13. 703.36 $in.^2$ 15. 123.84 ft^2 17. 6949.3 ft^2 19. 483.48 in.

Pages 357–359

Test 1

1. pi 3. 27.7 5. $(\text{radius})^2$ 7. 2

Test 2

1. T 3. F 5. T 7. F 9. F 11. F 13. F

Test 3

1. 8.5 $in.^2$ 3. 62.4 ft^2 5. 114 $in.^2$ 7. 157 ft

Pages 361–362

1. $\{x|x > -2\}$ 3. $\{-2, 0, 2\}$ 5. $\{\ \}$

7.

−3 −2 −1 0 1 2

9.

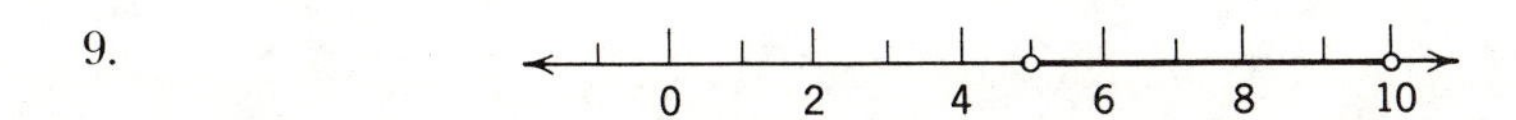

11.

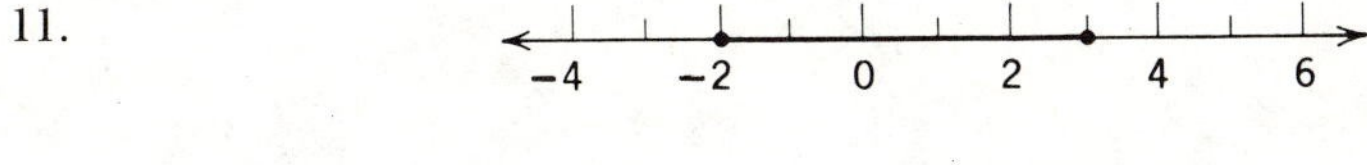

13.

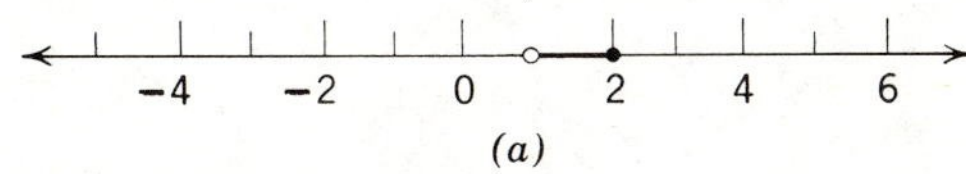

(a)

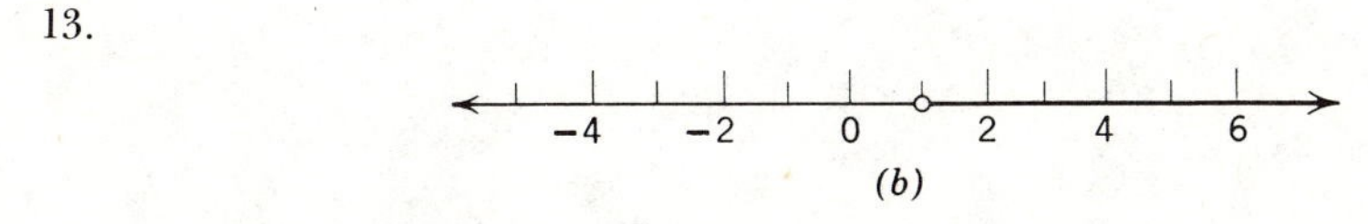

(b)

15.

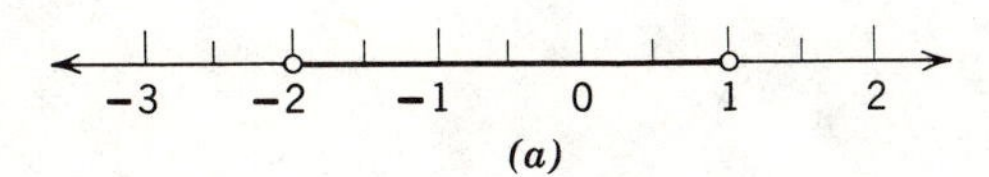

(a)

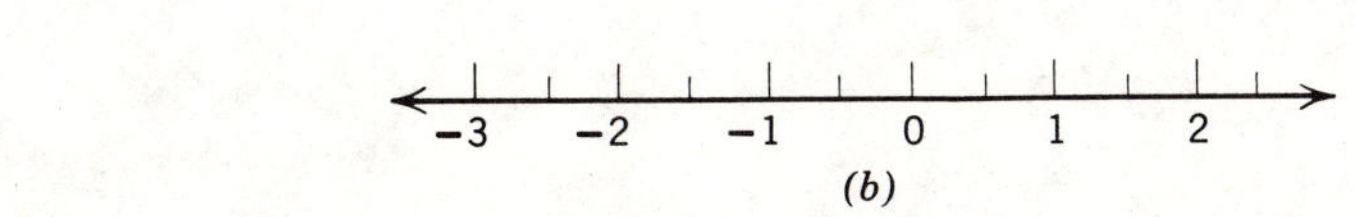

(b)

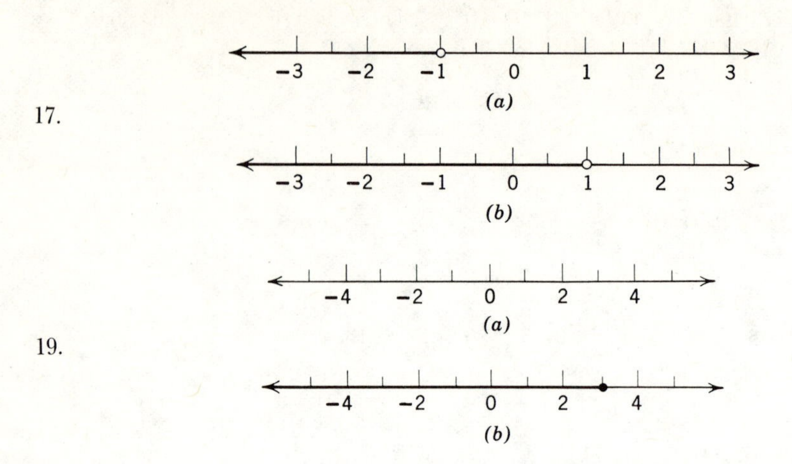

Pages 365–367

3. $A(1,0)$; $B(-\frac{3}{2},0)$; $C(2,0)$; $D(-\frac{5}{2},0)$; $E(0,\frac{3}{2})$; $P_1(2,\frac{3}{2})$; $P_2(-\frac{3}{2},1)$; $P_3(-\frac{5}{2},-2)$; $P_4(\frac{3}{2},-\frac{3}{2})$ 5. (a) 2; (b) 3; (c) 4; (d) 5; (e) 5; (f) 9; (g) 3; (h) 4; (i) 3; (j) 2; (k) 5; (l) 3 13. 5 15. y_2-y_1 17. (0, 4) 19. $\frac{x_1+x_2}{2}$

Page 370

1. (a) 5; (b) 13; (c) 17; (d) $\sqrt{113}$ (e) $8\sqrt{2}$ (f) $12\sqrt{2}$ (g) $2\sqrt{68}$ 3. 11.414 5. 24 11. (a) (3, 2); (c) $(-\frac{11}{2},-\frac{3}{2})$; $(b/2, a/2)$

Page 375

1. 2 3. −1 5. $-\frac{1}{5}$ 7. $\overline{AB} \parallel \overline{CD}$ 9. $\overline{AB} \parallel \overline{CD}$ 11. $\overline{AB} \perp \overline{CD}$

Page 379

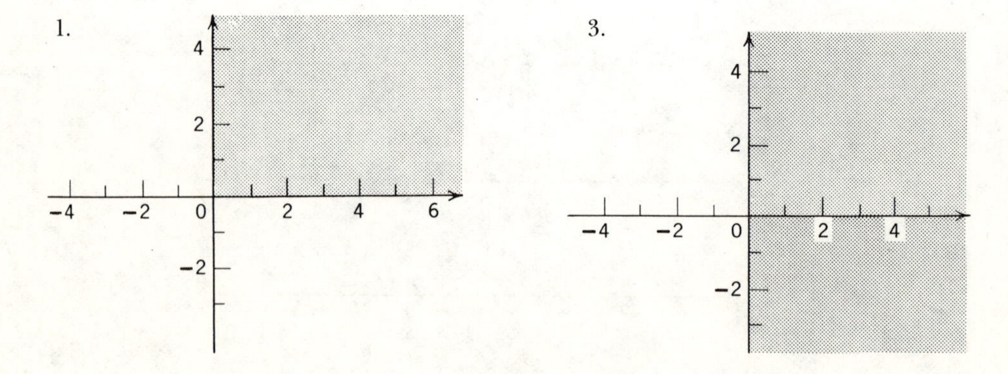

5.

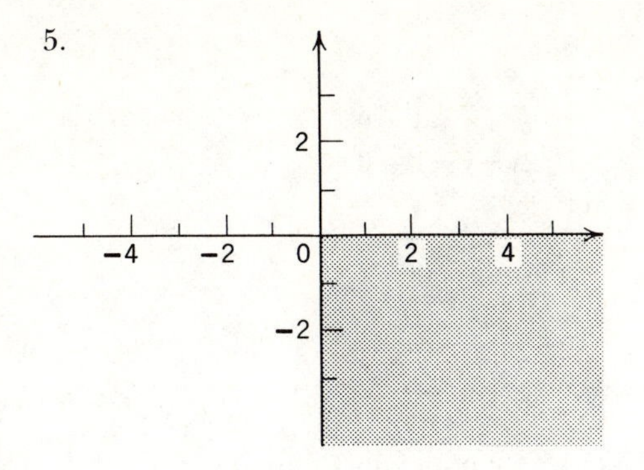
2
−4
−2
0
2
4
−2

7.

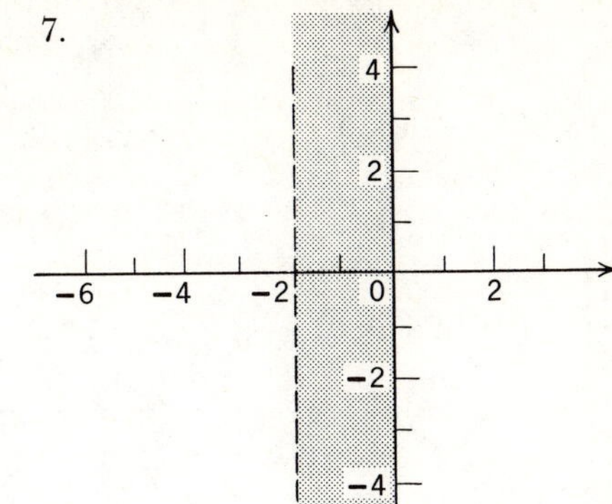
4
2
−6
−4
−2
0
2
−2
−4

9

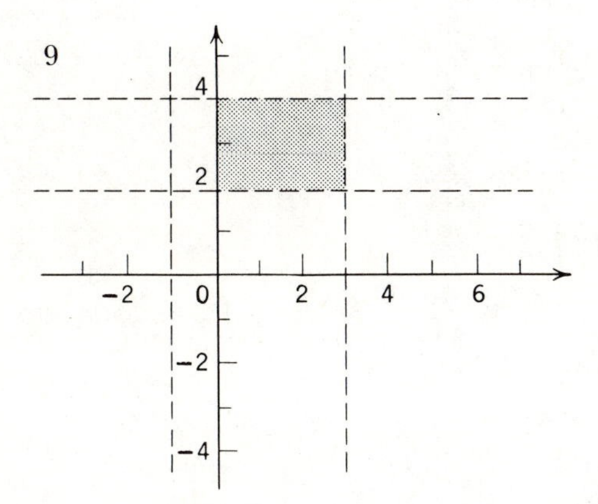
4
2
−2
0
2
4
6
−2
−4

11.

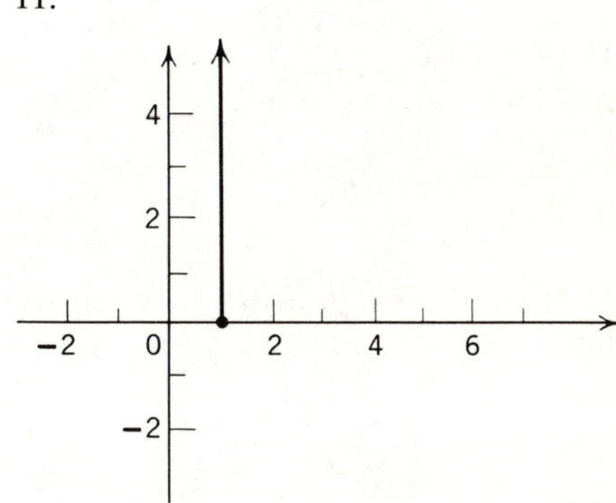
4
2
−2
0
2
4
6
−2

13.

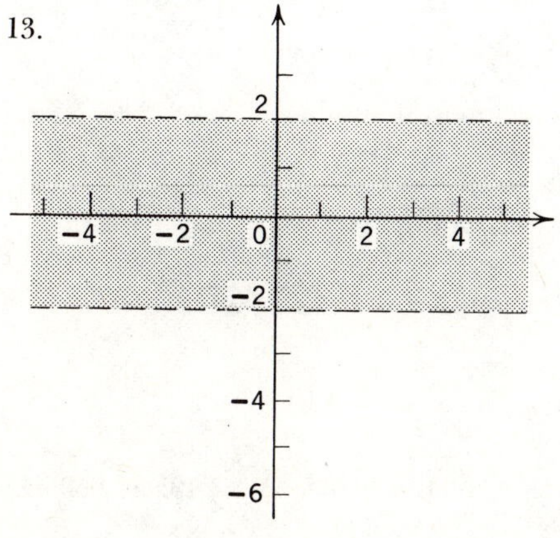
2
−4
−2
0
2
4
−2
−4
−6

15.

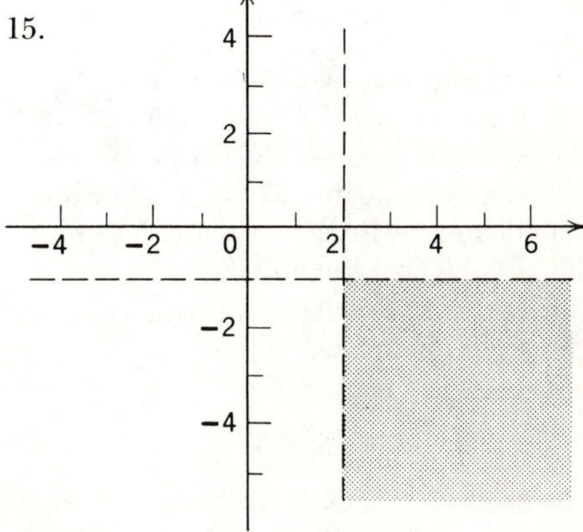
4
2
−4
−2
0
2
4
6
−2
−4

Pages 381–382

1. $4x-y-25=0$
3. $2x+y+4=0$
5. $y=-3$
7. $2x-y-1=0$
9. 3
11. $-\frac{5}{3}$
13. 2
15. 0
17. $2x+y=0$
19. $x-y=0$
21. $x-2y=0$

Page 385

1. 4; 3; $-\frac{3}{4}$
3. -5; none; not defined
5. $5x-y-7=0$
7. $2x+y-5=0$
9. $3x-y-9=0$
11. $2x-y+5=0$
13. $3x+y+4=0$
15. $x=0$
17. $x=3$
19. $x+y-8=0$
21. $3x-y+6=0$
23. $x+5y=0$

Pages 386–387

Test 1

1. (0, 0)
3. fourth
5. (0, 0)
7. 30
9. $-6\frac{1}{2}$
11. 2nd and 4th
13. (1, −2)
15. $-8y$

Test 2

1. $\sqrt{425}$
3. $\frac{15}{2}$
5. $14x-9y+12=0$
7. $15x-2y-140=0$
9. 28.3 (approx.)
11. $3x-y+6=0$
13. $2x-y+9=0$

Pages 394–396

1. any nonnegative integral number 3. yes 5. no 7. no
9. one 11. yes 13. yes 15. any nonnegative integral number
17. yes 19. yes (see Fig. 14.6) 21. one
23. not necessarily 27. no conclusion
29. Plane *MN* intersects plane *RS* in $\overleftrightarrow{EF}$. 31. $\overleftrightarrow{EJ} \parallel \overleftrightarrow{WZ}$
33. Point *Q* lies in plane *RS*.
35. $\overleftrightarrow{AB}$ and $\overleftrightarrow{EK}$ will not intersect, but need not be parallel. 37. $\overleftrightarrow{AB} \parallel \overleftrightarrow{KL}$
39. $\overleftrightarrow{PQ} \perp$ plane *KL*.

Pages 400–401

5. 384 in.^{2}
7. 2 ft^3
9. 60 ft^2
11. 480 in.^{2}
13. 12,960 in.^{3}
15. 4032 in.^{2}
17. 400/27 yd^3
19. 5.4 gals

Pages 403–404

1. 210 in.^{3} 3. 400 in.^{3} 5. 972 in.^{3} 7. 312 in.^{3} 9. 1164 in.^{2}

Pages 405–406

1. 471 in.^{3} 3. 1570 ft.^{2} 5. 938 in.^{3} 7. 405 in.^{2} 9. 127,000 ft^3 11. 15,400 ft^2

Pages 408–409

1. no 3. 829 in.^{2} 5. 5300 gals 7. 7850 ft^2 9. 206 lb

Pages 410–411

1. 904 in.^{3} 3. 924 in.^{2} 5. 4190 ft^3 7. 324 πft^2 $\approx$ 1020 ft^2 9. 1130 in.^{3}
11. $\frac{397}{6}\pi$ in.^{3} $\approx$ 208 in.^{3}

Pages 413–415

Test 1

1. line 3. $V = \frac{4}{3}\pi R^3$ 5. any nonnegative integral number 7. one 9. line
11. six 13. line 15. plane

Test 2

1. F 3. T 5. F 7. T 9. T 11. F 13. F 15. F 17. T
19. F 21. F 23. F 25. T

Problems Test

1. 41.25 ft^3 3. 9040 ft^2 5. 204 in.^{2} 7. 2790 in.^{3}

Index